Lecture Notes in Computer Science 6175

Commenced Publication in 1973
Founding and Former Series Editors:
Gerhard Goos, Juris Hartmanis, and Jan van Leeuwen

Ofer Strichman   Stefan Szeider (Eds.)

Theory and Applications of Satisfiability Testing – SAT 2010

13th International Conference, SAT 2010
Edinburgh, UK, July 11-14, 2010
Proceedings

 Springer

Volume Editors

Ofer Strichman
Technion
Technion City, Haifa 32000, Israel
E-mail: ofers@ie.technion.ac.il

Stefan Szeider
Vienna University of Technology
Favoritenstr. 9-11, 1040 Vienna, Austria
E-mail: stefan@szeider.net

Library of Congress Control Number: 2010929579

CR Subject Classification (1998): F.2, C.2.4, H.4, F.3, F.1, F.4.1

LNCS Sublibrary: SL 1 – Theoretical Computer Science and General Issues

ISSN   0302-9743
ISBN-10   3-642-14185-4 Springer Berlin Heidelberg New York
ISBN-13   978-3-642-14185-0 Springer Berlin Heidelberg New York

springer.com

© Springer-Verlag Berlin Heidelberg 2010
Printed in Germany

Typesetting: Camera-ready by author, data conversion by Scientific Publishing Services, Chennai, India
Printed on acid-free paper   06/3180

Preface

This volume contains the papers presented at SAT 2010, the 13th International Conference on Theory and Applications of Satisfiability Testing. SAT 2010 was held as part of the 2010 Federated Logic Conference (FLoC) and was hosted by the School of Informatics at the University of Edinburgh, Scotland. In addition to SAT, FLoC included the conferences CAV, CSF, ICLP, IJCAR, ITP, LICS, RTA, as well as over 50 workshops. Affiliated with SAT were the workshops LaSh (Logic and Search, co-affiliated with ICLP), LoCoCo (Logics for Component Configuration), POS (Pragmatics Of SAT), PPC (Propositional Proof Complexity: Theory and Practice), and SMT (Satisfiability Modulo Theories, co-affiliated with CAV). SAT featured three competitions: the MAX-SAT Evaluation 2010, the Pseudo-Boolean Competition 2010, and the SAT-Race 2010.

Many hard combinatorial problems such as problems arising in verification and planning can be naturally expressed within the framework of propositional satisfiability. Due to its wide applicability and enormous progress in the performance of solving methods, satisfiability has become one of today's most important core technologies. The SAT 2010 call for papers invited the submission of original practical and theoretical research on satisfiability. Topics included but were not limited to proof systems and proof complexity, search algorithms and heuristics, analysis of algorithms, combinatorial theory of satisfiability, random instances vs structured instances, problem encodings, industrial applications, applications to combinatorics, solvers, simplifiers and tools, case studies and empirical results, exact and parameterized algorithms. Satisfiability is considered in a rather broad sense: besides propositional satisfiability, it includes the domain of Quantified Boolean Formulae (QBF), Constraint Programming Techniques (CP) for word-level problems and their propositional encoding and particularly Satisfiability Modulo Theories (SMT).

The conference received 75 submissions, including 56 regular papers with a page limit of 14 pages and 21 short papers with a page limit of 6 pages. Each submission was reviewed by at least four members of the Program Committee. The committee decided to accept 21 regular papers and 14 short papers. Six out of the 14 submitted short papers were accepted; eight papers accepted as short papers had been submitted as regular papers. Three of the short papers were given a slightly larger page limit for the final version.

The program included 30-minute presentations of the accepted regular papers and 20-minute presentations of the accepted short papers. The program also included invited talks by Yehuda Naveh and Ramamohan Paturi (extended abstracts can be found in this volume) and presentations of the results of the three affiliated competitions. In addition, this year's program included an invited tutorial on SMT by Daniel Kroening and a joint session with the SMT workshop.

First and foremost we would like to thank the members of the Program Committee and the additional external reviewers for their careful and thorough work, without which it would not have been possible for us to put together such an outstanding conference program. We also wish to thank all the authors who submitted their work for our consideration.

We wish to thank the Workshop Chair Carsten Sinz for his excellent work, and all the organizers of the SAT affiliated workshops and competitions. Special thanks go to the organizers of FLoC, in particular to Moshe Vardi, for their great help and for coordinating the various conferences. We would like to thank Andrei Voronkov for his excellent EasyChair system that made many tasks easy, and Oliver Kullman, the Chair of SAT 2009, for his advice on running the conference. We also would like to acknowledge the support of our sponsors: EPSRC, NSF, Microsoft Research, Association for Symbolic Logic, CADE Inc., Google, Hewlett-Packard and Intel; Intel also sponsored SAT 2010 separately from their support of FLoC 2010.

May 2010   Ofer Strichman
                                                           Stefan Szeider

Conference Organization

Program Chairs

Ofer Strichman
Stefan Szeider

Program Committee

Dimitris Achlioptas
Fahiem Bacchus
Armin Biere
Nadia Creignou
Stefan Dantchev
Adnan Darwiche
John Franco
Enrico Giunchiglia
Kazuo Iwama
Hans Kleine Büning
Oliver Kullmann
Sais Lakhdar
Daniel Le Berre
Chu-Min Li
Inês Lynce
Hans van Maaren
Panagiotis Manolios
João Marques-Silva
David Mitchell
Alexander Nadel
Robert Nieuwenhuis
Albert Oliveras
Ramamohan Paturi
Igor Razgon
Karem Sakallah
Roberto Sebastiani
Laurent Simon
Carsten Sinz
Robert Sloan
Mirosław Truszczyński
Alasdair Urquhart
Allen Van Gelder
Toby Walsh
Emo Welzl

Lintao Zhang
Xishun Zhao

SAT Worshop Chair

Carsten Sinz

Organizing Committee (FLoC)

Seth Fogarty
Stephan Kreutzer
Leonid Libkin
Gordon Plotkin
Nicole Schweikardt
Philip Scott
Moshe Vardi

Local Organization (FLoC)

Claire David
Anuj Dawar
Kousha Etessami
Jacques Fleuriot
Floris Geerts
Paul Jackson
Bartek Klin
Stephan Kreutzer
Ian Stark
Perdita Stevens

External Reviewers

Uwe Bubeck
Doron Bustan
Lucas Cordeiro
Herve Daude
Jessica Davies
Gilles Dequen
Emanuele Di Rosa
Uwe Egly
Stephan Falke
Eugene Goldberg
Alexandra Goultiaeva
Ana Graça

Djamal Habet
Marijn Heule
Said Jabbour
Matti Järvisalo
Jiwei Jin
Ian Johnson
Daher Kaiss
Hadi Katebi
George Katsirelos
Zurab Khasidashvili
Lefteris Kirousis
Arist Kojevnikov

Theodor Lettmann
Han Lin
Christof Löding
Florian Lonsing
Michael Maher
Vasco Manquinho
Victor Marek
Paolo Marin
Barnaby Martin
Ruben Martins
Deepak Mehta
Robin A. Moser
Alexander Nadel
Nina Narodytska
M. A. Hakim Newton
Arlindo Oliveira
Richard Ostrowski
Vasileios Papavasileiou
Vasilis Papavasileiou

Cédric Piette
Knot Pipatsrisawat
Jordi Planes
Luca Pulina
Zhe Quan
Olivier Roussel
Philipp Rümmer
Vadim Ryvchin
Horst Samulowitz
Dominik Alban Scheder
Viktor Schuppan
Yuping Shen
Silvia Tomasi
Stefano Tonetta
Gyorgy Turan
Danny Vilenchik
Magnus Wahlström
Wanxia Wei
Ke Xu

Table of Contents

Part 3. Short Papers

The Big Deal: Applying Constraint Satisfaction Technologies Where It Makes the Difference

Yehuda Naveh

IBM Research – Haifa, Haifa University Campus, Haifa 31905, Israel
naveh@il.ibm.com

Abstract. In my talk, I will present a few industrial-scale applications of satisfaction technology (constraint programming and satisfiability), all of which are of prime importance to the respective business. The talk will focus on high-end solutions to unique but immense problems. This, as opposed to off-the-shelf solutions which are suitable for more commoditized problems. I argue that the former case is where cutting-edge satisfaction technology can bring the most significant impact. The following is an extended abstract of my talk.

1 Introduction

Constraint satisfaction technologies, including satisfiability (SAT) and its younger sibling constraint programming (CP), have fascinated the computer science community for decades. One of the most intriguing aspects is their declarative nature, bridging the gap between the front-end specification of the problem, and the back-end algorithm which solves it. Thus, the unique position of the discipline at the crossways of artificial intelligence, programming models, logic, algorithms, and theory accounts for much of the charm of this area.

Furthermore, the declarative nature of constraint satisfaction is also the basis of its strong practical importance, and provides the linkage to operations research areas, in particular linear and non-linear optimization. The ability of the user to specify the problem in a language which emerges from the actual application domain may be of critical importance, especially in domains which are complex, dynamic, and require a fast response.

The purpose of my talk is to present a few application domains which exhibit those criteria, and to show how constraint satisfaction is applied in these domains. The common theme of all applications I will describe is their immensity. In fact, all applications have only a small number of instances worldwide (for example, there are only a few high-end truck manufacturers in the world). However, each such instance is of a huge strategic importance to its company. My claim is that it is those cases in which it is beneficial for the company to invest a large effort (or a large amount of resources) in building a high-end constraint-satisfaction-based solution. This, in contrast to the more traditional operations research solutions, which are of a more commoditized nature, and which serve to solve problems more commonly exhibited in many small and medium sized businesses.

O. Strichman and S. Szeider (Eds.): SAT 2010, LNCS 6175, pp. 1–7, 2010.

2   Hardware Verification

Hardware verification is perhaps the prime example of the application of satisfaction technologies at a huge industrial scale. Here, the goal is to ensure that a hardware design works according to its specification while still at the design phase, before cast into silicon. The goal is so important that all large hardware manufacturers, as well as electronic design tool vendors, have for years invested a large amount of resources in the R&D of satisfaction technologies for this domain.

2.1   Model Checking

Model checking [1] for hardware verification is beyond the scope of this talk because the audience is intimately familiar with this topic. It is by far the largest and most important industrial application of SAT technology.

2.2   Stimuli Generation

While model checking and other formal verification techniques have their clear advantages, most notably the ability to formally prove functional correctness of the design, they can hardly cope with modern complex designs at the level of a single unit or larger. To this end, simulation-based verification, in which the design behavior is checked by simulating it over external inputs, accounts for roughly ninety percent of the overall verification efforts and resources.

The major challenge in such methods is in creating inputs, or 'stimuli', which are (1) valid according to the hardware specification and the simulation environment, (2) interesting in the sense that they are likely to excite prone to bugs areas of the design, and (3) diverse [2,3].

Item (1) is dealt with by modeling the entire hardware specification, as well as that of the simulation environment, as a set of mandatory constraints over the simulated variables (memory addresses, data transferred, processor instruction parameters, and so on). Item (2) is dealt with in two ways: first, generic expert knowledge is modeled as a set of soft, non-mandatory, constraints (for example, a soft constraint may require the result of operation $a+b$ to be zero, because this is a known prone-to-bugs area of the floating point processing unit). In addition, the verification engineer, who is directly responsible for creating the stimuli, may add mandatory and non-mandatory constraints to any particular run, directing the stimuli into required scenarios.

Figure 1 illustrates this scheme by considering two variables: the effective address and the real address of a 'load' instruction. These two variables are related by architectural constraints (complex translation scheme), expert knowledge (requirement to reuse cache rows), and specific verification scenarios.

Once modeled, this set of mandatory and non-mandatory constraints can be fed into a constraint solver, which comes up with a solution to the constraint problem in the form of a valid and interesting stimulus. In order to achieve target item (3), the solver typically has a built-in diversification mechanism,

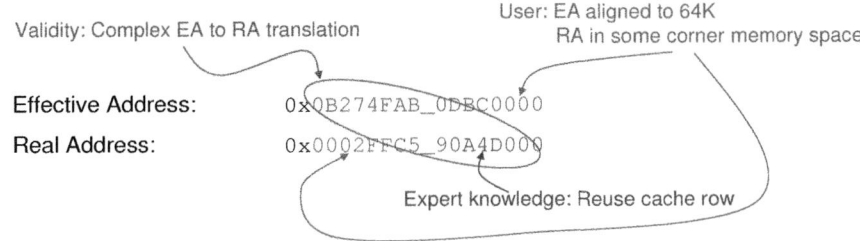

Fig. 1. Stimuli generation constraints on effective and real address values

meaning that each time it is called on the same input (same set of constraints), a sufficiently different output (stimuli) is returned.

3   Workforce Management

Not many companies have a professional services workforce on the scale of tens of thousands of professionals. However, those who are in this business face a critical challenge [4]. How do they identify and assign a team of professionals who best fit a specific customer engagement? Each of the professionals in the team must be skilled — but not over-skilled — to do the job, must be available at the area, or otherwise able to work remotely, must be free of their current engagement and not committed to further projects before the expected end of the work, and must have a personal affinity for the job. The team as a whole must have the correct distribution of skills and of experience levels, must conform to the budget requirements, and needs to be composed of professionals able to work together with each other. In addition, at any given time we need to staff as many projects as possible.

All those 'musts' and 'needs' better be met, or the projects would suffer the consequences of assigning under-qualified or over-qualified teams. These consequences include prolonged project durations, excessive compensation costs, fines and interests, unnecessary commute, and disruption to other projects. In addition to monetary losses, a poorly staffed project may result in an unsatisfied customer and demotivated professionals, leading eventually both to customer churn and employee attrition, which may become a death stroke for the business.

The above problem translates into a constraint problem, where some of the constraints have a clear mathematical foundation, while others are softer in the sense that they describe human attributes and as such are handled in a less strict manner.

The most obvious example of the first kind of constraints is that the same professional cannot be assigned to two different projects if the projects overlap in time. This consideration leads to a new type of global constraint, the `some-different` constraint [5]. The second type of constraints is best dealt

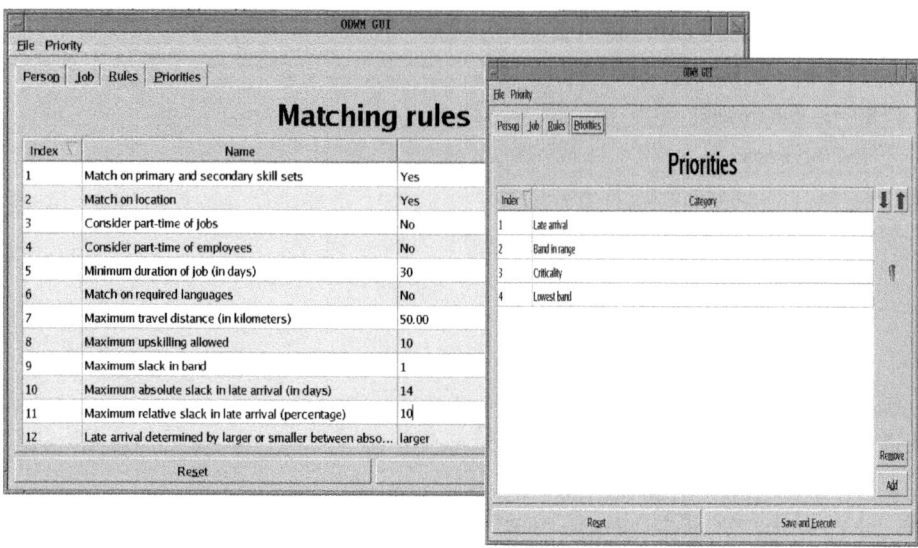

Fig. 2. User interface for specifying hard matching rules and soft priorities for the workforce management problem of service professionals

with as a set of preferences. For example, is it more important for the business to engage professionals with the exact required skill levels, or should geographical proximity of the professionals be the prime concern? similarly, is it more important to have best fits to individual projects, or is it better to maximize the overall number of engagements at the cost of compromising each individual engagement? These sets of rules and preferences are defined by the user in a list as in Fig. 2, and is then translated into a set of hard and soft constraints, respectively, which in turn are solved by a constraint solver.

The ability of constraint programming to account for the rigid mathematical constraints at the same level with the soft human aspects of the problem, is what provides for the unique business advantage of this approach.

4   Truck Configuration

Unlike regular automobiles, for which we enter our local neighborhood dealer and once we decide on a model we get a few options to choose from, large trucks, which cost a few hundred thousand dollars a piece, are highly configurable according to specific business needs of the customers. In fact, unless ordered as a batch from a single customer, there are no two identical large trucks on the road.

The customer, when ordering a truck, has some very specific needs in mind, and will not happily compromise on them. For example, a large dairy company may require a cargo area able to reach a specific temperature while driving in the desert, a drive-train suitable for mountainous terrain, a cabin with space and accessories to meet agreements with the drivers union, and a position for a

Jack is either telescopic 12 T, or regular 25 T

If fuel prefilter is with heated water separator, air-intake cannot be behind cab with round filter

If front axle design is straight, then front axle weight is 7.5 T, and there is no front override guard

Fig. 3. Examples of rules which must be met by any truck configuration

crane of up to half-ton leverage. Those requirements need to be all met, while still conforming to a multitude of engineering, manufacturing, marketing, and legal constraints. An inability to satisfy the customer's needs may result in the customer ordering the truck from a competitor. Conversely, conforming to the specified requirements, may lead to a happy customer and another million-dollar deal.

Given the thousands of configurable variables, and the tens of thousands of constraints on those variables (see some examples in Fig. 3), and given that the configuration problem is NP-complete, it may well be the case that a valid configuration which satisfies the customer's requirements exists, but is not found by the configurator at hand. Therefore, this is a classic example where a stronger technology, incorporating the best algorithms and heuristics available, can truly make the difference.

5 Systems Engineering

Complex systems design (the canonical example is that of an airplane) involves many different disciplines such as requirements engineering, system architecture, mechanical engineering, software engineering, electronic engineering, testing, parametric analysis, and more. In each of these disciplines a model of the product is managed, see Fig. 4. Today all of those models are managed separately. At best, there is an integration of two models, usually done by simple copy or export. This limits the possibilities to maintain traceability between the different teams and project parts, allow synchronization of the data, perform impact analysis when part of the model changes, and achieve optimal design.

One of the major obstacles limiting the ability to combine all sub-models into a single coherent model is the complexity of creating and maintaining a valid and consistent structure. The issue is the various validity rules which the models must conform to (for example, a given methodology may require that each functional feature must be associated with at least one, but not more than three, tests). Attempts to link the various models together may very fast lead to inconsistencies with respect to those rules.

Constraint satisfaction technology can ensure that the links are created in accordance with the design methodology, detect discrepancies between the

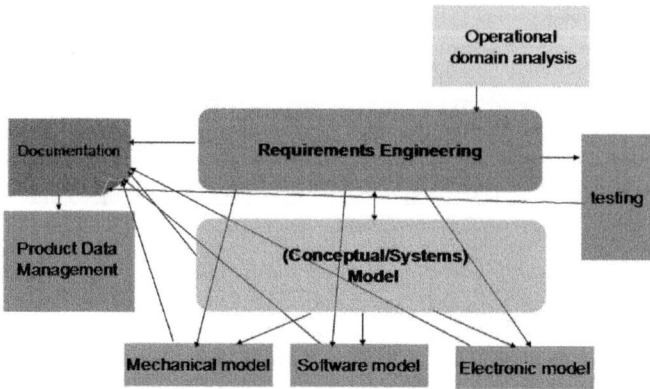

Fig. 4. Inter-relations between models of different disciplines in systems engineering

models, and deduce the existence or absence of links, thus assisting the creation of a unified model.

6   Additional Areas

Above I discussed thriving applications of satisfaction technologies. Space limitation allowed only mere description of the problems, and some flavor of their criticality to the business. It also forbids me from detailing other applications of similar nature, but which are currently only at various levels of prototyping. These include placement of virtual machines on physical hosts at data centers, job-scheduling for massively parallel processors (also known as supercomputers, e.g., IBM's BlueGene), pricing of services engagements, and variability management of product lines.

7   Conclusions

I hope to have delivered the message that in cases where a main part of the business is at stakes, companies would rightfully be willing to invest a large effort in R&D, and specifically in constraint satisfaction technologies. This is the case in hardware verification (shipping functionally correct hardware to the market), services project staffing (ensuring assignments of the correct professional teams), truck configuration (supplying the customer with the truck they need), systems engineering (better management of complex, airplane-size, models), and more. In all those huge-application cases, constraint satisfaction technology can be the means to achieve better, cutting edge, results, and thus provide the competitive advantage at the most critical aspect of the business.

Epilogue and Acknowledgments

The applications described above are all part of IBM's current activities. Obviously, only a strong team of motivated and experienced researchers can reach such achievements. The model at IBM Research is to develop a generic solver as a single technological core, while each application is then developed as an independent module using the same core solver as others. When an application requires an improved algorithm, the application and solver teams together find the right level of generalization of the requirements, and it is then implemented in the solver. This way, all other applications benefit from the original application's needs.

The number of past and present researchers who contributed to this work in the past fifteen years is too large to enumerate here. Many contributors to the stimuli generation activities can be found as authors of the references in [2,3]. As for the other applications discussed, the contributors generally come from the Constraint Satisfaction group at IBM Research – Haifa. The group consists of the solver team: Merav Aharoni, Ari Freund, Wesam Ibraheem, and Nathan Fridlyand; Optimatch (workforce management) team: Sigal Asaf, Michael Veksler, and Haggai Eran; truck configuration: Yael Ben-Haim; and systems engineering: Odellia Boni. In addition, Mage, our independent SAT Solver, is developed by Tanya Veksler and her team at the Formal Verification group. I thank all those contributors who are at the basis of this talk. I am also grateful to Odellia for her writeup of the System Engineering Section of this work.

References

1. Biere, A., Cimatti, A., Clarke, E.M., Fujita, M., Zhu, Y.: Symbolic model checking using sat procedures instead of bdds. In: DAC 1999: Proceedings of the 36th annual ACM/IEEE Design Automation Conference, pp. 317–320. ACM, New York (1999)
2. Naveh, Y., Rimon, M., Jaeger, I., Katz, Y., Vinov, M., Marcus, E., Shurek, G.: Constraint-based random stimuli generation for hardware verification. AI Magazine 28, 13–30 (2007)
3. Adir, A., Naveh, Y.: Stimuli generation for functional hardware verification with constraint programming. In: van Hentenryck, P., Milano, M. (eds.) Hybrid Optimization: the 10 Years of CPAIOR (to appear, 2010)
4. Naveh, Y., Richter, Y., Altshuler, Y., Gresh, D.L., Connors, D.P.: Workforce optimization: Identification and assignment of professional workers using constraint programming. IBM Journal of Research and Development 51, 263–280 (2007)
5. Richter, Y., Freund, A., Naveh, Y.: Generalizing alldifferent: The somedifferent constraint. In: Benhamou, F. (ed.) CP 2006. LNCS, vol. 4204, pp. 468–483. Springer, Heidelberg (2006)

Exact Algorithms and Complexity

Ramamohan Paturi[*]

Department of Computer Science and Engineering
University of California, San Diego
La Jolla, CA 92093-0404, USA

Over the past couple of decades, a series of exact exponential-time algorithms have been developed with improved run times for a number of problems including IndependentSet, k-SAT, and k-colorability using a variety of algorithmic techniques such as backtracking, dynamic programming, and inclusion-exclusion. The series of improvements are typically in the form of better exponents compared to exhaustive search. These improvements prompt several questions, chief among them is whether we can expect continued improvements in the exponent. Is there a limit beyond which one should not expect improvement? If we assume $\mathbf{NP} \neq \mathbf{P}$ or other appropriate complexity statement, what can we say about the likely exact complexities of various $\mathbf{NP}$-complete problems?

Besides the improvement in exponents, there are two other general aspects to the algorithmic developments. Problems seem to differ considerably in terms of the improvements in the exponents. Secondly, different algorithmic paradigms seem to work best for different problems. These aspects are particularly interesting given the well-known fact that all $\mathbf{NP}$-complete problems are equivalent as far as polynomial-time solvability is concerned. How do the best possible exponents differ for different problems? Can we explain the difference in terms of the structural properties of the problems? Are the likely complexities of various problems related? What is relative power of various algorithmic paradigms?

One approach would be to consider natural, though restricted, computational models. For example, consider the class $\mathbf{OPP}$ of one-sided error probabilistic polynomial-time algorithms. $\mathbf{OPP}$ captures a common design paradigm for randomized exact exponential-time algorithms: to repeat sufficiently many times a one-sided error probabilistic polynomial-time algorithm that is correct with an exponentially small probability so that the overall algorithm finds a witness with constant probability. This class includes Davis-Putnam-style backtracking algorithms developed in recent times to provide improved exponential-time upper bounds for a variety of $\mathbf{NP}$-hard problems. While the original versions of some of these algorithms are couched as exponential-time algorithms, one can observe from a formalization due to Eppstein that these algorithms can be converted into probabilistic polynomial-time algorithms whose success probability is the reciprocal of the best exponential-time bound. The class is interesting not just because of ubiquity, but because such algorithms are ideal from the point

[*] This research is supported by NSF grant CCF-0947262 from the Division of Computing and Communication Foundations.

O. Strichman and S. Szeider (Eds.): SAT 2010, LNCS 6175, pp. 8–9, 2010.
© Springer-Verlag Berlin Heidelberg 2010

of view of space efficiency, parallelization, and speed-up by quantum computation. What are the limitations of such algorithms for deciding **NP**-complete problems? Could the best algorithm for specific **NP**-complete problems be in this class?

On the other hand, the recent algorithms for k-colorability for $k \geq 3$ use inclusion-exclusion principle in combination with dynamic programming to achieve the bound of 2^n. This raises a natural question whether we can expect an **OPP** algorithm for k-colorability whose success probability is at least 2^{-n}. More generally, can we expect **OPP**-style optimal algorithms for k-colorability? Does there exist any **OPP** algorithm for k-colorability whose success probability is at least c^{-n} where c is independent of k? Negative answers (or evidence to that effect) for these questions would provide convincing proof (or evidence) that exponential-time inclusion-exclusion and dynamic programming paradigms are strictly more powerful than that of **OPP**. On the other hand, algorithmic results that would place k-colorability in the class **OPP** with c^{-n} success probability would be exciting.

The current state of the art in complexity theory is far from being able to resolve these questions, especially the question of best exponents, even under reasonable complexity assumptions. However, recent algorithmic and complexity results are interesting and they provide food for thought. In this talk, I will present key algorithmic results as well as our current understanding of the limitations.

Improving Stochastic Local Search for SAT with a New Probability Distribution

Adrian Balint and Andreas Fröhlich

Ulm University
Institute of Theoretical Computer Science
89069 Ulm, Germany
adrian.balint,andreas.froehlich@uni-ulm.de

Abstract. This paper introduces a new SLS-solver for the satisfiability problem. It is based on the solver *gNovelty+*. In contrast to *gNovelty+*, when our solver reaches a local minimum, it computes a probability distribution on the variables from an unsatisfied clause. It then flips a variable picked according to this distribution. Compared with other state-of-the-art SLS-solvers this distribution needs neither noise nor a random walk to escape efficiently from cycles. We compared this algorithm which we called *Sparrow* to the winners of the SAT 2009 competition on a broad range of 3-SAT instances. Our results show that *Sparrow* is significantly outperforming all of its competitors on the random 3-SAT problem.

1 Introduction

The propositional satisfiability problem (SAT) is one of the most studied $\mathcal{NP}$-complete problems in computer science. Given a propositional formula in conjunctive normal form (CNF) with variables $\{x_1, \ldots, x_N\}$ the SAT-problem consists in finding an assignment for the variables so that all clauses are satisfied. In this paper we will focus on SLS-solvers for SAT and describe how their performance can be improved with a new probability distribution.

SLS-solvers operate on complete assignments trying to find a solution by flipping variables according to a given heuristic. Most SLS-solvers are based on the following scheme: First a random assignment is chosen. If the formula is satisfied by the assignment the solution is found. If not, a variable is chosen according to a (possibly probabilistic) variable selection heuristic, which we further call *pickVar*. The variable is flipped and the process starts over again until a solution is found. Depending on the heuristic used in *pickVar*, SLS-solvers can be divided into three categories: GSAT, WalkSAT and dynamic local search (DLS). The currently best-performing solvers are mainly combinations of heuristics from these categories. For example the winner of the 2009 SAT Competition category random (*TNM*) uses in a first stage a GSAT-scheme. When reaching a local minimum a WalkSAT-like heuristic is used to escape. *gNovelty+2*, which won the second place in the 2009 SAT Competition, additionally uses an additive clause weighting scheme like DLS-solvers.

O. Strichman and S. Szeider (Eds.): SAT 2010, LNCS 6175, pp. 10–15, 2010.

Most SLS-solvers use different measures in their *pickVar*-heuristic. For example *Novelty+* uses the score and the age of variables (the number of steps since the variable was last flipped). The score of variable x_i is defined as the number of clauses that x_i will satisfy minus the number of clauses that will become unsatisfied by flipping x_i. To choose a variable *Novelty+* picks a random unsatisfied clause and then selects the best and the second best variable relative to their score. If the best variable is not the one with the lowest age-value then this variable is always chosen. Otherwise it is only chosen with probability $(1 - noise)$ and with probability *noise* the second best variable is selected. With probability *wp* a random walk is performed. Neither the difference between the scores nor the age-difference is taken into account. This lack of differentiation is a big disadvantage in our opinion.

In this paper we would like to address this weakness and improve a state-of-the-art solver like *gNovelty+*. We will replace the *adaptNovelty+*-heuristic from *gNovelty+* with a novel heuristic based on a probability distribution that takes into account the difference between the scores and the age of variables. We implemented these improvements in a solver called *Sparrow*. To show its superior performance we compare *Sparrow* with the winners of the last SAT-Competition on a wide range of 3-SAT formulas from the SAT 2009 random benchmark. *Sparrow* is able to outperform all winners of the SAT 2009 Competition.

2   *Sparrow*

2.1   *gNovelty+*

Because our solver *Sparrow* is based on *gNovelty+* we first want to describe it briefly. *gNovelty+* is one of the best performing SLS-solvers (winner of the SAT 2007 Competition category random). Our work is based on the SAT 2007 version of the solver which can be decomposed in three components.

1. gradient-walk like *G2WSAT* [4]
2. *adaptNovelty+* [3]
3. additive weigthing-scheme as used in *PAWS* [8]

The functionality of the components and their interactions are described in detail in [6]. *gNovelty+*'s *pickVar*-heuristic works in three phases, that all take into account the weights of the clauses. In the first phase it uses the gradient-walks until there are no more promising variables to flip. This state characterizes a local minimum. To escape from this local minimum a variable is chosen according to the *adaptNovelty+*-heuristic. *AdaptNovelty+* is a walk-SAT algorithm which behaves like *Novelty+* and additionally uses an adaptive scheme for the noise described in [3] and includes a random walk. Whenever an *adaptNovelty+*-step is performed the weights of the clauses are updated according to an additive weighting scheme.

The first two phases of *gNovelty+* without weights are the core of the *G2WSAT*-algorithms. It is well known that the performance of this kind of algorithm heavily relies on the *adaptNovelty+*-component.

2.2   Defining a Probability Distribution

One drawback of algorithms that use *adaptNovelty+*-like heuristics to escape from local minima is the lack of differentiation between the variables. While always selecting the best variable in gradient-steps seems to work very well, a more advanced heuristic is needed when a local minimum is reached. We therefore kept the gradient-walk as well as the adaptive weighting scheme but removed the *adaptNovelty+*-component and replaced it by a new heuristic which is based on a probability distribution over the variables from a random unsatisfied clauses.

When defining this probability distribution we focused on two aspects: First we wanted to keep the features that made *adaptNovelty+* a very well-performing solver. Second we wanted to perform better in those cases when *adaptNovelty+* fails to make the best decision. As we already discussed, we think that this is mainly due to insufficient differentiation.

Let $\{x_{i_1}, \ldots, x_{i_u}\}$ be the variables from a random unsatisfied clause. We now define the probability of choosing a variable x_{i_j} as

$$p(x_{i_j}) := \frac{p_s(x_{i_j}) * p_a(x_{i_j})}{\sum_{k=1}^{u} p_s(x_{i_k}) * p_a(x_{i_k})} \tag{1}$$

where $p_s(x_{i_j})$ is a function of $score(x_{i_j})$ and $p_a(x_{i_j})$ is a function of $age(x_{i_j})$.

We now have the possibility to directly let the score and the age of a variable influence its probability of getting flipped. This offers a much better differentiation between the individual variables than just by deciding whether they have the best score or the lowest age in their clause.

In particular we have chosen the following functions for our implementation:

$$p_s(x_{i_j}) := c_1^{score(x_{i_j})} \qquad p_a(x_{i_j}) := \left(\frac{age(x_{i_j})}{c_3}\right)^{c_2} + 1 \tag{2}$$

This yields a probability distribution p which grants bigger values to variables with higher score and age (like *adaptNovelty+* does). If we choose c_1 as a power of 2 we can efficiently implement p_s by binary shifts. p_a can also be efficiently implemented for $c_2 \in \mathbb{Z}$. As it can easily be seen from the formula, already small changes in score have a huge impact on the probability because of the exponential character of the function. On the other hand the age of a variable only slowly starts to influence the probability but is also able to have a great impact once an age of c_3 is exceeded. The degree of influence depends on c_2. We are going to specify values for these constants in our empirical results section.

Further we do not need any noise because repeating the same flips over and over again is automatically avoided by the growing age of variables that were not flipped due to other variables with higher score. We also don't need an explicit implementation of a random walk since our new scheme can efficiently escape cycles.

3   Empirical Results

3.1   Soft- and Hardware

The *gNovelty+2T* (the code from the competition uses a tabu scheme and we added a T in the name because of this), *TNM* and the *hybridGM3* code we used for the comparison were the ones submitted to the SAT 2009 Competition [7]. For our solver we used the following settings for all runs: $sp = 35$, $c_1 = 2$, $c_2 = 4$, $c_3 = 10^5$. The solvers were run on a part of the bwGrid [2], where we were provided with Intel Harpertown quad-core CPUs with 2.83 GHz.

3.2   The Benchmark Formulae

For our tests we set up a two benchmarks. The first one contains 64 instances from random large category of the SAT 2009 Competition ($2000 \leq$ #variables ≤ 18000). The second contains all formulas of the additional benchmark [7] of the same category ($20000 \leq$ #variables ≤ 26000). All instances have a clause/variable-ratio of 4.2. The solvers were run 100 times (50 for the second benchmark) on each instance and the mean values for the running time and the number of flips were calculated. The time limit was set to 1200 seconds (2400 for the second benchmark). On those instances where one of the compared solvers did not finish within the time limit in all 100 (50) runs the number of successful runs was plotted.

3.3   Results

The results of the benchmarks can be seen in figure 1. *TNM* and *hybridGM3* are compared to *Sparrow* on the regular benchmark first and on the additional benchmark in the following row. For *gNovelty+2T* we did not plot the additional benchmark.

In the first two rows, we compare *Sparrow* to *TNM*. As we can see, *Sparrow* is superior to *TNM* considering the number of flips as well as considering the runtime on all instances. This becomes even more obvious when we take a look at the success runs in the third column. Escpecially taking a look at the additional benchmark in row two, we can see that *TNM* had difficulties solving the more difficult instances while *Sparrow* still performs very well.

The next row shows *Sparrow* compared to *gNovelty+2T*. It is easy to see that *Sparrow* dominates *gNovelty+2T* on our benchmark in flipcount as well as in runtime. The success rates show that many instances of our benchmark indeed are very difficult to solve even for a state-of-the-art-solver. We did not include the results of *gNovelty+2T* on the additional benchmark in figure 1 since *gNovelty+2T* was not able to solve any of the instances in all runs and the success rate was constantly very low.

In the last two rows we compare *Sparrow* to *hybridGM3*, a hybrid SLS-solver using a DPLL-component [1]. *HybridGM3* can compete with *Sparrow* on some instances but also loses ground with increasing difficulty.

Altogether there were only 4 instances on which *Sparrow* did not finish within the time limit on all 100 runs. Except in 1 of them the success rate was nearly

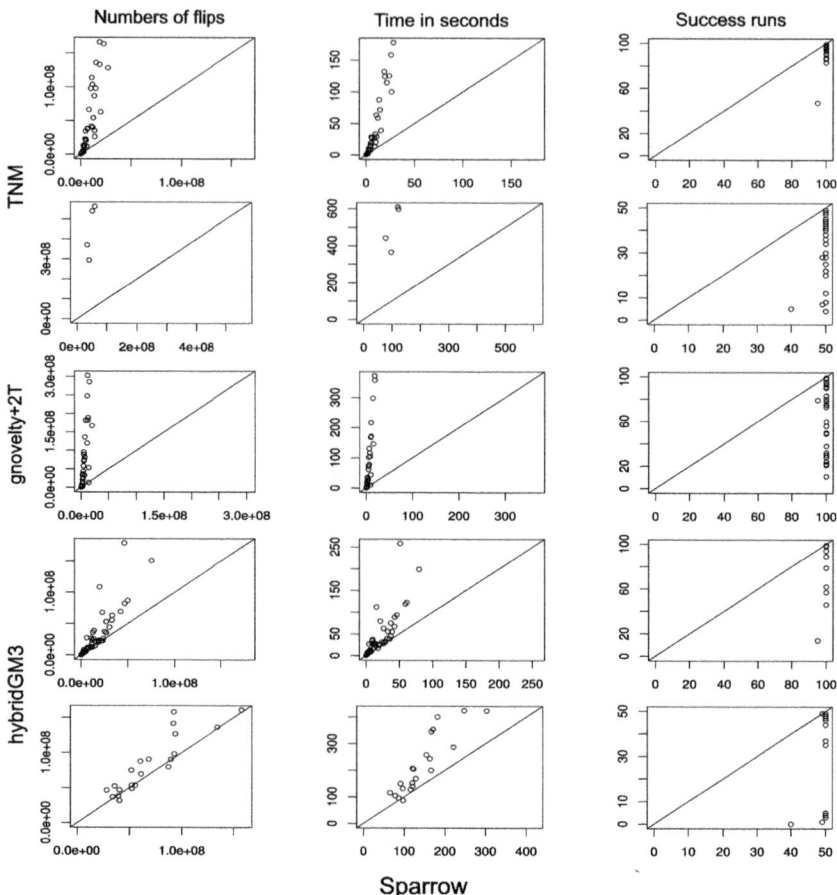

Fig. 1. *Sparrow* compared to *TNM*, *gNovelty+2T* and *hybridGM3* on 104 randomly selected large-size 3SAT-instances

100%. This however was also one of the most difficult instances for all solvers and their success rate was far below the one of *Sparrow*. We are confident that *Sparrow* would be able to solve even larger instances.

4   Related Work

There were many attempts to modify the *Novelty*-heuristic to increase its performance on different benchmarks. The first time the difference between the score of the variables was taken into account was in the solver *R-Novelty* by McAllester et. al. in [5]. However the variables taken into consideration were still only the first and the second best one. The third variable had always probability zero. The *Novelty++*-heuristic by Li [4] introduced a further parameter dp (diversification probability) to the *Novelty+*-heuristic to enable choosing the least flipped variable from a clause. This permits the heuristic to choose the third variable from a clause but there is no differentiation between the scores nor between the age of the variables. The solver *TNM* by Wei which won SAT Competition 2009 in the random category uses two noise mechanisms and switches between them whenever the weights of variables meet a given criteria. We are not aware of a heuristic that assigns probabilities to all variables depending on the difference between the score and between the age of variables.

5   Conclusions and Future Work

We presented in this paper a probability distribution that is a function of the score and the age of all variables from a random unsat clause and which takes into consideration the difference between these values. An advantage of such a heuristic is that it needs no noise nor a random walk, which are incorporated by definition. The parameters of the distribution are quite stable and do not need to be tuned manually.

We conducted an empirical study on different 3SAT-problems from the SAT 2009 Competition benchmark to show the good performance of this approach. As a future work it would be of interest to incorporate more information into the probability distribution. Also we would like to test our approach on other classes of instances including crafted and industrial instances.

Acknowledgment

We would like to thank bwGrid [2] for providing the computational resources and the students Borislav Junk and Raffael Bild for improving the code.

References

1. Balint, A., Henn, M., Gableske, O.: A novel approach to combine a SLS- and a DPLL-Solver for the satisfiability problem. In: Kullmann, O. (ed.) SAT 2009. LNCS, vol. 5584, pp. 284–297. Springer, Heidelberg (2009)
2. bwGRiD, member of the German D-Grid initiative, funded by the Ministry for Education and Research and the Ministry for Science, Research and Arts Baden-Württemberg, http://www.bw-grid.de
3. Hoos, H.H.: An adaptive noise mechanism for WalkSAT. In: Proceedings of AAAI 2002, pp. 635–660 (2002)
4. Li, C.M., Huang, W.Q.: Diversification and determinism in local search for satisfiability. In: Bacchus, F., Walsh, T. (eds.) SAT 2005. LNCS, vol. 3569, pp. 158–172. Springer, Heidelberg (2005)
5. McAllester, D., Selman, B., Kautz, H.: Evidence for invariant in local search. In: Proceedings of AAAI 1997, pp. 321–326 (1997)
6. Pham, D.N., Thornton, J.R., Gretton, C., Sattar, A.: Advances in local search for satisfiability. In: Australian Conference on Artificial Intelligence 2007, pp. 213–222 (2007)
7. The SAT Competition Homepage: http://www.satcompetition.org
8. Thornton, J., Pham, D.N., Bain, S., Ferreira Jr., V.: Additive versus multiplicative clause weighting for SAT. In: Proceedings of 19th AAAI 2004, pp. 191–196 (2004)

Lower Bounds for Width-Restricted Clause Learning on Small Width Formulas

Eli Ben-Sasson[1] and Jan Johannsen[2]

[1] Computer Science Department, Technion – Israel Institute of Technology,
Haifa, Israel
[2] Institut für Informatik, LMU München, Munich, Germany

Abstract. It has been observed empirically that clause learning does not significantly improve the performance of a SAT solver when restricted to learning clauses of small width only. This experience is supported by lower bound theorems. It is shown that lower bounds on the runtime of width-restricted clause learning follow from resolution width lower bounds. This yields the first lower bounds on width-restricted clause learning for formulas in 3-CNF.

1 Introduction

Most SAT solvers are based on extensions of the basic backtracking procedure known as the DLL algorithm [10]. The recursive procedure is called for a formula F in conjunctive normal form and a partial assignment α (which is empty in the outermost call). If α satisfies F, then it is returned, and if α causes a conflict, i.e., falsifies a clause in F, then the call fails. Otherwise a variable x unset by α is chosen according to some heuristic, and the procedure is called recursively twice, with α extended by $x := 1$ and by $x := 0$. If one recursive call returns a satisfying assignment, then it is returned, otherwise — if both recursive calls fail — the call fails.

Contemporary SAT solvers employ several refinements and extensions of the basic DLL algorithm. One of the most successful of these extensions is *clause learning* [22], which works as follows: When the procedure encounters a conflict, then a sub-assignment α' of α that suffices to cause this conflict is computed. This sub-assignment α', thought of as the reason for the conflict, can then be stored in form of a new clause C added to the formula, viz. the unique largest clause C falsified by α'. This way, when in a later branch of the search another partial assignment extending α' occurs, the procedure can backtrack earlier since then the added clause C becomes falsified and causes a conflict.

When clause learning is implemented, a heuristic is needed to decide which learnable clauses to keep in memory, because learning a large number of clauses leads to excessive consumption of memory, which slows the solver down rather than helping it. Many early heuristics for clause learning were such that the *width*, i.e., the number of literals, of learnable clauses was restricted, so that the solver learned only clauses whose width does not exceed a certain threshold.

O. Strichman and S. Szeider (Eds.): SAT 2010, LNCS 6175, pp. 16–29, 2010.

Experience has shown that such heuristics are not very helpful, i.e., learning only short clauses does not significantly improve the performance of a DLL algorithm for hard formulas. The present paper continues a line of work that aims at supporting this experience with rigorous mathematical analyses in the form of lower bound theorems.

The first lower bound for width-restricted clause learning was shown [9] for the well-known pigeonhole principle clauses PHP_n. These formulas require time $2^{\Omega(n \log n)}$ to solve when learning clauses of width up to $n/2$ only, whereas they can be solved in time $2^{O(n)}$ when learning arbitrary clauses. While this example in principle shows that learning wide clauses can yield a speed-up, it is not fully satisfactory, since even with arbitrary learning, the time required is exponential in n.

Another lower bound was shown [15] for a a set of clauses Ord_n based on the ordering principle on n elements. These formulas can be solved in polynomial time when learning arbitrary clauses, but require exponential time to solve when learning clauses of size up to $n/4$ only.

Both lower bounds are asymptotically the same as the known lower bounds [14,8] on the time for solving the respective formulas by DLL algorithms without clause learning.

In these previous lower bounds, the hard example formulas themselves contain clauses of large width. Since it is conceivable that the necessity to learn wide clauses is merely due to the presence of these wide initial clauses, the question arose whether similar lower bounds can be shown for formulas of small width. We answer this question by proving lower bounds on width-restricted clause learning for small width formulas.

The lower bounds are shown by proving the same lower bounds on the length of refutations in a certain resolution based propositional proof system called RTL (see Section 2). The relationship of this proof system to the DLL algorithm with clause learning has been established in several earlier works [9,12]. We will show that for formulas of small width, lower bounds for this proof system follow from lower bounds on the width of resolution proofs. This also gives an easier proof of a slightly weaker form of the previous lower bound [15] for the formulas Ord_n.

The lower bound for clause learning algorithms on formulas requiring large resolution width is somewhat dual to a result of Atserias et al. [4], who give a small polynomial upper bound on the runtime of a clause learning algorithm *with restarts* on formulas having resolution refutations of small width.

We will now informally describe our proof method, see Section 2 for precise definitions of the terms appearing below. Let F be an unsatisfiable formula in conjunctive normal form (CNF) over n variables. For the sake of simplicity assume that F is a 3-CNF, i.e., each clause in F contains at most 3 literals. Suppose furthermore that F requires large resolution refutation width, i.e., every resolution refutation of F must contain a clause with a large number w of literals, where e.g. $w \approx \sqrt{n}$. Finally, suppose we try to solve F using a DLL algorithm augmented with clause learning, where the width of learned clauses is limited to be less than w, say $w/3$. In other words, the maximal width of a learned clause is significantly smaller than the minimal refutation width of F.

Inspired by the combinatorial characterization of resolution width due to Atserias and Dalmau [3] we classify the learned clauses of small width into two categories. The *useless* clauses are those that can be derived from F via a resolution derivation of width less than w, whereas the *useful* clauses are those that can only be derived by going through a clause of width at least w. Our main observation (Theorem 6) roughly says that learning useless clauses will not significantly reduce the running time needed to obtain the first useful clause. In fact, we show that $2^{w/3}$ steps will be needed before our algorithm learns its first useful clause.

Using known constructions [8,21,3] of families of unsatisfiable 3-CNF formulas that have short resolution refutations of polynomial in n size, but which require large refutation width of about $w \approx \sqrt{n}$, we obtain in Section 5 a number of results which show that, in certain cases, width-restricted clause learning DLL algorithms will require exponentially longer running time than clause learning algorithms with unrestricted width.

2   Preliminaries

A *literal* a is a variable $a = x$ or a negated variable $a = \bar{x}$. A *clause* C is a disjunction $C = a_1 \vee \ldots \vee a_k$ of literals a_i. The *width* of C is k, the number of literals in C.

A formula in *conjunctive normal form* (CNF) is a conjunction $F = C_1 \wedge \ldots \wedge C_m$ of clauses, it is usually identified with the set of clauses $\{C_1, \ldots, C_m\}$. A formula F in CNF is in k-CNF if every clause C in F is of width $w(C) \leq k$.

We consider resolution-based refutation systems for formulas in CNF, which are known to be strongly related to DLL algorithms. These proof systems have two inference rules: the *weakening rule*, which allows to conclude a clause D from any clause C with $C \subseteq D$, and the *resolution rule*, which allows to infer the clause $C \vee D$ from the two clauses $C \vee x$ and $D \vee \bar{x}$, provided that the variable x does not occur in either C or D, pictorially:

$$\frac{C \vee x \qquad D \vee \bar{x}}{C \vee D}$$

We say that the variable x is *eliminated* in this inference.

A resolution derivation of a clause C from a CNF-formula F is a directed acyclic graph (dag) with a unique sink, in which every node has in-degree at most 2, where every node v is labeled with a clause C_v such that:

1. The sink is labeled with C.
2. If a node v has one predecessor u, then C_v follows from C_u by the weakening rule.
3. If a node v has two predecessors u_1, u_2, then C_v follows from C_{u_1} and C_{u_2} by the resolution rule.
4. A source node ν is labeled by a clause C in F.

A *resolution refutation* of F is a resolution derivation of the empty clause from F. Resolution is sound and complete: a CNF-formula F has a resolution refutation if and only if it is unsatisfiable.

We call a derivation *tree-like* if the underlying unlabeled dag is a tree, otherwise we may call it *dag-like* for emphasis. As usual, for a dag that is a tree we refer to the sink as the *root*, to the predecessors of a node as its *children* and to a source node as a *leaf*.

The *size* of a resolution derivation is the number of nodes in the dag. The *width* of a resolution refutation R is the maximal width of a clause occurring in R. The *resolution width* of F is the minimal width of a resolution refutation of F.

Ben-Sasson and Wigderson [7] have shown the following relation between resolution width and size of tree-like resolution:

Theorem 1. *If a d-CNF formula F requires resolution width at least w, then every tree-like resolution refutation of F is of size at least 2^{w-d}.*

In the literature, resolution proof systems are sometimes defined without the weakening rule, but since applications of this rule can be eliminated from a tree-like resolution refutation without increasing the size or width, all lower bounds shown for tree-like resolution without weakening apply to the system with weakening as well.

Let X be a set of variables. A *restriction* ρ of X is a partial assignment $X \to \{0,1\}$. A restriction ρ is extended to literals by setting

$$\rho(\bar{x}) := \begin{cases} 1 & \text{if } \rho(x) = 0 \\ 0 & \text{if } \rho(x) = 1 \end{cases}$$

For a clause C in variables X, we define

$$C\lceil\rho := \begin{cases} 1 & \text{if } \rho(a) = 1 \text{ for some } a \in C \\ \bigvee_{a \in C,\, \rho(a) \neq 0} a & \text{otherwise,} \end{cases}$$

where the empty disjunction is identified with the constant 0. For a CNF-formula F over X, we define

$$F\lceil\rho := \begin{cases} 0 & \text{if } C\lceil\rho = 0 \text{ for some } C \in F \\ \bigwedge_{C \in F,\, C\lceil\rho \neq 1} C\lceil\rho & \text{otherwise,} \end{cases}$$

where the empty conjunction is identified with 1.

Proposition 2. *Let R be a (tree-like) resolution derivation of C from F of size s, and ρ a restriction. Then there is a (tree-like) resolution derivation R' of $C\lceil\rho$ from $F\lceil\rho$ of size at most s.*

In particular, if $C\lceil\rho = 0$ then R' is a resolution refutation of $F\lceil\rho$. As usual, we denote the derivation R' by $R\lceil\rho$.

A resolution derivation is called *regular* if on every path through the dag, no variable is eliminated twice. This condition is inessential for tree-like resolution since minimal tree-like refutations are always regular [24], but regular dag-like refutations can necessarily be exponentially longer than general ones [1].

Tree-like resolution exactly corresponds to the DLL algorithm by the following well-known correspondence: the search tree produced by the run of a DLL algorithm on an unsatisfiable formula F forms a tree-like resolution refutation of F, and from a given tree-like regular resolution refutation of F one can construct a run of a DLL algorithm showing the unsatisfiability of F that produces essentially the given search tree.

In order to define proof systems that correspond to the DLL algorithm with clause learning in the same way, we define *resolution trees with lemmas* (RTL). In these proof systems, the order of branches in the proof tree is significant, thus the underlying trees need to be ordered.

An *ordered binary tree* is a rooted tree in which every node has at most 2 children, and where every node with 2 children has a distinguished *left* and *right* child. The post-ordering $\prec$ of an ordered binary tree is the order in which the nodes of the tree are visited by a post-order traversal, i.e., $u \prec v$ holds for nodes u, v if u is a descendant of v, or if there is a common ancestor w of u and v such that u is a descendant of the left child of w and v is a descendant of the right child of w.

An RTL-derivation of a clause C from a CNF-formula F is an ordered binary tree, in which every node v is labeled with a clause C_v such that:

1. The root is labeled with C.
2. If a node v has one child u, then C_v follows from C_u by the weakening rule.
3. If a node v has two children u_1, u_2, then C_v follows from C_{u_1} and C_{u_2} by the resolution rule.
4. A leaf v is labeled by a clause D in F, or by a clause C labeling some node $u \prec v$. In the latter case we call C a *lemma*.

An RTL-derivation is an RTL(k)-derivation if every lemma C is of width $w(C) \leq k$. An RTL-refutation of F is an RTL-derivation of the empty clause from F.

A subsystem WRTI of RTL was defined by Buss et al. [9], which exactly corresponds to a general formulation of the DLL algorithm with clause learning: the size of a refutation of an unsatisfiable formula F in WRTI has been shown [9] to be polynomially related to the runtime of a schematic algorithm DLL-L-UP on F. This schema DLL-L-UP subsumes most clause learning strategies commonly used in practice, including *first-UIP* [22], *all-UIP*, *decision* [25] and *rel-sat* [5]. A variant of DLL-L-UP which incorporates these learning strategies and also allows for non-chronological backtracking [5] was described by Hoffmann [13] and shown to be likewise simulated by WRTI.

In addition to clause learning, most state-of-the-art satisfiability solvers also use *restarts* [11], therefore their performance is not modeled by RTL. The runtime of a DLL algorithm with clause learning and restarts was shown to be polynomially related to the size of general dag-like resolution refutations, for certain particular learning strategies [6] and more recently also for most natural learning strategies [20]. However, these simulations of general dag-like resolution proofs, as well as the clause learning algorithm of Atserias et al. [4] that simulates resolution proofs of small width, use a particular restart policy: they perform a restart after every conflict. An interesting question is whether general resolution proofs can be simulated with more natural restart policies.

It follows from the mentioned results of Buss et al. [9] that if an unsatisfiable formula F can be solved by a DLL algorithm with clause learning in time t, then it has an RTL-refutation of size polynomial in t. Moreover, if the algorithm learns only clauses of width at most k, then the refutation is in RTL(k). In this work we prove lower bounds on the size of RTL(k)-refutations, which thus yield lower bounds on the runtime of DLL algorithms with width-restricted clause-learning.

3   Resolution Width and Systems of Restrictions

Let X be a set of variables, and $w \in \mathbb{N}$ with $w \leq |X|$. A *w-system of restrictions* over X is a non-empty set $\mathcal{H}$ of restrictions with the following properties:

- $|\rho| \leq w$ for all $\rho \in \mathcal{H}$,
- downward closure: if $\rho \in \mathcal{H}$ and $\rho' \subseteq \rho$, then $\rho' \in \mathcal{H}$,
- the extension property: if $\rho \in \mathcal{H}$ with $|\rho| < w$, and $x \in X \setminus \operatorname{dom} \rho$, then there is $\rho' \in \mathcal{H}$ with $\rho' \supseteq \rho$ and $x \in \operatorname{dom} \rho'$.

We say that $\mathcal{H}$ *avoids* a clause C if $C \lceil \rho \neq 0$ for every $\rho \in \mathcal{H}$, and $\mathcal{H}$ avoids a formula F if $\mathcal{H}$ avoids every clause $C \in F$.

The notion was introduced by Atserias and Dalmau [3], who showed the following characterization of resolution width:

Theorem 3. *A formula F requires resolution width at least w if and only if there is a w-system of restrictions over $var(F)$ that avoids F.*

Atserias and Dalmau [3] called a w-system of restrictions avoiding F a *winning strategy for the Duplicator in the Boolean existential w-pebble game on F*, which is explained by the origin of the notion in the existential k-pebble game [16] in finite model theory. Since we make no use of the model-theoretic background, we chose to use a shorter name for the concept.

For our application we shall use the concept of a system of restrictions being restricted by one of its elements, which we define now.

Lemma 4. *If $\mathcal{H}$ is a w-system of restrictions over X, and $\rho \in \mathcal{H}$ with $|\rho| = r < w$, then the set*

$$\mathcal{H} \lceil \rho := \big\{ \sigma \; ; \; \operatorname{dom} \sigma \subseteq X \setminus \operatorname{dom} \rho \text{ and } \sigma \cup \rho \in \mathcal{H} \text{ and } |\sigma| \leq w - r \big\}$$

is a $(w - r)$-system of restrictions over $X \setminus \operatorname{dom} \rho$.

Note that $\mathcal{H} \lceil \rho$ would be empty, and hence not a system of restrictions in the sense of the definition, if the definition were extended to restrictions $\rho \notin \mathcal{H}$: if there is a $\sigma \in \mathcal{H} \lceil \rho$, then by definition $\sigma \cup \rho \in \mathcal{H}$, and by downward closure $\rho \in \mathcal{H}$.

Proof. Every $\sigma \in \mathcal{H}\lceil\rho$ has $|\sigma| \leq w - r$ by definition. If $\sigma \in \mathcal{H}\lceil\rho$ and $\sigma' \subseteq \sigma$, then $\sigma' \cup \rho \subseteq \sigma \cup \rho$, and thus by downward closure of $\mathcal{H}$ we have $\sigma' \cup \rho \in \mathcal{H}$. Therefore $\sigma' \in \mathcal{H}\lceil\rho$, hence $\mathcal{H}\lceil\rho$ is downward closed.

If $\sigma \in \mathcal{H}\lceil\rho$ is a restriction with $|\sigma| < w - r$ and $x \in X \setminus \rho$ is a variable with $x \notin \operatorname{dom}\sigma$, then $|\sigma \cup \rho| < w$, and hence by the extension property of $\mathcal{H}$ there is $\sigma' \supseteq \sigma \cup \rho$ in $\mathcal{H}$ with $x \in \operatorname{dom}\sigma'$. Then $\sigma' \setminus \rho \supseteq \sigma$ is in $\mathcal{H}\lceil\rho$, and $x \in \operatorname{dom}(\sigma' \setminus \rho)$. Therefore $\mathcal{H}\lceil\rho$ has the extension property, and hence is a $(w - r)$-system of restrictions over $X \setminus \operatorname{dom}\rho$. $\qquad\square$

Lemma 5. *If $\mathcal{H}$ is a w-system of restrictions that avoids F, and $\rho \in \mathcal{H}$, then $\mathcal{H}\lceil\rho$ avoids $F\lceil\rho$.*

Proof. Assume that $\mathcal{H}\lceil\rho$ does not avoid $F\lceil\rho$, i.e., there is a clause C in $F\lceil\rho$ and a restriction $\sigma \in \mathcal{H}\lceil\rho$ such that $C\lceil\sigma = 0$. Since C is in $F\lceil\rho$, there is a clause D with $D\lceil\rho = 0$ such that $C \vee D \in F$. By definition, $\sigma' = \sigma \cup \rho \in \mathcal{H}$ and $(C \vee D)\lceil\sigma' = C\lceil\sigma \vee D\lceil\rho = 0$, hence $\mathcal{H}$ does not avoid F, in contradiction to the hypothesis. $\qquad\square$

4   The Lower Bound

We now prove our main theorem, which shows that lower bounds for RTL(k)-refutations of F follow from lower bounds on the resolution width of F, for formulas F of sufficiently small width.

Theorem 6. *If F is a d-CNF that requires resolution width at least w to refute, then for any k, every RTL(k)-refutation of F is of size at least*

$$2^{w - (k + \max\{d,k\})} \geq 2^{w - (2k+d)}.$$

Proof. Let R be an RTL(k)-refutation of F. By Theorem 3, there is a w-system of restrictions $\mathcal{H}$ that avoids F.

Let C be the first clause in R of small enough width $w(C) \leq k$ to be used as a lemma, and that is not avoided by $\mathcal{H}$. In particular, every lemma in R derived before C *is* avoided by $\mathcal{H}$. Let ρ be the smallest restriction in $\mathcal{H}$ with $C\lceil\rho = 0$, so that we have $r := |\rho| = w(C) \leq k$.

Let R_C be the subtree of R below C, so R_C is an RTL(k)-derivation of C from F. Let G be the set of lemmas used in R_C, so R_C is a tree-like resolution derivation of C from $F \wedge G$, and thus $R' := R_C\lceil\rho$ is a tree-like resolution refutation of $F' := (F \wedge G)\lceil\rho$. Note that every clause in F is of width d, and every clause in G is of width k, therefore $w(F') \leq w(F \wedge G) \leq \max\{d, k\}$.

By the choice of C we know that $\mathcal{H}$ avoids every clause in G, and hence $\mathcal{H}$ avoids $F \wedge G$. It follows by the lemmas above that $\mathcal{H}\lceil\rho$ is a $(w - r)$-system of restrictions that avoids F'.

Therefore, by Theorem 3, F' requires resolution width $w - r \geq w - k$, and thus by Theorem 1, the refutation $R_C\lceil\rho$, and therefore R, is of size at least $2^{(w-k) - w(F')} \geq 2^{w - (k + \max\{d,k\})}$ as claimed. $\qquad\square$

5 Applications

We now instantiate our general lower bound to prove several lower bounds for RTL(k)-refutations of certain concrete formulas.

Ordering Principle

The ordering principle expresses the fact that every finite linear ordering has a maximal element. Its negation is expressed in propositional logic by the following set of clauses Ord_n over the variables $x_{i,j}$ for $1 \leq i, j \leq n$ with $i \neq j$:

$$
\begin{array}{lll}
\bar{x}_{i,j} \vee \bar{x}_{j,i} & \text{for } 1 \leq i < j \leq n & (A_{i,j}) \\[4pt]
x_{i,j} \vee x_{j,i} & \text{for } 1 \leq i < j \leq n & (T_{i,j}) \\[4pt]
\bar{x}_{i,j} \vee \bar{x}_{j,k} \vee \bar{x}_{k,i} & \text{for } 1 \leq i, j, k \leq n \text{ pairwise distinct} & (\Delta_{i,j,k}) \\[4pt]
\bigvee_{1 \leq j \leq n, j \neq i} x_{i,j} & \text{for } 1 \leq i \leq n & (M_i)
\end{array}
$$

The clauses $A_{i,j}$, $T_{i,j}$ and $\Delta_{i,j,k}$ state that in a satisfying assignment, the values of the variables define a linear ordering on n points. The clause M_i expresses that i is not a maximum in this ordering, therefore the formula Ord_n is unsatisfiable.

The formulas Ord_n were introduced by Krishnamurthy [17] as potential hard example formulas for resolution, but short regular resolution refutations for them were constructed by Stålmarck [23].

Proposition 7. *There are dag-like regular resolution refutations of* Ord_n *of size* $O(n^3)$.

Note that the size of the formula Ord_n is $\Theta(n^3)$, so the size of these refutations is linear in the size of the formula. A general simulation of regular resolution by WRTI [9] yields WRTI-refutations of Ord_n of polynomial size. On the other hand, a lower bound for RTL(k)-refutations of Ord_n was shown by the second author [15]:

Theorem 8. *For* $k < n/4$, *every* RTL(k)-*refutation of* Ord_n *is of size* $2^{\Omega(n)}$.

Thus this lower bound shows the necessity to use wide lemmas to refute them efficiently. But since the formula Ord_n itself contains wide clauses, it is conceivable that it is these wide clauses that cause this necessity. We therefore apply our general lower bound to derive similar lower bounds for variants of the ordering principle formulas having small width. The most straightforward way to obtain a formula of small width from any formula is to expand it into a 3-CNF, as described below:

For a CNF-formula F, the 3-CNF-expansion $E_3(F)$ of F is obtained as follows: for every clause $C = a_1 \vee \ldots \vee a_k$ in F of width $w(C) = k \geq 4$, introduce $k+1$ new *extension variables* $y_{C,0}, \ldots, y_{C,k}$, and replace C by the clauses:

$$
y_{C,0} \qquad \bar{y}_{C,i-1} \vee a_i \vee y_{C,i} \quad \text{for } 1 \leq i \leq k \qquad \bar{y}_{C,k}
$$

The formula $E_3(F)$ is obviously in 3-CNF and is satisfiable if and only if F is satisfiable.

Bonet and Galesi [8] show a lower bound of $n/6$ on the resolution width of the 3-CNF expansion $E_3(\mathrm{Ord}_n)$ of the ordering principle. We show a larger lower bound by exhibiting a suitable system of restrictions:

Theorem 9. *The formula $E_3(\mathrm{Ord}_n)$ requires resolution width at least $n/2$.*

For ease of notation, we denote the clauses in the 3-CNF expansion $E_3(M_i)$ of the formula M_i as follows:

$$y_{i,0} \quad \cdots \quad \bar{y}_{i,i-1} \vee x_{i,i+1} \vee y_{i,i+1} \quad \cdots \quad \bar{y}_{i,n}$$

For a non-empty set $D \subseteq \{1, \ldots, n\}$, a total ordering $\prec$ on D and a partial mapping $s : D \rightarrow D$ with the properties

- $s(i)$ is defined for every $i \in D$ except $\max_{\prec} S$,
- $i \prec s(i)$ for every $i \in \mathrm{dom}\, s$,

we define a restriction $\rho(D, \prec, s)$ as follows:

$$x_{i,j} \mapsto \begin{cases} 1 & \text{if } i, j \in D \text{ and } i \prec j \\ 0 & \text{if } i, j \in D \text{ and } j \prec i \end{cases}$$

$$y_{i,j} \mapsto \begin{cases} 1 & \text{if } i \in D, \; s(i) \text{ is defined and } j \geq s(i) \\ 0 & \text{if } i \in D, \; s(i) \text{ is defined and } j < s(i) \end{cases}$$

and is undefined in all other cases. Now let $\mathcal{H}_{\mathrm{ord}}$ be the set of restrictions σ such that $|\sigma| \leq n/2$ and $\sigma \subseteq \rho(D, \prec, s)$ for some subset $D \subseteq \{1, \ldots, n\}$, ordering $\prec$ on D and suitable mapping $s : D \rightarrow D$. Theorem 9 now follows immediately from the following lemma by Theorem 3.

Lemma 10. *$\mathcal{H}_{\mathrm{ord}}$ is an $n/2$-system of restrictions that avoids $E_3(\mathrm{Ord}_n)$.*

Proof. Obviously, $\mathcal{H}_{\mathrm{ord}}$ is non-empty, and the size bound $|\sigma| \leq n/2$ for all $\sigma \in \mathcal{H}_{\mathrm{ord}}$ and downward closure hold by definition. The clauses $A_{i,j}$, $T_{i,j}$ and $\Delta_{i,j,k}$ are avoided since the variables $x_{i,j}$ are set according to the ordering $\prec$. The clauses in $E_3(M_i)$ containing a variable $x_{i,j}$ for $j \neq s(i)$ are avoided since both extension variables are set to the same value, and one of them occurs positively and the other negatively. The clause in $E_3(M_i)$ containing $x_{i,s(i)}$ is avoided since this variable cannot be set to 0.

It remains to show that $\mathcal{H}_{\mathrm{ord}}$ has the extension property. If $\sigma \in \sigma \in \mathcal{H}_{\mathrm{ord}}$ is of size $|\sigma| < n/2$, then there is a set D of size $|D| \leq n - 2$, an ordering $\prec$ on D and a mapping $s : D \rightarrow D$ such that $\sigma \subseteq \rho := \rho(D, \prec, s)$. Let $v \notin \mathrm{dom}\, \sigma$ be a variable left unset by σ.

If $v = x_{i,j}$, then we set $D' := D \cup \{i, j\}$. If $D' = D$, then $\rho(v)$ is already defined, and we set $\prec' = \prec$ and $s' = s$. Otherwise, if $i \in D' \setminus D$, we extend $\prec$ and s to $\prec'$ and s' by setting $i \prec' k$ for every $k \in D$ and $s'(i) := \min_{\prec} D$, and similarly for j.

If $v = y_{i,j}$ and $i \in D$, then $\rho(v)$ is already defined unless $i = \max_\prec D$. In the latter case, we pick an arbitrary $k \notin D$ and set $D' := D \cup \{k\}$, extend $\prec$ to $\prec'$ by setting $i \prec' k$ and s to s' by setting $s'(i) = k$.

If $v = y_{i,j}$ and $i \notin D$, then we set $D' := D \cup \{i\}$, and we extend $\prec$ to $\prec'$ by setting $i \prec' k$ for all $k \in D$ and s to s' by setting $s'(i) = \min_\prec D$.

In all cases $\rho' := \rho(D', \prec', s')$ is an extension $\rho' \supseteq \rho$ with $v \in \operatorname{dom}\rho'$. Let $\sigma' := \sigma \cup \{(v, \rho'(v))\}$, then we have $|\sigma'| \leq n/2$, and $\sigma' \subseteq \rho'$, hence we have found $\sigma' \in \mathcal{H}_{\mathrm{ord}}$ with $v \in \operatorname{dom}\sigma'$. $\qquad\square$

By Theorem 6, a lower bound for $\mathrm{RTL}(k)$-refutations of $E_3(\mathrm{Ord}_n)$ follows from Theorem 9: by choosing $k = n/6$ and observing that for $n \geq 18$ we get $k \geq 3$, we obtain from Theorem 6 a lower bound of $2^{n/2-2n/6} = 2^{n/6}$.

Corollary 11. *For $n \geq 18$, every $\mathrm{RTL}(n/6)$-refutation of $E_3(\mathrm{Ord}_n)$ is of size $2^{n/6}$.*

It follows that a DLL algorithm with clause learning requires exponential time to solve the formulas $E_3(\mathrm{Ord}_n)$ when only clauses of width $n/6$ are learned. On the other hand, from the short regular resolution refutations of Ord_n, short regular refutations of $E_3(\mathrm{Ord}_n)$ are obtained easily. From those, one can construct a run of a DLL algorithm with arbitrary clause learning on $E_3(\mathrm{Ord}_n)$ in polynomial time. Hence we have an example of 3-CNF formulas for which learning wide clauses is necessary to solve them efficiently.

Since the clauses M_i have tree-like derivations from $E_3(M_i)$ of size n, an $\mathrm{RTL}(k)$-refutation of Ord_n of size s can be converted into an $\mathrm{RTL}(k)$-refutation of $E_3(\mathrm{Ord}_n)$ of size sn. Hence the Corollary 11 also yields an easier proof of a slightly weaker variant of the lower bound from Theorem 8: every $\mathrm{RTL}(n/6)$-refutation of Ord_n is of size at least $2^{n/6-\log n}$.

Graph Ordering Principle

A different way to obtain a small width formula from the ordering principle is to consider the restriction of it to a graph, as introduced by Segerlind et al. [21]. The only wide clauses in Ord_n are the clauses M_i stating that there is an element larger than i, for every i. A formula of small width can be obtained by defining for every i a small set of elements and requiring that one element in this set is larger than i.

For a graph $G = (V, E)$ on n vertices $V = \{1, \dots, n\}$, the formula $\mathrm{Ord}(G)$ consists of the clauses $A_{i,j}$, $T_{i,j}$ and $\Delta_{i,j,k}$ of Ord_n, plus the following restricted version of the clauses M_i:

$$\bigvee_{j \in N(i)} x_{i,j} \qquad \text{for } 1 \leq i \leq n \qquad\qquad (M_i')$$

Here $N(i)$ denotes the neighborhood of i in G, i.e.., the set $\{j \in V; \{i, j\} \in E\}$. The formula requires that for every vertex, there is a larger one in the ordering among its neighbors. Thus in this notation, the formula Ord_n is $\mathrm{Ord}(K_n)$ for the complete graph K_n on n vertices. If the graph G has maximum degree $d \geq 3$, then $\mathrm{Ord}(G)$ is a d-CNF.

A graph G on n vertices is called ϵ-*neighborly*, if for all pairs of disjoint subsets $A, B \subseteq V$ with $|A|, |B| \geq \epsilon n$ there is an edge $\{a, b\} \in E$ with $a \in A$ and $b \in B$. A lower bound on the resolution width of $\mathrm{Ord}(G)$ depending on the neighborliness of G was shown by Segerlind et al. [21]:

Theorem 12. *If G is a connected graph on n vertices that is ϵ-neighborly for $0 < \epsilon < 1/3$, then $\mathrm{Ord}(G)$ requires resolution width at least $(\frac{1-3\epsilon}{6})n$.*

The following lemma follows from known results about expander graphs that can e.g. be found in the book of Alon and Spencer [2, Section 9.2]. In what follows a family of graphs $\{G_n; n \in \mathbb{N}\}$ is said to be *explicitly constructible* if there exists a polynomial time Turing machine that on input 1^n outputs a description (say, by means of its adjacency matrix) of the graph G_n.

Lemma 13. *For every $0 < \epsilon < 1/3$ there is a constant $d = O(1/\epsilon^2)$ such that there is an explicitly constructible family of ϵ-neighborly graphs $\{G_n; n \in \mathbb{N}\}$ on n vertices of maximal degree d.*

Proof. As explained in the mentioned book [2, Section 9.2] (and using the notation there), the works of Lubotzky et al. [18] and of Margulis [19] explicitly construct for every integer d of the form $d = p + 1$, where p is a prime congruent to 1 modulo 4, and for every sufficiently large n, a d-regular expander graph G_n, with all eigenvalues of the adjacency matrix except for the largest being bounded in absolute value by $2\sqrt{d-1}$ (such graphs are known as "Ramanujan" expander graphs). For such graphs one may apply Corollary 9.2.5 in the book [2] which implies that every two disjoint subsets of the vertices of G_n of size at least $\frac{2n}{\sqrt{d}}$ must be connected by an edge, i.e., G_n is $\frac{2}{\sqrt{d}}$-neighborly. □

The lemma yields, e.g., a family of graphs G_n on n vertices of maximal degree $d = 150$ that are $1/6$-neighborly, and for these graphs G_n the formula $\mathrm{Ord}(G_n)$ is a d-CNF that requires resolution width $n/12$. By invoking Theorem 6 with $k = n/36$ we obtain the following lower bound for n large enough that $k \geq 150$:

Corollary 14. *For sufficiently large n, every $\mathrm{RTL}(n/36)$-refutation of $\mathrm{Ord}(G_n)$ for the graphs G_n is of size at least $2^{n/36}$.*

As above, it follows that a DLL algorithm with clause learning requires exponential time to solve $\mathrm{Ord}(G_n)$ when only clauses of width $n/36$ are learned. On the other hand, short regular resolution refutations of $\mathrm{Ord}(G_n)$ are contained in the refutations of Ord_n. From those, one can again construct a run of a DLL algorithm with arbitrary clause learning on $\mathrm{Ord}(G_n)$ in polynomial time. Hence the formulas $\mathrm{Ord}(G_n)$ are another example of formulas of small width for which learning wide clauses is necessary to solve them efficiently.

Dense Linear Ordering Principle

The dense linear ordering principle yields another family of formulas that have short regular resolution refutations, but require large resolution width. It says that a finite linear ordering cannot be dense. It gives rise to an unsatisfiable set of clauses DLO_n, in the variables $x_{i,j}$ representing the ordering as in Ord_n, and additional variables $z_{i,j,k}$ intended to express that j is between i and k in the ordering. It consists of the clauses $A_{i,j}$, $T_{i,j}$ and $\Delta_{i,j,k}$ of Ord_n, plus new clauses containing the variables $z_{i,j,k}$:

$$\bar{x}_{i,j} \vee \bar{x}_{j,k} \vee z_{i,j,k} \qquad \text{for } 1 \leq i, j, k \leq n \text{ pairwise distinct}$$

$$\bar{z}_{i,j,k} \vee x_{i,j} \qquad \text{for } 1 \leq i, j, k \leq n \text{ pairwise distinct}$$

$$\bar{z}_{i,j,k} \vee x_{j,k} \qquad \text{for } 1 \leq i, j, k \leq n \text{ pairwise distinct}$$

$$\bar{x}_{i,k} \vee \bigvee_{1 \leq j \leq n, j \neq i,k} z_{i,j,k} \qquad \text{for } 1 \leq i, k \leq n \text{ with } i \neq k \qquad (D_{i,k})$$

The first three groups of clauses enforce that the values of the variables $z_{i,j,k}$ define the relation "j is between i and k", and the clause $D_{i,k}$ states that if $i \prec k$, then there exists an element between i and k. Therefore the formula DLO_n expresses that there is a dense linear ordering on n points, and is thus unsatisfiable.

Atserias and Dalmau [3] show a lower bound on the resolution width of the 3-CNF expansion $E_3(\mathrm{DLO}_n)$ of the dense linear ordering principle:

Theorem 15. *The formula $E_3(\mathrm{DLO}_n)$ requires resolution width at least $n/3$.*

Using Theorem 6 with $k = n/9$, it follows:

Corollary 16. *For $n \geq 27$, every $\mathrm{RTL}(n/9)$-refutation of $E_3(\mathrm{DLO}_n)$ is of size at least $2^{n/9}$.*

Again, it follows that a DLL algorithm with clause learning requires exponential time to solve $E_3(\mathrm{DLO}_n)$ when only clauses of width $n/9$ are learned. On the other hand, short resolution refutations of DLO_n and of $E_3(\mathrm{DLO}_n)$ are given by Atserias and Dalmau [3], and these refutations are easily seen to be regular. Hence there is a run of a DLL algorithm with arbitrary clause learning on $E_3(\mathrm{DLO}_n)$ in polynomial time, and thus learning wide clauses is necessary to solve these formulas efficiently.

References

1. Alekhnovich, M., Johannsen, J., Pitassi, T., Urquhart, A.: An exponential separation between regular and general resolution. Theory of Computing 3, 81–102 (2007); Preliminary Version in Proc. 34th ACM Symposium on Theory of Computing (2002)
2. Alon, N., Spencer, J.: The Probabilistic Method. John Wiley and Sons, Chichester (2002)
3. Atserias, A., Dalmau, V.: A combinatorial characterization of resolution width. Journal of Computer and System Sciences 74, 323–334 (2008); Preliminary version in Proc. 18th IEEE Conference on Computational Complexity (2003)

4. Atserias, A., Fichte, J.K., Thurley, M.: Clause learning algorithms with many restarts and bounded-width resolution. In: Kullmann, O. (ed.) SAT 2009. LNCS, vol. 5584, pp. 114–127. Springer, Heidelberg (2009)
5. Bayardo Jr., R.J., Schrag, R.C.: Using CSP look-back techniques to solve real-world SAT instances. In: Proc. 14th Natl. Conference on Artificial Intelligence, pp. 203–208 (1997)
6. Beame, P., Kautz, H.A., Sabharwal, A.: Towards understanding and harnessing the potential of clause learning. Journal of Artificial Intelligence Research 22, 319–351 (2004)
7. Ben-Sasson, E., Wigderson, A.: Short proofs are narrow — resolution made simple. Journal of the ACM 48 (2001); Preliminary Version in Proc. 31st ACM Symposium on Theory of Computing (1999)
8. Bonet, M.L., Galesi, N.: Optimality of size-width tradeoffs for resolution. Computational Complexity 10(4), 261–276 (2001)
9. Buss, S.R., Hoffmann, J., Johannsen, J.: Resolution trees with lemmas: Resolution refinements that characterize DLL algorithms with clause learning. Logical Methods in Computer Science 4(4) (2008)
10. Davis, M., Logemann, G., Loveland, D.W.: A machine program for theorem-proving. Communications of the ACM 5(7), 394–397 (1962)
11. Gomes, C.P., Selman, B., Crato, N.: Heavy-tailed distributions in combinatorial search. In: Smolka, G. (ed.) CP 1997. LNCS, vol. 1330. Springer, Heidelberg (1997)
12. Hertel, P., Bacchus, F., Pitassi, T., van Gelder, A.: Clause learning can effectively p-simulate general propositional resolution. In: Fox, D., Gomes, C.P. (eds.) Proceedings of the 23rd AAAI Conference on Artificial Intelligence, AAAI 2008, pp. 283–290. AAAI Press, Menlo Park (2008)
13. Hoffmann, J.: Resolution proofs and DLL-algorithms with clause learning. Diploma Thesis, LMU München (2007)
14. Iwama, K., Miyazaki, S.: Tree-like resolution is superpolynomially slower than dag-like resolution for the pigeonhole principle. In: Aggarwal, A.K., Pandu Rangan, C. (eds.) ISAAC 1999. LNCS, vol. 1741, p. 133. Springer, Heidelberg (1999)
15. Johannsen, J.: An exponential lower bound for width-restricted clause learning. In: Kullmann, O. (ed.) SAT 2009. LNCS, vol. 5584, pp. 128–140. Springer, Heidelberg (2009)
16. Kolaitis, P.G., Vardi, M.Y.: On the expressive power of Datalog: Tools and a case study. Journal of Computer and System Sciences 51(1), 110–134 (1990); Preliminary version in Proc. 9th ACM Symposium on Principles of Database Systems (1990)
17. Krishnamurthy, B.: Short proofs for tricky formulas. Acta Informatica 22, 253–274 (1985)
18. Lubotzky, A., Phillips, R., Sarnak, P.: Ramanujan graphs. Combinatorica 8(3), 261–277 (1988)
19. Margulis, G.A.: Explicit group-theoretical constructions of combinatorial schemes and their application to the design of expanders and concentrators. Problems of Information Transmission 24, 39–46 (1988)
20. Pipatsrisawat, K., Darwiche, A.: On the power of clause-learning SAT solvers with restarts. In: Gent, I.P. (ed.) CP 2009. LNCS, vol. 5732, pp. 654–668. Springer, Heidelberg (2009)

21. Segerlind, N., Buss, S.R., Impagliazzo, R.: A switching lemma for small restrictions and lower bounds for k-DNF resolution. SIAM Journal on Computing 33(5), 1171–1200 (2004); Preliminary version in Proc. 43rd IEEE Symposium on Foundations of Computer Science (2002)
22. Silva, J.P.M., Sakallah, K.A.: GRASP - a new search algorithm for satisfiability. In: Proc. IEEE/ACM International Conference on Computer Aided Design (ICCAD), pp. 220–227 (1996)
23. Stålmarck, G.: Short resolution proofs for a sequence of tricky formulas. Acta Informatica 33, 277–280 (1996)
24. Tseitin, G.: On the complexity of derivation in propositional calculus. Studies in Constructive Mathematics and Mathematical Logic, Part 2, pp. 115–125 (1968)
25. Zhang, L., Madigan, C.F., Moskewicz, M.W., Malik, S.: Efficient conflict driven learning in a Boolean satisfiability solver. In: Proc. IEEE/ACM International Conference on Computer Aided Design (ICCAD), pp. 279–285 (2001)

Proof Complexity of Propositional Default Logic[*]

Olaf Beyersdorff[1], Arne Meier[2], Sebastian Müller[3],
Michael Thomas[2], and Heribert Vollmer[2]

[1] Institute of Computer Science, Humboldt University Berlin, Germany
[2] Institute of Theoretical Computer Science, Leibniz University Hanover, Germany
[3] Faculty of Mathematics and Physics, Charles University Prague, Czech Republic

Abstract. Default logic is one of the most popular and successful formalisms for non-monotonic reasoning. In 2002, Bonatti and Olivetti introduced several sequent calculi for credulous and skeptical reasoning in propositional default logic. In this paper we examine these calculi from a proof-complexity perspective. In particular, we show that the calculus for credulous reasoning obeys almost the same bounds on the proof size as Gentzen's system *LK*. Hence proving lower bounds for credulous reasoning will be as hard as proving lower bounds for *LK*. On the other hand, we show an exponential lower bound to the proof size in Bonatti and Olivetti's enhanced calculus for skeptical default reasoning.

1 Introduction

Trying to understand the nature of human reasoning has been one of the most fascinating adventures since ancient times. It has long been argued that due to its monotonicity, classical logic is not adequate to express the flexibility of commonsense reasoning. To overcome this deficiency, a number of formalisms have been introduced (cf. [19]), of which Reiter's default logic [20] is one of the most popular and widely used systems. Default logic extends the usual logical (first-order or propositional) derivations by patterns for default assumptions. These are of the form "in the absence of contrary information, assume ...". Reiter argued that his logic adequately formalizes human reasoning under the *closed world assumption*. Today default logic is widely used in artificial intelligence and computational logic.

The semantics and the complexity of default logic have been intensively studied during the last decades (cf. [6] for a survey). In particular, Gottlob [12] has identified and studied two reasoning tasks for propositional default logic: the *credulous* and the *skeptical* reasoning problem which can be understood as analogues of the classical problems SAT and TAUT. Due to the stronger expressibility of default logic, however, credulous and skeptical reasoning become

---

[*] Research supported in part by DFG grants KO 1053/5-2 and VO 630/6-1, by a grant from the John Templeton Foundation, and by the Marie Curie FP7 Initial Training Network MALOA (no. 238381).

O. Strichman and S. Szeider (Eds.): SAT 2010, LNCS 6175, pp. 30–43, 2010.

harder than their classical counterparts—they are complete for the second level Σ_2^p and Π_2^p of the polynomial hierarchy, respectively [12].

Less is known about the complexity of proofs in default logic. While there is a rich body of results for propositional proof systems (cf. [16]), proof complexity of non-classical logics has only recently attracted more attention, and a number of exciting results have been obtained for modal and intuitionistic logics [13–15]. Starting with Reiter's work [20], several proof-theoretic methods have been developed for default logic (cf. [1, 10, 17, 18, 21] and [8] for a survey). However, most of these formalisms employ external constraints to model non-monotonic deduction and thus cannot be considered purely axiomatic (cf. [9] for an argument). This was achieved by Bonatti and Olivetti [3] who designed simple and elegant sequent calculi for credulous and skeptical default reasoning. Subsequently, Egly and Tompits [9] extended Bonatti and Olivetti's calculi to first-order default logic and showed a speed-up of these calculi over classical first-order logic, *i.e.*, they construct sequences of first-order formulae which need long classical proofs but have short derivations using default rules.

In the present paper we investigate the original calculi of Bonatti and Olivetti [3] from a proof-complexity perspective. Apart from some preliminary observations in [3], this comprises, to our knowledge, the first comprehensive study of lengths of proofs in propositional default logic. Our results can be summarized as follows. Bonatti and Olivetti's *credulous default calculus* BO_{cred} obeys almost the same bounds to the proof size as Gentzen's propositional sequent calculus LK, *i.e.*, we show that upper bounds to the proof size in both calculi are polynomially related. The same result also holds for the proof length (the number of steps in the system). Thus, proving lower bounds to the size of BO_{cred} will be as hard as proving lower bounds to LK (or, equivalently, to Frege systems), which constitutes a major challenge in propositional proof complexity [4, 16]. This result also has implications for automated theorem proving. Namely, we transfer the non-automatizability result of Bonet, Pitassi, and Raz [5] for Frege systems to default logic: BO_{cred}-proofs cannot be efficiently generated, unless factoring integers is possible in polynomial time.

While already BO_{cred} appears to be a strong proof system for credulous default reasoning, admitting very concise proofs, we also exhibit a general method of how to construct a proof system $Cred(P)$ for credulous reasoning from a propositional proof system P. This system $Cred(P)$ bears the same relation to P with respect to proof size as BO_{cred} does to LK. Thus, choosing for example P as extended Frege might lead to stronger proof systems for credulous reasoning.

For *skeptical reasoning*, the situation is different. Bonatti and Olivetti [3] construct two proof systems for this task. While they already show an exponential lower bound for their first skeptical calculus, we obtain also an exponential lower bound to the proof length in their enhanced skeptical calculus.

This paper is organized as follows. In Sect. 2 we start with some background information on proof systems and default logic. The calculi of Bonatti and Olivetti [3] consist of four main ingredients: classical sequents, antisequents to refute non-tautologies, a residual calculus, and default rules. Thus we start our investigation in Sect. 3 by analyzing the preliminary antisequent and residual calculi. Our main results on the proof complexity of credulous and skeptical default reasoning follow in Sects. 4 and 5, respectively. In Sect. 6, we conclude with a discussion and some open questions.

Due to space constraints, some proofs are only sketched.

2   Preliminaries

We assume familiarity with propositional logic and basic notions from complexity theory (cf.[16]). By $\mathcal{L}$ we denote the set of all propositional formulae over some fixed standard set of connectives. For $T \subseteq \mathcal{L}$, the set of all logical consequences of T will be denoted by $Th(T)$.

2.1   Proof Systems

Cook and Reckhow [7] defined the notion of a *proof system* for an arbitrary language L as a polynomial-time computable function f with range L. A string w with $f(w) = x$ is called an *f-proof* for $x \in L$. Proof systems for $L = $ TAUT are called *propositional proof systems*. The sequent calculus LK of Gentzen [11] is one of the most important and best studied propositional proof systems. It is well known that LK and Frege systems mutually p-simulate each other(cf. [16]).

There are two measures which are of primary interest in proof complexity. The first is the minimal *size* of an f-proof for some given element $x \in L$. To make this precise, let $s_f(x) = \min\{|w| \mid f(w) = x\}$ and $s_f(n) = \max\{s_f(x) \mid |x| \leq n\}$. We say that the proof system f is *t-bounded* if $s_f(n) \leq t(n)$ for all $n \in \mathbb{N}$. If t is a polynomial, then f is called *polynomially bounded*. Another interesting parameter of a proof is the *length* defined as the number of proof steps. This measure only makes sense for proof systems where proofs consist of lines containing formulae or sequents. This is the case for LK and most systems studied in this paper. For such a system f, we let $t_f(\varphi) = \min\{k \mid f(\pi) = \varphi$ and π uses k steps$\}$ and $t_f(n) = \max\{t_f(\varphi) \mid |\varphi| \leq n\}$. Obviously, it holds that $t_f(n) \leq s_f(n)$, but the two measures are even polynomially related for a number of natural systems as extended Frege (cf. [16]).

For sequent calculi one distinguishes between dag-like and tree-like proofs where in the latter notion each derived sequent can be used at most once as a prerequisite of a rule. While for LK these two measures are equivalent [16], we will concentrate here only on the stronger dag-like model.

2.2   Default Logic

Default logic is an extension of classical logic that has been proposed by Reiter [20]. The logic is non-monotonic in the sense that an increase in information may decrease the number of consequences. A *default theory* $\langle W, D \rangle$ consists of a set W of propositional sentences and a set D of *defaults*. A default (rule) δ is an inference rule of the form $\dfrac{\alpha : \beta}{\gamma}$, where α and γ are propositional formulae and β is a set of propositional formulae. The *prerequisite* α is also referred to as $p(\delta)$, the formulae in β are called *justifications* (referred to as $j(\delta)$), and γ is the *conclusion* that is referred to as $c(\delta)$. *Stable extensions* are originally defined in terms of a fixed-point equation [20], but we use the following characterization as a starting definition:

Theorem 1 (Reiter [20]). *Let $E \subseteq \mathcal{L}$ be a set of formulae and $\langle W, D \rangle$ be a default theory. Furthermore let $E_0 = W$, and*

$$E_{i+1} = Th(E_i) \cup \{c(\delta) \mid \delta \in D, E_i \vdash p(\delta), \neg j(\delta) \cap E = \emptyset\},$$

where $\neg j(\delta)$ denotes the set of all negated sentences contained in $j(\delta)$. Then E is a (stable) extension of $\langle W, D \rangle$ if and only if $E = \bigcup_{i \in \mathbb{N}} E_i$.

A default theory $\langle W, D \rangle$ can have none or several stable extensions (cf. [12] for examples). A sentence $\psi \in \mathcal{L}$ is *credulously* entailed by $\langle W, D \rangle$ if ψ holds in *some* stable extension of $\langle W, D \rangle$. If ψ holds in *every* extension of $\langle W, D \rangle$, then ψ is *skeptically* entailed by $\langle W, D \rangle$.

Default rules with empty justification are called *residues*. We use the notation $\mathcal{L}^{res} = \mathcal{L} \cup \left\{ \frac{\alpha}{\gamma} \mid \alpha, \gamma \in \mathcal{L} \right\}$ for the set of all formulae and residues. Residues can be used to alternatively characterize stable extensions. For a set D of defaults and $E \subseteq \mathcal{L}$ let $RES(D, E) = \left\{ \frac{p(\delta)}{c(\delta)} \mid \delta \in D, E \cap \neg j(\delta) = \emptyset \right\}$. Apparently, $RES(D, E)$ is a set of residues. We can then build stable extensions via the following closure operator. For a set R of residues we define $Cl_0(W, R) = W$ and $Cl_{i+1}(W, R) = Th(Cl_i(W, R)) \cup \left\{ \gamma \mid \frac{\alpha}{\gamma} \in R, \alpha \in Th(Cl_i(W, R)) \right\}$. Let $Cl(W, R) = \bigcup_{i=0}^{\infty} Cl_i(W, R)$. Then we obtain for the sets E_i from Theorem 1:

Proposition 1 (Bonatti, Olivetti [3]). *Let $\langle W, D \rangle$ be a default theory and let $E \subseteq \mathcal{L}$. Then $E_i = Cl_i(W, RES(D, E))$ for all $i \in \mathbb{N}$. In particular, E is a stable extension of $\langle W, D \rangle$ if and only if $E = Cl(W, RES(D, E))$.*

If D only contains residues, then there is an easier way of characterizing Cl:

Lemma 1 (Bonatti, Olivetti [3]). *For $D \subseteq \mathcal{L}^{res} \setminus \mathcal{L}$, $W \subseteq \mathcal{L}$, and for $i \in \mathbb{N}$ let $C_0 = W$ and $C_{i+1} = C_i \cup \left\{ \gamma \mid \frac{\alpha}{\gamma} \in D, \alpha \in Th(C_i) \right\}$. Then $\gamma \in Cl(W, D)$ if and only if there exists $k \in \mathbb{N}$ with $\gamma \in Th(C_k)$.*

3   Complexity of the Antisequent and Residual Calculi

Bonatti and Olivetti's calculi for default logic use four main ingredients: usual propositional sequents and rules of LK, antisequents to refute formulae, residual rules, and default rules. In this section we will investigate the complexity of the antisequent calculus AC and the residual calculus RC.

We start with the definition of Bonatti's *antisequent calculus AC* from [2]. A related refutation calculus for first-order logic was previously developed by Tiomkin [22]. In AC we use *antisequents* $\Gamma \nvdash \Delta$, where $\Gamma, \Delta \subseteq \mathcal{L}$. Intuitively,

$$\frac{\Gamma \nvdash \Sigma, \alpha}{\Gamma, \neg\alpha \nvdash \Sigma} \ (\neg\nvdash) \qquad\qquad \frac{\Gamma, \alpha \nvdash \Sigma}{\Gamma \nvdash \Sigma, \neg\alpha} \ (\nvdash \neg)$$

$$\frac{\Gamma, \alpha, \beta \nvdash \Sigma}{\Gamma, \alpha \wedge \beta \nvdash \Sigma} \ (\wedge \nvdash) \quad \frac{\Gamma \nvdash \Sigma, \alpha}{\Gamma \nvdash \Sigma, \alpha \wedge \beta} \ (\nvdash \bullet\wedge) \quad \frac{\Gamma \nvdash \Sigma, \beta}{\Gamma \nvdash \Sigma, \alpha \wedge \beta} \ (\nvdash \wedge\bullet)$$

$$\frac{\Gamma \nvdash \Sigma, \alpha, \beta}{\Gamma \nvdash \Sigma, \alpha \vee \beta} \ (\nvdash \vee) \quad \frac{\Gamma, \alpha \nvdash \Sigma}{\Gamma, \alpha \vee \beta \nvdash \Sigma} \ (\bullet\vee \nvdash) \quad \frac{\Gamma, \beta \nvdash \Sigma}{\Gamma, \alpha \vee \beta \nvdash \Sigma} \ (\vee\bullet \nvdash)$$

$$\frac{\Gamma, \alpha \nvdash \Sigma, \beta}{\Gamma \nvdash \Sigma, \alpha \rightarrow \beta} \ (\nvdash\rightarrow) \quad \frac{\Gamma \nvdash \Sigma, \alpha}{\Gamma, \alpha \rightarrow \beta \nvdash \Sigma} \ (\bullet \rightarrow\nvdash) \quad \frac{\Gamma, \beta \nvdash \Sigma}{\Gamma, \alpha \rightarrow \beta \nvdash \Sigma} \ (\rightarrow \bullet \nvdash)$$

Fig. 1. Inference rules of the antisequent calculus AC

$\Gamma \nvdash \Delta$ means that $\bigvee \Delta$ does not follow from $\bigwedge \Gamma$. Axioms of AC are all sequents $\Gamma \nvdash \Delta$, where Γ and Δ are disjoint sets of propositional variables. The inference rules of AC are shown in Fig. 1. For this calculus, Bonatti [2] shows:

Theorem 2 (Bonatti [2]). *The calculus AC is sound and complete.*

Concerning the size of proofs in the antisequent calculus we observe:

Proposition 2. *The antisequent calculus AC is polynomially bounded.*

Proof. Observe that the calculus contains only unary inference rules, each of which reduces the logical complexity of one of the contained formulae (if perceived bottom-up). Thus each use of an inference rule decrements the size of the formulae by at least one. After a linear number of steps we end up with only propositional variables which we cannot reduce any further. Each antisequent is of linear size, hence the complete derivation has quadratic size. □

The above observation is not very astounding, since, to verify $\Gamma \nvdash \Delta$ we could alternatively guess assignments to the propositional variables in Γ and Δ and thereby verify antisequents in NP.

We now turn to the *residual calculus RC* of Bonatti and Olivetti [3]. Its objects are *residual sequents* $\langle W, R \rangle \vdash \Delta$ and *residual antisequents* $\langle W, R \rangle \nvdash \Delta$ where $W, \Delta \subseteq \mathcal{L}$ and $R \subseteq \mathcal{L}^{res}$. The intuitive meaning is that Δ does (respectively does not) follow from W using the residues R. The rules of RC comprise of the inference rules from Fig. 2 together with the rules of LK and AC. However, the use of rules from LK and AC is restricted to purely propositional (anti)sequents. For this calculus, Bonatti and Olivetti [3] showed:

Theorem 3 (Bonatti, Olivetti [3]). *The residual calculus RC is sound and complete, i.e., for all default theories $\langle W, R \rangle$ with $R \subseteq \mathcal{L}^{res}$ and all $\Delta \subseteq \mathcal{L}$,*

1. $\langle W, R \rangle \vdash \Delta$ is derivable in RC if and only if $\bigvee \Delta \in Cl(W, R)$;

$$(\textbf{Re1}) \ \frac{\Gamma \vdash \Delta}{\Gamma, \frac{\alpha}{\gamma} \vdash \Delta} \qquad\qquad (\textbf{Re2}) \ \frac{\Gamma \vdash \alpha \quad \Gamma, \gamma \vdash \Delta}{\Gamma, \frac{\alpha}{\gamma} \vdash \Delta}$$

$$(\textbf{Re3}) \ \frac{\Gamma \nvdash \Delta \quad \Gamma \nvdash \alpha}{\Gamma, \frac{\alpha}{\gamma} \nvdash \Delta} \qquad\qquad (\textbf{Re4}) \ \frac{\Gamma, \gamma \nvdash \Delta}{\Gamma, \frac{\alpha}{\gamma} \nvdash \Delta}$$

Fig. 2. Inference rules of the residual calculus RC

2. $\langle W, R \rangle \nvdash \Delta$ *is derivable in* RC *if and only if* $\bigvee \Delta \notin Cl(W, R)$.

To bound the lengths of proofs in this calculus we exploit the property that residues only have to be used at a certain level and are not used to deduce any formulae afterwards (cf. Lemma 1). Using this we prove that the complexity of RC is tightly linked to that of LK.

Lemma 2. *There exist a polynomial p and a constant c such that $s_{RC}(n) \leq p(n) \cdot s_{LK}(cn)$ and $t_{RC}(n) \leq p(n) \cdot t_{LK}(cn)$.*

Proof. The proof consists of two parts. First we will show the bounds stated above for sequents. In the second part we will then show that antisequents even admit polynomial-size proofs in RC.

Assume first that we want to derive the sequent $\langle W, R \rangle \vdash \Delta$, where $W, \Delta \subseteq \mathcal{L}$ and $R = \{r_1, \ldots, r_k\}$ is a set of residues with $r_i = \frac{\alpha_i}{\gamma_i}$. Let $R' \subseteq R$ be minimal with respect to the size $|R'|$ such that $\langle W, R' \rangle \vdash \Delta$. We may w.l.o.g. assume that $R' = \{r_1, \ldots, r_{k'}\}$ and $k' \leq k$. Furthermore, by Lemma 1, we may assume that the rules r_i are ordered in the way they are applied when computing the sets C_i. In particular, this means that for each $i = 1, \ldots, k'$,

$$W \cup \{\gamma_1, \ldots, \gamma_{i-1}\} \vdash \alpha_i$$

is a true propositional sequent for which we fix an LK-proof Π_i. We augment Π_i by $k' - i$ applications of rule (**Re1**) to obtain

$$\langle W \cup \{\gamma_1, \ldots, \gamma_{i-1}\}, \{r_{i+1}, \ldots, r_{k'}\} \rangle \vdash \alpha_i \ .$$

Let us call the proof of this sequent Π_i'.

The proof tree depicted in Fig. 3 for deriving $\langle W, R \rangle \vdash \Delta$ unfurls as follows. We start with an LK-proof for the sequent $W \cup \{\gamma_1, \ldots, \gamma_{k'}\} \vdash \Delta$ and then apply k'-times the rule (**Re2**) in the step

$$\frac{\langle W \cup \{\gamma_1, \ldots, \gamma_{i-1}\}, \{r_{i+1}, \ldots, r_{k'}\} \rangle \vdash \alpha_i \quad \langle W \cup \{\gamma_1, \ldots, \gamma_i\}, \{r_{i+1}, \ldots, r_{k'}\} \rangle \vdash \Delta}{\langle W \cup \{\gamma_1, \ldots, \gamma_{i-1}\}, \{r_i, \ldots, r_{k'}\} \rangle \vdash \Delta}$$

to reach $\langle W, R' \rangle \vdash \Delta$. To derive the left prerequisite we use the proof Π_i'. Finally we use $k - k'$ applications of the rule (**Re1**) to get $\langle W, R \rangle \vdash \Delta$.

$$\cfrac{\Pi'_1 \quad \cfrac{\Pi'_2 \quad \cfrac{\Pi'_{k'} \quad \langle W \cup \{\gamma_1, \dots, \gamma_{k'}\}, \emptyset \rangle \vdash \Delta}{\vdots}(\textbf{Re2})}{\cfrac{\langle W \cup \{\gamma_1, \gamma_2\}, \{r_3, \dots, r_{k'}\} \rangle \vdash \Delta}{\langle W \cup \{\gamma_1\}, \{r_2, \dots, r_{k'}\} \rangle \vdash \Delta}(\textbf{Re2})}(\textbf{Re2})}{\cfrac{\langle W, R' \rangle \vdash \Delta}{\vdots}(\textbf{Re1})}$$

$$\langle W, R \rangle \vdash \Delta$$

Fig. 3. Proof tree for the sequent $\langle W, R \rangle \vdash \Delta$ in the residual calculus

Our proof for $\langle W, R \rangle \vdash \Delta$ uses at most $(k'+1) \cdot t_{LK}(n) + \frac{k'(k'+1)}{2} + k$ steps, i.e., $t_{RC}(n) \leq \mathcal{O}(n \cdot t_{LK}(n) + n^2)$. Each sequent is of linear size. Hence, $s_{RC}(n) \leq p(n) \cdot s_{LK}(n)$ for some polynomial p.

In the second part of the proof we have to show that any true antisequent has an RC-proof of polynomial size. We omit the details. $\square$

Let us remark that while the RC-proof of $\langle W, R \rangle \vdash \Delta$ in Fig. 3 is tree-like, this is not true for our dag-like RC-proof of $\langle W, R \rangle \nvdash \Delta$ constructed in the second part of the proof of Lemma 2.

4   Proof Complexity of Credulous Default Reasoning

Now we turn to the analysis of Bonatti and Olivetti's calculus for credulous default reasoning. An essential ingredient of the calculus are *provability constraints* which resemble a necessity modality. Provability constraints are of the form $\mathbf{L}\alpha$ or $\neg \mathbf{L}\alpha$ with $\alpha \in \mathcal{L}$. A set $E \subseteq \mathcal{L}$ satisfies a constraint $\mathbf{L}\alpha$ if $\alpha \in Th(E)$. Similarly, E satisfies $\neg \mathbf{L}\alpha$ if $\alpha \notin Th(E)$.

We can now describe the calculus BO_{cred} of Bonatti and Olivetti [3] for credulous default reasoning. A *credulous default sequent* is a 3-tuple $\langle \Sigma, \Gamma, \Delta \rangle$, denoted by $\Sigma; \Gamma \vdash \Delta$, where $\Gamma = \langle W, D \rangle$ is a default theory, Σ is a set of provability constraints and Δ is a set of propositional sentences. Semantically, the sequent $\Sigma; \Gamma \vdash \Delta$ is true, if there exists a stable extension E of Γ which satisfies all of the constraints in Σ and $\bigvee \Delta \in E$. The calculus BO_{cred} uses such sequents and extends LK, AC, and RC by the inference rules in Fig. 4.

For this calculus Bonatti and Olivetti [3] show the following:

Theorem 4 (Bonatti, Olivetti [3]). *BO_{cred} is sound and complete, i.e., a credulous default sequent is true if and only if it is derivable in BO_{cred}.*

We now investigate lengths of proofs in BO_{cred}. Our next lemma shows that upper bounds on the proof size of RC can be transferred to BO_{cred}.

Lemma 3. *For any function $t(n)$, if RC is $t(n)$-bounded, then BO_{cred} is $p(n) \cdot t(n)$-bounded for some polynomial p. The same relation holds for the number of steps in RC and BO_{cred}.*

$$(\textbf{cD1})\ \frac{\Gamma \vdash \Delta}{;\ \Gamma \hspace{-2pt}\sim\hspace{-2pt}\Delta} \qquad (\textbf{cD2})\ \frac{\Gamma \vdash \alpha \qquad \Sigma;\ \Gamma \hspace{-2pt}\sim\hspace{-2pt}\Delta}{\textbf{L}\alpha,\ \Sigma;\ \Gamma \hspace{-2pt}\sim\hspace{-2pt}\Delta} \qquad (\textbf{cD3})\ \frac{\Gamma \nvdash \alpha \qquad \Sigma;\ \Gamma \hspace{-2pt}\sim\hspace{-2pt}\Delta}{\neg\textbf{L}\alpha,\ \Sigma;\ \Gamma \hspace{-2pt}\sim\hspace{-2pt}\Delta}$$

where $\Gamma \subseteq \mathcal{L}^{res}$ in rules $(\textbf{cD1})$, $(\textbf{cD2})$, and $(\textbf{cD3})$

$$(\textbf{cD4})\ \frac{\textbf{L}\neg\beta_i,\ \Sigma;\ \Gamma \hspace{-2pt}\sim\hspace{-2pt}\Delta}{\Sigma;\ \Gamma,\ \frac{\alpha:\ \beta_1\ldots\beta_n}{\gamma} \hspace{-2pt}\sim\hspace{-2pt}\Delta} \qquad (\textbf{cD5})\ \frac{\neg\textbf{L}\neg\beta_1 \ldots \neg\textbf{L}\neg\beta_n,\ \Sigma;\ \Gamma,\ \frac{\alpha}{\gamma} \hspace{-2pt}\sim\hspace{-2pt}\Delta}{\Sigma;\ \Gamma,\ \frac{\alpha:\ \beta_1\ldots\beta_n}{\gamma} \hspace{-2pt}\sim\hspace{-2pt}\Delta}$$

Fig. 4. Inference rules for the credulous default calculus BO_{cred}

Proof. Let $\Sigma; \Gamma \hspace{-2pt}\sim\hspace{-2pt}\Delta$ be a true credulous default sequent. We will construct a BO_{cred}-derivation of $\Sigma; \Gamma \hspace{-2pt}\sim\hspace{-2pt}\Delta$ starting from the bottom with the given sequent. Observe that we cannot use any of the rules $(\textbf{cD1})$ through $(\textbf{cD3})$ as long as Γ contains proper defaults with nonempty justification. Thus we first have to reduce all defaults to residues plus some set of constraints using $(\textbf{cD4})$ or $(\textbf{cD5})$. As one of these rules has to be applied exactly once for each appearance of some default in Γ we end up with $\Sigma'; \Gamma' \hspace{-2pt}\sim\hspace{-2pt}\Delta$, where $|\Sigma'|$ is polynomial in $|\Gamma \cup \Sigma|$ and Γ' is equal to Γ on its propositional part and contains some of the corresponding residues instead of the defaults from Γ. From this point on we can only use rules $(\textbf{cD2})$ and $(\textbf{cD3})$ until we have eliminated all constraints and then finally apply rule $(\textbf{cD1})$ once. Thus, BO_{cred}-proofs look as shown in Fig. 5 where RC indicates

$$\frac{RC \qquad \dfrac{\dfrac{RC}{\Gamma' \hspace{-2pt}\sim\hspace{-2pt}\Delta}\ (\textbf{cD1})}{\sigma; \Gamma' \hspace{-2pt}\sim\hspace{-2pt}\Delta}\ (\textbf{cD2})\ \text{or}\ (\textbf{cD3})}{}\ (\textbf{cD2})\ \text{or}\ (\textbf{cD3})$$

$$\vdots$$

$$\frac{RC \qquad \Sigma''; \Gamma' \hspace{-2pt}\sim\hspace{-2pt}\Delta}{\Sigma'; \Gamma' \hspace{-2pt}\sim\hspace{-2pt}\Delta}\ (\textbf{cD2})\ \text{or}\ (\textbf{cD3})$$

$$\phantom{\frac{\Sigma'; \Gamma'}{}}\ (\textbf{cD4})\ \text{or}\ (\textbf{cD5})$$

$$\vdots$$

$$\Sigma; \Gamma \hspace{-2pt}\sim\hspace{-2pt}\Delta$$

Fig. 5. The structure of the BO_{cred}-proof in Lemma 3

a derivation in the residual calculus and σ is the remaining constraint from Σ after applications of $(\textbf{cD2})$ or $(\textbf{cD3})$. Hence we obtain the bounds on $s_{BO_{cred}}$ and $t_{BO_{cred}}$. $\qquad\square$

Combining Lemmas 2 and 3 we obtain our main result in this section stating a tight connection between the proof complexity of LK and BO_{cred}.

Theorem 5. *There exist a polynomial p and a constant c such that $s_{LK}(n) \leq s_{BO_{cred}}(n) \leq p(n) \cdot s_{LK}(cn)$ and $t_{LK}(n) \leq t_{BO_{cred}}(n) \leq p(n) \cdot t_{LK}(cn)$.*

In the light of this result, proving either non-trivial lower or upper bounds to the proof size of BO_{cred} seems very difficult—as such a result would mean a major breakthrough in propositional proof complexity (cf. [2, 16]).

4.1   On the Automatizability of BO_{cred}

Practitioners are not only interested in the size of a proof, but face the more complicated problem to actually construct a proof for a given instance. Of course, in the presence of super-polynomial lower bounds to the proof size this cannot be done in polynomial time. Thus, in proof search the best one can hope for is the following notion of automatizability:

Definition 1 (Bonet, Pitassi, Raz [5]). *A proof system P for a language L is* automatizable *if there exists a deterministic procedure that takes as input a string x and outputs a P-proof of x in time polynomial in the size of the shortest P-proof of x if $x \in L$. If $x \notin L$, then the behaviour of the algorithm is unspecified.*

For practical purposes automatizable systems would be very desirable. Searching for a proof we may not find the shortest one, but we are guaranteed to find one that is only polynomially longer. Unfortunately, for BO_{cred} there are strong limitations towards this goal as our next result shows:

Theorem 6. *BO_{cred} is not automatizable unless factoring integers is possible in polynomial time.*

Proof. First we observe that automatizability of BO_{cred} implies automatizability of Frege systems. For this let φ be a propositional tautology. By assumption, we can construct a BO_{cred}-proof of $\emptyset \hspace{0.3em}\vert\hspace{-0.3em}\sim \varphi$. This BO_{cred}-proof contains an LK-proof of $\emptyset \vdash \varphi$ by rule (**cD1**). As LK is polynomially equivalent to Frege systems [16], we can construct from this LK-proof a Frege proof of φ in polynomial time. By a result of Bonet, Pitassi, and Raz [5], Frege systems are not automatizable unless Blum integers can be factored in polynomial time (a Blum integer is the product of two primes which are both congruent 3 modulo 4).   □

4.2   A General Construction of Proof Systems for Credulous Default Reasoning

In this section we will explain a general method how to construct proof systems for credulous default reasoning. These proof systems arise from the canonical Σ_2^p algorithm for credulous default reasoning (Algorithm 1). Algorithm 1 first guesses a generating set G_{ext} for a potential stable extension and then verifies by the stage construction from Theorem 1 that G_{ext} indeed generates a stable extension which moreover contains the formula φ. Algorithm 1 is a Σ_2^p procedure, *i.e.*, it can be executed by a nondeterministic polynomial-time Turing machine M with access to a coNP-oracle. The nondeterminism solely lies in line 1 and

---

Algorithm 1. A Σ_2^p procedure for credulous default reasoning

---

Require: $\langle W, D \rangle, \varphi$

1: guess $D_0 \subseteq D$ and let $G_{\text{ext}} \leftarrow W \cup \left\{ \gamma \mid \frac{\alpha:\beta}{\gamma} \in D_0 \right\}$

2: $G_{\text{new}} \leftarrow W$

3: **repeat**

4:     $G_{\text{old}} \leftarrow G_{\text{new}}$

5:     **for all** $\frac{\alpha:\beta}{\gamma} \in D$ **do**

6:         **if** $G_{\text{old}} \models \alpha$ and $G_{\text{ext}} \not\models \neg\beta$ **then**

7:             $G_{\text{new}} \leftarrow G_{\text{new}} \cup \{\gamma\}$

8:         **end if**

9:     **end for**

10: **until** $G_{\text{new}} = G_{\text{old}}$

11: **if** $G_{\text{new}} = G_{\text{ext}}$ and $G_{\text{ext}} \models \varphi$ **then**

12:     **return true**

13: **else**

14:     **return false**

15: **end if**

---

the oracle queries are made in lines 6 and 11 to the coNP-complete problem of propositional implication IMP $= \{\langle \Psi, \varphi \rangle \mid \Psi \subseteq \mathcal{L}, \ \varphi \in \mathcal{L}, \text{ and } \Psi \models \varphi\}$.

Algorithm 1 can be converted into a proof system for credulous default reasoning as follows. We fix a propositional proof system P and define a proof system $Cred(P)$ for credulous default reasoning where proofs are of the form

$$\langle W, D, \varphi, comp, q_1, \ldots, q_k, a_1, \ldots, a_k \rangle \ .$$

Here $comp$ is a computation of M on input $\langle W, D, \varphi \rangle$ and $q_1, \ldots, q_k$ are the queries to IMP during this computation. If the IMP-query $q_i = \langle \Psi_i, \varphi_i \rangle$ is answered positively, then a_i is a P-proof of $\left(\bigwedge_{\psi \in \Psi_i} \psi \right) \to \varphi_i$, otherwise a_i is an assignment falsifying this formula. For this proof system we obtain the following bounds:

Theorem 7. *Let P be a propositional proof system. Then $Cred(P)$ is a proof system for credulous default reasoning with $s_P(n) \leq s_{Cred(P)}(n) \leq \mathcal{O}(n^2 s_P(n))$.*

Proof. The first inequality holds because we can use $Cred(P)$ to prove propositional tautologies φ by choosing $W = D = \emptyset$.

For the second inequality, we observe that Algorithm 1 has quadratic running time. In particular, a computation of Algorithm 1 contains at most a quadratic number of queries to IMP. Each of these queries is of linear size because it only consists of formulae from the input. If the query is answered positively, then we have to supply a P-proof and there exists such a P-proof of size $\leq s_P(n)$. For a negative answer we just include an assignment of linear size. This yields $s_{Cred(P)}(n) \leq \mathcal{O}(n^2 s_P(n))$. $\qquad\square$

Theorem 7 tells us that proving lower bounds for proof systems for credulous default reasoning is more or less the same as proving lower bounds to propositional proof systems. In particular, we get:

Corollary 1. *There exists a polynomially bounded proof system for credulous default reasoning if and only if there exists a polynomially bounded propositional proof system.*

5   Lower Bounds for Skeptical Default Reasoning

Bonatti and Olivetti [3] introduce two calculi for skeptical default reasoning. As before, objects are sequents of the form $\Sigma; \Gamma \hspace{-0.3em}\sim\hspace{-0.3em} \Delta$, where Σ is a set of constraints, Γ is a propositional default theory, and Δ is a set of propositional formulae. But now, the sequent $\Sigma; \Gamma \hspace{-0.3em}\sim\hspace{-0.3em} \Delta$ is true, if $\bigvee \Delta$ holds in *all* extensions of Γ satisfying the constraints in Σ.

The first calculus BO_{skep} consists of the defining axioms of LK and AC, the inference rules of LK, AC, RC, and the rules from Fig. 6. Bonatti and Olivetti

$$(\textbf{sD1})\ \frac{\Gamma \vdash \Delta}{\Sigma; \Gamma \hspace{-0.3em}\sim\hspace{-0.3em} \Delta} \qquad (\textbf{sD2})\ \frac{\Gamma \vdash \alpha}{\neg \textbf{L}\alpha, \Sigma; \Gamma \hspace{-0.3em}\sim\hspace{-0.3em} \Delta} \qquad (\textbf{sD3})\ \frac{\Gamma \nvdash \alpha}{\textbf{L}\alpha, \Sigma; \Gamma \hspace{-0.3em}\sim\hspace{-0.3em} \Delta}$$

where $\Gamma \subseteq \mathcal{L}^{res}$ in rules (**sD1**), (**sD2**), and (**sD3**)

$$(\textbf{sD4})\ \frac{\neg \textbf{L}\neg\beta_1, \ldots, \neg \textbf{L}\neg\beta_n, \Sigma; \Gamma, \frac{\alpha}{\gamma} \hspace{-0.3em}\sim\hspace{-0.3em} \Delta \qquad \textbf{L}\neg\beta_1, \Sigma; \Gamma \hspace{-0.3em}\sim\hspace{-0.3em} \Delta \ \ldots\ \textbf{L}\neg\beta_n, \Sigma; \Gamma \hspace{-0.3em}\sim\hspace{-0.3em} \Delta}{\Sigma; \Gamma, \frac{\alpha:\beta_1\ldots\beta_n}{\gamma} \hspace{-0.3em}\sim\hspace{-0.3em} \Delta}$$

Fig. 6. Inference rules for the skeptical default calculus BO_{skep}

show that each true sequent is derivable in BO_{skep}, *i.e.*, the calculus is sound and complete. However, they already remark that proofs in BO_{skep} are of exponential size in the number of default rules in the sequent. This is due to the residual rules for they cannot be applied unless all defaults with nonempty justifications have been eliminated using rule (**sD4**).

To get more concise proofs, Bonatti and Olivetti [3] suggest an enhanced calculus BO'_{skep} where the rules (**sD1**) to (**sD3**) are replaced by rules (**sD1'**) to (**sD3'**) and rule (**sD4**) is kept (see Fig. 7). Bonatti and Olivetti prove soundness and completeness for BO'_{skep}. Moreover, they show that BO'_{skep} is exponentially separated from BO_{skep}, *i.e.*, there exist sequents $(S_n)_{n \geq 1}$ which require exponential-size proofs in BO_{skep} but have linear-size derivations in BO'_{skep}. In our next result we will show an exponential lower bound to the proof length (and therefore also to the proof size) in the enhanced skeptical calculus BO'_{skep}.

Theorem 8. *The calculus BO'_{skep} has exponential lower bounds to the lengths of proofs. More precisely, there exist sequents S_n of size $\mathcal{O}(n)$ such that every BO'_{skep}-proof of S_n uses $2^{\Omega(n)}$ steps. Therefore, $s_{BO'_{skep}}(n), t_{BO'_{skep}}(n) \in 2^{\Omega(n)}$.*

$$(\text{sD1'}) \ \frac{\Sigma', \Gamma' \vdash \Delta}{\Sigma; \Gamma \hspace{-2pt}\sim\hspace{-2pt} \Delta} \qquad (\text{sD2'}) \ \frac{\Sigma; \Gamma \hspace{-2pt}\sim\hspace{-2pt} \alpha}{\neg \mathbf{L}\alpha, \Sigma; \Gamma \hspace{-2pt}\sim\hspace{-2pt} \Delta} \qquad (\text{sD3'}) \ \frac{\Gamma'' \not\vdash \alpha}{\mathbf{L}\alpha, \Sigma; \Gamma \hspace{-2pt}\sim\hspace{-2pt} \Delta}$$

$$(\text{sD4}) \ \frac{\neg \mathbf{L}\neg\beta_1, \ldots, \neg \mathbf{L}\neg\beta_n, \Sigma; \Gamma, \frac{\alpha}{\gamma} \hspace{-2pt}\sim\hspace{-2pt} \Delta \qquad \mathbf{L}\neg\beta_1, \Sigma; \Gamma \hspace{-2pt}\sim\hspace{-2pt} \Delta \ \ldots \ \mathbf{L}\neg\beta_n, \Sigma; \Gamma \hspace{-2pt}\sim\hspace{-2pt} \Delta}{\Sigma; \Gamma, \frac{\alpha:\beta_1\ldots\beta_n}{\gamma} \hspace{-2pt}\sim\hspace{-2pt} \Delta}$$

where $\Sigma' \subseteq \{\alpha \mid \mathbf{L}\alpha \in \Sigma\}$, $\Gamma' \subseteq \Gamma \cap \mathcal{L}^{res}$, and $\Gamma'' = (\Gamma \cap \mathcal{L}) \cup \left\{ \frac{p(\delta)}{c(\delta)} \,\middle|\, \delta \in \Gamma \right\}$.

Fig. 7. Inference rules for the enhanced skeptical default calculus BO'_{skep}

Proof. (Sketch) We construct a sequence $(S_n)_{n \geq 1} = (\Sigma_n; \Gamma_n \hspace{-2pt}\sim\hspace{-2pt} \psi_n)_{n \geq 1}$ such that for some constant c, every BO'_{skep}-proof of S_n has length at least $2^{\Omega(n)}$. We choose $\Sigma_n = \emptyset$, $\psi_n = x_{2n}$, and $\Gamma_n = \langle \emptyset, D_{2n} \rangle$, where D_{2n} consists of the defaults listed in Fig. 8. The default theory Γ_n possesses 2^{n+1} stable extensions. Observe that each of these contains x_{2n}, but that each pair of stable extensions differs in truth assigned to the propositional variables $x_0, \ldots, x_n$. We claim that every proof of S_n has exponential length in n. More precisely, we show that rule (**sD4**) has to be applied an exponential number of times.

$$\frac{: \ x_0}{x_0} \qquad \frac{: \ \neg x_0}{\neg x_0}$$

$$\frac{x_i \ : \ x_{i+1}}{x_{i+1}} \qquad \frac{\neg x_i \ : \ x_{i+1}}{x_{i+1}} \qquad \frac{x_i \ : \ \neg x_{i+1}}{\neg x_{i+1}} \qquad \frac{\neg x_i \ : \ \neg x_{i+1}}{\neg x_{i+1}}$$

$$\frac{x_{n+j} \ : \ x_{n-j-1}}{x_{n+j+1}} \qquad \frac{\neg x_{n+j} \ : \ x_{n-j-1}}{x_{n+j+1}} \qquad \frac{x_{n+j} \ : \ \neg x_{n-j-1}}{\neg x_{n+j+1}} \qquad \frac{\neg x_{n+j} \ : \ \neg x_{n-j-1}}{\neg x_{n+j+1}}$$

for $i = 0, \ldots, n-1$ and $j = 0, \ldots, n-2$

$$\frac{x_{2n-1} \ : \ x_0}{x_{2n}} \qquad \frac{\neg x_{2n-1} \ : \ x_0}{x_{2n}} \qquad \frac{x_{2n-1} \ : \ \neg x_0}{x_{2n}} \qquad \frac{\neg x_{2n-1} \ : \ \neg x_0}{x_{2n}}$$

Fig. 8. The defaults in D_{2n} in the proof of Theorem 8

We point out that our argument does not only work against tree-like proofs, but also rules out the possibility of sub-exponential dag-like derivations for $D_{2n} \hspace{-2pt}\sim\hspace{-2pt} x_{2n}$. The lower bound is obtained from the fact that to derive x_{2n}, we have to derive x_i and $\neg x_i$ for each $n < i < 2n$, each of which can only be achieved from ancestors with mutually different proof constraints. This, by definition of BO_{skep}, leads to mutually disjoint sets of ancestor sequents. $\square$

6   Conclusion

In this paper we have shown that with respect to lengths of proofs, proof systems for credulous default reasoning and for propositional logic are very close to each other. Although deciding credulous default sequents is presumably harder than deciding tautologies (the former is Σ_2^p-complete [12], while the latter is complete for coNP), the difference disappears when we want to prove these objects (Sect. 4.2).

For skeptical reasoning this is less clear. While skeptical default reasoning has polynomially bounded proof systems if and only if this holds for TAUT, we leave open whether this equivalence extends to other bounds. However, in the light of our exponential lower bound for BO'_{skep} (Theorem 8), searching for natural proof systems for skeptical default reasoning with more concise proofs will be a rewarding task for future research.

In this direction Bonatti and Olivetti [3] themselves introduced two rules to supplement their enhanced calculus. These are the cut rule

$$\frac{\Sigma; \Gamma \vdash \alpha \qquad \Sigma; \Gamma, \alpha \vdash \Delta}{\Sigma; \Gamma \vdash \Delta} \text{ (\textbf{Cut})}$$

and the following version of the rule (**sD4**)

$$\frac{\Sigma_0, \Sigma; \Gamma, \frac{\alpha}{\gamma} \vdash \Delta \qquad \Sigma_1, \Sigma; \Gamma \vdash \Delta \quad \ldots \quad \Sigma_n, \Sigma; \Gamma \vdash \Delta}{\Sigma; \Gamma, \frac{\alpha:\beta_1\ldots\beta_n}{\gamma} \vdash \Delta} \text{ (\textbf{sD4}')}$$

where $\Sigma_i = \mathbf{L}\neg\beta_{\pi(i)}, \neg\mathbf{L}\neg\beta_{\pi(i+1)}, \ldots, \neg\mathbf{L}\neg\beta_{\pi(n)}$ for an arbitrary permutation π of $\{1, \ldots, n\}$. While it is not hard to see that our lower bound in Theorem 8 still remains true if we add (**sD4**') to BO'_{skep}, we leave open the problem to show super-polynomial lower bounds in the presence of the cut rule.

Acknowledgements

The first author wishes to thank Neil Thapen for interesting discussions on the topic of this paper during a research visit to Prague. We also thank the anonymous referees for helpful comments.

References

1. Amati, G., Aiello, L.C., Gabbay, D.M., Pirri, F.: A proof theoretical approach to default reasoning I: Tableaux for default logic. Journal of Logic and Computation 6(2), 205–231 (1996)
2. Bonatti, P.A.: A Gentzen system for non-theorems. Technical Report CD/TR 93/52, Christian Doppler Labor für Expertensysteme (1993)
3. Bonatti, P.A., Olivetti, N.: Sequent calculi for propositional nonmonotonic logics. ACM Transactions on Computational Logic 3(2), 226–278 (2002)

4. Bonet, M.L., Buss, S.R., Pitassi, T.: Are there hard examples for Frege systems? In: Clote, P., Remmel, J. (eds.) Feasible Mathematics II, pp. 30–56. Birkhäuser, Basel (1995)
5. Bonet, M.L., Pitassi, T., Raz, R.: On interpolation and automatization for Frege systems. SIAM Journal on Computing 29(6), 1939–1967 (2000)
6. Cadoli, M., Schaerf, M.: A survey of complexity results for nonmonotonic logics. Journal of Logic Programming 17(2-4), 127–160 (1993)
7. Cook, S.A., Reckhow, R.A.: The relative efficiency of propositional proof systems. The Journal of Symbolic Logic 44(1), 36–50 (1979)
8. Dix, J., Furbach, U., Niemelä, I.: Nonmonotonic reasoning: Towards efficient calculi and implementations. In: Handbook of Automated Reasoning, pp. 1241–1354. Elsevier/MIT Press (2001)
9. Egly, U., Tompits, H.: Proof-complexity results for nonmonotonic reasoning. ACM Transactions on Computational Logic 2(3), 340–387 (2001)
10. Gabbay, D.: Theoretical foundations of non-monotonic reasoning in expert systems. In: Logics and Models of Concurrent Systems, pp. 439–457. Springer, Heidelberg (1985)
11. Gentzen, G.: Untersuchungen über das logische Schließen. Mathematische Zeitschrift 39, 68–131 (1935)
12. Gottlob, G.: Complexity results for nonmonotonic logics. Journal of Logic and Computation 2(3), 397–425 (1992)
13. Hrubeš, P.: On lengths of proofs in non-classical logics. Annals of Pure and Applied Logic 157(2-3), 194–205 (2009)
14. Jeřábek, E.: Frege systems for extensible modal logics. Annals of Pure and Applied Logic 142, 366–379 (2006)
15. Jeřábek, E.: Substitution Frege and extended Frege proof systems in non-classical logics. Annals of Pure and Applied Logic 159(1-2), 1–48 (2009)
16. Krajíček, J.: *Bounded Arithmetic, Propositional Logic, and Complexity Theory*. Encyclopedia of Mathematics and Its Applications, vol. 60. Cambridge University Press, Cambridge (1995)
17. Kraus, S., Lehmann, D.J., Magidor, M.: Nonmonotonic reasoning, preferential models and cumulative logics. Artificial Intelligence 44(1-2), 167–207 (1990)
18. Makinson, D.: General theory of cumulative inference. In: Reinfrank, M., Ginsberg, M.L., de Kleer, J., Sandewall, E. (eds.) Non-Monotonic Reasoning 1988. LNCS, vol. 346, pp. 1–18. Springer, Heidelberg (1988)
19. Marek, V.W., Truszczyński, M.: Nonmonotonic Logics—Context-Dependent Reasoning. Springer, Heidelberg (1993)
20. Reiter, R.: A logic for default reasoning. Artificial Intelligence 13, 81–132 (1980)
21. Risch, V., Schwind, C.: Tableaux-based characterization and theorem proving for default logic. Journal of Automated Reasoning 13(2), 223–242 (1994)
22. Tiomkin, M.L.: Proving unprovability. In: Proc. 3rd Annual Symposium on Logic in Computer Science, pp. 22–26 (1988)

Automated Testing and Debugging of SAT and QBF Solvers

Robert Brummayer, Florian Lonsing, and Armin Biere

Institute for Formal Models and Verification
Johannes Kepler University Linz, Austria

Abstract. Robustness and correctness are essential criteria for SAT and QBF solvers. We develop automated testing and debugging techniques designed and optimized for SAT and QBF solver development. Our fuzz testing techniques are able to find critical solver defects that lead to crashes, invalid satisfying assignments and incorrect satisfiability results. Moreover, we show that sequential and concurrent delta debugging techniques are highly effective in minimizing failure-inducing inputs.

1 Introduction

Satisfiability solving has been shown to be a competitive problem solving technique that is used in many different domains such as verification, test case generation, scheduling, computational biology and artificial intelligence. For a recent survey on satisfiability solving we refer the reader to [8]. Recent advances of propositional satisfiability (SAT) solvers and quantified boolean formula (QBF) solvers are driven by competitions and real industrial applications such as formal hardware and software verification.

Essential criteria of satisfiability solvers are robustness and correctness. SAT and QBF solvers are used as core decision engines and the clients heavily depend on these important criteria. For instance, an incorrect SAT solver used as decision engine in a formal verification framework may lead to incorrect verification results, i.e. either the system may be spuriously proven to be correct or the verification framework generates a spurious counter-example. Moreover, wrong satisfying assignments (models) may be mapped to spurious verification counter-examples that hinder the overall verification process.

While a large part of current research focuses on speeding up SAT and QBF solving with various techniques such as improved decision heuristics and low-level optimizations, there are, to the best of our knowledge, no rigorous scientific publications about automated testing and debugging techniques for SAT and QBF solvers. This paper tries to improve this situation by introducing automated state-of-the-art testing and (multi-threaded) delta debugging techniques, designed and optimized for SAT and QBF solvers. Our experimental results are available at `http://fmv.jku.at/brummayer/fuzz-dd-sat-qbf.tar.7z`. Every tool is available at `http://fmv.jku.at/software/`.

O. Strichman and S. Szeider (Eds.): SAT 2010, LNCS 6175, pp. 44–57, 2010.
© Springer-Verlag Berlin Heidelberg 2010

2 Fuzzing

Fuzzing is an automated negative testing technique, typically used in software security and quality assurance [45,46]. The original idea is to treat software as a black-box and repeatedly "attack" it with random inputs in order to find critical defects, e.g. buffer overflows. Fuzz testing methods such as "monkey testing" were already used around 1980 [46]. Miller, one of the fuzzing pioneers, demonstrated that fuzz testing could find many critical defects in UNIX applications [36]. The lack of a formal model and the brute force nature of the approach lead to the situation that papers about fuzz testing were often offended. Miller simply responded that he was just trying to find bugs [46], which is also exactly what we want to achieve with our fuzzing techniques, explicitly designed and optimized for SAT and QBF solvers.

The goal of previous work on random generation of SAT and QBF instances was to study the phase transition phenomenon [22,40,23,16] and to generate hard instances [27,48,1]. However, our work focuses on generating random instances in order to *find defects* in current state-of-the-art solver implementations. We propose to use *grammar-based black-box fuzzing* in order to test SAT and QBF solvers. A fuzzer repeatedly generates syntactically valid inputs. Solvers are treated as black-boxes, which makes our approach highly flexible. They are run on the generated inputs in order to detect critical defects such as segmentation faults and aborts. Moreover, reported satisfiability results are validated in order to find defects that lead to incorrect results and models.

One of the main success factors of fuzz testing is a high test throughput, e.g. testing a solver with five instances per hour is unlikely to be successful. Therefore, generating hard instances solely is counter-productive. On the other hand, trivial instances are unlikely to trigger interesting defects. Ideally, a fuzzer should be able to generate a variety of different inputs that lead to the execution of different paths in the tested solver. The majority of the generated instances should be easy to solve in order to maintain a sufficiently high test throughput. The combination of automation, diversity and high throughput makes fuzz testing an effective negative testing technique. Our experiments in section 2.3 show that this technique can be successfully applied to SAT and QBF solvers.

In the following we describe our novel fuzzing techniques for SAT and QBF solvers, implemented in our fuzzers CNFuzz, FuzzSAT and QBFuzz. Due to the probabilistic nature of fuzzing, our fuzzers use magic constants found through direct experimentation. All fuzzing approaches use a random number generator. We assume that picks during fuzzing are performed uniformly at random.

2.1 SAT Fuzzing

3SATGen. Easy to solve instances do not exercise solvers enough. Therefore, it is unlikely to find interesting defects with easy instances alone. However, as research [22] on the phase transition in random 3-SAT suggests, it is straightforward to write a random CNF generator that generates reasonable hard instances. In our view, this is an important application of [22].

Our 3SAT generator 3SATGen is based on this technique and works as follows. First, the number of variables m is picked, typically between 10 and 400 variables. The next step is to determine a clause variable ratio r, which should be around the hardness threshold, typically between 3 and 5. Finally, $m \cdot r$ random ternary clauses are generated, where each literal is picked uniformly.

CNFuzz. Random 3-SAT formulas are lacking structure. However, the success of SAT solvers in industry seems to rely on their ability to use structure, at least implicitly, even though we do not actually know how to describe this connection in a more formal way. This rather vague argument implies that 3SAT does not exercise all the interesting features of an industrial SAT solver. Therefore, we were looking for other ways to generate "more structured" instances. Our fuzzer CNFuzz enforces certain locality restrictions and thus generates instances that contain more internal structure than the simple 3SAT approach.

The CNF generated by CNFuzz consists of $l \in [1, 20]$ layers of maximum width $w \in [10, 70]$. Both numbers are picked randomly within these ranges. The i-th layer with $i \in [1, l]$ introduces $n_i \in [10, w]$ new variables, again chosen randomly. Each layer is associated with a separately "picked-clause-variable-ratio" $r_i \in [3, 4.5]$ from which the number $c_i = r_i w_i$ of clauses in layer i is calculated. Clauses are at least ternary, and with exponentially decreasing probability longer: $2/3$ are expected to be ternary, $1/3 \cdot 2/3$ of length 4, $(1/3)^2 \cdot 2/3$ of length 5, etc. Variables are picked either from the same or from smaller layers. The layer from which a variable is picked is determined in a similar way as the length of clauses. A variable from layer i is picked with probability $1/2$, from layer $i - 1$ with probability $1/4$, from layer $i - 2$ with probability $1/8$, etc., down to the first layer, which accumulates the remaining probabilities. As a further refinement of the iterative clause generation process, variables that have not been selected are preferred within the same layer.

FuzzSAT. The even more structured approach of FuzzSAT is based on the translation of boolean circuits into CNF. To be more precise, a directed acyclic graph (DAG) representation of a random boolean circuit is generated. The generated DAG is converted into CNF by using the Tseitin transformation [47] afterwards.

The boolean circuit DAG is constructed as follows. First, $v \in [1, 100]$ boolean input nodes are generated and inserted into a global set n, which is a container for all nodes generated during the construction process. Then, in the core routine of our DAG generation approach, we randomly select a boolean operator op from the set of operators $O = \{\text{AND, OR, XOR, IFF}\}$. Moreover, we select two operands o_1 and o_2 from n, negate each operand with probability $1/2$, generate the new operator node, and insert it into n. This process is repeated until each original input variable is referenced at least t times, where t is usually 1.

Then, we take the set r of all boolean roots, i.e. generated operators that are not referenced by other operators, and combine them to one boolean root as follows. We select a boolean operator op from O, select two operands r_1 and r_2 from r, negate each of them with probability $1/2$, remove r_1 and r_2 from r, generate the new root, and insert it into r. This process is repeated until there is only one root left. Then, we perform the Tseitin transformation on this root.

Let c be the number of clauses generated so far and let p be a probability $\in [0.01, 0.1]$. Finally, $c \cdot p$ random clauses of varying size $s \in [2, 6]$ are added to the CNF in order to increase diversity. The size of the additional random clauses is picked for each clause individually.

2.2 QBF Fuzzing

A quantified boolean formula (QBF) $F = B_1 \ldots B_n.\phi$ in *prenex conjunctive normal form (PCNF)* consists of a propositional formula ϕ in CNF over a set of variables V and a *quantifier prefix* $B_1 \ldots B_n$. The quantifier prefix is a linearly ordered set of *blocks* B_i where $B_1 < \ldots < B_n$, forming a partition on V.

A block B_i is *existential* ($q(B_i) = \exists$) if it is associated with an existential quantifier and *universal* ($q(B_i) = \forall$) otherwise. For two adjacent blocks B_i and B_{i+1} where $1 \le i < n$, $q(B_i) \ne q(B_{i+1})$.

A clause C is *forall-reduced* [12] if for every literal $l \in C$ with $l \in B_i$ and $q(B_i) = \forall$ there is a literal $k \in C$ with $k \in B_j$ and $q(B_j) = \exists$ and $i < k$.

In the following we describe two different approaches for QBF fuzz testing. All generated formulas are in PCNF and contain forall-reduced clauses only.

Random QBF model. We have implemented a QBF fuzzer BlocksQBF which generates random QBFs in PCNF according to the model described in [14]. This model is an extension of an approach originally introduced in [13], which was further improved in [23].

The model [14] used in our fuzzer has the following parameters: the number of clauses n_c, the number of quantifier blocks n_b in the prefix $B_1 \ldots B_{n_b}$, the number of variables $n_{v,1}, \ldots, n_{v,i}, \ldots, n_{v,n_b}$ in each block B_i for $1 \le i \le n_b$ and the number of literals $n_{l,1}, \ldots, n_{l,i}, \ldots, n_{l,n_b}$ taken from a block B_i to appear in each clause, where $n_{l,i} \le n_{v,i}$ for $1 \le i \le n_b$. By convention, always $q(B_{n_b}) = \exists$, i.e. all clauses are forall-reduced by construction and have the same length.

We generate exactly n_c *distinct* clauses one after the other as follows. From each block B_i for $1 \le i \le n_b$ we select and negate exactly $n_{l,i}$ literals where $n_{l,i} \le n_{v,i}$. Different from the description given in [14], complementary or duplicate literals in a clause are *always* discarded until a new literal is generated which can be added to the clause. This is possible as we never add more literals from a block B_i than there are variables in B_i (since $n_{l,i} \le n_{v,i}$ for $1 \le i \le n_b$).

Newly generated clauses are added to the formula only if there is no duplicate clause already present. Otherwise the new clause is discarded and another attempt is carried out. This process continues until exactly n_c distinct clauses are generated, which is different from [14]. For improper parameter settings such as big n_c and very small $n_{v,i}$ it can be impossible to generate exactly n_c distinct clauses, but this was avoided in our experiments, where we used the following settings: $n_c = 160$, $n_b = 3$ (i.e. quantifier prefix $\exists\forall\exists$), block sizes $n_{v,1} = 15$, $n_{v,2} = 10$, $n_{v,3} = 25$ and $n_{l,1} = n_{l,2} = 2$, $n_{l,3} = 1$ literals taken from each block.

QBFuzz. The second QBF fuzzer `QBFuzz` we used in our experiments generates QBFs in PCNF which do not follow an exact model such as [13,23,14], leading to a higher diversity. The following parameters are *maximum* values: number of clauses n_c, number of variables n_v and number of blocks n_b. Further, minimum *min* and maximum number *max* of literals in a clause and the ratio $r \in [0, 1]$ of existential variables in the formulas and in each clause is specified.

Formulas according to the given setting are generated as follows: first a quantifier prefix is selected according to values of n_b, n_v and r, where the number of variables per block is selected at random. Next n_c clauses are generated of length $len \in [min, max]$ and each containing $r.len$ existential variables. Different from `BlocksQBF`, literals are selected from *any* block and are negated uniformly at random. As described above, duplicate and complementary literals are discarded. After generated clauses have been forall-reduced, duplicate clauses and unused variables are removed from the formula. We used the following settings in our experiments: $n_c = 80$, $n_v = 40$, $n_b = 15$, $min = 5$, $max = 15$ and $r = 0.4$.

2.3   Experiments

In order to evaluate our fuzzing techniques, we performed fuzz testing experiments with a selected subset of complete SAT solvers that participated in the SAT competition 2007 and 2009. Moreover, we fuzz tested several state-of-the-art QBF solvers. We ran our experiments under Ubuntu Linux on an Intel Core 2 Quad machine with 2.66 GHz and 8 GB RAM. Our fuzzing test framework used each of the four cores. The results of our fuzzing experiments with SAT solvers are shown in Tab. 1 and Tab. 2. The QBF results are shown in Tab. 3.

The results of our fuzz testing experiments with SAT solvers in Tab. 1 and Tab. 2 clearly show the overall effectiveness of our fuzz testing techniques. We were able to find serious defects such as segmentation faults, aborts, assertion failures, invalid models and incorrect results. We classified the defects into the following categories. Unexpected termination without providing a result was classified as an *error*. Cases where solvers reported an incorrect satisfiability status, i.e. a solver reported that an instance is unsatisfiable although the instance is provably satisfiable, were classified as *incorrect*. Finally, providing the correct satisfiability status but an invalid satisfying assignment was classified as *invalid model*, labeled *model* in tables Tab. 1 and Tab. 2. Notice that multiple observable failures may be caused by the same solver defect.

We used our tool `PrecoCheck` to validate models. Cases where we could not fully decide which satisfiability status is correct, e.g. some solvers claim that the instance is unsatisfiable and some others claim that instance is satisfiable, but provide an invalid model, did not occur. If all solvers agreed that the current instance is unsatisfiable, we did not further validate the unsatisfiability status as it is highly unlikely that all solvers are wrong.

We were able to find six defective solvers that participated in the SAT competition 2007. Notice that we did not test all solvers. We selected only a subset of the most competitive complete SAT solvers in order to demonstrate the effectiveness of our fuzzing techniques. Moreover, in order to keep our set of solvers small, we did not test incomplete and portfolio-based solvers. Notice that our fuzz testing and delta debugging techniques can be applied to any kind of solver.

Five of the six SAT competition 2007 solvers shown in Tab. 1 have defects that lead to incorrect results, which we consider as the worst case that can happen. Incorrect results reported by the multi-threaded SAT solver MiraXTv3 are non-deterministic. Depending on the thread scheduling and the actual utilization of the individual processing cores, MiraXTv3 either reports that an instance is satisfiable or unsatisfiable. Moreover, our fuzzers detected that two SAT solvers generate invalid models. Notice that RSat respectively PicoSAT, were ranked first respectively second in the industrial category (satisfiable and unsatisfiable instances). Moreover, notice that March_ks was the second best solver in the random category (satisfiable and unsatisfiable instances).

Our fuzzing techniques were able to find three defective solvers that participated in the SAT competition 2009. We found critical defects causing segmentation faults in MiniSat-9z, the winner of the MiniSat hack track. Moreover, we found non-deterministic crashes in ManySat, which was the winner of the parallel solver application track. Finally, our fuzzer FuzzSAT was able to reveal that Mirch_hi, second best solver (SAT + UNSAT) and best solver (UNSAT) in the random track, sometimes generates invalid models.

None of the fuzzing techniques is clearly superior to the others, except that CNFuzz and FuzzSAT were able to find more varying defects as the simple 3SAT generator 3SATGen. The restriction to 3SAT CNF instances may miss failures that occur if the input contains clauses of arbitrary size. Nevertheless, the 3SAT generator was still able to find defects in three of the six solvers, which is rather surprising as SAT solvers are typically tested with 3SAT instances. Interestingly, while CNFuzz was the only fuzzer that found defects in Barcelogic-fixed and incorrect results of Barcelogic, FuzzSAT was the only fuzzer that was able to

Table 1. Experimental results of fuzz testing SAT solvers from SAT competition 2007. The 3SAT generator 3SATGen and our fuzzers CNFuzz and FuzzSAT generated 10000 CNF instances, respectively. We fuzz tested Barcelogic [9] and Barcelogic-fixed [9], CMUSAT [30], March_ks [26], MiniSat [18], MiraXTv3 [34], MXC [10], PicoSAT [7], RSat [41], Sat7 [32], SAT4J [4], Spear [2] and Tinisat [29]. All solver binaries were taken from the SAT competition 2007. Only solvers for which defects have been found are shown in the table. The testing time was about two hours for the 3SAT generator and FuzzSAT, respectively, and one hour for CNFuzz. For each solver and each CNF instance a time limit of thirty seconds was used.

	3SATGen			CNFuzz			FuzzSAT		
solver	error	incorrect	model	error	incorrect	model	error	incorrect	model
Barcelogic	0	0	0	1	**3**	1	1	0	1
Barcelogic-fixed	0	0	0	0	**1**	**1**	0	0	0
March_ks	**24**	2	0	5	0	0	2	2	0
MiraXTv3	26	7	0	91	**13**	0	**286**	2	0
PicoSAT	0	0	0	0	0	0	0	**2**	0
RSat	**56**	0	0	27	0	0	3	0	0

Table 2. Experimental results of fuzz testing SAT solvers from SAT competition 2009. The 3SAT generator 3SATGen and our fuzzers CNFuzz and FuzzSAT generated 10000 CNF instances, respectively. We fuzz tested CirCUs [5], Clasp [20], Cumr_p [5], Glucose [5], LySATi [5], ManySAT [25], March_hi [26], MiniSat [18], MiniSat-9z [5], MXC [5], PicoSAT [7], PrecoSAT [5], RSat [5], SApperloT-base [5], SAT4J [4] and Varsat-industrial [28]. All solvers binaries were taken from the SAT competition 2009. Only solvers for which defects have been found are shown in the table. No discrepancies were found, i.e. all solvers agreed on the satisfiability status of each CNF instance. The testing time was about two hours for the 3SAT generator, three hours for FuzzSAT and one hour and thirty minutes for CNFuzz. For each solver and each instance a time limit of thirty seconds was used.

solver	3SATGen		CNFuzz		FuzzSAT	
	error	model	error	model	error	model
ManySat	2	0	56	0	**836**	0
March_hi	0	0	0	0	0	**24**
MiniSat-9z	2	0	58	0	**852**	0

Table 3. Experimental results of fuzz testing QBF solvers with BlocksQBF and QBFuzz. Both generated 10000 CNF instances, respectively. We fuzz tested an internal version of DepQBF [35], MiniQBF-090608 [43], QMRES [39], Quantor-3.0 [6], QuBE6.0 [24], QuBE6.5 [24], QuBE6.6 [24], Semprop-010604 [33], sKizzo-0.8.2 [3], SQBF-1.0 [44], Squolem-1.03 [31] and yQuaffle-021006 [53]. The fuzz testing time was one hour and fifteen minutes for QBFuzz and one hour and twenty minutes for BlocksQBF. Only solvers for which defects have been found are shown in the table. For each solver and each instance a time limit of thirty seconds was used.

solver	BlocksQBF		QBFuzz	
	error	incorrect	error	incorrect
Quantor	0	0	1	0
QuBE 6.0	0	684	5	7
QuBE 6.5	0	0	4	0
sKizzo	0	0	2	29
SQBF	0	0	35	0
yQuaffle	0	0	94	0

generate instances on which PicoSAT reports an incorrect satisfiability status. Moreover, the Barcelogic errors found by CNFuzz and FuzzSAT are different. Additionally, FuzzSAT was able to find an assertion failure of March_ks, which the other fuzzers were not able find. Our experimental results suggest that a portfolio of fuzzers should be used in order to find different solver defects. Notice that we listed the defects that each of our fuzzers were able to find in only *one* hour, which shows the impressing effectiveness of fuzz testing. Moreover, a portfolio of fuzzers could be run on a cluster for days or even weeks, which would strongly increase the probability of finding defects that could not be found so far.

The fuzz testing results for QBF solvers listed in Tab. 3 shows that also our QBF fuzzing techniques were able to find many critical defects in state-of-the-art QBF solvers. As validating QBF solver results is much harder than validating SAT solvers, we used a majority voting in order to determine the correct result. If at least 90% of the QBF solvers agreed on the satisfiability status, then all solvers reporting the opposite were classified as incorrect.

Our QBF fuzzer QBFuzz is clearly superior to the QBF generator BlocksQBF. The higher diversity of instances generated by QBFuzz enabled finding defects that BlocksQBF was not able to detect.

3 Delta Debugging

The overall goal of delta debugging [51,50,15,37] is to minimize failure-inducing inputs. Typically, minimized inputs simplify the debugging process as irrelevant input parts have been removed. In principle, delta debugging SAT and QBF solvers works as follows. First, the delta debugger runs the solver on the original failure-inducing input in order to observe the failure induced by the original input, e.g. the solver crashes or reports an incorrect satisfiability status. Then, the delta debugger repeatedly tries to simplify the failure-inducing input. After each simplification, the delta debugger runs the solver on the simplified input. If the solver shows the same observable behavior, the delta debugger treats the simplification as success and continues simplifying the reduced input. Otherwise, the delta debugger undoes the last simplification, and continues with other simplifications. The delta debugger repeats this process until a given time limit or fix-point is reached.

In general, it is not guaranteed that delta debugging generates a minimal failure-inducing input. However, this feature is rarely needed in practice. Instead, greedy minimization techniques are used to simplify the input as much and as fast as possible in order to generate a small failure-inducing input that can be used for effective debugging. In the following we present our delta debugging techniques for SAT and QBF.

3.1 SAT Delta Debugging

Our first CNF delta-debugger cnfdd is based on a variant of the algorithm described in [50]. With increasing granularity it iteratively tries to remove subsets of the whole clause set, without changing the exit code of the solver on the reduced formula. Eventually the delta-debugger will try to remove individual clauses. Thus cnfdd applied to solving an unsatisfiable instance, using a sound SAT solver of course, simply simulates a binary search for minimal unsatisfiable cores. In contrast to [50], complements of subsets are not considered to be removed, and cnfdd is also not restarted after a successful removal of a subset of clauses. This makes cnfdd greedier than the original DDMIN approach [51]. These changes lead to a reduction of the actual number of calls to the SAT solver during delta debugging, leading to improved performance.

However, and this is a key insight, only removing clauses, will just make the formula easier to satisfy. This will rarely lead to sufficient overall reduction. It is essential, to also strengthen the formula, of course without removing the failure. Our current version tries to remove individual literal occurrences, which is rather costly and an opportunity for future improvement. After this phase of removing individual literals, and if at least one literal was removed, the delta debugger tries to reduce the variable range, and the whole procedure is restarted.

There is also a multi-threaded version `mtcnfdd` which tries to remove clauses and literals in parallel. In the clause removal phase all sets of clauses of the current granularity are split into as many parts as threads are available. Each thread checks in parallel whether some subsets of the clauses of its part can be removed. For the clause removal phase, the worker threads are synchronized after all subsets of the current granularity have been tried. Successful removals are merged sequentially by the master thread, starting with the local view of a thread that was able to remove the largest number of clauses. In the literal removal phase, which is far less frequently successful than clause removal, clauses are split among the threads as well. Successful literal removal attempts will be tried to be merged immediately. They become permanent if the attempt of a worker thread to merge its reduced local view with the global view succeeded. Otherwise the global view takes precedence and is copied as local view.

3.2   QBF Delta Debugging

Our tool `qbfdd` is a highly configurable delta debugger for QBF instances in PCNF. It supports different variants of delta debugging strategies such as the original DDMIN [51] approach (default), DDMIN with complements only [50], and a simple strategy based on one-by-one elimination. Similar to `cnfdd`, it tries to remove subsets of the whole clause set. Then, it tries to remove individual literals. Optionally, it can move variables between quantifier sets, which may enable further simplifications. If any simplification was possible, the delta debugger continues with a new simplification round, and terminates otherwise.

3.3   Experiments

We ran our experiments on the same hardware as our fuzzing experiments. The results of our delta debugging experiments for SAT solvers are shown in Tab. 4. The experimental results clearly show the overall effectiveness of our delta debugging techniques in shrinking failure-inducing CNF instances. With the exception of RSat, our delta debugger could eliminate huge parts of the original failure-inducing parts. In the case of PicoSAT the delta debugger was able to shrink the original failure-inducing CNF instances containing more than one thousand clauses to a tiny CNF with only a few clauses as shown in Fig. 1. The defects found for RSat, which are aborts and segmentation faults, could not be minimized significantly. This in contrast to delta debugging crash-inducing instances of other solvers. For instance, segmentation faults for MiniSat-9z could be delta debugged efficiently with an average reduction of 98.8%. Therefore, we suppose that the defects of RSat are non-trivial and cannot be triggered by a small CNF easily. For example, the failures could need a minimum number of unit propagations in order to occur.

During our experimental evaluation we observed that RSat and March_hi sometimes needed an unexpected long time (several hours) to solve instances generated during delta debugging. For instance, March_hi generated an invalid solution for the original failure-inducing instance almost immediately, but it needed hours to solve simplified instances proposed by the delta debugger. In order to speed up delta debugging, we used a time limit as proposed in [11]. We used a time limit of ten seconds to each call to Rsat and March_hi during delta debugging. If the solver exceeds the limit the delta debugger simply treats this case as if the current failure-inducing input does not lead to the same observable failure as the original input, i.e. the current simplification was not successful.

Our multi-threaded delta debugger mtcnfdd clearly outperforms our single-threaded delta debugger cnfdd. It is significantly faster on the failure-inducing instances of Barcelogic, Barcelogic-fixed, March_hi and RSat. Moreover, mtcnfdd tends to generate smaller instances than cnfdd.

Notice that we did not show experimental results of delta debugging failure-inducing inputs for ManySAT and MiraXTv3 as they showed non-deterministic behavior. For instance, MiraXTv3 reported different satisfiability results when all four cores of our computer were utilized. Due to space constraints we omit our preliminary results on delta debugging non-deterministic solvers.

In order to delta debug incorrect results, we used MiniSAT from SAT competition 2009 for SAT and Qube6.6 for QBF as reference solvers. The delta debugger calls a wrapper script instead of calling the incorrect solver directly. The script calls the reference solver and the incorrect solver on the current instance proposed by the delta debugger. If both solver agree on the satisfiability status, the script returns 1, and 0 otherwise. The possibility of calling scripts instead of solvers directly makes our delta debuggers highly flexible. Optionally, satisfiability results could be validated with techniques as proposed in [21,52].

In order to illustrate the success of our delta debuggers, we show some selected examples of minimized instances in Fig. 1. PicoSAT from SAT competition 2007 prints the solution 1 2 3 -4 for the first instance shown left, although it is obviously unsatisfiable. March_ks from SAT competition 2007 prints the solution 1 2 3 for the second unsatisfiable example. Moreover, it claims that the solution has been verified, which shows the demand for external checking tools such as [52] for the unsatisfiable case. The QBF solver yQuaffle aborts with an assertion failure when run on the third instance. QuBE 6.0 claims that the fourth instance (shown right) is satisfiable although it contains a universal unit clause.

3.4 Related Work

The work most closely related is [11]. The authors showed that fuzz testing and delta debugging techniques can be successfully applied to Satisfiability Modulo Theories (SMT) solvers. In this paper, we introduce techniques that have been explicitly designed and optimized for pure SAT and QBF. The SMT-LIB

Table 4. Experimental results of delta debugging SAT solvers from SAT competition 2007 and 2009. We evaluated our single-threaded delta debugger `cnfdd` and our multi-threaded delta debugger `mtcnfdd`, configured to use six threads. From left to right, the table shows the solver name (`solver`), the number of failure-inducing files (`files`), the number of bug classes (`classes`), the average delta debugging time (`time`) in seconds, the average file size (`size`) of the reduced instances in bytes and the average file size reduction (`red`) achieved by the delta debugger. Notice that the delta debugging time includes the time needed for the solver calls. We used a time limit of three hours for delta debugging each CNF instance. The delta debugger `cnfdd` exceeded this time limit three times (one instance of march_hi and two instances of RSat). The multi-threaded delta debugger `mtcnfdd` exceeded the time limit two times (the same RSat instances as `cnfdd`). Moreover, we used a time limit of ten seconds for each call to RSat and March_hi during delta debugging.

			cnfdd			mtcnfdd		
solver	files	classes	time	size	red	time	size	red
Barcelogic	7	4	39	432	95.8%	**20**	**378**	**96.4%**
Barcelogic-fixed	2	2	41	361	99.0%	**29**	**360**	99.0%
March_hi	24	1	638	**1982**	**88.4%**	**277**	2507	85.4%
March_ks	35	3	4	147	97.8%	**3**	**130**	**98.0%**
MiniSat-9z	912	1	<1	10	98.8%	<1	10	98.8%
PicoSAT	2	1	2	**39**	99.8%	2	40	99.8%
RSat	86	2	1478	17068	32.5%	**762**	**16971**	**32.9%**

Table 5. Experimental results of delta debugging QBF solvers. The columns have the same meaning as in Tab. 4. The delta debugging time includes the time needed for the solver calls. In order to effectively delta debug failure-inducing inputs on which SQBF crashed almost immediately, we used a time limit of two seconds for each call to SQBF during delta debugging.

		qbfdd			
solver	files	classes	time	size	red
Quantor	1	1	35	446	83.0%
QuBE 6.0	696	2	150	33	99.0%
QuBE 6.5	4	1	84	363	83.8%
sKizzo	31	2	330	497	76.2%
SQBF	35	1	57	289	86.7%
yQuaffle	94	1	26	31	98.8%

format [42] is much more complex than the DIMACS and QDIMACS format as it supports specifying formulas in several fragments of first order logic. However, in contrast to the flat CNF in SAT and QBF instances, the structural information in SMT-LIB instances can be used to apply Hierarchical Delta Debugging [37] (HDD), which is hardly possible in SAT as hierarchical information is typically lost during the translation to CNF.

```
c Picosat07        c March_ks07        c yQuaffle09        c QuBE6.0
p cnf 4 4          p cnf 3 5           p cnf 1 2           p cnf 2 2
-2 -1 0            1 0                 e 1                 a 1 0
-2 1 0             2 -1 -3 0           1 0                 e 2 0
2 0                -2 -1 3 0           -1 0                2 0
-3 -4 0            -3 -2 0                                 1 0
                   3 2 0
```

Fig. 1. Examples of delta debugged failure-inducing inputs for SAT and QBF

Freeman mentions in his thesis [19] that he uses a 3SAT generator to test his SAT solver. However, to the best of our knowledge, there does not exist any rigorous scientific publication about automated testing and debugging SAT and QBF solvers. Nevertheless, there are a few publications that treat the problem of validating solvers. For instance, in [21,52,49] the authors instrument DPLL-based [17] solvers in order to verify unsatisfiability claims by checking traces. Recent work focuses on QBF solver validation with the help of certificates [38,31].

4   Conclusion

Essential criteria of SAT and QBF solvers are robustness and correctness. We have demonstrated that our fuzzing techniques were able to find critical defects that lead to crashes, incorrect results and invalid models in state-of-the-art SAT and QBF solvers. In particular, our fuzzers detected critical defects in top-ranked solvers at the SAT competition 2007 and 2009. Therefore, we propose to use fuzz testing in an extra qualification phase in SAT and QBF competitions in order to increase the reliability of competition results. Moreover, we showed that our delta debugging techniques are very effective in minimizing failure-inducing inputs for SAT and QBF solvers. All tools are available as open source and provide support for automated testing and debugging of SAT and QBF solvers.

Acknowledgements. We would like to thank T. Hribernig, M. Preiner and A. Niemetz for the implementations of `mtcnfdd`, `QBFuzz` and `qbfdd`.

References

1. Ansótegui, C., Béjar, R., Fernández, C., Mateu, C.: Generating Hard SAT/CSP Instances Using Expander Graphs. In: AAAI (2008)
2. Babić, D.: Exploiting Structure for Scalable Software Verification. PhD thesis, University of British Columbia, Vancouver, Canada (2008)
3. Benedetti, M.: sKizzo: A Suite to Evaluate and Certify QBFs. In: Nieuwenhuis, R. (ed.) CADE 2005. LNCS (LNAI), vol. 3632, pp. 369–376. Springer, Heidelberg (2005)
4. Le Berre, D.: SAT4J: Bringing the Power of SAT Technology to the Java Platform, http://www.sat4j.org

5. Le Berre, D., Roussel, O., Simon, L.: SAT 2009 Competitive Events Booklet - Preliminary Version. In: SAT competition solver descriptions (September 2009), http://www.cril.univ-artois.fr/SAT09/solvers/booklet.pdf
6. Biere, A.: Resolve and Expand. In: SAT (Selected Papers) (2004)
7. Biere, A.: PicoSAT Essentials. JSAT 4 (2008)
8. Biere, A., Heule, M., van Maaren, H., Walsh, T. (eds.): Handbook of Satisfiability. IOS Press, Amsterdam (2009)
9. Bofill, M., Nieuwenhuis, R., Oliveras, A., Rodríguez-Carbonell, E., Rubio, A.: The Barcelogic SMT Solver. In: Gupta, A., Malik, S. (eds.) CAV 2008. LNCS, vol. 5123, pp. 294–298. Springer, Heidelberg (2008)
10. Bregman, D., Mitchell, D.: The SAT solver MXC, Version 0.5., SAT competition solver description (2007)
11. Brummayer, R., Biere, A.: Fuzzing and Delta-Debugging SMT Solvers. In: SMT. ACM International Conference Proceedings Series. ACM, New York (2009)
12. Kleine Büning, H., Karpinski, M., Flögel, A.: Resolution for Quantified Boolean Formulas. Inf. Comput. 117(1), 12–18 (1995)
13. Cadoli, M., Schaerf, M., Giovanardi, A., Giovanardi, M.: An Algorithm to Evaluate Quantified Boolean Formulae and Its Experimental Evaluation. J. Autom. Reasoning 28 (2002)
14. Chen, H., Interian, Y.: A Model for Generating Random Quantified Boolean Formulas. In: IJCAI (2005)
15. Claessen, K., Hughes, J.: QuickCheck: a Lightweight Tool for Random Testing of Haskell Programs. ICFP 35(9) (2000)
16. Creignou, N., Daudé, H., Egly, U., Rossignol, R.: New Results on the Phase Transition for Random Quantified Boolean Formulas. In: Kleine Büning, H., Zhao, X. (eds.) SAT 2008. LNCS, vol. 4996, pp. 34–47. Springer, Heidelberg (2008)
17. Davis, M., Logemann, G., Loveland, D.: A Machine Program for Theorem-Proving. ACM Commun. 5 (1962)
18. Eén, N., Sörensson, N.: An extensible SAT-solver. In: Giunchiglia, E., Tacchella, A. (eds.) SAT 2003. LNCS, vol. 2919, pp. 502–518. Springer, Heidelberg (2004)
19. Freeman, J.W.: Improvements to Propositional Satisfiability Search Algorithms. PhD thesis, Depart. of Comp. and Inf. Science, University of Pennsylvania (1995)
20. Gebser, M., Kaufmann, B., Schaub, T.: The Conflict-Driven Answer Set Solver clasp: Progress Report. In: Erdem, E., Lin, F., Schaub, T. (eds.) LPNMR 2009. LNCS, vol. 5753, pp. 509–514. Springer, Heidelberg (2009)
21. Van Gelder, A.: Extracting (Easily) Checkable Proofs from a Satisfiability Solver that Employs both Preorder and Postorder Resolution. In: AMAI (2002)
22. Gent, I., Walsh, T.: The SAT Phase Transition. In: ECAI (1994)
23. Gent, I., Walsh, T.: Beyond NP: the QSAT phase transition. In: AAAI/IAAI (1999)
24. Giunchiglia, E., Narizzano, M., Tacchella, A.: QuBE++: An Efficient QBF Solver. In: Hu, A.J., Martin, A.K. (eds.) FMCAD 2004. LNCS, vol. 3312, pp. 201–213. Springer, Heidelberg (2004)
25. Hamadi, Y., Jabbour, S.: ManySAT: a Parallel SAT Solver. JSAT 6 (2009)
26. Heule, M.: SmArT Solving: Tools and Techniques for Satisfiability Solvers. PhD thesis, TU Delft (2008)
27. Horie, S., Watanabe, O.: Hard instance generation for SAT. CoRR, cs.CC/9809117 (1998)
28. Hsu, E., McIlraith, S.: VARSAT: Integrating Novel Probabilistic Interference Techniques with DPLL Search. In: Kullmann, O. (ed.) SAT 2009. LNCS, vol. 5584, pp. 377–390. Springer, Heidelberg (2009)
29. Huang, J.: TINISAT in SAT Competition 2007. SAT competition solver description (2007)

30. Jain, H., Clarke, E.: SAT Solver Descriptions: CMUSAT-Base and CMUSAT. SAT competition solver description (2007)
31. Jussila, T., Biere, A., Sinz, C., Kröning, D., Wintersteiger, C.: A First Step Towards a Unified Proof Checker for QBF. In: Marques-Silva, J., Sakallah, K.A. (eds.) SAT 2007. LNCS, vol. 4501, pp. 201–214. Springer, Heidelberg (2007)
32. Kern, C., Khaleghi, M., Kugele, S., Schallhart, C., Tautschnig, M., Weis, A.: SAT 7 - Engineering a Modular SAT-Solver. SAT competition solver description (2007)
33. Letz, R.: Lemma and Model Caching in Decision Procedures for Quantified Boolean Formulas. In: Egly, U., Fermüller, C. (eds.) TABLEAUX 2002. LNCS (LNAI), vol. 2381, p. 160. Springer, Heidelberg (2002)
34. Lewis, M., Schubert, T., Becker, B.: Multithreaded SAT Solving. In: Asia and South Pacific DAC (2007)
35. Lonsing, F.: DepQBF 0.1 Source Code (2010), http://fmv.jku.at/depqbf/
36. Miller, B., Koski, D., Lee, C., Maganty, V., Murthy, R., Natarajan, A., Steidl, J.: Fuzz Revisited: A Re-examination of the Reliability of UNIX Utilities and Services. Technical Report CS-TR-1995-1268, University of Wisconsin, Madison (1995)
37. Misherghi, G., Su, Z.: HDD: Hierarchical Delta Debugging. In: ICSE, pp. 142–151. ACM, New York (2006)
38. Narizzano, M., Peschiera, C., Pulina, L., Tacchella, A.: Evaluating and Certifying QBFs: A Comparison of State-of-the-Art Tools. AI Commun. 22 (2009)
39. Pan, G., Vardi, M.: Symbolic Decision Procedures for QBF. In: Wallace, M. (ed.) CP 2004. LNCS, vol. 3258, pp. 453–467. Springer, Heidelberg (2004)
40. Pennock, D., Stout, Q.: Exploiting a Theory of Phase Transitions in Three-Satisfiability Problems. In: AAAI/IAAI, vol. 1 (1996)
41. Pipatsrisawat, K., Darwiche, A.: RSat 2.0: SAT Solver Description. Technical Report D–153, Automated Reasoning Group, CSD, UCLA (2007)
42. Ranise, S., Tinelli, C.: The SMT-LIB Standard: Version 1.2. Technical report, Department of Computer Science, The University of Iowa (2006)
43. Samulowitz, H.: MiniQBF Solver (2010), http://miniqbf.spaces.live.com/
44. Samulowitz, H., Bacchus, F.: Using SAT in QBF. In: van Beek, P. (ed.) CP 2005. LNCS, vol. 3709, pp. 578–592. Springer, Heidelberg (2005)
45. Sutton, M., Greene, A., Amini, P.: Fuzzing - Brute Force Vulnerability Discovery. Pearson Ed., London (2007)
46. Takanen, A., Demott, J., Miller, C.: Fuzzing for Software Security Testing and Quality Assurance. Artech House (2008)
47. Tseitin, G.: On the Complexity of Proofs in Propositional Logics. Automation of Reasoning: Classical Papers in Computational Logic 1967-1970 2 (1983)
48. Xu, K., Boussemart, F., Hemery, F., Lecoutre, C.: A Simple Model to Generate Hard Satisfiable Instances. CoRR, abs/cs/0509032 (2005)
49. Yu, Y., Malik, S.: Validating the Result of a Quantified Boolean Formula (QBF) Solver: Theory and Practice. In: Asia and South Pacific DAC (2005)
50. Zeller, A.: Why Programs Fail. A Guide to Systematic Debugging. Morgan Kaufmann, San Francisco (2005)
51. Zeller, A., Hildebrandt, R.: Simplifying and Isolating Failure-Inducing Input. IEEE Transactions on Software Engineering 28(2), 183–200 (2002)
52. Zhang, L., Malik, S.: Validating SAT Solvers Using an Independent Resolution-Based Checker: Practical Implementations and Other Applications. In: DATE 2003 (2003)
53. Zhang, L., Malik, S.: Towards Symmetric Treatment of Conflicts And Satisfaction in Quantified Boolean Satisfiability Solver. In: Van Hentenryck, P. (ed.) CP 2002. LNCS, vol. 2470, p. 200. Springer, Heidelberg (2002)

Rewriting (Dependency-)Quantified 2-CNF with Arbitrary Free Literals into Existential 2-HORN

Uwe Bubeck and Hans Kleine Büning[*]

Universität Paderborn, Germany
{bubeck,kbcsl}@upb.de

Abstract. We extend quantified 2-CNF formulas by also allowing literals over free variables which are exempt from the 2-CNF restriction. That means we consider quantified CNF formulas with clauses that contain at most two bound literals and an arbitrary number of free literals. We show that these $Q2\text{-}CNF^b$ formulas can be transformed in polynomial time into purely existentially quantified CNF formulas in which the bound literals are in 2-HORN ($\exists 2\text{-}HORN^b$).

Our result still holds if we allow Henkin-style quantifiers with explicit dependencies. In general, dependency quantified Boolean formulas ($DQBF$) are assumed to be more succinct at the price of a higher complexity. This paper shows that $DQ2\text{-}CNF^b$ has a similar expressive power and complexity as $\exists 2\text{-}HORN^b$. In the special case that the 2-CNF restriction is also applied to the free variables ($DQ2\text{-}CNF^*$), the satisfiability can be decided in linear time.

1 Introduction

Quantified Boolean formulas (QBF) generalize propositional formulas by allowing variables to be quantified universally or existentially. In this paper, we also allow free variables which are not quantified and indicate this with a star (QBF^*). An interesting property of quantified Boolean formulas with free variables is that it is possible to define an equivalence between such formulas and propositional formulas. We say that $\Phi \in QBF^*$ is equivalent to $\psi \in PROP$ ($\Phi \approx \psi$) if and only if the free variables in Φ correspond to the propositional variables in ψ and both formulas have the same truth value for each assignment to the free/propositional variables. This means that quantified variables inside of Φ are not taken into consideration here, so these can be thought of as local or auxiliary variables. An important application of auxiliary variables is to introduce abbreviations for repeating parts in a given formula, such as multiple copies of transition or reachability relations in verification problems [9, 14]. Accordingly, QBF^* representations are often much more compact than equivalent propositional encodings, in addition to the advantage that many problems have a natural forall-exists semantics which can elegantly be modeled by quantifiers [20].

---

[*] Partially supported by the German Research Foundation (DFG), grant KL 529/QBF.

Unfortunately, quantified Boolean formulas appear to be much harder to solve than propositional formulas, with QBF and QBF^* satisfiability being $PSPACE$-complete. This makes it worthwhile to investigate subclasses with a lower decision complexity. An interesting idea is to consider QBF^* formulas in clausal form with additional restrictions only on the quantified literals. Let $\Phi = Q \bigwedge_i (\phi_i^b \vee \phi_i^f)$ be a quantified Boolean formula with quantifiers Q, such that ϕ_i^b is a clause over bound variables (called *bound part*) and ϕ_i^f a clause over free variables (the *free part*). Then we require that $Q \bigwedge_i \phi_i^b \in QK$ for a formula class QK, while the free parts ϕ_i^f may have arbitrary structure. Such formulas, which we call QK^b for a base class QK, can be surprisingly powerful.

For example, $QHORN^b$ denotes quantified Horn formulas in which the Horn property is only enforced on the quantified variables, which means each clause has at most one positive and arbitrarily many negative literals over quantified variables, but an arbitrary number of free literals with arbitrary polarity. Obviously, every propositional CNF formula is also a $QHORN^b$ formula, but this class is significantly more capable. For example, $QHORN^b$ formulas can compactly encode Boolean circuits with arbitrary fan-out (and vice versa) [1, 15], while it is generally assumed that there exist circuits with fan-out greater than 1 for which every equivalent propositional formula is exponentially larger. Furthermore, while there are propositional formulas for which every equivalent CNF formula is exponential, every propositional formula has a poly-size equivalent $QHORN^b$ formula, e.g. by the one-sided Tseitin transformation [22, 19] when the newly introduced variables are bound by existential quantifiers. In fact, such poly-size CNF transformations can even be accomplished with $\exists 2\text{-}HORN^b$ formulas, that is, existentially quantified formulas in clausal form with at most two bound literals per clause, one of which may be positive [8]. At the same time, $QHORN^b$ satisfiability is not significantly more difficult than propositional satisfiability, because the universal quantifiers can easily be eliminated [15], which makes $QHORN^b$ satisfiability NP-complete.

Besides $HORN$, another standard restriction on the structure of clauses is $2\text{-}CNF$. The goal of this paper is to investigate the implications of enforcing a $2\text{-}CNF$ restriction on the bound parts of QBF^* formulas in clausal form. That means we have clauses with at most two bound and arbitrarily many free literals, called $Q2\text{-}CNF^b$ in line with the above terminology. This class is surprisingly powerful and indeed exponentially more expressive than propositional CNF because of the above remark about $\exists 2\text{-}HORN^b \subseteq Q2\text{-}CNF^b$ formulas being sufficient for poly-size CNF transformation.

Normally, $2\text{-}CNF$ formulas are not more difficult than $HORN$ formulas. In the propositional case, it is well known that the satisfiability problem for both classes can be solved in linear time ([11, 2] and [13, 10]). For quantified $2\text{-}CNF$ formulas with free variables, the satisfiability problem is still linear [2], whereas the best known algorithms for determining the satisfiability of a quantified Horn formula Φ with $|\forall|$ universal quantifiers require time $O(|\forall| \cdot |\Phi|)$ [12] ($|\Phi|$ is the length of Φ, counting all occurrences of variables, including those in quantifier definitions).

Is it possible to make similar statements about the complexity and expressive power of $Q2\text{-}CNF^b$ in comparison to $QHORN^b$ formulas? Our goal is to show that $Q2\text{-}CNF^b$ formulas can be transformed in polynomial time into equivalent $\exists 2\text{-}HORN^b$ formulas. This immediately implies that $Q2\text{-}CNF^b$ satisfiability is NP-complete, like $QHORN^b$ satisfiability.

An intermediate result that we present is the elimination of all universal quantifiers from a $Q2\text{-}CNF^b$ formula Φ in time and space $O(|\forall|^2 |\Phi|)$. This might be useful for QBF solvers, since a $Q2\text{-}CNF^b$ formula can be embedded as a subformula in a QBF formula if we consider variables which are bound by preceding quantifiers as free variables. For example, let $\Phi = Q((Q'\ \phi) \wedge \varphi) \approx QQ'\ (\phi \wedge \varphi)$ be a QBF formula in CNF where each clause in ϕ contains at most two literals over variables that are bound in Q', whereas the variables from Q can appear without restrictions in ϕ and φ. Then the transformation presented below allows the elimination of all universals in the $Q2\text{-}CNF^b$ formula $Q'\ \phi$.

2   Dependency Quantified Boolean Formulas

In QBF, an existentially quantified variable can have different values depending on the values of universal variables whose quantifiers occur further outside. This imposes an ordering on the quantifiers where each existentially quantified variable depends on all preceding universal variables. Even if we waive the usual requirement that all quantifiers have to appear at the beginning in a dedicated quantifier prefix, it is not possible for two existential variables which occur in common clauses to depend on disjoint non-empty sets of universally quantified variables. Dependency quantified Boolean formulas ($DQBF$ or $DQBF^*$ with free variables) [18] make this possible by explicitly stating for each existentially quantified variable on which universals it depends. For example, $\Phi = \forall x_1 \forall x_2 \exists y_1(x_1) \exists y_2(x_2) \exists y_3(x_1, x_2)\ \phi(x_1, x_2, y_1, y_2, y_3)$ is a $DQBF$ formula in which y_1 depends only on x_1, y_2 only on x_2 and y_3 on both x_1 and x_2.

Can we apply our poly-time transformation from $Q2\text{-}CNF^b$ to $\exists 2\text{-}CNF^b$ also to $DQ2\text{-}CNF^b$ formulas, which means formulas with dependency quantifiers as in the example above and at most two bound literals per clause? The fact that universal variables can be eliminated cheaply from $Q2\text{-}CNF^b$ formulas implies that $2\text{-}CNF$ is such a strong restriction that the ordering of quantifiers in the prefix loses much of its relevance. For $DQHORN^b$, the situation is similar: it is indeed possible to eliminate all dependency quantifiers with less than quadratic formula growth [6] (that proof is for $DQHORN^*$, but it also applies to $DQHORN^b$, since it does not rely on a particular structure of the free variables).

In general, however, $DQBF^*$ encodings are assumed to be exponentially more compact in the best case than QBF^* encodings. Whereas QBF can be seen as a two-player game with an existential player reacting to moves of a universal player, $DQBF$ corresponds to a three-player game where a universal player challenges two existential players with different inputs. Disjoint dependencies, like for y_1 and y_2 in the example above, guarantee that both existential players work independently. Such variables can still occur together in the same clauses, which is a vital feature that is not possible with *QBF*, even in non-prenex form. It allows the universal player to compare the results of independent existential players. This corresponds to a multi-prover interactive proof system [4], which is a very powerful concept, but also causes another jump in complexity, with $DQBF^*$ satisfiability being *NEXPTIME*-complete [18].

Before we can develop a transformation from $DQ2\text{-}CNF^b$ to $\exists 2\text{-}CNF^b$, we need a few basics. We require $DQBF^*$ formulas to be in prenex form with a quantifier-free matrix, as negations of existential dependency quantifiers would be problematic. Because of the explicit dependencies, $DQBF^*$ formulas can always be written with a $\forall^*\exists^*$ prefix. To quickly enumerate the dependencies of a given existential variable y_i, we use indices $d_{i,1}, ..., d_{i,n_i}$ which point to the n_i universals on which y_i depends. For example, given the existential quantifier $\exists y_4(x_3, x_5)$, we say that y_4 depends on $x_{d_{4,1}}$ and $x_{d_{4,2}}$ with $d_{4,1} = 3$ and $d_{4,2} = 5$. We also use a shorter notation $\exists y_i(\mathbf{x_{d_i}})$ where we abbreviate $\mathbf{x_{d_i}} := (x_{d_{i,1}}, ..., x_{d_{i,n_i}})$. It is allowed to have empty dependencies with $n_i = 0$, i.e. existential quantifiers $\exists y_i()$ that do not depend on any universals.

The semantics of *DQBF* and $DQBF^*$ is defined by associating dependency quantified existentials $\exists y_i(x_{d_{i,1}}, ..., x_{d_{i,n_i}})$ with functions $f_{y_i}(x_{d_{i,1}}, ..., x_{d_{i,n_i}})$:

Definition 1. *(Satisfiability Model)*
For $\Phi \in DQBF$ with existential variables $\mathbf{y} = (y_1, ..., y_m)$, let $M = (f_{y_1}, ..., f_{y_m})$ map each existential y_i to a propositional formula f_{y_i} over the universal variables $x_{d_{i,1}}, ..., x_{d_{i,n_i}}$ on which y_i depends.
*M is a **satisfiability model** for Φ if and only if $\Phi[\mathbf{y}/M] := \Phi[y_1/f_{y_1}, ..., y_m/f_{y_m}]$ is true, i.e. if a tautological formula is obtained when simultaneously each existential variable y_i is replaced with f_{y_i} and the existential quantifiers are dropped from the prefix.*

Definition 2. *(DQBF and DQBF* Semantics)*
*A DQBF formula Φ is **true** if and only if it has a satisfiability model.*
A DQBF formula $\Psi(\mathbf{z})$ with free variables $\mathbf{z} = (z_1, ..., z_r)$ is **satisfiable** if and only if there exists a truth assignment $\tau(\mathbf{z}) = (\tau(z_1), ..., \tau(z_r)) \in \{0,1\}^r$ to the free variables such that $\Psi(\tau(\mathbf{z})) \in DQBF$ is true, i.e. replacing all occurrences of free variables with their assigned truth value produces a true formula.*

3   Transformation from $DQ2\text{-}CNF^b$ to $\exists 2\text{-}CNF^b$

There are two powerful concepts that we need for transforming $DQ2\text{-}CNF^b$ formulas into $\exists 2\text{-}CNF^b$: universal expansion and minimal falsity/unsatisfiability.

Universal expansion in QBF^* is the elimination of universal quantifiers by the well-known equivalence $\forall x\ \Phi(x, \mathbf{z}) \approx \Phi(0, \mathbf{z}) \wedge \Phi(1, \mathbf{z})$, an operation which has been used successfully in various solvers, e.g. [3, 5, 7]. Care must be taken to duplicate also subsequent existential quantifiers which are in the scope of the expanded quantifier, in order to retain the ability to assign different values to an existential for different values of a preceding universal. In general, repeated application of this method obviously produces exponential formulas, even though the amount of duplication can often be significantly reduced in practice [5, 7, 17, 21]. We are going to show that the *2-CNF* restriction on the bound variables allows us to always apply universal expansion in a tractable fashion.

Universal expansion also works for *DQBF**, and the dependency lists immediately indicate which existentials must be duplicated when a universal variable is expanded. The correctness of universal expansion is bit more difficult to verify for *DQBF** because of the more implicit semantics definition by using model functions.

Lemma 1. *(Correctness of Universal Expansion for DQBF*)*
Let Φ be a DQBF* formula in which we want to expand the universal quantifier $\forall x_n$. Without loss of generality, assume that the existentials are arranged in two blocks, depending on whether they are dominated by x_n or not:

$$\Phi(\mathbf{z}) = \forall x_1...\forall x_n \exists y_1(\mathbf{x_{d_1}})...\exists y_k(\mathbf{x_{d_k}}) \exists y_{k+1}(\mathbf{x_{d_{k+1}}}, x_n)...\exists y_m(\mathbf{x_{d_m}}, x_n)$$
$$\phi(x_1, ..., x_n, y_1, ..., y_m, \mathbf{z})$$

with $x_n \notin \mathbf{x_{d_i}}$ for all $1 \le i \le m$. Then $\Phi(\mathbf{z}) \approx \Phi'(\mathbf{z})$ for the expanded formula

$$\Phi'(\mathbf{z}) = \forall x_1...\forall x_{n-1} \exists y_1(\mathbf{x_{d_1}})...\exists y_k(\mathbf{x_{d_k}})$$
$$\exists y_{k+1,(0)}, y_{k+1,(1)}(\mathbf{x_{d_{k+1}}})...\exists y_{m,(0)}, y_{m,(1)}(\mathbf{x_{d_m}})$$
$$\phi(x_1, ..., x_{n-1}, 0, y_1, ..., y_k, y_{k+1,(0)}, ..., y_{m,(0)}, \mathbf{z}) \wedge$$
$$\phi(x_1, ..., x_{n-1}, 1, y_1, ..., y_k, y_{k+1,(1)}, ..., y_{m,(1)}, \mathbf{z}) .$$

Proof. We must prove that $\Phi(\tau(\mathbf{z})) = 1 \Leftrightarrow \Phi'(\tau(\mathbf{z})) = 1$ for any truth assignment $\tau(\mathbf{z}) := (\tau(z_1), ..., \tau(z_r)) \in \{0,1\}^r$ to the free variables $\mathbf{z} = (z_1, ..., z_r)$. For fixed $\tau(\mathbf{z})$, we can consider $\Phi(\tau(\mathbf{z}))$ and $\Phi'(\tau(\mathbf{z}))$ as closed *DQBF* formulas.

From left to right: let $M = (f_{y_1}, ..., f_{y_m})$ be a satisfiability model for $\Phi(\tau(\mathbf{z}))$. Define $G_{(0)} := (g_{y_1}, ..., g_{y_k}, g_{y_{k+1,(0)}}, ..., g_{y_{m,(0)}})$ with $g_{y_i} := f_{y_i}$ for $i = 1, ..., k$ and $g_{y_{i,(0)}}(x_{d_{i,1}}, ..., x_{d_{i,n_i}}) := f_{y_i}(x_{d_{i,1}}, ..., x_{d_{i,n_i}}, 0)$ for $i = k+1, ..., m$. Then $\forall x_1...\forall x_{n-1} \phi(x_1, ..., x_{n-1}, 0, g_{y_1}, ..., g_{y_{m,(0)}}, \tau(\mathbf{z})) = 1$. With an analogous definition of $G_{(1)}$ with functions $g_{y_{i,(1)}}(x_{d_{i,1}}, ..., x_{d_{i,n_i}}) := f_{y_i}(x_{d_{i,1}}, ..., x_{d_{i,n_i}}, 1)$ for $i = k+1, ..., m$, $G = (g_{y_1}, ..., g_{y_k}, g_{y_{k+1,(0)}}, g_{y_{k+1,(1)}}, ..., g_{y_{m,(0)}}, g_{y_{m,(1)}})$ is a satisfiability model for $\Phi'(\tau(\mathbf{z}))$.

From right to left: let $G = (g_{y_1}, ..., g_{y_k}, g_{y_{k+1,(0)}}, g_{y_{k+1,(1)}}, ..., g_{y_{m,(0)}}, g_{y_{m,(1)}})$ be a satisfiability model for $\Phi'(\tau(\mathbf{z}))$. We now construct a model $M = (f_{y_1}, ..., f_{y_m})$ that satisfies $\Phi(\tau(\mathbf{z}))$. Let $f_{y_i} := g_{y_i}$ for $i = 1, ..., k$, and for $i = k+1, ..., m$, let

$$f_{y_i}(x_{d_{i,1}}, ..., x_{d_{i,n_i}}, x_n) := (x_n \vee g_{y_{i,(0)}}(\mathbf{x_{d_i}}) \wedge (\neg x_n \vee g_{y_{i,(1)}}(\mathbf{x_{d_i}}))$$

such that $f_{y_i}[x_n/0] := f_{y_i}(x_{d_{i,1}}, ..., x_{d_{i,n_i}}, 0) \approx g_{y_{i,(0)}}(x_{d_{i,1}}, ..., x_{d_{i,n_i}})$, and thus:

$$\forall x_1...\forall x_{n-1} \phi(x_1, ..., x_{n-1}, 0, f_{y_1}, ..., f_{y_k}, f_{y_{k+1}}[x_n/0], ..., f_{y_m}[x_n/0], \tau(\mathbf{z}))$$
$$\approx \forall x_1...\forall x_{n-1} \phi(x_1, ..., x_{n-1}, 0, g_{y_1}, ..., g_{y_k}, g_{y_{k+1,(0)}}, ..., g_{y_{m,(0)}}, \tau(\mathbf{z}))$$

The latter is true, since G is a satisfiability model for $\Phi'(\tau(\mathbf{z}))$. The case $x_n = 1$ is analogous, so $\forall x_1...\forall x_{n-1} \forall x_n \phi(x_1, ..., x_{n-1}, x_n, f_{y_1}, ..., f_{y_m}, t(\mathbf{z})) = 1$. $\square$

The expressive power of $DQBF^*$ and QBF^* formulas in clausal form depends essentially on the structure of the minimal unsatisfiable subformulas of the bound part of the matrix, so we first recall some well-known properties. A CNF formula ϕ is called *minimal unsatisfiable* if and only if ϕ is unsatisfiable and the removal of an arbitrary clause produces a satisfiable formula. A (dependency) quantified Boolean formula $\Phi = Q \bigwedge_{1 \leq i \leq q} \phi_i$ with CNF matrix and without free variables is called *minimal false* if and only if Φ is false and removing an arbitrary clause ϕ_i leads to a true formula. If Φ is purely existentially quantified, it is minimal false if and only if the matrix is minimal unsatisfiable. A clause $L \vee K$ is called an *∃-unit clause* for a formula $\Phi \in DQ2\text{-}CNF$ if and only if L is a literal over an existentially quantified variable and either $L = K$ or K is a universally quantified literal.

A well-known fact about minimal unsatisfiable propositional 2-CNF formulas is that they contain at most two unit clauses (see, e.g., [16]). This result can be lifted to minimal false $DQ2$-CNF formulas:

Lemma 2. *(Number of ∃-unit clauses)*

1. *A minimal unsatisfiable 2-CNF formula contains at most two unit clauses.*
2. *A minimal false DQ2-CNF formula contains at most two ∃-unit clauses.*

Proof. Ad 1: Suppose there is some minimal unsatisfiable formula α with at least three unit clauses, say L_1, L_2 and L_3. Then there are clauses $\neg L_1 \vee P_1^{j_1}, \neg L_2 \vee P_2^{j_2}, \neg L_3 \vee P_3^{j_3}$ for $1 \leq j_1 \leq t_1, 1 \leq j_2 \leq t_2, 1 \leq j_3 \leq t_3$. Please notice that α contains no complementary unit clause $\neg L_i$ and no clauses $L_i \vee K_i$ for some literal K_i. Furthermore, all the literals $P_1^1, ..., P_3^{t_3}$ must be distinct. Let α be such a formula with a minimal number of variables. After applying unit resolution on the L_i and removing the parent clauses, we obtain a minimal unsatisfiable formula with at least three unit clauses. These are the clauses $P_1^1, ..., P_3^{t_3}$. The variables of L_i do not occur in the resulting formula, which is a contradiction to our initial assumption that α has a minimal number of variables.

Ad 2: Let $\Phi = \forall x_1 ... \forall x_n \exists y_1(\mathbf{x_{d_1}}) ... \exists y_m(\mathbf{x_{d_m}}) \, \phi$ be a minimal false formula in $DQ2$-CNF with at least three ∃-unit clauses. By expansion of the universal variables, we obtain an existentially quantified formula $\exists \mathbf{y}' \phi' \in \exists 2$-$CNF$ whose matrix ϕ' is unsatisfiable. A subset of the clauses in ϕ' forms a minimal unsatisfiable formula ϕ''. From the first part of the lemma, we know that ϕ'' contains at most two unit clauses, say L_1 and L_2. These literals are unit clauses in the original formula or come from clauses $U_1 \vee L_1$ or $U_2 \vee L_2$ with universal literals U_1, U_2. That means two ∃-unit clauses in ϕ are sufficient to produce two unit clauses L_1 and L_2 in ϕ''. All the other ∃-unit clauses in ϕ can be removed without making the formula satisfiable, which contradicts our initial assumption that ϕ is minimal false. □

Subsequently, we assume that all $DQ2$-CNF^b formulas are normalized to have no clauses without an existentially quantified literal. This is justified by the fact that clauses without bound variables can be moved in front of the prefix while preserving the equivalence. And in a 2-clause that contains a universal and a free variable, the universal variable can be omitted. Obviously, clauses consisting only of universal variables are unsatisfiable. We can also assume that there are no clauses ϕ_i without free literals. Otherwise, we could replace such a clause with clauses $\phi_i \vee z$ and $\phi_i \vee \neg z$ for a free variable z that already occurs in the formula. But the transformations also work if we assume $\phi_i^f := 0$ for such clauses without free literals.

The following lemma introduces a handy representation in which the minimal false subsets of the quantified bound parts determine which combinations of the free parts must be true.

Lemma 3. *(MF Skeleton)*
Let $\Phi = Q \bigwedge_{1 \leq i \leq q}(\phi_i^b \vee \phi_i^f)$ be a formula in DQ2-CNF^b with non-empty bound parts ϕ_i^b and free parts ϕ_i^f. Let

$$S(\Phi) := \{\Phi' \mid \Phi' = Q\phi_{i_1}^b \wedge \ldots \wedge \phi_{i_r}^b \text{ is minimal false}, 1 \leq i_1, ..., i_r \leq q\}$$

be the set of minimal false subformulas of the quantified bound parts of Φ. Then we have the following equivalence:

$$\Phi \approx \bigwedge_{(Q\phi_{i_1}^b \wedge \ldots \wedge \phi_{i_r}^b) \in S(\Phi)} (\phi_{i_1}^f \vee \ldots \vee \phi_{i_r}^f)$$

Proof. Let $M(\Phi) := \bigwedge_{(Q\phi_{i_1}^b \wedge \ldots \wedge \phi_{i_r}^b) \in S(\Phi)}(\phi_{i_1}^f \vee \ldots \vee \phi_{i_r}^f)$ be the right side of the equivalence. From right to left, let $M(\Phi)$ be true for a truth assignment τ to the free variables. Suppose $\tau(\Phi)$ is false. Let $Q\phi' := Q(\phi_{i_1}^b \wedge \ldots \wedge \phi_{i_r}^b)$ be the quantified bound parts for which $\tau(\phi_{i_k}^f)$ is false for $1 \leq k \leq r$. Under the assumption that $\tau(\Phi)$ is false, $Q\phi'$ is also false and contains therefore a minimal false subformula, say $Q\phi^* := Q(\phi_{j_1}^b \wedge \ldots \wedge \phi_{j_t}^b)$. Since $\tau(M(\Phi))$ is true, one of the free parts $\phi_{j_1}^f, ..., \phi_{j_t}^f$ must be true for τ. That is a contradiction.
From left to right, let Φ be true for a truth assignment τ to the free variables. Suppose $\tau(M(\Phi))$ is false. Then there is a clause $\phi' := (\phi_{i_1}^f \vee \ldots \vee \phi_{i_r}^f)$ in $M(\Phi)$ for which $\tau(\phi_{i_k}^f)$ is false for $1 \leq k \leq r$. Since $Q(\phi_{i_1}^b \wedge \ldots \wedge \phi_{i_r}^b)$ is minimal false, we can conclude that $\tau(\Phi)$ is false in contradiction to our assumption.   □

On the basis of Lemmas 2 and 3, we now establish a poly-time transformation from $DQ2$-CNF^b to $\exists 2$-CNF^b.

Theorem 1. *(DQ2-CNF^b $=_{poly-time}$ $\exists 2$-CNF^b)*
Every DQ2-CNF^b formula Φ can be transformed in time $O(|\forall|^2|\Phi|)$ into an equivalent $\exists 2$-CNF^b formula of length at most $O(|\forall|^2|\Phi|)$, where $|\forall|$ is the number of universal quantifiers in Φ.

Proof. In the following, we treat conjunctions of clauses as sets of clauses. Let $\Phi = Q\{(\phi_i^b \vee \phi_i^f) \mid 1 \leq i \leq q\}$ be a formula in $DQ2$-CNF^b with non-empty bound parts ϕ_i^b and free parts ϕ_i^f. We assume that Φ is forall-reduced, which means each clause contains at most one literal over a universal variable. For universal variables u_1, u_2 (not necessarily distinct), we let $\Phi|u_1, u_2$ denote the formula which contains only those clauses of Φ in which the universal literal is over u_1 or u_2 and those clauses without universals:

$$\Phi|u_1, u_2 := Q\{(\phi_i^b \vee \phi_i^f) \mid \text{every universal literal in } \phi_i^b \text{ is } u_1 \text{ or } u_2, 1 \leq i \leq q\}$$

According to Lemma 3, we have

$$\Phi \approx \bigwedge_{(Q\phi_{i_1}^b \wedge \ldots \wedge \phi_{i_r}^b) \in S(\Phi)} (\phi_{i_1}^f \vee \ldots \vee \phi_{i_r}^f)$$

where $S(\Phi)$ is the set of minimal false subformulas of the quantified bound parts of Φ. Lemma 2 implies that each minimal false formula in $S(\Phi)$ has at most two $\exists$-unit clauses. In analogy to the above notation, we let $S(\Phi)|u_1, u_2 \subseteq S(\Phi)$ be those minimal false formulas in which every $\exists$-unit clause with a universal literal contains either a literal over u_1 or a literal over u_2. Then the union of $S(\Phi)|u_1, u_2$ for all pairs of universals u_1, u_2 equals $S(\Phi)$:

$$\Phi \approx \bigwedge_{u_1, u_2 \in \forall var(\Phi)} \bigwedge_{(Q\phi_{i_1}^b \wedge \ldots \wedge \phi_{i_r}^b) \in S(\Phi)|u_1, u_2} (\phi_{i_1}^f \vee \ldots \vee \phi_{i_r}^f)$$

It is not difficult to see that $S(\Phi)|u_1, u_2 = S(\Phi|u_1, u_2)$. Then by applying Lemma 3 backwards, we obtain:

$$\Phi \approx \bigwedge_{u_1, u_2 \in \forall var(\Phi)} \bigwedge_{(Q\phi_{i_1}^b \wedge \ldots \wedge \phi_{i_r}^b) \in S(\Phi)|u_1, u_2} (\phi_{i_1}^f \vee \ldots \vee \phi_{i_r}^f)$$

$$\approx \bigwedge_{u_1, u_2 \in \forall var(\Phi)} \bigwedge_{(Q\phi_{i_1}^b \wedge \ldots \wedge \phi_{i_r}^b) \in S(\Phi|u_1, u_2)} (\phi_{i_1}^f \vee \ldots \vee \phi_{i_r}^f)$$

$$\approx \bigwedge_{u_1, u_2 \in \forall var(\Phi)} \Phi|u_1, u_2$$

The prefix of each formula $\Phi|u_1, u_2$ can be simplified, because only u_1 and u_2 occur as universal variables in the matrix, so the other universal quantifiers can be dropped. By universal expansion of u_1, u_2 in $\Phi|u_1, u_2$, we obtain an equivalent existentially quantified formula. Its size is at most four times the length of $\Phi|u_1, u_2$. We perform this expansion for every formula $\Phi|u_1, u_2$ and rename the bound variables, such that different pairs of universal variables u_1, u_2 have distinct bound variables. Now, all existential variables can be moved up front, and the result is an equivalent formula in $\exists 2\text{-}CNF^b$. Since there are at most $|\forall|^2$ pairs of $|\forall|$ universal variables, the resulting formula has a length of $O(|\forall|^2|\Phi|)$. $\square$

If the whole formula matrix, including the free variables, is in $2\text{-}CNF$, we write $DQ2\text{-}CNF^*$ instead of $DQ2\text{-}CNF^b$. In this special case, the above transformation produces an existentially quantified formula with matrix in $2\text{-}HORN$, which can easily be solved in linear time. Together with the costs of the transformation, we would have a complexity of $O(|\forall|^2|\Phi|)$ for determining the satisfiability of a $DQ2\text{-}CNF^*$ formula. There is, however, a faster way to solve such formulas without the above transformation. Without loss of generality, we can focus on $DQ2$-CNF formulas without free variables, because a $DQBF^*$ formula with prefix $Q = \forall x_1 ... \forall x_n \exists y_1 (x_{d_1}) ... \exists y_m (x_{d_m})$ and free variables $z_1, ..., z_r$ is satisfiable if and only if the formula with prefix $Q' = \forall x_1 ... \forall x_n \exists z_1 () ... \exists z_r () \exists y_1 (x_{d_1}) ... \exists y_m (x_{d_m})$ and the same matrix is true.

As outlined in [2], a quantified 2-CNF formula Φ can be represented as a directed graph $G(\Phi)$. The idea is to associate with every clause $L \vee K$ the edges $\neg L \to K$ and $\neg K \to L$ for the nodes $L, \neg L, K$ and $\neg K$. Nodes are called existential or universal if the corresponding variable is existentially or universally quantified. For a unit clause L, we introduce the edge $\neg L \to L$. By computing the strongly connected components of the resulting graph, the satisfiability of the formula can be determined in linear time: it is unsatisfiable if and only if one of the following conditions holds:

1. There is a complementary pair of existential nodes, say y and $\neg y$, in some strongly connected component, which is equivalent to the graph having a path from y to $\neg y$ and a path from $\neg y$ to y.

2. A universal node over x is in the same strong component as an existential node over y, and $\exists y$ precedes $\forall x$ in the prefix of Φ.

3. There exists a path from one universal node to another universal node (possibly both over the same variable).

This idea can also be applied to $DQ2$-CNF formulas. The only necessary modification is to replace condition 2 with the following condition 2': "A universal node over x is in the same strong component as an existential node over y, and y does not depend on x."

For $\Phi = Q\phi \in DQ2$-CNF, notice that if $L_1 \to L_2$ is a path in $G(\Phi)$ then Φ is true if and only if $Q(\phi \wedge (\neg L_1 \vee L_2))$ is true. This can be shown by induction on the path length with the observation that for two clauses $\neg L \vee V$ and $\neg V \vee K$ (both not purely universal) in ϕ, we have $Q \phi = 1$ if and only if $Q (\phi \wedge (\neg L \vee K)) = 1$, where V may be a universal or an existential literal. Then it is easy to see that each of the conditions implies the unsatisfiability of the given formula.

To show the satisfiability of the formula if none of the above conditions hold, the same marking algorithm as in [2] can be used, with the only modification that we stop for condition 2' instead of 2 if a strong component contains both a universal and an existential node. Then it follows that the marking has the same properties as the one in [2], except that a component containing a universal node over some x contains only existential nodes over variables that depend on x. Then it is clear that we can satisfy the formula in the same way as in the original proof by assigning 0 or 1 to existential variables in purely existential components. The truth value of the other existential variables is derived only from those universals on which they depend, so the quantifier dependencies are respected, and we immediately have the following theorem.

Theorem 2. $DQ2$-CNF^* satisfiability is solvable in linear time.

4   Transformation from $\exists 2\text{-}CNF^b$ to $\exists 2\text{-}HORN^b$

In the following, we consider graphs with the structure from the last section also for $\exists 2\text{-}CNF^b$ formulas $\Phi = \exists y_1...\exists y_m \bigwedge_{1\le i\le q}(\phi_i^b \vee \phi_i^f)$. The idea is to associate only the bound literals with nodes in the graph, whereas the free parts become the labels of the corresponding edges. A clause $L \vee K \vee \phi_i^f$ with bound part $\phi_i^b = L \vee K$ and free part ϕ_i^f is then associated with the labeled edges $\neg L \xrightarrow{\phi_i^f} K$ and $\neg K \xrightarrow{\phi_i^f} L$. A clause $L \vee \phi_i^f$ where the bound part is a unit literal is mapped to an edge $\neg L \xrightarrow{\phi_i^f} L$. Figure 1 (left) shows the graph for the following example:

$$\Phi = \exists a \exists b \, (a \vee b \vee z_1) \wedge (\neg a \vee b \vee z_2) \wedge (a \vee \neg b \vee z_3) \wedge (\neg a \vee \neg b \vee z_4).$$

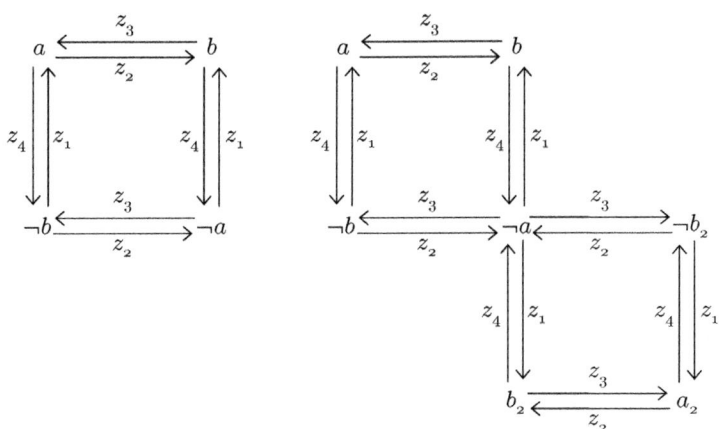

Fig. 1. Example graph (left) and unfolding for variable a (right)

We now translate such graphs into $\exists 2\text{-}HORN^b$ clauses by mapping an edge $L \xrightarrow{\phi_i^f} K$ to a clause $\neg L \vee K \vee \phi_i^f$. The input formula Φ is satisfiable if and only if there exists an assignment of truth values to the free variables such that for all paths from a node y_k to $\neg y_k$ and back to y_k, at least one edge label evaluates to true. The trick is now to encode this check separately for each quantified variable. That means we create a subformula which is false if and only if there is a path from y_1 to $\neg y_1$ and back to y_1 with all edge labels being false, another subformula for paths from y_2 to $\neg y_2$ and back to y_2, and so on. Furthermore, we unfold the graph for each y_k by "mirroring" it around $\neg y_k$, so that instead of checking for a cycle, it is sufficient to detect a simple path from y_k to $\neg y_k$ and from there to the mirrored copy of y_k. Suitable renamings make sure that all nodes in the unfolded graph have unique names. Figure 1 (right) shows how the graph for the previous example is unfolded for the variable a.

Theorem 3. $(\exists 2\text{-}CNF^b =_{poly-time} \exists 2\text{-}HORN^b)$
Every $\exists 2\text{-}CNF^b$ formula Φ with $|\exists|$ existential quantifiers can be transformed in time and space $O(|\exists| \cdot |\Phi|)$ into an equivalent $\exists 2\text{-}HORN^b$ formula.

Proof. Let $\Phi = \exists y_1...\exists y_m \bigwedge_{1 \le i \le q}(\phi_i^b \vee \phi_i^f) \in \exists 2\text{-}CNF^b$. In addition to the previously stated assumption that the bound parts ϕ_i^b are not empty, we also assume that the quantified bound parts $\exists y_1...\exists y_m \bigwedge_{1 \le i \le q} \phi_i^b$ yield an unsatisfiable formula. Otherwise, Φ would be true for any truth assignment to the free variables and therefore be a tautology. Furthermore, we do not allow multiple occurrences of identical bound parts. If the formula contains clauses $L \vee K \vee \phi_i^f$ and $L \vee K \vee \phi_j^f$ with the same bound part $L \vee K$, we can replace the first clause with the clauses $L \vee y \vee \phi_i^f$ and $\neg y \vee K \vee \phi_i^f$ for a new existentially quantified variable y.

Let G be the graph associated with Φ as outlined above. The following procedure transforms G into a formula $\Phi^* \in \exists 2\text{-}HORN^b$:

For all bound variables y, compute the graphs $G(y)$ and $G(\neg y)$ by the following renamings with new names $a_y, a_{\neg y}, b_y$:

$G(y)$ is obtained from G by renaming y into a_y and $\neg y$ into $a_{\neg y}$, all the other nodes are given new unique names.

$G(\neg y)$ is obtained from G by renaming $\neg y$ into $a_{\neg y}$ and y into b_y, all the other nodes are given new unique names.

For all bound variables y,

compute the combined graph $H(y) := G(y) \cup G(\neg y)$,

with $\mathbf{v}_y$ being the set of names of all nodes in $H(y)$,

build the formula $F(y) := \exists \mathbf{v}_y\, a_y \wedge \neg b_y \wedge \bigwedge_{(L \xrightarrow{\sigma} K) \in H(y)}(\neg L \vee K \vee \sigma)$.

Combine the formulas $F(y_i)$ for the bound variables $y_1, ..., y_m$ in Φ into $\Psi := \exists \mathbf{v}_{y_1}...\exists \mathbf{v}_{y_m} F(y_1) \wedge ... \wedge F(y_m)$. Clearly, $\Psi \in \exists 2\text{-}HORN^b$.

In order to prove that $\Phi \approx \Psi$, we use the equivalent representations from Lemma 3:

$$M(\Phi) := \bigwedge_{(\exists \phi_{i_1}^b \wedge...\wedge \phi_{i_r}^b) \in S(\Phi)}(\phi_{i_1}^f \vee ... \vee \phi_{i_r}^f) \approx \Phi$$

$$M(\Psi) := \bigwedge_{(\exists \psi_{j_1}^b \wedge...\wedge \psi_{j_s}^b) \in S(\Psi)}(\psi_{j_1}^f \vee ... \vee \psi_{j_s}^f) \approx \Psi$$

Since the matrix of an existentially quantified minimal false formula is minimal unsatisfiable, we represent the formulas $M(\Phi)$ and $M(\Psi)$ as follows:

$M(\Phi) = \{(\phi_{i_1}^f \vee ... \vee \phi_{i_r}^f) \mid (\phi_{i_1}^b \wedge ... \wedge \phi_{i_r}^b)$ minimal unsat, $\phi_{i_k}^b$ bound part in $\Phi\}$

$M(\Psi) = \{(\psi_{j_1}^f \vee ... \vee \psi_{j_s}^f) \mid (\psi_{j_1}^b \wedge ... \wedge \psi_{j_s}^b)$ minimal unsat, $\psi_{j_l}^b$ bound part in $\Psi\}$

Ad $M(\Psi) \models M(\Phi)$: Let $\varphi := \phi_{i_1}^f \vee ... \vee \phi_{i_r}^f$ be a clause in $M(\Phi)$. Then $\beta := \phi_{i_1}^b \wedge ... \wedge \phi_{i_r}^b$ is minimal unsatisfiable, and according to [2], there must be some variable y in β for which a path from y to $\neg y$ and from $\neg y$ to y exists in the graph representing the propositional 2-CNF formula β. Since the graph G has the same structure, it contains the same path. For fixed y, this path must be unique and have length r, because β would not be minimal unsatisfiable otherwise. Accordingly, there is also exactly one path of length r from a_y to $a_{\neg y}$ and then to b_y in $H(y)$, which implies that the corresponding bound parts of the formula $F(y)$ are minimal unsatisfiable and thus define a clause in $M(\Psi)$. By construction, the path in $H(y)$ is labeled with the same free parts as the corresponding path in G, namely $\{\phi_{i_1}^f, ..., \phi_{i_r}^f\}$. This shows that $M(\Psi)$ contains the clause φ, so $M(\Psi) \models M(\Phi)$.

Ad $M(\Phi) \models M(\Psi)$: This direction is essentially the inverse of the preceding case. Let $\varphi := \psi_{j_1}^f \vee ... \vee \psi_{j_s}^f$ be a clause in $M(\Psi)$. Due to the unique node names, a minimal unsatisfiable subset of bound parts $\psi_{j_1}^b \wedge ... \wedge \psi_{j_s}^b$ in Ψ can only arise within a single formula $F(y)$ for some variable y. The existence of such a minimal unsatisfiable subset of bound parts implies a path of length s from a_y to $a_{\neg y}$ and to b_y in $H(y)$. The path is labeled with $\{\psi_{j_1}^f, ..., \psi_{j_s}^f\}$ and corresponds to a path from y to $\neg y$ and back to y in G with the same edge labels. Such a path implies that there is an unsatisfiable set of bound parts $\phi_{i_1}^b \wedge ... \wedge \phi_{i_s}^b$ in Φ. A subset of these is minimal unsatisfiable, and the corresponding free parts are a subset of the edge labels on the path. It follows that a subset of each clause φ in Ψ is a clause in Φ, and thus $M(\Phi) \models M(\Psi)$. □

5 Conclusion

We have shown that the formula class $DQ2\text{-}CNF^b$ is not significantly more expressive than $\exists 2\text{-}HORN^b$ and that $DQ2\text{-}CNF^b$ satisfiability is also NP-complete. An important intermediate result was a poly-time elimination of all universal quantifiers in a $DQ2\text{-}CNF^b$ or $Q2\text{-}CNF^b$ formula, which might also be useful for QBF solvers fighting against the exponential blowup caused by universal expansion in the general case. Along the lines, we have also shown that $DQ2\text{-}CNF^*$ satisfiability can be decided in linear time and that universal expansion is also correct for $DQBF^*$.

While there are formulas for which $DQ2\text{-}CNF^b$ and $\exists 2\text{-}HORN^b$ are known to be exponentially more concise than propositional CNF, the relationship between $\exists 2\text{-}HORN^b$ and $\exists HORN^b$ remains unclear. The latter class has the same expressive power as Boolean circuits with arbitrary fan-out, which are assumed to be more powerful than propositional formulas. It is not known whether such circuits can also be compactly encoded as poly-size $\exists 2\text{-}HORN^b$ formulas or, equivalently, whether every $\exists HORN^b$ formula has an equivalent $\exists 2\text{-}HORN^b$ formula of polynomial length. Perhaps, the poly-time equivalence between $(D)Q2\text{-}CNF^b$ and $\exists 2\text{-}HORN^b$ can help to shed some light onto this problem. It would also be interesting to investigate whether the transformation between the two classes can be carried out with lower costs than with the procedure that is presented here.

References

[1] Aanderaa, S., Börger, E.: The Horn Complexity of Boolean Functions and Cook's Problem. In: Proc. 5th Scandinavian Logic Symposium 1979, pp. 231–256. Aalborg University Press (1979)

[2] Aspvall, B., Plass, M., Tarjan, R.: A Linear-Time Algorithm for Testing the Truth of Certain Quantified Boolean Formulas. Information Processing Letters 8(3), 121–123 (1979)

[3] Ayari, A., Basin, D.: QUBOS: Deciding Quantified Boolean Logic using Propositional Satisfiability Solvers. In: Aagaard, M.D., O'Leary, J.W. (eds.) FMCAD 2002. LNCS, vol. 2517, pp. 187–201. Springer, Heidelberg (2002)

[4] Babai, L., Fortnow, L., Lund, C.: Non-Deterministic Exponential Time has Two-Prover Interactive Protocols. Journal Computational Complexity 1(1), 3–40 (1991)

[5] Biere, A.: Resolve and Expand. In: Hoos, H.H., Mitchell, D.G. (eds.) SAT 2004. LNCS, vol. 3542, pp. 59–70. Springer, Heidelberg (2005)

[6] Bubeck, U., Kleine Büning, H.: Dependency Quantified Horn Formulas: Models and Complexity. In: Biere, A., Gomes, C.P. (eds.) SAT 2006. LNCS, vol. 4121, pp. 198–211. Springer, Heidelberg (2006)

[7] Bubeck, U., Kleine Büning, H.: Bounded Universal Expansion for Preprocessing QBF. In: Marques-Silva, J., Sakallah, K.A. (eds.) SAT 2007. LNCS, vol. 4501, pp. 244–257. Springer, Heidelberg (2007)

[8] Bubeck, U., Kleine Büning, H.: A New 3-CNF Transformation by Parallel-Serial Graphs. Journal Information Processing Letters 109(7), 376–379 (2009)

[9] Dershowitz, N., Hanna, Z., Katz, J.: Bounded Model Checking with QBF. In: Bacchus, F., Walsh, T. (eds.) SAT 2005. LNCS, vol. 3569, pp. 408–414. Springer, Heidelberg (2005)

[10] Dowling, W., Gallier, J.: Linear-Time Algorithms for Testing the Satisfiability of Propositional Horn Formulae. Journal of Logic Programming 1(3), 267–284 (1984)

[11] Even, S., Itai, A., Shamir, A.: On the Complexity of Timetable and Multi-Commodity Flow Problems. SIAM Journal on Computing 5(4), 691–703 (1976)

[12] Flögel, A., Karpinski, M., Kleine Büning, H.: Resolution for Quantified Boolean Formulas. Information and Computation 117(1), 12–18 (1995)

[13] Itai, A., Makowsky, J.: On the Complexity of Herbrand's Theorem. Technical Report 243, Technion, Haifa (1982)

[14] Jussila, T., Biere, A.: Compressing BMC Encodings with QBF. Electronic Notes in Theoretical Computer Science 174(3), 45–56 (2007)

[15] Kleine Büning, H., Zhao, X., Bubeck, U.: Resolution and Expressiveness of Subclasses of Quantified Boolean Formulas and Circuits. In: Kullmann, O. (ed.) SAT 2009. LNCS, vol. 5584, pp. 391–397. Springer, Heidelberg (2009)

[16] Liberatore, P.: Redundancy in Logic II: 2-CNF and Horn Propositional Formulae. Artificial Intelligence 172(2-3), 265–299 (2008)

[17] Lonsing, F., Biere, A.: A Compact Representation for Syntactic Dependencies in QBFs. In: Kullmann, O. (ed.) SAT 2009. LNCS, vol. 5584, pp. 398–411. Springer, Heidelberg (2009)

[18] Peterson, G., Reif, J., Azhar, S.: Lower Bounds for Multiplayer Non-Cooperative Games of Incomplete Information. Computers and Mathematics with Applications 41(7-8), 957–992 (2001)

[19] Plaisted, D., Greenbaum, S.: A Structure-preserving Clause Form Translation. Journal of Symbolic Computation 2(3), 293–304 (1986)

[20] Sabharwal, A., Ansotegui, C., Gomes, C., Hart, J., Selman, B.: QBF Modeling: Exploiting Player Symmetry for Simplicity and Efficiency. In: Biere, A., Gomes, C.P. (eds.) SAT 2006. LNCS, vol. 4121, pp. 382–395. Springer, Heidelberg (2006)

[21] Samer, M., Szeider, S.: Backdoor Sets of Quantified Boolean Formulas. In: Marques-Silva, J., Sakallah, K.A. (eds.) SAT 2007. LNCS, vol. 4501, pp. 230–243. Springer, Heidelberg (2007)

[22] Tseitin, G.: On the Complexity of Derivation in Propositional Calculus. In: Silenko, A. (ed.) Studies in Constructive Mathematics and Mathematical Logic, Part II, pp. 115–125 (1970)

Synthesizing Shortest Linear Straight-Line Programs over GF(2) Using SAT[*]

Carsten Fuhs[1] and Peter Schneider-Kamp[2]

[1] LuFG Informatik 2, RWTH Aachen University, Germany
fuhs@informatik.rwth-aachen.de
[2] IMADA, University of Southern Denmark, Denmark
petersk@imada.sdu.dk

Abstract. Non-trivial linear straight-line programs over the Galois field of two elements occur frequently in applications such as encryption or high-performance computing. Finding the shortest linear straight-line program for a given set of linear forms is known to be MaxSNP-complete, i.e., there is no ϵ-approximation for the problem unless $P = NP$.

This paper presents a non-approximative approach for finding the shortest linear straight-line program. In other words, we show how to search for a circuit of XOR gates with the minimal number of such gates. The approach is based on a reduction of the associated decision problem ("Is there a program of length k?") to satisfiability of propositional logic. Using modern SAT solvers, optimal solutions to interesting problem instances can be obtained.

1   Introduction

Straight-line programs over the Galois field of two elements, often denoted GF(2), have many practically relevant applications. The most prominent ones are probably in high performance computing (inversion of sparse binary matrices), networking and storage (error detection by checksumming), and encryption (hashing, symmetric ciphers).

In this paper, we focus on linear straight-line programs over GF(2) with applications in cryptography. The motivation behind this choice is that modern symmetric ciphers like AES can be implemented by lookup tables and addition in GF(2). Multiplication and addition in GF(2) correspond to the Boolean AND and XOR operations, respectively. In other words, we are looking at straight-line programs composed of array lookups and sequences of XOR operations.

The goal of this paper is, given a specification of a linear function from a number of inputs to a number of outputs, to find the shortest linear straight-line program over GF(2) that satisfies the specification. In other words, we show how to find a XOR circuit with the minimal number of gates that connects inputs to outputs. Finding such shortest programs is obviously interesting both for software and for hardware implementations of, for example, the symmetric cipher Advanced Encryption Standard (AES) [13].

---

[*] Supported by the G.I.F. grant 966-116.6 and the Danish Natural Science Research Council.

O. Strichman and S. Szeider (Eds.): SAT 2010, LNCS 6175, pp. 71–84, 2010.

While there are heuristic methods for finding short straight-line linear programs [4] (see also [3] for the corresponding patent application), to the best of our knowledge, there is no feasible method for finding an optimal solution. In this paper, we present an approach based on reducing the associated decision problem ("Is there a program of length k?") to satisfiability of propositional logic. The reduction is performed in a way that every model found by the SAT solver represents a solution. Recent work [11] has shown that reductions to satisfiability problems are a promising approach for circuit synthesis. By restricting our attention to linear functions, we now obtain a polynomial-size encoding.

The structure of this paper is as follows. In Section 2, we formally introduce our optimization problem and show how linear straight-line programs can be used to compute a given set of linear forms. Section 3 presents a novel encoding for the associated decision problem to SAT. Then, we discuss in Section 4 how to tackle our optimization problem by reducing it to the associated decision problem using a customized search for k.

In Section 5 we present an empirical case study where we try to optimize an important component of AES. To prove optimality of the solution found, the case study prompts us to improve the performance of our encoding for the decision problem in the unsatisfiable case. For this, we discuss different approaches in Section 6. We conclude with a summary of our contributions in Section 7.

2   Linear Straight-Line Programs

In this paper, we assume that we have n inputs $x_1, \ldots, x_n$ and m outputs $y_1, \ldots, y_m$. Then the linear function to be computed can be specified by m equations of the following form:

$$y_1 = a_{1,1} \cdot x_1 \oplus a_{1,2} \cdot x_2 \oplus \ldots \oplus a_{1,n} \cdot x_n$$
$$y_2 = a_{2,1} \cdot x_1 \oplus a_{2,2} \cdot x_2 \oplus \ldots \oplus a_{2,n} \cdot x_n$$
$$\ldots$$
$$y_m = a_{m,1} \cdot x_1 \oplus a_{m,2} \cdot x_2 \oplus \ldots \oplus a_{m,n} \cdot x_n$$

We call each equation a *linear form*. Note that each $a_{\ell,j}$ is a constant from GF(2) $= \{0, 1\}$, each x_j is a variable over GF(2), and $\oplus$ and $\cdot$ denote standard addition and multiplication on GF(2), respectively. In this paper, we always assume that the linear forms are pairwise different.

Our goal is to come up with an algorithm that computes these linear forms given $x_1, \ldots, x_n$ as inputs. More specifically, we would like to express this algorithm via a *linear straight-line program* (or, for brevity, just *program*). Here, every line of the program has the shape $u := e \cdot v \oplus f \cdot w$ with $e, f \in$ GF(2) and v, w variables. Some lines of the program will contain the output, i.e., assign the value of one of the desired linear forms to a variable. The *length* of a program is the number of lines the program contains. Without loss of generality, we perform write operations only to fresh variables, so no input is overwritten and no intermediate variable is written to twice. A program is *optimal* if there is no shorter program that computes the same linear forms.

Example 1. Consider the following linear forms:

$$
\begin{aligned}
y_1 &= x_1 \oplus x_2 \oplus x_3 \oplus x_4 \oplus x_5 \\
y_2 &= x_1 \oplus x_2 \oplus x_3 \oplus x_4 \\
y_3 &= x_1 \oplus x_2 \oplus x_3 \qquad\quad \oplus x_5 \\
y_4 &= \qquad\qquad\qquad x_3 \oplus x_4 \oplus x_5 \\
y_5 &= x_1 \qquad\qquad\qquad\qquad\quad \oplus x_5
\end{aligned}
$$

A shortest linear program for computing these linear forms has length 6. The following linear program is an optimal solution for this example.

$$
\begin{aligned}
v_1 &= x_1 \oplus x_5 & [y_5] \\
v_2 &= x_2 \oplus v_1 \\
v_3 &= x_3 \oplus v_2 & [y_3] \\
v_4 &= x_4 \oplus v_3 & [y_1] \\
v_5 &= x_5 \oplus v_4 & [y_2] \\
v_6 &= v_2 \oplus v_5 & [y_4]
\end{aligned}
$$

It is easy to check that for each output y_ℓ there is a variable v_i that contains the linear form for y_ℓ. In the above program, this mapping from intermediate variables to outputs is given by annotating the program lines with the associated output in square brackets.

Note that finding the shortest program over GF(2) is *not* an instance of the common subexpression elimination problem known from program optimization. The above shortest program makes extensive use of *cancellation*, i.e., of the fact that for all x in GF(2), we have $x \oplus x = 0$. For example, the output y_4 is computed by adding v_2 and v_5. These two variables are described by the linear forms $x_1 \oplus x_2 \oplus x_3 \oplus x_4$ and $x_1 \oplus x_2 \oplus x_5$, respectively. By adding these two linear forms, we obtain the desired $x_3 \oplus x_4 \oplus x_5$ since $x_1 \oplus x_1 \oplus x_2 \oplus x_2 = 0$ for all values of x_1 and x_2. Without cancellations, a shortest linear straight-line program has length 8, i.e., it uses 25% more XOR gates.

The goal that we are now pursuing in this paper is to synthesize an optimal linear straight-line program for a given set of linear forms both automatically and efficiently. Formally, this problem can be described as follows:

> Given n variables $x_1, \ldots, x_n$ over GF(2) and m linear forms $y_\ell = a_{\ell,1} \cdot x_1 \oplus \ldots \oplus a_{\ell,n} \cdot x_n$, find the shortest linear program that computes all y_ℓ.

Note that here we are aiming at a (provably) *optimal* solution. This is opposed to allowing approximations with more lines than actually necessary, which is currently the state of the art [2].

As a step towards solving this optimization problem, first let us consider the corresponding decision problem:

> Given n variables $x_1, \ldots, x_n$ over GF(2), m linear forms $y_\ell = a_{\ell,1} \cdot x_1 \oplus \ldots \oplus a_{\ell,n} \cdot x_n$ and a natural number k, decide if there exists a linear program of length k that computes all y_ℓ.

Of course, if the answer to this question is "Yes", we do not only wish to get this answer, but we would also like to obtain a corresponding program of length (at most) k. In line i, the variable v_i is defined as the sum of two other variables. Here, one may read from the variables $x_1, \ldots, x_n$ and also from the intermediate variables $v_1, \ldots, v_j$ with $j < i$, i.e., from those intermediate variables that have been defined before.

To facilitate the description of our encoding in the following section, we reformulate the problem via matrices over GF(2). Here, given a natural number k, we represent the given coefficients of the m linear forms over n inputs with $y_\ell = a_{\ell,1} \cdot x_1 \oplus a_{\ell,2} \cdot x_2 \oplus \ldots \oplus a_{\ell,n} \cdot x_n$ $(1 \leq \ell \leq m)$ as rows of an $m \times n$-matrix A. The ℓ-th row thus consists of the entries $a_{\ell,1} a_{\ell,2} \ldots a_{\ell,n}$ from GF(2).

Likewise, we can also express the resulting program via two matrices:

- A matrix $B = (b_{i,j})_{k \times n}$ over GF(2), where $b_{i,j} = 1$ iff in line i of the program the input variable x_j is read.
- A matrix $C = (c_{k,k})_{k \times k}$ over GF(2) where $c_{i,j} = 1$ iff in line i of the program the intermediate variable v_j is read.

To represent for example the program line $v_3 = x_3 \oplus v_2$ from Example 1, all $b_{3,j}$ except for $b_{3,3}$ and all $c_{3,j}$ except for $c_{3,2}$ have to be 0. Thus, the third row in B is $\begin{pmatrix} 0\,0\,1\,0\,0 \end{pmatrix}$ while in C it is $\begin{pmatrix} 0\,1\,0\,0\,0\,0 \end{pmatrix}$.

Now, for the matrices B and C to actually represent a legal linear straight-line program, for any row i there must be exactly two non-zero entries in the combined i-th row of B and C. That is, the vector $\begin{pmatrix} b_{i,1} \ldots b_{i,n} \; c_{i,1} \ldots c_{i,k} \end{pmatrix}$ must contain exactly two 1s.

Furthermore, for the represented program to actually compute our linear forms, we have to demand that for each desired output y_ℓ, there is a line i in the program (and the matrices) such that $v_i = y_\ell$ where $y_\ell = a_{\ell,1} \cdot x_1 \oplus \ldots \oplus a_{\ell,n} \cdot x_n$ and $v_i = b_{i,1} \cdot x_1 \oplus \ldots \oplus b_{i,n} \cdot x_n \oplus c_{i,1} \cdot v_1 \oplus \ldots \oplus c_{i,i-1} \cdot v_{i-1}$. Note that we only use the lower triangular matrix as a program may only read intermediate values that have already been written. To represent the mapping of intermediate variables to outputs, we use a function $f : \{1, \ldots, m\} \mapsto \{1, \ldots, k\}$.

Example 2. Consider again the linear forms from Example 1. They are represented by the following matrix A. Likewise, the program is represented by the matrices B and C and the function f.

$$A = \begin{pmatrix} 1\,1\,1\,1\,1 \\ 1\,1\,1\,1\,0 \\ 1\,1\,1\,0\,1 \\ 0\,0\,1\,1\,1 \\ 1\,0\,0\,0\,1 \end{pmatrix} \quad B = \begin{pmatrix} 1\,0\,0\,0\,1 \\ 0\,1\,0\,0\,0 \\ 0\,0\,1\,0\,0 \\ 0\,0\,0\,1\,0 \\ 0\,0\,0\,0\,1 \\ 0\,0\,0\,0\,0 \end{pmatrix} \quad C = \begin{pmatrix} 0\,0\,0\,0\,0\,0 \\ 1\,0\,0\,0\,0\,0 \\ 0\,1\,0\,0\,0\,0 \\ 0\,0\,1\,0\,0\,0 \\ 0\,0\,0\,1\,0\,0 \\ 0\,1\,0\,0\,1\,0 \end{pmatrix} \quad f = \begin{cases} 1 \mapsto 4 \\ 2 \mapsto 5 \\ 3 \mapsto 3 \\ 4 \mapsto 6 \\ 5 \mapsto 1 \end{cases}$$

Obviously, all combined rows of B and C contain exactly two non-zero elements. Furthermore, by computing the v_i and the y_ℓ, we can see that each of the linear forms described by A is computed by the program represented by B and C.

3   Encoding to Propositional Logic

Now that the scenario has been set up and the matrix formulation has been introduced, we start by giving a high-level encoding of the decision problem as a logical formula in second order logic. Then we perform a stepwise refinement of that encoding where in each step we eliminate some elements that cannot directly be expressed by satisfiability of propositional logic.

For our first encoding, the carrier is the set of natural numbers, and we use predicates over indices to represent the matrices A, B, and C as well as the vectors x, y, and v. We also use a function f to map indices of outputs from y to indices of intermediate variables from v. Finally, we make use of cardinality constraints by predicates exactly$_k$ that take a list of variables and check that the number of variables that are assigned 1 is exactly k.

First, we need to ensure that B and C represent a legal linear straight-line program. This is encoded by the following formula α_1:

$$\alpha_1 = \bigwedge_{1 \leq i \leq k} \text{exactly}_2(B(i,1), \dots, B(i,n), C(i,1), \dots, C(i, i-1))$$

Second, we demand that the values for the intermediate variables from v are computed by using the values from B and C:

$$\alpha_2 = \bigwedge_{1 \leq i \leq k} \left(v(i) \leftrightarrow \bigoplus_{1 \leq j \leq n} B(i,j) \wedge x(j) \oplus \bigoplus_{1 \leq p < i} C(i,p) \wedge v(p) \right)$$

Third, we ensure that the value of the intermediate variable determined by f for the ℓ-th output actually takes the same value as the ℓ-th linear form:

$$\alpha_3(\ell) = v(f(\ell)) \leftrightarrow \bigoplus_{1 \leq j \leq n} A(\ell, j) \wedge x(j)$$

Here, $v(f(\ell))$ denotes the intermediate variable which stores the result of the linear form $y(\ell)$. In other words, the (existentially quantified) function f maps the index ℓ of the linear form y_ℓ to the index $i = f(\ell)$ of the variable v_i in v which contains the result of y_ℓ.

Now we can give our first encoding by the following formula α:

$$\alpha = \exists B. \exists C. \exists f. \forall x. \exists v.\ \alpha_1 \wedge \alpha_2 \wedge \bigwedge_{1 \leq \ell \leq m} \alpha_3(\ell)$$

Note that we indeed use the expressivity of second order logic as all our quantifications are over predicates and functions. Fortunately, all these only need to be defined on finite domains. In order not to have to deal with quantification over predicates representing matrices and vectors, we can just introduce a finite number of Boolean variables to represent the elements of the matrices and vectors and work on these directly. For example, for the $k \times n$ matrix B we introduce the $k \cdot n$ Boolean variables $b_{1,1} \dots, b_{k,n}$.

Similarly, for the function f we introduce $m \cdot k$ Boolean variables $f_{\ell,i}$ that denote that the ℓ-th linear form is computed by the i-th intermediate variable. To make sure that these variables actually represent a function, we need to encode well-definedness: for each ℓ there must be exactly one i with $f_{\ell,i}$.

We obtain the refined overall constraint β, which is a formula from QBF:

$$\beta_1 = \bigwedge_{1 \leq i \leq k} \text{exactly}_2(b_{i,1}, \ldots, b_{i,n}, c_{i,1}, \ldots, c_{i,i-1})$$

$$\beta_2 = \bigwedge_{1 \leq i \leq k} \left(v_i \leftrightarrow \bigoplus_{1 \leq j \leq n} b_{i,j} \wedge x_j \oplus \bigoplus_{1 \leq p < i} c_{i,p} \wedge v_p \right)$$

$$\beta_3(\ell) = \bigwedge_{1 \leq i \leq k} \left(f_{\ell,i} \rightarrow \left(v_i \leftrightarrow \bigoplus_{1 \leq j \leq n} a_{\ell,j} \wedge x_j \right) \right) \wedge \text{exactly}_1(f_{\ell,1}, \ldots, f_{\ell,k})$$

$$\beta = \exists b_{1,1}.\ldots.\exists b_{k,n}.\exists c_{1,1}.\ldots.\exists c_{k,k}.\exists f_{1,1}.\ldots.\exists f_{m,k}.\forall x_1.\ldots.\forall x_n.\exists v_1.\ldots.\exists v_k.$$
$$\beta_1 \wedge \beta_2 \wedge \bigwedge_{1 \leq \ell \leq m} \beta_3(\ell)$$

The above formula β is in prenex normal form and has a quantifier prefix of the shape $\exists^+\forall^+\exists^+$. This precludes us from using a SAT solver on β directly. For this, we would need to have a quantifier prefix of the shape $\exists^+$ alone. Thus, unless we want to use a QBF solver, we need to get rid of the $\forall^+\exists^+$ suffix of the quantifier prefix of β. In other words, we need to get rid of the quantifications over $x_1, \ldots, x_n$ and $v_1, \ldots, v_k$.

We observe that β explicitly contains the computed values v_i of the intermediate variables. We can eliminate them by unrolling the defining equations of an intermediate variable v_i to be expressed directly via $x_1, \ldots, x_n$. In other words, we do not regard the intermediate variables for "computing" the result of the linear forms y_ℓ, but we directly use a closed expression that depends on the $b_{i,j}$ and the $c_{i,p}$. Here, we introduce the auxiliary formulae $\varphi(i)$ for $1 \leq i \leq k$ whose truth value should correspond to the value taken by the corresponding v_i:

$$\varphi(i) = (\bigoplus_{1 \leq j \leq n} b_{i,j} \wedge x_j) \oplus (\bigoplus_{1 \leq p < i} c_{i,p} \wedge \varphi(p))$$

We now reformulate β to obtain a refined encoding γ. Note that we do not need to redefine β_1 and we do not need an equivalent of β_2 as we unroll the definition of the v_i into γ_3 using $\varphi(i)$.

$$\gamma_3(\ell) = \bigwedge_{1 \leq i \leq k} \left(f_{\ell,i} \rightarrow \left(\varphi(i) \leftrightarrow \bigoplus_{1 \leq j \leq n} a_{\ell,j} \wedge x_j \right) \right) \wedge \text{exactly}_1(f_{\ell,1}, \ldots, f_{\ell,k})$$

$$\gamma = \exists b_{1,1}.\ldots.\exists b_{k,n}.\exists c_{1,1}.\ldots.\exists c_{k,k}.\exists f_{1,1}.\ldots.\exists f_{m,k}.\forall x_1.\ldots.\forall x_n.\beta_1 \wedge \bigwedge_{1 \leq \ell \leq m} \gamma_3(\ell)$$

Note that it looks as though for each i we had obtained many redundant copies of the subformulae $\varphi(i)$, which would entail a blow-up in formula size. However, in practical implementations it is beneficial to represent propositional formulae not as trees, but as directed acyclic graphs with sharing of common subformulae. This technique is also known as *structural hashing* [6]. We perform standard Boolean simplifications (e.g., $\varphi \wedge 1 = \varphi$), we share Boolean junctor applications modulo commutativity and idempotence (where applicable), and we use varyadic $\wedge$ and $\vee$. In contrast, the junctors $\leftrightarrow$ and $\oplus$ are binary and associate to the left.

Nevertheless, we still have universal quantification over the inputs as part of our encoding. This states that regardless of the input values for $x_1, \ldots, x_n$, our program should yield the correct result. Fortunately, we can now benefit from linearity of the operation $\oplus$ on GF(2), which means that the absolute positiveness criterion for polynomials [10] (a simple technique commonly used in automated termination provers, cf. e.g. [7]) is not only sound, but also complete. Essentially, the idea is that two linear forms compute the same function iff their coefficients are identical. In this way, we can now drop the inputs $x_1, \ldots, x_n$.

For $1 \leq j \leq n$ and $1 \leq i \leq k$, we introduce the auxiliary formulae $\psi(j, i)$, which should denote the dependence of the value for v_i with respect to x_j (i.e., whether the value of v_i toggles if x_j changes or not):

$$\psi(j, i) = b_{i,j} \oplus \bigoplus_{1 \leq p < i} c_{i,p} \wedge \psi(j, p)$$

We finally get an encoding δ in prenex normal form that can be used as input for a SAT solver (by dropping explicit existential quantification, encoding cardinality constraints using [5,1], and performing Tseitin's transformation [14]).

$$\delta_3(\ell) = \bigwedge_{1 \leq i \leq k} \left(f_{\ell,i} \rightarrow \bigwedge_{1 \leq j \leq n} (\psi(j, i) \leftrightarrow a_{\ell,j}) \right) \wedge \text{exactly}_1(f_{\ell,1}, \ldots, f_{\ell,k})$$

$$\delta = \exists b_{1,1}. \ldots . \exists b_{k,n}. \exists c_{1,1}. \ldots \ldots . \exists c_{k,k}. \exists f_{1,1}. \ldots . \exists f_{m,k}. \; \beta_1 \wedge \bigwedge_{1 \leq \ell \leq m} \delta_3(\ell)$$

For the implementation of δ we used the SAT framework in the verification environment AProVE [8] and the Tseitin implementation from SAT4J [12].

3.1   Size of the Encoding

Given a decision problem with an $m \times n$ matrix and a natural number k (where w.l.o.g. $m \leq k$ holds since for $m > k$, we could just set $\delta = 0$), our encoding δ has size $\mathcal{O}(n \cdot k^2)$ if the cardinality constraints are encoded in space linear in the number of arguments [5]. To see this, consider the following size estimation for δ where due to the use of structural hashing we can look at δ_3 and ψ separately.

$$|\delta| = \mathcal{O}(k \cdot n + k \cdot k + m \cdot k + |\beta_1| + m \cdot |\delta_3| + n \cdot k \cdot |\psi|)$$

For β_1 and δ_3 we obtain the following estimations where g is a function describing the size of the cardinality constraint:

$$|\beta_1| = \mathcal{O}(k \cdot g(n + k)) \qquad\qquad |\delta_3| = \mathcal{O}(k \cdot n + g(k))$$

For ψ we immediately obtain the size estimation $|\psi| = \mathcal{O}(k)$. Now, we can simplify the estimation for δ by using $m \leq k$:

$$|\delta| = \mathcal{O}(k \cdot n + k \cdot k + m \cdot k + k \cdot g(n+k) + m \cdot (k \cdot n + g(k)) + n \cdot k \cdot k)$$
$$= \mathcal{O}(n \cdot k^2 + k \cdot g(n+k) + m \cdot g(k))$$

3.2   Tuning the Encoding

The models of the encoding δ from this section are all linear straight-line programs of length k that compute the m linear forms $y_1, \ldots, y_m$. The programs can be decoded from a satisfying assignment of the propositional formula by simply reconstructing the matrices B and C.

In this paper, we are interested in finding short programs. Thus, we can exclude many programs that perform redundant computations. We do so by adding further conjuncts that exclude those undesired programs. While we change the set of models, note that we do not change the satisfiability of the decision problem. That is, if there is a program that computes the given linear forms in k steps, we will find one which does not perform these kinds of redundant computation.

The first kind of redundant programs are programs that compute the same linear form twice, i.e., there are two different intermediate variables that contain the same linear form. We exclude such programs by demanding that for all distinct pairs of intermediate variables v_i and v_p, there is also some x_j that influences exactly one of the two variables:

$$\bigwedge_{1 \leq i \leq k} \bigwedge_{1 \leq p < i} \bigvee_{1 \leq j \leq n} (\psi(j,p) \oplus \psi(j,i))$$

The second kind of redundant programs are programs that compute the constant 0 or a linear form just depending on the value of one input variable. To exclude such programs, we add cardinality constraints stating that each compute linear form must depend on at least two input variables.

$$\bigwedge_{1 \leq i \leq k} \text{atLeast}_2(\psi(1,i), \ldots, \psi(n,i))$$

In fact, statements that compute linear forms that only depend on two input variables can be restricted not to use any other intermediate variables (as they could be computed in one step from the inputs).

$$\bigvee_{1 \leq j < i} c_{i,j} \to \bigwedge_{1 \leq i \leq k} \text{atLeast}_3(\psi(1,i), \ldots, \psi(n,i))$$

Apart from disallowing redundant programs, we additionally include implied conjuncts to further constrain the search space. In this way, the SAT solver becomes more efficient as unit propagation can be employed in more situations.

As stated in Section 2, we require that the input does not contain duplicate linear forms. Consequently, we may require f to be injective, i.e., any intermediate variable covers at most one linear form.

$$\bigwedge_{1 \leq i \leq k} \text{atMost}_1(f_{1,i}, \dots, f_{m,i})$$

Often, CDCL-based SAT solvers are not very good at solving the pigeonhole problem. Additional constraints facilitate better unit propagation in these cases. Since f maps from $\{1, \dots, m\}$ to $\{1, \dots, k\}$, only at most k of the $f_{\ell,i}$ may become true.

$$\text{atMost}_k(f_{1,1}, \dots, f_{m,k})$$

Similarly, we can even state that at least m of the $f_{\ell,i}$ need to become true as we have to compute all given (distinct) m linear forms.

$$\text{atLeast}_m(f_{1,1}, \dots, f_{m,k})$$

4   From Decision Problem to Optimization

A simple approach for solving an optimization problem given a decision procedure for the associated decision problem is to search for the parameter to be optimized by repeatedly calling the decision procedure.

In our case, for minimizing the length k of the synthesized linear straight-line program, we start by observing that this minimal length must be at least the number of linear forms. At the same time, if we compute each linear form separately, we obtain an upper bound for the minimal length. More precisely, we know that the minimal length k_{min} is in the closed interval from m to $|A|_1 - m$ where $|\cdot|_1$ denotes the number of 1s in a matrix.

Without further heuristic knowledge about the typical length of shortest programs, the obvious thing to do is to use a bisecting approach for refining the interval. That is, one selects the middle element of the current interval and calls a decision procedure based on our encoding from Section 3 for this parameter. If there is a model, the interval is restricted to the lower half of the previous interval and we continue bisecting. If there is no model and δ is unsatisfiable, the interval is restricted to the upper half of the previous interval. When the interval becomes empty, the lower limit indicates the minimal parameter k_{min}.

The above approach requires a logarithmic number of calls to the decision procedure, approximately half of which will return the result "unsatisfiable". This approach is very efficient if we can assume that our decision procedure takes approximately the same time for a positive answer as for a negative answer. As we will see in the case study of the following section, though, for realistic problem instances the negative answers may require orders of magnitude more time.

Thus, to minimize the number of calls to the decision procedure resulting in a negative answer, we propose the following algorithm for refining the length k.

1. Start with $k := |A|_1 - m - 1$.
2. Call the decision procedure with k.
3. If UNSAT, return $k + 1$ and exit.
4. If SAT, compute used length k_{used} from B and C.
5. Set $k := k_{used} - 1$ and go to Step 2.

Here, the *used length* of a program is the number of variables that are needed directly or indirectly to compute the m linear forms. For given matrices B and C and a function f, the set of used variables *used* is the least set such that:

- if $f(\ell) = i$, then $v_i \in$ *used* and
- if $v_i \in$ *used* and $c_{i,j} = 1$, then $v_j \in$ *used*.

The used length can then be obtained as the cardinality of the set *used*.

This algorithm obviously only results in exactly one call to UNSAT – directly before finding the minimal solution. The price we pay for this is that in the worst case we have to call the decision procedure a linear number of times. In practice, though, for $k > k_{min}$, there are many solutions and the solution returned by the SAT solver typically has $k_{used} < k$. Consequently, at the beginning the algorithm typically approaches k_{min} in rather large steps.

While it seems natural to use MaxSAT for our optimization problem instead of calling the SAT solver repeatedly, the decision problems close to the optimum are already so hard that solving these as part of a larger instance seems infeasible.

5   Case Study: Advanced Encryption Standard

As mentioned in the introduction, a major motivation for our work is the minimization of circuits for implementing cryptographic algorithms. In this section, we study how our contributions can be applied to optimize an important component of the Advanced Encryption Standard (AES) [13].

The AES algorithm consists of the (repeated) application of four steps. The main step for introducing non-linearity is the SubBytes step that is based on a so-called S-box. This S-box is a transformation based on multiplicative inverses in $GF(2^8)$ combined with an invertible affine transformation. This step can be decomposed into two linear parts and a minimal non-linear part.

For our case study, we consider the first of the linear parts (called the "top matrix" in [4]) which is represented by the following 21×8 matrix A:

$$A = \begin{pmatrix} 0&1&1&0&0&0&0&1 \\ 1&1&1&0&0&0&0&1 \\ 1&1&1&0&0&1&1&1 \\ 0&1&1&1&0&0&0&1 \\ 0&1&1&0&0&0&1&1 \\ 1&0&0&1&1&0&1&1 \\ 0&1&0&0&1&1&1&1 \\ 1&0&0&0&0&1&0&0 \\ 1&0&0&1&0&0&0&0 \\ 1&1&1&1&1&0&1&0 \\ 0&1&0&0&1&1&1&0 \\ 1&0&0&1&0&1&1&0 \\ 1&0&0&0&0&0&1&0 \\ 0&0&0&1&0&1&0&0 \\ 1&0&0&1&1&0&1&0 \\ 0&0&1&0&1&1&1&0 \\ 1&0&1&1&0&1&0&0 \\ 1&0&1&0&1&1&1&0 \\ 0&1&1&1&1&1&1&0 \\ 1&1&0&1&1&1&1&0 \\ 1&0&1&0&1&1&0&0 \end{pmatrix} \quad B = \begin{pmatrix} 0&0&1&0&0&0&0&1 \\ 1&0&0&0&0&1&0&0 \\ 0&0&1&0&0&0&0&0 \\ 0&0&0&0&1&0&0&0 \\ 0&1&0&0&0&0&0&0 \\ 1&0&0&0&0&0&1&0 \\ 1&0&0&1&0&0&0&0 \\ 0&0&0&0&0&0&1&0 \\ 0&0&1&0&0&0&0&0 \\ 0&0&0&0&0&0&0&0 \\ 0&0&0&0&0&0&0&0 \\ 0&0&0&0&0&0&0&0 \\ 0&0&1&0&0&1&0&0 \\ 0&0&0&1&0&0&0&0 \\ 0&0&0&1&0&0&0&0 \\ 0&0&0&0&0&0&0&0 \\ 0&0&0&0&0&0&1&0 \\ 0&0&0&0&0&0&0&0 \\ 0&0&0&0&0&0&0&1 \\ 0&0&0&0&0&0&0&0 \\ 0&0&0&0&0&0&0&0 \end{pmatrix}$$

$$C = \begin{pmatrix} 0&0&0&0&0&0&0&0&0&0&0&0&0&0&0&0&0&0&0&0&0&0&0&0&0&0&0&0&0&0&0&0 \\ 0&0&0&0&0&0&0&0&0&0&0&0&0&0&0&0&0&0&0&0&0&0&0&0&0&0&0&0&0&0&0&0 \\ 0&1&0&0&0&0&0&0&0&0&0&0&0&0&0&0&0&0&0&0&0&0&0&0&0&0&0&0&0&0&0&0 \\ 0&0&1&0&0&0&0&0&0&0&0&0&0&0&0&0&0&0&0&0&0&0&0&0&0&0&0&0&0&0&0&0 \\ 1&0&0&0&0&0&0&0&0&0&0&0&0&0&0&0&0&0&0&0&0&0&0&0&0&0&0&0&0&0&0&0 \\ 0&0&0&0&0&0&0&0&0&0&0&0&0&0&0&0&0&0&0&0&0&0&0&0&0&0&0&0&0&0&0&0 \\ 0&0&0&0&0&0&0&0&0&0&0&0&0&0&0&0&0&0&0&0&0&0&0&0&0&0&0&0&0&0&0&0 \\ 0&0&1&0&0&0&0&0&0&0&0&0&0&0&0&0&0&0&0&0&0&0&0&0&0&0&0&0&0&0&0&0 \\ 0&0&1&0&0&0&0&0&0&0&0&0&0&0&0&0&0&0&0&0&0&0&0&0&0&0&0&0&0&0&0&0 \\ 0&0&0&1&0&1&0&0&0&0&0&0&0&0&0&0&0&0&0&0&0&0&0&0&0&0&0&0&0&0&0&0 \\ 0&0&0&0&0&0&0&1&1&0&0&0&0&0&0&0&0&0&0&0&0&0&0&0&0&0&0&0&0&0&0&0 \\ 0&0&0&0&1&0&0&0&0&1&0&0&0&0&0&0&0&0&0&0&0&0&0&0&0&0&0&0&0&0&0&0 \\ 0&0&0&0&0&0&0&0&0&0&0&0&0&0&0&0&0&0&0&0&0&0&0&0&0&0&0&0&0&0&0&0 \\ 0&0&0&0&0&1&0&0&0&0&0&0&1&0&0&0&0&0&0&0&0&0&0&0&0&0&0&0&0&0&0&0 \\ 0&0&0&0&0&1&0&0&0&0&0&0&0&0&0&0&0&0&0&0&0&0&0&0&0&0&0&0&0&0&0&0 \\ 0&0&0&0&0&0&0&1&0&0&0&1&0&0&0&0&0&0&0&0&0&0&0&0&0&0&0&0&0&0&0&0 \\ 0&0&0&0&1&0&0&0&0&0&0&0&0&0&0&0&0&0&0&0&0&0&0&0&0&0&0&0&0&0&0&0 \\ 0&0&0&0&0&0&0&0&0&0&0&1&1&0&0&0&0&0&0&0&0&0&0&0&0&0&0&0&0&0&0&0 \\ 0&0&0&0&0&0&0&1&0&0&0&0&0&0&0&0&0&0&0&0&0&1&0&0&0&0&0&0&0&0&0&0 \\ 0&0&0&0&0&1&0&0&0&0&0&0&0&0&0&0&0&0&0&0&0&1&0&0&0&0&0&0&0&0&0&0 \\ 0&1&0&0&0&0&0&0&0&0&0&0&0&0&0&0&0&0&0&0&0&1&0&0&0 \\ 0&0&0&1&0&0&0&0&0&0&0&0&0&0&0&0&0&0&0&1&0&0&0 \end{pmatrix}$$

Here, the matrices B and C represent a solution with length $k = 23$. This solution was found in less than one minute using our decision procedure from

Section 3 with MiniSAT v2.1 as backend on a 2.67 GHz Intel Core i7. We strongly conjecture that $k_{min} = 23$ and, indeed, the shortest known linear straight-line program for the linear forms described by the matrix A has length $k = 23$ [4]. This shows that our SAT-based optimization method is able to find very good solutions in reasonable time. The UNSAT case is harder, though. For $k = 20$ (which is trivially unsatisfiable due to the pigeonhole problem), without the tunings from Section 3 we cannot show unsatisfiability within 4 days. But with the tunings enabled we can show unsatisfiability in less than one second.

Unfortunately, proving the unsatisfiability for $k = 22$ proves to be much more challenging. Indeed, we have run many different SAT solvers (including but not limited to glucose, ManySat, MiniSat, MiraXT with 8 threads, OKsolver, PicoSAT, PrecoSAT, RSat, SAT4J) on the CNF file for this instance of our decision problem. Some of the more promising solvers for this instance were run for more than 40 days without returning either SAT or UNSAT.

In an effort to prove unsatisfiability of this instance and thereby prove optimality of the solution with $k = 23$, we have asked for and received a lot of support and good advice from the SAT community (see the Acknowledgements at the end of this paper). Still, to this day the unsatisfiability of this instance remains a conjecture. Using pre-processing techniques, the number of variables of this instance can be reduced from more than 45000 to less than 5000 in a matter of minutes. The remaining SAT problem seems to be very hard, though.[1]

To analyze how difficult these problems really are, we consider a small subset of the linear forms to be computed for the top matrix. The table to the right shows how the run-times in seconds of the SAT solver are affected by the choice of k for the case that we consider only the first 8 out of 21 linear forms from A. In order to keep runtimes manageable we already incorporated the symmetry breaking improvement described in Section 6. Note that unsatisfiability for $k = 12$ is still much harder to show than satisfiability for $k_{min} = 13$.

To conclude this case study, we see that while finding (potentially) minimal solutions is obviously feasible, proving their optimality (i.e., unsatisfiability of the associated decision problem for $k = k_{min}-1$) is challenging. This observation confirms observations made in [11]. In the following section we present some of our attempts to improve the efficiency of our encoding for the UNSAT case.

k	result	time
8	UNSAT	0.4
9	UNSAT	0.5
10	UNSAT	1.2
11	UNSAT	5.0
12	UNSAT	76.8
13	SAT	1.0
14	SAT	3.4
15	SAT	2.8
16	SAT	1.5
17	SAT	4.3
18	SAT	2.7
19	SAT	2.5
20	SAT	3.0
21	SAT	3.0
22	SAT	3.5
23	SAT	3.6
24	SAT	5.5
25	SAT	5.9

6   Towards Handling the UNSAT Case

Satisfiability of propositional logic is an NP-complete problem and, thus, we can expect that at least some instances are computationally expensive. While

---

[1] The reader is cordially invited to try his favorite SAT solver on one of the instances available from: http://aprove.informatik.rwth-aachen.de/eval/slp.zip

SAT solvers have proven to be a Swiss army knife for solving practically relevant instances of many different NP-complete problems, our kind of program synthesis problems seems to be a major challenge for today's SAT solvers even on instances with "just" 1500 variables.

In this section we discuss three different approaches based on unary SAT encodings, on Pseudo-Boolean satisfiability, and on symmetry breaking.

6.1   Unary Encodings

As remarked by [9], encoding arithmetic in unary representation instead of the more common binary (CPU-like) representation can be very beneficial for the performance of modern conflict-driven SAT solvers on the resulting instances. Unfortunately, encoding the computations not via XOR on GF(2), but rather in unary representation on $\mathbb{Z}$ with a deferred parity check turned out to be prohibitively expensive as the (integer) values for the i-th line are bounded only by $\mathcal{O}(fib(i))$ where fib is the Fibonacci function.

6.2   Encoding to Pseudo-Boolean Constraints

Instead of optimizing and tuning our encoding to SAT, we also implemented a straight-forward encoding to Pseudo-Boolean constraints. The hope was that, e.g., cutting plane approaches could be useful for showing unsatisfiability.

We experimented with MiniSat+, Pueblo, SAT4J, and SCIP but were not able to obtain any improvements for e.g. the first 8 linear forms of the top matrix.

6.3   Symmetry Breaking

In general, having many solutions is considered good for SAT instances as the SAT solver is more likely to "stumble" upon one of them. For UNSAT instances, though, having many potential solutions usually means that the search space to exhaust is very large.

One of the main reasons for having many solutions is symmetry. For example, it does not matter if we first compute $v_1 = x_1 \oplus x_2$ and then $v_2 = x_3 \oplus x_4$ or the other way around. Limiting these kinds of symmetries can be expected to significantly reduce the runtimes for UNSAT instances.

In our concrete setting, being able to reorder independent program lines is one of the major sources of symmetry. Two outputs in a straight-line program are said to be *independent* if neither of them depends on the other (directly through the matrix C or indirectly).

Now, the idea for breaking symmetry is to impose an order on these lines: the line which computes the "smaller" linear form (w.r.t. a total order on linear forms, which can e.g. be obtained by lexicographic comparison of the coefficient vectors) must occur before the line which computes the greater linear form. We can encode the direct dependence of v_i on v_p:

$$\bigwedge_{1 \leq i \leq k} \bigwedge_{1 \leq p < i} c(i,p) \rightarrow dep(i,p)$$

Likewise, the indirect dependence of v_i on v_p can be encoded by transitivity:

$$\bigwedge_{1\leq i\leq k} \bigwedge_{1\leq p<i} \bigwedge_{p<q<i} c(i,q) \wedge dep(q,p) \rightarrow dep(i,p)$$

We also need to encode the reverse direction, i.e.:

$$\bigwedge_{1\leq i\leq k} \bigwedge_{1\leq p<i} \left(dep(i,p) \rightarrow \left(c(i,p) \vee \bigvee_{p<q<i} (c(i,q) \wedge dep(q,p)) \right) \right)$$

Now we can enforce that for $i > p$, the output v_i depends on the output v_p or v_i encodes a greater linear form than v_p:

$$\bigwedge_{1\leq i\leq k} \bigwedge_{1\leq p<i} (dep(i,p) \vee [\psi(1,i),\ldots,\psi(n,i)] >_{lex} [\psi(1,p),\ldots,\psi(n,p)])$$

Here lexicographic comparison of formula tuples is encoded in the usual way (see for example the encodings in [7,5]).

While this approach eliminates some otherwise valid solutions of length k and thus reduces the set of admissible solutions, obviously there is at least one solution of length k which satisfies our constraints whenever solutions of length k exist at all. This way, we greatly reduce the search space by breaking symmetries that are not relevant for the result, but may slow down the search considerably.

Consider again the restriction of our S-box top matrix to the first 8 linear forms. With symmetry breaking, we can show unsatisfiability for the "hard" case $k = 12$ in 76.8 seconds. In contrast, without symmetry breaking, we cannot show unsatisfiability within 4 days.

7   Conclusion

In this paper we have shown how shortest linear straight-line programs for given linear forms can be synthesized using SAT solvers. To this end we have presented a novel polynomial-size encoding of the associated decision problem to SAT and a customized white-box method for again turning this decision procedure into an optimization algorithm.

We have evaluated the feasibility of this approach by a case study where we minimize an important part of the S-box for the Advanced Encryption Standard. This study shows that our SAT-based approach is indeed able to synthesize shortest-known programs for realistic problem settings within reasonable time.

Proving the optimality of the programs found by showing unsatisfiability of the associated decision problem leads to very challenging SAT problems. To improve the performance for the UNSAT case, we discussed three approaches based on unary encodings, on a port to Pseudo-Boolean satisfiability, and on symmetry breaking. We have shown that symmetry breaking significantly reduces runtimes in the UNSAT case.

In future work, we consider to apply our method to other problems from cryptography. Also, we plan to further enhance our encoding and specialize existing SAT solvers to further improve performance in the UNSAT case.

Acknowledgements

Our sincere thanks go to Erika Ábrahám, Daniel Le Berre, Armin Biere, Youssef Hamadi, Oliver Kullmann, Matthew Lewis, Lakhdar Saïs, and Laurent Simon for input on and help with the experiments. Furthermore, we thank Joan Boyar and René Peralta for providing us with information on their work and Michael Codish for pointing out similarities to common subexpression elimination.

References

1. Asín, R., Nieuwenhuis, R., Oliveras, A., Rodríguez-Carbonell, E.: Cardinality networks and their applications. In: Kullmann, O. (ed.) SAT 2009. LNCS, vol. 5584, pp. 167–180. Springer, Heidelberg (2009)
2. Boyar, J., Matthews, P., Peralta, R.: On the shortest linear straight-line program for computing linear forms. In: Ochmański, E., Tyszkiewicz, J. (eds.) MFCS 2008. LNCS, vol. 5162, pp. 168–179. Springer, Heidelberg (2008)
3. Boyar, J., Peralta, R.: A new technique for combinational circuit optimization and a new circuit for the S-Box for AES. In: Patent Application Number 61089998 filed with the U.S. Patent and Trademark Office (2009)
4. Boyar, J., Peralta, R.: A new combinational logic minimization technique with applications to cryptology. In: Festa, P. (ed.) SEA 2010. LNCS, vol. 6049, pp. 178–189. Springer, Heidelberg (2010)
5. Codish, M., Lagoon, V., Stuckey, P.: Solving partial order constraints for LPO termination. Journal on Satisfiability, Boolean Modeling and Computation (JSAT) 5, 193–215 (2008)
6. Eén, N., Sörensson, N.: Translating pseudo-boolean constraints into SAT. Journal on Satisfiability, Boolean Modelling and Computation (JSAT) 2(1-4), 1–26 (2006)
7. Fuhs, C., Giesl, J., Middeldorp, A., Thiemann, R., Schneider-Kamp, P., Zankl, H.: SAT solving for termination analysis with polynomial interpretations. In: Marques-Silva, J., Sakallah, K.A. (eds.) SAT 2007. LNCS, vol. 4501, pp. 340–354. Springer, Heidelberg (2007)
8. Giesl, J., Schneider-Kamp, P., Thiemann, R.: AProVE 1.2: Automatic termination proofs in the dependency pair framework. In: Furbach, U., Shankar, N. (eds.) IJCAR 2006. LNCS (LNAI), vol. 4130, pp. 281–286. Springer, Heidelberg (2006)
9. Grinchtein, O., Leucker, M., Piterman, N.: Inferring network invariants automatically. In: Furbach, U., Shankar, N. (eds.) IJCAR 2006. LNCS (LNAI), vol. 4130, pp. 483–497. Springer, Heidelberg (2006)
10. Hong, H., Jakuš, D.: Testing positiveness of polynomials. Journal of Automated Reasoning (JAR) 21(1), 23–38 (1998)
11. Kojevnikov, A., Kulikov, A.S., Yaroslavtsev, G.: Finding efficient circuits using SAT-solvers. In: Kullmann, O. (ed.) SAT 2009. LNCS, vol. 5584, pp. 32–44. Springer, Heidelberg (2009)
12. Le Berre, D., Parrain, A.: SAT4J, http://www.sat4j.org
13. Federal Information Processing Standard 197. The advanced encryption standard. Technical report, National Institute of Standards and Technology (2001)
14. Tseitin, G.: On the complexity of derivation in propositional calculus. Studies in Constructive Mathematics and Mathematical Logic, pp. 115–125 (1968); Reprinted in Automation of Reasoning 2, 466–483 (1983)

sQueezeBF: An Effective Preprocessor for QBFs Based on Equivalence Reasoning

Enrico Giunchiglia, Paolo Marin, and Massimo Narizzano

DIST - Università di Genova
Viale Causa 13, 16145 Genova, Italy
name.lastname@unige.it

Abstract. In this paper we present sQueezeBF, an effective preprocessor for QBFs that combines various techniques for eliminating variables and/or redundant clauses. In particular sQueezeBF combines (*i*) variable elimination via Q-resolution, (*ii*) variable elimination via *equivalence substitution* and (*iii*) equivalence breaking via *equivalence rewriting*. The experimental analysis shows that sQueezeBF can produce significant reductions in the number of clauses and/or variables - up to the point that some instances are solved directly by sQueezeBF - and that it can significantly improve the efficiency of a range of state-of-the-art QBF solvers - up to the point that some instances cannot be solved without sQueezeBF preprocessing.

1 Introduction

Quantified Boolean Formulas are a powerful extension of the Satisfiability (SAT) problem in which variables are universally as well as existentially quantified. Adding the quantification makes QBFs a more compact language than SAT, but this comes with a price: QBFs are in practice much harder to solve than SAT formulas. Many different problems can be efficiently encoded into QBF instances —such as in Verification [1,17], Planning (Synthesis) [5,14], and Reasoning about Knowledge [13]— and recently there has been great interest and progress in developing efficient solvers for effectively dealing with such instances. Preprocessing formulas has been proven to be very effective for solving SAT instances: it can reduce the size of the formula considerably and decrease the solving time substantially, even taking into account the time spent on preprocessing (see for example [7,18]). Recently two preprocessors have been presented for QBFs: *preQuel* [15,16] and *proverbox* [4]. *preQuel* derives as many binary clauses as possible via Q-resolution [12] and then it eliminates variables and clauses applying binary equality reduction and clause subsumption. *proverbox*, instead, eliminates universal variables by expansion and existential variables at the innermost quantification level by Q-resolution [12]. In order to limit the size of the resulting formula, only a conveniently selected subset of universally quantified variables with bounded expansion costs, is expanded by *proverbox*.

In this paper we present *sQueezeBF*, an effective preprocessor for QBFs that combines: (i) variable elimination via Q-resolution, (ii) variable elimination via

O. Strichman and S. Szeider (Eds.): SAT 2010, LNCS 6175, pp. 85–98, 2010.

equivalence substitution and (iii) equivalence breaking via *equivalence rewriting*. Variable elimination via Q-resolution aims to eliminate a variable v Q-resolving [12] all the clauses where v occurs positively with all the clauses where v occurs negatively. The resulting clauses, if not tautological, are learned while the original clauses where v occurs are deleted. This can be performed only when the resulting clauses verify a size criterion. Variable elimination via Q-resolution in *sQueezeBF* is an extension of the one used by *proverbox* [4]: In *proverbox* only the variables at the innermost quantifier level are eliminated, while in *sQueezeBF* variables at any quantifier level can be eliminated. Variable elimination via *equivalence substitution* replaces each defined literal in the formula with its definition. This well known technique in SAT (see, e.g., [7,18]), has been implemented also in QBF solvers but only for binary equivalences (see, e.g., [2,3]). However, if the size of the resulting formula is greater than the original one, then the substitution does not take place. When substitution increases the size of the formula, we apply *equivalence rewriting*: This technique introduces a new variable aiming to break the equivalence, the size of the resulting formula is increased of at most a binary clause, and —under particular conditions— it allows to eliminate half of the equivalence thus decreasing the size of the resulting formula. As we will see, *equivalence rewriting* can be very effective, especially when coupled with pure literal detection.

The experimental analysis shows that *sQueezeBF* significantly improves the efficiency of various state-of-the-art QBF solvers. In particular, *sQueezeBF*: (*i*) reduces the size of the preprocessed formula, (*ii*) resolves by itself some instances and (*iii*) when coupled with a QBF solver, is able to improve the solver efficiency significantly. The experimental analysis also shows that equivalence rewriting among all the preprocessing techniques implemented in *sQueezeBF* is the one leading to the greatest increase in the number of problems solved. However, the best overall performances are obtained when variable elimination via Q-resolution and equivalence substitution are also activated. Finally, comparing *sQueezeBF* with *preQuel* and *proverbox* we see that most of the time *sQueezeBF* outperforms them both in terms of size reduction of the resulting formula, and number of problem solved when coupled with various state-of-the-art QBF solvers.

This paper is organized as follows. First we review the basics of QBF satisfiability. Then we discuss the algorithm of *sQueezeBF*. We end the paper with the experimental analysis and the conclusions.

2   Basic Definitions

Consider a set P of *variables*. A *literal* is a variable or the negation of a variable. In the following, for any literal l,

- $|l|$ is the variable in l; and
- $\bar{l}$ is $\bar{l}$ if l is a variable, and is $|l|$ otherwise.

A *clause* C is an n-ary ($n \geq 0$) disjunction of literals such that, for any two distinct disjuncts l, l' in C, it is not the case that $|l| = |l'|$. A *propositional formula* is a k-ary ($k \geq 0$) conjunction of clauses.

A *(closed) QBF* is an expression of the form

$$Q_1 z_1 \ldots Q_n z_n \Phi \qquad (n \geq 0) \qquad\qquad (1)$$

where

- every Q_i ($1 \leq i \leq n$) is a quantifier, either existential $\exists$ or universal $\forall$,
- $z_1, \ldots, z_n$ are distinct variables in P, and
- Φ is a propositional formula in the variables $z_1, \ldots, z_n$.

In (1), $Q_1 z_1 \ldots Q_n z_n$ is the *prefix*, Φ is the *matrix*, and Q_i is the *binding quantifier* of z_i. Further, we say that a literal l is *existential* if $\exists |l|$ belongs to the prefix, and is *universal* otherwise. In the following, we will use TRUE and FALSE as abbreviations for the empty conjunction and the empty disjunction respectively.

We define

- the *level of a variable* z_i in the prefix $Q_j z_j Q_{j+1} z_{j+1}$ with $j \geq i$ and $Q_j \neq Q_{j+1}$ is the number of alternating quantifier blocks from left to right (starting with 1);
- the *level of a literal* l, to be the level of $|l|$;
- the *level of the formula* (1), to be the level of z_1.

If φ is a QBF and l is a literal, φ_l is the QBF

1. whose matrix Φ is obtained from the matrix of φ by deleting the clauses C such that $l \in C$, and removing $\bar{l}$ from the others, and
2. whose prefix is obtained from the prefix of φ by deleting each variable and corresponding bounding quantifier not occurring in Φ.

The semantics of a QBF φ can be defined recursively as follows. If the prefix is empty, then φ's satisfiability is defined according to the truth tables of propositional logic. If φ is $\exists x \psi$ (respectively $\forall x \psi$), φ is satisfiable if and only if φ_x or (respectively and) $\varphi_{\bar{x}}$ are satisfiable. If $\varphi = Q x \psi$ is a QBF and l is a literal, φ_l is the QBF obtained from ψ by substituting l with TRUE and $\bar{l}$ with FALSE. It is easy to see that if φ is a QBF without universal quantifiers, the problem of deciding the satisfiability of φ reduces to SAT.

A literal l,

1. *occurs positively* in (1), if l is a disjunct of a clause in Φ,
2. *occurs negatively* in (1), if $\bar{l}$ is a disjunct of a clause in Φ,
3. *occurs* in (1), if l occurs positively or negatively in (1).

Further, in (1), we say that a literal l is

- *Unit* if l is existential, and, for some $m \geq 0$,
  - a clause $(l \vee l_1 \vee \ldots \vee l_m)$ belongs to Φ, and
  - each literal l_i ($1 \leq i \leq m$) is universal and has a lower level than l.

```
0 function sQueezeBF(φ)
1     do
2        φ' = φ
3        φ = Simplify(φ)
4        φ = Eq-Subs(φ)
5        φ = Eq-Rw(φ)
6        φ = Q-resolution(φ)
7        if φ ≡ TRUE  return φ
8        if φ ≡ FALSE  return φ
9     while φ' ≠ φ
10    return φ
```

Fig. 1. The algorithm of *sQueezeBF*

- *Pure* (or *monotone*) if
  - either l is existential and l does not negatively occur in Φ,
  - or l is universal and l does not positively occur in Φ.

Given a set of clauses S we define the *size* $|S|$ of S as the number of literals in S. Finally given a QBF φ, a literal l and a set of clauses α, $\varphi(l/\alpha)$ is the QBF obtained from φ by substituting each occurrence of l in Φ with α[1].

3   *sQueezeBF*

In Fig. 1 we present the main algorithm of *sQueezeBF*. The algorithm takes as input a QBF, and it returns a simplified QBF, that can either be empty (i.e., equivalent to TRUE) or contains an empty clause (i.e., equivalent to FALSE). *sQueezeBF* starts saving the current state of the formula at line 2, and then it applies four operations sequentially, i.e., Simplify(φ), Eq-Subs(φ), Eq-Rw(φ), Q-resolution(φ), till no further simplification is possible (line 9).

Simplify(φ) (line 3) gets in input the formula and simplifies it propagating all the unit and pure literals. Given a formula φ and a literal l, which is unit or pure in φ, φ is equivalent to φ_l. Moreover, it also eliminates *subsumed* clauses, i.e. clauses that are a superset of another clause in φ (see for more details [21]), or *self-subsumed* clauses as in [7]. Eq-Subs(φ), Eq-Rw(φ) and Q-resolution(φ) are explained in details in the next subsections.

From now on, we abuse notation and write QBFs with arbitrary matrix, not necessarily in CNF, with the intended meaning to represent the CNF obtained by applying standard rewriting rules without introducing additional variables. For example, given a CNF α, we write $(l \vee \alpha)$ as an abbreviation for the CNF obtained adding l to each disjunct of α.

---

[1] Strictly speaking the result is not a QBF. We assume that the resulting expression is suitably converted into a QBF without introducing additional variables. For instance, in $\varphi(l/\alpha)$ each clause $C \vee l$ in φ is substituted by the clauses in $\{C \vee C' : C' \in \alpha\}$.

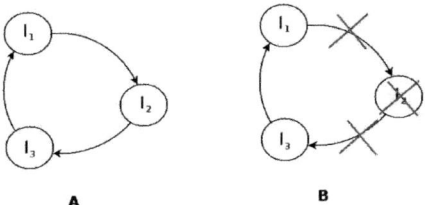

Fig. 2. Dependency graph

3.1 Variable Elimination via Equivalence Substitution

The variable elimination via equivalence checking was introduced for SAT in [7]. It is an algorithm that works in two steps: (i) identification of *equivalences* and (ii) literal substitution. In the first step Eq-Subs looks for the following set of clauses in the formula:

$$(\bar{l} \vee l_1 \vee \ldots \vee l_{n-1} \vee l_n) \wedge (l \vee \bar{l}_1) \wedge \ldots \wedge (l \vee \bar{l}_{n-1}) \wedge (l \vee \bar{l}_n) \qquad (2)$$

$$(l \vee l_1 \vee l_2) \wedge (l \vee \bar{l}_1 \vee \bar{l}_2) \wedge (\bar{l} \vee \bar{l}_1 \vee l_2) \wedge (\bar{l} \vee l_1 \vee \bar{l}_2) \qquad (3)$$

where in both the formulas (2) and (3) l is existential and with level lower than the level of each $|l_i|$ occurring in the equivalence. Notice that the set (2) corresponds to the equivalence:

$$l \Leftrightarrow (l_1 \circ \ldots \circ l_{n-1} \circ l_n) \qquad \text{where } \circ \in \{\vee, \wedge\} \text{ and } n \geq 1 \qquad (4)$$

and the set (3) corresponds to the equivalence:

$$l \Leftrightarrow l_1 \Leftrightarrow l_2 \qquad (5)$$

Once Eq-Subs finds an equivalence of the form (4) or (5), then it eliminates the equivalence and it substitutes each occurrence of l with its definition. Notice that after a substitution of a literal l with its definition α in a formula φ, the resulting formula $\varphi(l/\alpha)$ is not guaranteed to be in CNF, and a CNF conversion step may be necessary, using standard distribution laws[2]. Notice that it is well known that the size of $\varphi(l/\alpha)$ can be significantly greater than the size of φ. However, by eliminating redundant literals or clauses in the produced set of clauses, such increase often does not happen. In case it does, we discard the changes, and the substitution does not take place. This ensures the formula to never increase in size after each equivalence substitution. Indeed, for binary equivalences of the form $l \Leftrightarrow l_1$, l can be safely replaced by l_1 and the size does not increase. Binary equivalences are substituted as soon as they are detected.

For non-binary equivalences some care as to be taken. Indeed, not all such equivalences can be substituted. Take for example:

---

[2] Indeed, other CNF conversion methods are possible, e.g., based on renaming, but these methods would introduce back new variables.

$$\varphi = \alpha \wedge (l_1 \Leftrightarrow l_2 \vee l_4) \wedge (l_2 \Leftrightarrow l_3 \vee l_4) \wedge (l_3 \Leftrightarrow l_1 \vee l_5)$$

if l_1 is eliminated first, then when l_3 is substituted in the formula, l_1 is reintroduced, and an analogous phenomenon happens no matter which variable is substituted first. This problem arises since at least a variable occurs in the definitions of the others (notice that this can only happen if the literals are all at the same level). In order to solve the problem of circularity, Eq-Subs constructs a dependency graph where each node represents a defined variable, and each edge, connecting two nodes, represents the dependency between their definitions. So for example, Fig. 2:A represents the dependency graph of the example above, where the edge from the node l_3 and the node l_1 depict the fact that l_1 occurs in the definition of l_3. Notice that the edges have a direction, meaning that the pointed node belongs to the definition of the starting one. In Fig. 2:A, l_1 occurs in the definition of l_3. After the graph is created, then the algorithm looks for circular paths and, if any is found, then the path is cut removing one of the definition from it. Looking at Fig. 2:A, in order to eliminate the circular path the algorithm deletes one of the three definitions, for example l_2 (Fig. 2:B). At this point we can substitute first l_3, and then l_1 and the formula is simplified eliminating two variables. Notice that the order matters, since if we substitute l_1 first, when l_3 is substituted l_1 will be reintroduced: Substituting literals starting from the ones with no entering edges solves the problem.

3.2   Equivalence Breaking via Equivalence Rewriting

Equivalence substitution may increase substantially the size of the QBF. For this reason not all the equivalences are substituted in a QBF, but a bound on the new size of the formula is always considered. Yet, keeping equivalences in a QBF may slow down the search. For example consider a QBF with matrix:

$$\varphi = (l \vee \alpha) \wedge (l \Leftrightarrow \gamma) \wedge \phi. \tag{6}$$

where α, γ and ϕ are CNFs in which l does not occur. During the search whenever α becomes true, then the only occurrency of l is in the definition of the equivalence, and thus the equivalence can be safely eliminated. However, (6) is equivalent to $(l \vee \alpha) \wedge (l \Rightarrow \gamma) \wedge \phi$ and thus $(l \Rightarrow \gamma)$ can be eliminated by pure literal detection as soon as α becomes true. The fact that $(l \Leftrightarrow \gamma)$ can be replaced by $(l \Rightarrow \gamma)$ in (6) is formally stated by the following theorem.

Theorem 1. *Let φ be a QBF with matrix*

$$\Phi = (l \vee \alpha) \wedge (l \Leftrightarrow \gamma) \wedge \phi \tag{7}$$

where

1. *l is an existential literal which does not occur in α, γ and ϕ, and*
2. *l has prefix level lower than the prefix level of the literals in γ.*

φ is equivalent to the QBF obtained by substituting $(l \Leftrightarrow \gamma)$ with $(l \Rightarrow \gamma)$.

In the hypothesis of the theorem, rewriting equivalences gives two advantages to the solver:

- Less clauses to deal with, i.e. smaller QBFs: Suppose that $\gamma \equiv l_1 \circ \ldots \circ l_n$ then after rewriting the set of clauses $\gamma \rightarrow l$ is deleted. If $\circ$ is $\wedge$, a single but long clause is deleted. If $\circ$ is $\vee$, n binary clauses are deleted.
- More pruning during the search: Whenever α becomes true l occurs both positively and negatively in (7), while in the rewritten formula l occurs only negatively and thus can be simplified by pure literal detection.

The theorem above can be generalized to the case in which the l occurs also negatively in φ.

Theorem 2. *Let φ be a QBF with matrix*

$$\Phi = (l \vee \alpha) \wedge (\bar{l} \vee \beta) \wedge (l \Leftrightarrow \gamma) \wedge \phi \tag{8}$$

where

1. l *is an existential literal which does not occur in* α, γ, β *and* ϕ, *and*
2. l *has prefix level lower than the prefix level of the literals in* γ.

φ *is equivalent to the QBF* φ' *obtained from* φ *substituting:*

1. $(l \Leftrightarrow \gamma)$ *with* $(l \Rightarrow \gamma) \wedge (\gamma \Rightarrow \bar{l}')$, *and*
2. $\bar{l} \vee \beta$ *with* $l' \vee \beta$,

where l' *is a new existentially quantified variable introduced at the same quantifier level of* l.

The matrix of the QBF rewritten after Theorem 2 has the same size of the original one, but still gives the benefit due to the pure literal detection:

1. Whenever α (resp. β) becomes true, the matrix of φ can be simplified to $(\bar{l} \vee \beta) \wedge (l \Leftrightarrow \gamma) \wedge \phi$ (resp. $(l \vee \alpha) \wedge (l \Leftrightarrow \gamma) \wedge \phi$) while in the matrix Φ' of φ', l (resp. $\bar{l}'$) becomes pure, and Φ' can be simplified to $(l' \vee \beta) \wedge (\gamma \Rightarrow \bar{l}') \wedge \phi$ (resp. $(l \vee \alpha) \wedge (l \Rightarrow \gamma) \wedge \phi$).
2. Whenever both α and β become true, the matrix of φ can be simplified to $(l \Leftrightarrow \gamma) \wedge \phi$, while in the matrix Φ' of φ', l and $\bar{l}'$ become pure, and Φ' can be simplified to ϕ.

The last item shows that we obtain an effect analogous to the "don't care propagation" in non-CNF reasoning (see, e.g., [19]): Indeed, there is more than an analogy and it can be proved that given a non-CNF formula φ, if φ' is the CNF formula obtained by first converting φ using Tseitin conversion [20] and then applying equivalence rewriting to the result, if a subformula ψ of φ gets assigned by "don't care propagation", the clauses corresponding to ψ in φ' get assigned by pure literal detection [9]. Also notice that whenever γ becomes true (resp. false), in the rewritten formula l (resp. $\bar{l}'$) becomes pure, and the matrix can be simplified to β (resp. α) as in the original case. However, when α (resp. β) becomes false, the matrix of φ can be simplified to $\beta \wedge \gamma \wedge \phi$, (resp. $\alpha \wedge \gamma \wedge \phi$), while in the matrix Φ' of φ', l (resp. l') is unit, and Φ' can be simplified to $(l' \vee \beta) \wedge \gamma \wedge (\gamma \Rightarrow \bar{l}') \wedge \phi$. (resp. $(l \vee \alpha) \wedge (\neg\gamma) \wedge (l \Rightarrow \gamma) \wedge \phi$). In particular, from $\gamma \wedge (\gamma \Rightarrow \overline{l'})$, it might the case that $\overline{l'}$ is not derived by unit propagation. In order to obtain such a propagation, we may add the *efficiency clause* $\overline{l} \vee \overline{l'}$ to the matrix of φ', as sanctioned by the following Corollary.

Corollary 1. *In the hypothesis of Theorem 2, φ' is equivalent to the QBF obtained adding the new clause $(\overline{l} \vee \overline{l'})$ to the matrix of φ'.*

Indeed, the efficiency clause is entailed by $(l \Rightarrow \gamma) \wedge (\gamma \Rightarrow \overline{l'})$.

Finally, Eq-Rw applies Theorem 1 when a defined literal l occurs only positively, and the Corollary of Theorem 2 when a defined literal l occurs both positively and negatively.

3.3   Variable Elimination via Q-Resolution

Variable elimination via Q-resolution is a technique used in many state-of-the-art SAT solvers, first introduced in [6] during the search and also used in [18] as preprocessor. In QBF variable elimination via Q-resolution has been first introduced by *Quantor* [3]: During the search *Quantor* eliminates by Q-resolution the existential variables with the lowest prefix level. *sQueezeBF* instead can remove any existential variable, and starts from those quantified at the lowest level.

Given an existential variable x and two clauses $C_1 = \{x \vee l_1 \vee \ldots \vee l_n\}$ and $C_2 = \{\overline{x} \vee l'_1 \vee \ldots \vee l'_m\}$ such that $\overline{l_i} \neq l'_j$ when l_i or l'_j has a lower level than x, the clause $C = \{l_1 \vee \ldots \vee l_n \vee l'_1 \vee \ldots \vee l'_m\}$ is called the *resolvent of C_1 and C_2 (on the variable x)*, and is denoted by $C_1 \otimes C_2$. If S_x (resp. $S_{\overline{x}}$) is the set of clauses in which x (resp. $\overline{x}$) occurs, we define Q-resolution between S_x and $S_{\overline{x}}$ as the set of clauses

$$S_x \otimes S_{\overline{x}} : \{C_x \otimes C_{\overline{x}} | C_x \in S_x, C_{\overline{x}} \in S_{\overline{x}}\}. \tag{9}$$

Assuming we can perform the resolution of each clause in S_x with each clause in $S_{\overline{x}}$, we can replace the clauses in $S_x \cup S_{\overline{x}}$ with the clause in $S_x \otimes S_{\overline{x}}$, and delete $|l|$ and its quantifier from the prefix, resulting in an equivalent problem.

As an example of how the algorithm works, consider

$$S_x = \{\{x \vee \overline{a}\}, \{x \vee a \vee c\}, \{x \vee \overline{d}\}\}$$

and

$$S_{\overline{x}} = \{\{\overline{x} \vee b\}, \{\overline{x} \vee b \vee d\}\}.$$

Then the resolvent between the two set of clauses is

$$S_x \otimes S_{\overline{x}} = \{\{\overline{a} \vee b\}, \{\overline{a} \vee b \vee d\}, \{a \vee c \vee b\}, \{a \vee c \vee b \vee d\}, \{b \vee \overline{d}\}\}.$$

In order to eliminate the variable x from φ, all the clauses in $S_x \cup S_{\overline{x}}$ have to be deleted from φ, adding all the clauses in $S_x \otimes S_{\overline{x}}$ and obtaining the formula φ'. Notice that in the example, the size of φ' is greater then the size of φ, since

$|S_x \cup S_{\overline{x}}| = 12$, and $|S_x \otimes S_{\overline{x}}| = 14$. In order to avoid an increase in the size of the formula, *sQueezeBF* eliminates a variable x by Q-resolution if and only if

$$|S_x \cup S_{\overline{x}}| \geq |S_x \otimes S_{\overline{x}}| \tag{10}$$

Notice that the size of the resolvent can not be calculated in advance on the basis of $|S_x|$ and $|S_{\overline{x}}|$: Indeed, $S_x \otimes S_{\overline{x}}$ may contain tautological clauses which can be discarded, or a same literal l can belong to a clause $C_1 \in S_x$ and a clause $C_2 \in S_{\overline{x}}$. For these reasons, *sQueezeBF* first calculates the size of the original set of clauses, and then it computes the resolvent, discarding it if (10) is not satisfied.

4 Experimental Analysis

As environment, we use a cluster made of 4 IBM HS21 computing nodes, each with 2 Quad Core Xeon 2.5 GHz, 16 GB RAM, running Linux CentOS 5; the time limit was set to 600 s and the memory limit to 2 GB, where for each node we run 4 processes at the time. We compare *sQueezeBF* with *preQuel* and *proverbox* on the pool of (3326) fixed-structure QBF instances selected for the QBF Evaluation 2008 from QBFLIB [10]. We first compare their effectiveness comparing the size of each formula before and after the preprocessing. This is reported in Table 1, where for each row[3] is reported (i) in the first column the family name and its number of problems and (ii) a group of three columns with the averages number of the variables (V), clauses (C), and literals per clause (L/C) of the original formula, and (iii) for each preprocessor, a group of three columns, i.e. the average variation in the number of variables (V%) and clauses (C%), with respect to the column "original", and the average literals per clause (L/C). Notice that a negative value stands for a decrease in the QBF with respect to the original one. For example for the first row, "Abduction" is the family name which contains 300 instances, having, on average 1632 variables, 4529 clauses each with 2.51 literals on average. After *proverbox* is run on the "Abduction" family, the resulting QBFs have 14.78% less variables, 16.76% less clauses, but each clause now consists of 4.84 literals on average.

Considering only the preprocessors, we can say that:

– *proverbox* was able to solve by itself 141/3326 problems and on 112 instances was not able to terminate due to Memory or Time Out. On the remaining QBFs, as shown in Table 1, *proverbox* results to be the least effective preprocessor: In fact even if the number of literals per clause most of the time decreases, the number of families with a size reduction is very low: on 22/32 families *proverbox* fails to reduce both the number of variables and clauses;

---

[3] For the sake of compactness, in some cases families are grouped into larger sets. This is the case of "k_x_n" (respectively "k_x_p") that includes the families k_branch_n, k_d4_n, k_dum_n, k_grz_n, k_lin_n, k_path_n, k_ph_n, k_poly_n and k_t4p_n (k_branch_p, k_d4_p, k_dum_p, k_grz_p, k_lin_p, k_path_p, k_ph_p, k_poly_p, k_t4p_p), and "Scholl-Becker" that includes C432, C499, C5315, C6288, C880, comp, term1, and z4ml.

Table 1. Size-reduction comparison between different preprocessors. "V", "C" and "L/C" denote the average number of variables, clauses, and literals per clause. *sQueezeBF* is the preprocessor with all the techniques enabled. In the first column the number written in parentheses is the number of instances in the family.

Family	original			proverbox			preQuel			sQueezeBF		
	V	C	L/C	V%	C%	L/C	V%	C%	L/C	V%	C%	L/C
Abduction *(300)*	1632	4529	2.51	-14.78	-16.76	4.84	-2.54	-3.53	2.48	-5.88	-8.18	2.84
Adder *(31)*	5196	7052	5.54	-46.93	14.93	8.29	-11.65	-0.42	5.54	-65.57	-46.41	6.19
blackbox_design *(28)*	4082	9496	2.34	-64.55	-16.71	3.71	-53.83	-21.07	2.33	-80.37	-78.57	3.3
blackbox-01X-QBF *(295)*	20823	54282	2.32	-80.25	-25.13	4.76	-74.07	-30.58	2.33	-87.48	-83.89	3.24
Blocks *(13)*	485	6810	2.97	11.70	9.73	2.98	-41.73	-58.39	2.93	-28.65	-56.35	3.05
BMC *(58)*	18824	53922	3.31	-57.66	5.99	3.94	-44.42	-34.72	3.64	-56.32	-50.84	4.06
circuits *(45)*	15885	22367	2.58	-57.04	-3.56	4.37	-48.61	-11.40	2.6	-60.42	-50.97	3.27
Conformant Planning *(19)*	1467	21738	5.77	91.05	110.92	4.41	-1.91	-1.03	5.78	-33.36	-29.26	6.14
Counter *(22)*	6205	16518	3	-1.74	31.33	3.26	-3.13	-2.48	2.99	-5.38	-6.44	3.16
evader-pursuer-4x4-l *(7)*	1961	12973	3.72	14.08	60.21	3.97	-16.20	-13.77	3.87	-24.64	-26.71	3.87
evader-pursuer-4x4-s *(7)*	7454	67639	3.25	43.63	44.00	3.41	-4.41	-12.79	3.35	-28.24	-50.28	3.76
evader-pursuer-6x6-l *(8)*	5121	41564	3.9	7.08	71.34	4.52	-16.22	-13.55	4.1	-25.04	-28.03	4.07
evader-pursuer-6x6-s *(1)*	16385	212391	2.98	83.41	29.67	3.17	-4.24	-24.35	3.15	-52.48	-88.79	4.56
evader-pursuer-8x8-l *(8)*	9633	85906	3.96	12.47	72.49	4.66	-18.97	-16.89	4.19	-24.92	-28.51	4.15
FPGA_PLB_FIT_FAST *(5)*	69	658	6.52	36.24	72.70	5.41	-0.46	-0.11	6.39	0.73	-1.88	6.09
FPGA_PLB_FIT_SLOW *(3)*	72	522	6.59	36.63	89.70	5.6	-0.95	-0.30	6.52	-0.51	-0.92	6.22
irqlkeapcite *(46)*	9635	37515	2.53	-37.89	44.95	4.39	-14.50	-7.60	2.52	-37.54	-33.84	2.8
jmc_quant *(10)*	1441	3893	2.61	-3.59	27.72	3.06	-7.70	-6.58	2.61	-48.24	-65.73	3.54
jmc_quant_squaring *(10)*	1604	4264	2.62	4.47	35.43	3.01	-6.96	-6.03	2.62	-46.44	-65.98	3.6
k_x_n *(161)*	1915	14454	2.62	87.16	333.03	2.92	62.45	165.07	2.62	-4.38	46.56	2.83
k_x_p *(138)*	1678	11154	2.58	37.35	67.89	2.83	1.18	11.43	2.58	-38.07	-29.54	2.83
Mutex *(6)*	3744	4847	2.46	-63.85	-6.80	5.38	-0.11	-0.08	2.46	-79.88	-66.69	3.87
Sakallah *(141)*	14060	35871	2.64	-22.86	-23.44	3.25	-16.27	-16.99	2.66	-25.25	-46.30	3.27
Scholl-Becker *(64)*	1769	5001	2.36	-6.37	34.43	2.89	-48.75	-22.45	2.41	-35.91	-38.96	2.88
Sorting_networks *(84)*	2902	5935	2.57	-26.14	13.58	3.48	-10.82	-1.78	2.58	-37.86	-26.18	2.79
SzymanskiP *(6)*	69879	84653	2.6	-15.94	-13.98	3.37	-0.25	-0.26	2.6	-69.30	-57.42	3.42
terminator *(564)*	3568	11837	2.47	-14.17	37.97	3.85	-46.13	-2.70	2.46	-84.05	-87.75	3.02
tipdiam *(169)*	6286	18144	2.33	-30.61	17.57	3.26	-42.41	-34.95	2.32	-64.99	-67.71	3.15
tipfixpoint *(387)*	8391	24077	2.33	-49.49	17.69	3.55	-26.24	-19.53	2.32	-60.17	-61.35	2.92
Toilet *(7)*	638	2816	2.37	-18.76	-3.66	2.66	-30.07	-21.60	2.4	-47.15	-35.21	2.79
uclid *(3)*	2900	7869	2.31	-60.22	-21.77	3.76	-10.32	-7.96	2.33	-77.77	-77.29	3.32
wmiforward *(62)*	1125	3675	2.5	7.71	35.51	2.87	-43.51	-18.15	2.49	-59.85	-58.50	2.74

- *preQuel* solves by itself 273/3326 instances and did not terminate on 70 instances. It also reduces the size of almost each formula, in terms of number of variables and clauses. *preQuel* also keeps low the ratio between the number of literals and the number of clauses;

- *sQueezeBF* solves by itself 287/3326 instances and was not able to terminate on 73 instances. Its performances are comparable to those of *preQuel* when considering the number of families having a reduction in the number of variables and clauses: only 2 families were not reduced by *sQueezeBF*. However, in terms of percentage of variables and clause reduction, *sQueezeBF* is the most effective preprocessor, since it eliminates half of the formula on average.

Table 2. Number of instances solved and cumulative time (in parentheses) by using different preprocessors. Time (in thousands of seconds) includes both preprocessing and solving : a penalty of 600s is added in case the solver or the preprocessor ran out of time or memory. The first column refers to solving without the help of any external preprocessing. *sQueezeBF* labels mean: *sQueezeBF* is the preprocessor with all its techniques enabled, "*–Eq-**", "–Eq-Rw" and "–Eq-Subs" stand for *sQueezeBF* without both equivalence substitution and equivalence rewriting, without equivalence rewriting, or without equivalence substitution, respectively.

Solver	no-prepro	*preQuel*	*proverbox*	*sQueezeBF*			
				–Eq-*	–Eq-Rw	–Eq-Subs	–
QuBE	1008 (1407)	1094 (1358)	1208 (1311)	1223 (1285)	1294 (1241)	1849 (923)	2195 (716)
QuBE-np	737 (1565)	755 (1557)	938 (1459)	930 (1464)	988 (1427)	1014 (1415)	1100 (1357)
yQuaffle	937 (1449)	996 (1410)	781 (1549)	976 (1427)	1014 (1406)	1147 (1326)	1176 (1309)
sSolve	1503 (1117)	1096 (1355)	890 (1487)	1155 (1323)	1216 (1292)	1717 (996)	1790 (949)
Quantor	767 (1542)	1002 (1406)	778 (1541)	849 (1497)	901 (1472)	1354 (1196)	1429 (1151)
sKizzo	1585 (1081)	1494 (1132)	2005 (872)	1397 (1199)	1365 (1211)	1352 (1224)	1712 (1000)

In Table 2 we present the results of the preprocessor when coupled with different state-of-the-art QBF solvers. The first column (Solver) shows the name of the solver, namely *QuBE* (Release QuBE6.6 where its own preprocessor has been disabled) [11], *yQuaffle* (Version 021006) [23,22], and *sSolve* (Version *sSolveC* from QBFEVAL 08) [8] as search-based solvers; *Quantor* (Version 3.0.) [3] as a resolution based solver; and *sKizzo* (Version *sKizzo*-0.10-qck from QBFEVAL 07) [2] as a symbolic skolemization based solver. We have also run a version of *QuBE* in which the pure literal detection was disabled (*QuBE*-np in the tables). The second column, (no-prepro) presents the results of the solver when no preprocessor is applied. The third and fourth columns present the results for *preQuel* and *proverbox*, respectively. The last four columns present the results when *sQueezeBF* is applied in different versions: in the last column "-"), we show the results for the full-featured preprocessor (i.e. with no technique disabled), columns "–Eq-Subs", "–Eq-Rw" and "–Eq-*" represent a version of *sQueezeBF* featuring all the techniques but variable elimination via equivalence substitution, equivalence breaking via equivalence rewriting, or without both, respectively, i.e. performing only Q-resolution. The table reports the number of problems solved and the cumulative solving time, for each solver (on the rows) when coupled with the corresponding preprocessor written in the column. Table 2 witnesses that *sQueezeBF* is the only one able to constantly improve the efficiency of a range of state-of-the-art QBF Solvers. In particular the column "–" shows that the use of *sQueezeBF* improves up to a factor two the number of problems solved by a given solver. Using *sQueezeBF* as a preprocessor affects also the solving time of each solver, decreasing it substantially. Moreover, Table 2 also shows which technique has more impact on which solver. For example, it looks like that disabling variable elimination by equivalence rewriting, *QuBE* and all the other solvers but *sKizzo*, can no longer solve many problems. It is also interesting that *sQueezeBF* on this set of benchmarks, makes *QuBE* the most effective solver. On the other hand, it may be the

Table 3. Size-reduction comparison between different versions of *sQueezeBF*. The header has the same meaning of the one in Table 1, but the reference values are those of *sQueezeBF*.

Family	sQueezeBF			− Eq-Subs			− Eq-Rw			− Eq-*		
	V	C	L/C	V%	C%	L/C	V%	C%	L/C	V%	C%	L/C
Abduction *(300)*	1505	4162	2.84	0.45	0.65	2.82	-0.96	-0.17	2.83	-1.26	-0.47	2.82
Adder *(31)*	1313	3264	6.19	-2.42	-1.13	6.2	-1.44	-0.69	6.19	-1.81	-0.88	6.18
blackbox_design *(28)*	795	1970	3.3	8.16	11.46	3.1	9.07	53.29	3.24	16.03	57.93	3.06
blackbox-01X-QBF *(295)*	2630	8640	3.24	4.03	0.94	3.16	42.67	98.09	3.19	64.54	115.44	2.97
Blocks *(13)*	360	3271	3.05	-0.18	-0.05	3.04	-21.83	-4.11	3.08	-1.21	31.93	3.05
BMC *(58)*	7545	25108	4.06	5.06	4.77	3.97	-28.02	-11.64	4.25	-14.68	-0.89	4.08
circuits *(45)*	4060	7336	3.27	-0.74	-1.00	3.42	-5.35	-4.72	3.37	-5.37	-4.73	3.43
Conformant Planning *(19)*	1005	16956	6.14	1.13	0.08	6.13	-3.05	2.47	6.1	-3.10	2.13	6.09
Counter *(22)*	6177	16368	3.16	4.35	2.54	3.11	-3.56	-1.70	3.18	0.10	1.26	3.09
evader-pursuer-6x6-s *(1)*	7786	23805	4.56	0.03	0.01	4.56	0.00	29.83	3.96	0.01	29.84	3.96
evader-pursuer-4x4-l *(7)*	1509	9867	3.87	0.16	0.04	3.87	-4.92	0.36	3.84	-4.42	0.52	3.84
evader-pursuer-4x4-s *(7)*	5675	37477	3.76	0.05	0.02	3.76	0.00	13.94	3.53	0.03	13.95	3.53
evader-pursuer-6x6-l *(8)*	3928	31360	4.07	0.06	0.01	4.07	-4.33	0.16	4.04	-4.14	0.21	4.04
evader-pursuer-8x8-l *(8)*	7402	64521	4.15	0.03	0.01	4.15	-4.07	0.54	4.13	-3.96	0.57	4.13
FPGA_PLB_FIT_FAST *(5)*	69	646	6.09	0.00	0.00	6.06	-1.17	-0.27	6.07	-1.17	-0.16	6.14
FPGA_PLB_FIT_SLOW *(3)*	71	518	6.22	0.00	0.00	6.2	-0.44	-0.05	6.21	-0.44	-0.05	6.18
irqlkeapcite *(46)*	6105	25189	2.8	7.78	3.45	2.73	-16.76	-0.69	2.85	-18.75	-3.16	2.85
jmc_quant *(10)*	696	1436	3.54	1.34	2.73	3.63	-6.41	65.71	3.36	-0.05	75.72	3.12
jmc_quant_squaring *(10)*	797	1539	3.6	2.83	3.64	3.5	-5.65	71.91	3.34	0.74	80.13	3.13
k_x_n *(161)*	1136	7882	2.83	32.77	156.76	3.08	19.39	26.91	2.9	83.45	274.02	2.93
k_x_p *(138)*	959	5009	2.83	100.17	190.65	3.03	24.94	32.18	2.86	91.64	275.44	2.99
Mutex *(6)*	757	1681	3.87	-0.07	-0.03	3.84	0.71	0.40	3.84	0.45	0.24	3.81
Sakallah *(141)*	11278	20941	3.27	4.55	4.86	3.46	-5.97	29.32	3.35	8.31	50.42	3.54
Scholl-Becker *(64)*	1142	3022	2.88	1.80	5.47	2.86	-33.38	-13.68	2.92	-27.19	-7.90	2.9
Sorting_networks *(84)*	2036	4640	2.79	3.51	0.95	2.74	-1.75	3.41	2.84	-1.58	3.47	2.84
SzymanskiP *(6)*	21313	36083	3.42	-2.23	-1.42	3.44	-2.64	-1.66	3.44	-2.88	-1.75	3.44
terminator *(564)*	585	1521	3.02	108.96	119.67	2.76	250.52	476.68	2.75	302.49	576.40	2.67
tipdiam *(169)*	1599	4392	3.15	24.67	20.13	2.96	-24.95	5.55	3.18	-12.29	28.54	3.04
tipfixpoint *(387)*	3301	9197	2.92	26.58	29.87	2.9	-34.55	-7.73	3.02	-26.38	11.08	2.97
Toilet *(7)*	369	2109	2.79	4.81	-1.57	2.73	-4.43	0.71	2.83	22.85	11.40	2.86
uclid *(3)*	632	1852	3.32	3.96	2.01	3.2	27.15	66.41	3.22	65.04	97.45	3.03
wmiforward *(62)*	524	1734	2.74	6.27	1.86	2.68	-21.81	-5.17	2.92	-7.61	9.03	2.82

case that a solver performs worse when coupled with a preprocessor. This is not that surprising: given that many solvers have a built-in preprocesor, or are tuned to perform best when dealing with "plain encoded" formulas. In order to better understand the impact that the three techniques implemented in *sQueezeBF* have on QBFs we also collect the data presented in Table 1 disabling one at the time the three techniques in *sQueezeBF*.

In Table 3 we show the results where in each column, instead of a preprocessor, we show a variant of *sQueezeBF*. In particular, the first column (*sQueezeBF*) represents the full-featured preprocessor (i.e. no technique disabled); the suffixes "–Eq-Subs", "–Eq-Rw" and "–Eq-*" represent a version of *sQueezeBF* featuring (*i*) all the techniques but variable elimination via equivalence substitution; (*ii*) all the techniques but equivalence breaking via equivalence rewriting; and (*iii*) all the techniques but equivalence substitution and rewriting. For example, looking at row "Abduction" and column *sQueezeBF*, the three values (V, C, L/C)= (1505, 4162, 2.84) represent respectively the number of variables and clauses on average, and the literals per clause on average. The next columns state the variations with respect to the first one, i.e. looking at the first row (Abduction), the column –Eq-Subs shows the values (V%, C%, L/C) = (0.45%, 0.65%, 2.82), representing the fact that running the preprocessor without variable elimination via equivalence substitution makes the formula larger, having 0.45% more variables (i.e. almost 1511 variables on average) and 0.65% more clauses (i.e. almost 4189 clauses on average).

Table 3 shows that disabling one technique at the time usually leads to a greater formula, as witnessed by the small number of negative numbers in the table. Further, looking at the single families, we see that the reduction of variables is mostly affected by Equivalence Substitution, and that the number of variables decreases also in a few cases when disabling Equivalence Rewriting: This is not surprising because, at least in principle, the goal of Equivalence Substitution is to eliminate the defined variables, while Equivalence Rewriting can introduce additional variables (which can then later be eliminated, e.g., by pure literal detection).

About the time needed by the different preprocessors, these are cumulatively presented in Table 2, where it is shown also the effect of the different preprocessors (*sQueezeBF*, *proverbox*, *preQuel*) when coupled with different solvers (*QuBE*, sSolve, yQuaffle, sKizzo and Quantor). In general, the preprocessing time is negligible wrt the whole task of preprocess and solve, but for some large instances it can be more expensive to try to simplify the formula rather than solving it. The cumulative preprocessing times are approximately 199, 105s for *proverbox*, 98, 338s for *preQuel*, and 70, 271s for *sQueezeBF*.

5   Conclusions

In this paper we presented *sQueezeBF*, a very effective preprocessor for QBFs. We took into account many benchmarks from different families and two other different preprocessing tools, *preQuel* and *proverbox*. We shaw that *sQueezeBF* is much more effective in terms of formula reduction, since most of the times it decreases the size of the formula preprocessed, and never increases the size of the formula, while this is not always true for *preQuel* and *proverbox*. We also compare five different state-of-the-art solvers: The proposed techniques offer robust improvements across the different solvers on all the tested benchmark families. To the best of our knowledge thanks to *sQueezeBF* the solvers are able to solve 136 problems that have never been resolved before.

References

1. Ayari, A., Basin, D.A.: Bounded model construction for monadic second-order logics. In: Emerson, E.A., Sistla, A.P. (eds.) CAV 2000. LNCS, vol. 1855, pp. 99–112. Springer, Heidelberg (2000)
2. Benedetti, M.: sKizzo: A suite to evaluate and certify QBFs. In: Nieuwenhuis, R. (ed.) CADE 2005. LNCS (LNAI), vol. 3632, pp. 369–376. Springer, Heidelberg (2005)

3. Biere, A.: Resolve and expand. In: Hoos, H., Mitchell, D.G. (eds.) SAT 2004. LNCS, vol. 3542, pp. 59–70. Springer, Heidelberg (2005)
4. Bubeck, U., Büning, H.K.: Bounded universal expansion for preprocessing QBF. In: Marques-Silva, J., Sakallah, K.A. (eds.) SAT 2007. LNCS, vol. 4501, pp. 244–257. Springer, Heidelberg (2007)
5. Castellini, C., Giunchiglia, E., Tacchella, A.: Improvements to SAT-based conformant planning. In: Proc. ECP (2001)
6. Davis, M., Putnam, H.: A computing procedure for quantification theory. Journal of the ACM 7, 201–215 (1960)
7. Eén, N., Biere, A.: Effective preprocessing in SAT through variable and clause elimination. In: Bacchus, F., Walsh, T. (eds.) SAT 2005. LNCS, vol. 3569, pp. 61–75. Springer, Heidelberg (2005)
8. Feldmann, R., Monien, B., Schamberger, S.: A distributed algorithm to evaluate Quantified Boolean Formulae. In: Proc. AAAI (2000)
9. Giunchiglia, E., Marin, P., Narizzano, M.: Don't care propagation via equivalence rewriting and pure literal detection. Technical report (2010)
10. Giunchiglia, E., Narizzano, M., Tacchella, A.: Quantified Boolean Formulas satisfiability library, QBFLIB (2001), http://www.qbflib.org
11. Giunchiglia, E., Narizzano, M., Tacchella, A.: Clause/term resolution and learning in the evaluation of quantified Boolean formulas. Journal of Artificial Intelligence Research (JAIR) 26, 371–416 (2006)
12. Kleine Büning, H., Karpinski, M., Flögel, A.: Resolution for quantified Boolean formulas. Information and Computation 117(1), 12–18 (1995)
13. Pan, G., Sattler, U., Vardi, M.Y.: Bdd-based decision procedures for k. In: Voronkov, A. (ed.) CADE 2002. LNCS (LNAI), vol. 2392, pp. 16–30. Springer, Heidelberg (2002)
14. Rintanen, J.: Constructing conditional plans by a theorem prover. Journal of Artificial Intelligence Research 10, 323–352 (1999)
15. Samulowitz, H., Davies, J., Bacchus, F.: Preprocessing QBF. In: Benhamou, F. (ed.) CP 2006. LNCS, vol. 4204, pp. 514–529. Springer, Heidelberg (2006)
16. Samulowitz, H., Davies, J., Bacchus, F.: QBF Preprocessor Prequel (2006), http://www.cs.toronto.edu/~fbacchus/sat.html
17. Scholl, C., Becker, B.: Checking equivalence for partial implementations. In: Proceedings of the 38th Design Automation Conference, pp. 238–243 (2001)
18. Subbarayan, S., Pradhan, D.K.: NiVER: Non-increasing variable elimination resolution for preprocessing SAT instances. In: Hoos, H.H., Mitchell, D.G. (eds.) SAT 2004. LNCS, vol. 3542, pp. 276–291. Springer, Heidelberg (2005)
19. Thiffault, C., Bacchus, F., Walsh, T.: Solving non-clausal formulas with DPLL search. In: Wallace, M. (ed.) CP 2004. LNCS, vol. 3258, pp. 663–678. Springer, Heidelberg (2004)
20. Tseitin, G.: On the complexity of proofs in propositional logics. Seminars in Mathematics 8 (1970)
21. Zhang, L.: On subsumption removal and on-the-fly CNF simplification. In: Bacchus, F., Walsh, T. (eds.) SAT 2005. LNCS, vol. 3569, pp. 482–489. Springer, Heidelberg (2005)
22. Zhang, L., Malik, S.: Conflict driven learning in a quantified Boolean satisfiability solver. In: Proc. of International Conference on Computer Aided Design, ICCAD 2002 (2002)
23. Zhang, L., Malik, S.: Towards a symmetric treatment of satisfaction and conflicts in quantified Boolean formula evaluation. In: Van Hentenryck, P. (ed.) CP 2002. LNCS, vol. 2470, p. 200. Springer, Heidelberg (2002)

Non Uniform Selection of Solutions for Upper Bounding the 3-SAT Threshold

Yacine Boufkhad and Thomas Hugel

LIAFA - Université Denis Diderot Paris 7 - CNRS*
Case 7014
F-75205 Paris Cedex 13
{boufkhad,thomas.hugel}@liafa.jussieu.fr

Abstract. We give a new insight into the upper bounding of the 3-SAT threshold by the first moment method. The best criteria developed so far to select the solutions to be counted discriminate among neighboring solutions on the basis of *uniform* information about each individual free variable. What we mean by *uniform* information, is information which does not depend on the solution: e.g. the number of positive/negative occurrences of the considered variable. What is new in our approach is that we use *non uniform* information about variables. Thus we are able to make a more precise tuning, resulting in a slight improvement on upper bounding the 3-SAT threshold for various models of formulas defined by their distributions.

1 Introduction

We consider the phase transition phenomenon that occurs in some random satisfiability problems, where the probability of satisfiability for a random formula suddenly goes from 1 to 0 at a given ratio $\frac{\#\text{clauses}}{\#\text{variables}}$. It was first experimentally observed that this transition would occur at a ratio near 4.25 for the standard 3-SAT model (see [1]). The same kind of transition was also observed in some variants of the standard model, e.g. when occurrences and signs of variables are balanced (see [2]).

The first important step towards the quest of the threshold is the work of Friedgut and Bourgain [3] establishing that the width of the transition window tends to zero as the number of variables tends to infinity.

An important breakthrough was then made by Achlioptas and Peres [4]: using a sophisticated technique based on the second moment method they located asymptotically the threshold of k-SAT for large constant k at $2^k \ln 2 - O(k)$. However in the particular case of 3-SAT, there remains a gap between established lower and upper bounds.

The cornerstone method used for 25 years in order to establish upper bounds of the 3-SAT threshold is the so called first moment method. Indeed we are

---

* This work was partially supported by GANG project of INRIA.

O. Strichman and S. Szeider (Eds.): SAT 2010, LNCS 6175, pp. 99–112, 2010.

interested in the probability that a formula has some solutions, but that probability is currently out of reach of human-tractable calculations; however the moments under this probability are much easier to estimate. The first moment method consists in bounding the probability we are interested in by the first moment of a certain quantity X under this probability. The simplest quantity X one can imagine as a candidate for the first moment method is the number of solutions. This gives an upper bound of 5.191 [5], which is far above the experimentally observed threshold at around 4.25. There has been ever since lots of efforts [6,7,8,9,10] intended to lower this upper bound by removing as many solutions as possible from the counted quantity X, the only requirement of the first moment method being to count at least 1 solution whenever a formula is satisfiable; thus the technique is to count only particular solutions, designed to be present whenever there is a solution, and not too complicated to count.

We obtain some new upper bounds in a variety of models of 3-CNF formulas (which we introduce later in section 2.1). In the particular case of the standard model we get an upper bound of 4.500. We must mention here the work of Díaz et al. [11]; gathering the technique of [10,12] with a pure literal elimination and a filtering on the typicality of clauses, they got an upper bound of 4.490. The fact is that our new technique is quite compatible with the pure literal elimination and the filtering on the typicality of clauses, but we only aim at emphasizing the positive effect of our new technique for selecting solutions, by comparing it to previous analogous techniques in several models of formulas.

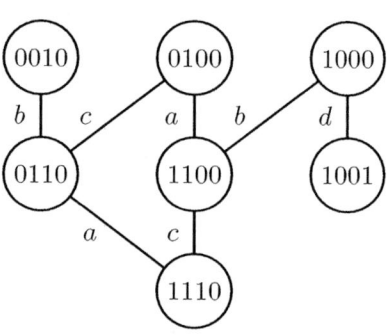

Fig. 1. Graph of solutions for formula F. The label of an edge is the name of the variable differing between both solutions.

The best implementations of the first moment method approximating the threshold of 3-SAT use local relationships between solutions, which involves solutions agreeing on the values of all variables but a constant number of them, in general one variable [8] or two [9].

We shall consider the set of solutions with local relationship as a graph which nodes are the solutions and an edge exists between two solutions if and only if both solutions agree on the values of all variables except one. Each edge will be labelled by the variable differing between both solutions.

For example the formula

$$\Phi = \left\{ a \vee b \vee c, a \vee c \vee \overline{d}, a \vee \overline{c} \vee \overline{d}, a \vee \overline{b} \vee \overline{d}, \overline{b} \vee c \vee \overline{d}, \overline{a} \vee \overline{b} \vee \overline{d}, \overline{a} \vee b \vee \overline{c} \right\}$$

has 7 solutions that can be represented by the non oriented graph of figure 1.

The techniques used so far amount to making an acyclic orientation of the above graph and to counting only the minimal solutions (those that do not have outgoing edges). The least is the number of minimal solutions the best is the upper bound obtained. In general, any graph can be oriented so as to obtain only one minimal element for every connected component (e.g. by a depth first search), but this orientation is obtained thanks to a sophisticated algorithm that is aware of the whole graph while in our case, the orientation must be decided locally.

The very first orientation [8,9] consisted in orienting an edge from the solution where the label variable is assigned 0 to the one where it is 1 regardless of which variable is considered. Later, in [12,11], an edge is oriented towards the value that makes true the most literals and this can be known thanks to the syntactic property of the number of occurrences of each variable in the formula. In both these types of orientation, the edges having the same labels are oriented the same way (e.g. from 0 to 1) anywhere in the graph. So we call such orientations *uniform* (see Figure 2(a)).

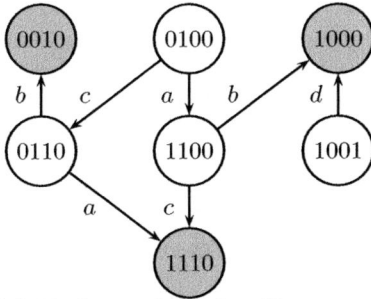

(a) Uniform orientation. For example b has 2 positive occurrences and 3 negative ones, so every edge labeled by b is oriented from 1 to 0.

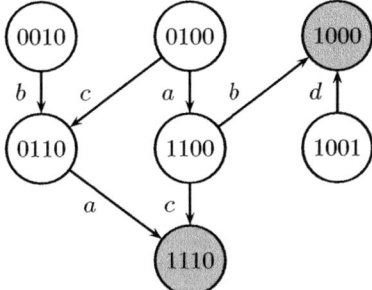

(b) Non uniform orientation, obtained in this example by minimizing $4\beta_1 + 2\beta_2 + \beta_3$ (see definition in section 2). Both edges labeled by b are oriented differently (i.e. from 0 to 1 as well as from 1 to 0).

Fig. 2. Two different orientations for the solutions of formula Φ. Minimal solutions are in gray.

The orientation that we use in this paper is less rigid: two edges labelled with the same variable can be oriented differently depending on the solutions involved (that is what we call *non uniform* orientation, see Figure 2(b)). Indeed we keep track of a set of 5 numbers associated with each variable and use it to discriminate among neighboring solutions. These 5 numbers provide information on the repartition of true and false occurrences of each variable in each type of clauses (clauses having 1, 2 or 3 true literals). Our intuition is that we should select solutions in which the least occurrences of true literals are *critical*. The less a clause has true literals, the more its true literals are critical. Such a property is by nature non uniform.

We develop our technique in a general framework allowing us to apply it to a wide variety of 3-CNF models of formulas defined by their distributions; thus we derive new bounds for some known models of formulas [2]. The existence of other non uniform orientations that may give a smaller number of minimal elements and then better bounds remains to be investigated.

In section 2 we present our framework and four different models of formulas; in section 3 we show how we make our non uniform selection of solutions, and sum up the bounds we obtain for each model. We give details on the calculation of the first moment and its constraints in section 4, as well as some hints on what led us to the weights we took for our non uniform selection.

2   Definitions and Notations

We consider a generic random model of 3-CNF formulas having n variables and cn clauses. Models are parametrized by a probability distribution $(d_{p,q})_{p,q \in \mathbf{N}}$ such that $\sum_{p,q \in \mathbf{N}} d_{p,q} = 3c$. In each model a satisfiability threshold will appear for a specific value of c we want to estimate. Before we get formulas we draw *configurations* as follows:

1. each of the n variables is given p labelled positive occurrences and q labelled negative occurrences in a way that the overall proportion of variables with p positive occurrences and q negatives occurrences is $d_{p,q}$;
2. a configuration can be seen as a matrix of $3cn$ bins containing literals occurrences; the repartition of literals into the $3cn$ bins is drawn uniformly among all $(3cn)!$ permutations of labelled literals occurrences.

A legal *formula* is a configuration where occurrences are unlabelled and each clause contains at most one occurrence of each variable. For the models we consider in this paper and described in section 2.1, it was shown that an upper bound on the satisfiability threshold obtained for configurations also applies to legal formulas (see [11] for the standard model and [2] for models where p and q are bounded). So we shall work on configurations all along this paper.

2.1   Overview of Models

Standard Model: all literals are drawn uniformly and independently; it was shown in [12,11] that the resulting distribution is the 2D Poisson distribution: $d_{p,q} = \binom{p+q}{p} \frac{e^{-3c}}{(p+q)!} \left(\frac{3c}{2} \right)^{p+q}$.

By analogy with the standard model we now define several other models where we force an equilibrium between variables occurrences and/or signs. These can be seen as *regular* variants of 3-SAT (just like regular graphs). The equilibrium cannot be perfect because of parity or truncation reasons, but we circumvent it as follows. Of course one can check that all of these distributions sum up to 1 and have an average of $3c$.

Model with Almost Balanced Signs: every variable appear with (almost) the same number of positive and negative occurrences; we define $d_{p,q}$ by $d_{p,p} = \frac{e^{-3c}(3c)^{2p}}{(2p)!}$ and $d_{p+1,p} = d_{p,p+1} = \frac{1}{2} \frac{e^{-3c}(3c)^{2p+1}}{(2p+1)!}$ (and zero elsewhere).

Model with Almost Balanced Occurrences: every variable appear with (almost) the same number occurrences; let $t^* = \lfloor 3c \rfloor$ and $r^* = 3c - t^*$; we define $d_{p,q}$ by $d_{p,t^*-p} = (1 - r^*) \frac{\binom{t^*}{p}}{2^{t^*}}$ and $d_{p,t^*+1-p} = r^* \frac{\binom{t^*+1}{p}}{2^{t^*+1}}$ (and zero elsewhere).

Model with Almost Balanced Signs and Occurrences: every variable appear with (almost) the same number occurrences and have strictly the same number of positive as negative occurrences (this model was examined in [2]); let $p^* = \lfloor \frac{3c}{2} \rfloor$ and $r^* = \frac{3c}{2} - p^*$. We define $d_{p,q}$ by $d_{p^*,p^*} = 1 - r^*$ and $d_{p^*+1,p^*+1} = r^*$ (and zero elsewhere).

2.2   Types of Clauses and Variables

Our selection method is based on different types of clauses: given any assignment, we call clause of *type t* a clause having t true literals under this assignment, and β_t the proportion of clauses of type t.

Moreover we want to have some control on the number of occurrences of variables in the different types of clauses; to do so we need 6 numbers per variable, so we say that a variable is of *type* (i, j, k, l, m, v) if it is assigned v and has:

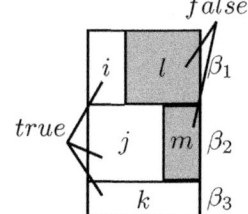

i true occurrences in clauses of type 1;
j true occurrences in clauses of type 2;
k true occurrences in clauses of type 3;
l false occurrences in clauses of type 1;
m false occurrences in clauses of type 2;

Remark 1. For each variable we have $i + j + k = p$ and $l + m = q$ or vice versa (according to the value v assigned to the variable).

Then we put some weights onto the solutions as follows: in a given solution each variable of type (i, j, k, l, m, v) receives a weight $\omega_{i,j,k,l,m,v}$. The weight of a solution will be the product of the weights of all variables. It turns out that in the end we shall take binary weights, yielding in fact an orientation between solutions. We explain the choice of the weights in sections 3 and 4.4. Then we apply the first moment method to the random variable X equal to the sum of the weights of the solutions.

3   Selection of Solutions

Let us recall how the first moment method works: we want to show that $\Pr(Y \geq a)$ is small but we don't have access to $\Pr(Y \geq a)$. Instead we use some EX. It suffices then to ensure that $\Pr(Y \geq a) \leq EX$. For our problem 3-SAT, Y is the number of solutions, $a = 1$ and X is the total weight on the solutions. Since $X \geq 0$, Markov's inequality yields that $\Pr(X \geq 1) \leq EX$; so if we choose X such that $Y \geq 1$ implies $X \geq 1$, we have $\Pr(Y \geq 1) \leq \Pr(X \geq 1) \leq EX$.

Then our goal will be to tune the weights so that $EX \to 0$ for the least ratio $c = \frac{\#\text{clauses}}{\#\text{variables}}$.

3.1   Construction of a Correct Weighting Scheme

Of course we must put some constraints onto the weights in order that the weighting scheme can be correct for the first moment method: namely the sum of the weights of the solutions of a satisfiable formula must be at least 1. However the constraints we choose here might not be necessary for the first moment method to hold.

Let us recall that given a solution, a variable is called *free* when the assignment obtained by inverting its value (0/1) remains a solution. Thus in our framework, a variable is free iff its i number is 0. How does the tuple $(0, j, k, l, m, v)$ for a free variable x behave when the value v is inverted to $1 - v$? $i(x) \leftarrow 0$, $j(x) \leftrightarrow l(x), k(x) \leftrightarrow m(x)$ and $v(x) \leftarrow 1 - v(x)$.

1. the first constraint we put is that $\omega_{i,j,k,l,m,v} = 1$ as soon as $i \geq 1$; that is, we put significant weights only onto free variables. The reason for this is that free variables allow to move between solutions.
2. the second constraint is that

$$\omega_{0,j,k,l,m,v} + \omega_{0,l,m,j,k,1-v} = 1 \; ; \tag{1}$$

   that is, the sum of the weights of a free variable in a couple of solutions differing only on that variable is 1. We impose this condition by analogy with the conditions on weights given in [13].

As suggested by the analysis given in section 4.4, we shall take $\omega_{0,j,k,l,m,v} = \mathbf{1}_{P(j,k,l,m,v)}$ for a certain predicate $P(j, k, l, m, v)$ linked with the sign of $\alpha_1 \rho_{j,l} + \alpha_3 \rho_{k,m}$ (where α_1 and α_3 are any real constants and ρ is an operator defined as $\rho_{a,b} = a - b$).

The fact that we imposed $\omega_{0,j,k,l,m,v} + \omega_{0,l,m,j,k,1-v} = 1$ tells us that given a solution and a free variable x at the value v, the predicate P is satisfied by x at the value v or (exclusively) by x at the value $1 - v$. Thus we are able to define an orientation between neighboring solutions.

Let us say that variable x is *obedient* when P is satisfied. We put an arc between 2 solutions differing only on 1 (free) variable x from the solution S_d (where x is disobedient) to the solution S_o (where x is obedient), and we call that relation $S_d > S_o$. The notation $>$ is not randomly chosen.

Namely our weighting scheme counts 1 for a solution when it does not have any disobedient free variables, and 0 otherwise; but what can ensure that whenever there is a solution, there is also a solution where all free variables are obedient? It suffices that the relation $>$ is circuit-free. Then the transitive closure of $>$ is an order, and we are precisely counting the minimal solutions in that order. Minimal solutions exist because the set of all solutions is finite. So let us see how we can make the relation $>$ circuit-free.

Recapitulation of Existing Methods

All Solutions: This method consists in computing the first moment on all solutions: $P(j, k, l, m, v) \equiv 1$.

Negatively Prime Solutions (NPS): This method consists in counting only solutions which free variables are assigned 1. That is $P(j, k, l, m, v) \equiv v > 0$. This method was introduced in [8].

NPS with Imbalance: This method was introduced in [12] and combined to some other ingredients in [11]. This method consists in allowing free variables to take only a value such that the number of true occurrences is larger than the number of negative occurrences of this variable (and in case of equality, ties are broken in favor of the value 1). In other words $P(j, k, l, m, v) \equiv (\rho_{j,l} + \rho_{k,m}, v) >_{\text{lex}} (0, 0)$, where $>_{\text{lex}}$ denotes the lexicographical order.

Our Method. May we choose arbitrary real coefficients α_1 and α_3 in the expression of $\alpha_1 \rho_{j,l} + \alpha_3 \rho_{k,m}$ in order that the first moment method should hold? It turns out that it is the case, and here is a proof of it.

We make the following observation: how does the population of the 3 different types of clauses evolve when a free variable x is flipped? $\beta_1 + = \rho_{j,l}(x)$, $\beta_2 + = (\rho_{k,m} - \rho_{j,l})(x)$ and $\beta_3 + = -\rho_{k,m}(x)$.

Thus $\alpha_1 \rho_{j,l} + \alpha_3 \rho_{k,m}$ is the variation of $\alpha_1 \beta_1 - \alpha_3 \beta_3$; so we may define our predicate P in the following way: $P(j, k, l, m, v) \equiv (\alpha_1 \rho_{j,l} + \alpha_3 \rho_{k,m}, v) >_{\text{lex}} (0, 0)$; thanks to v we break ties when $\alpha_1 \rho_{j,l} + \alpha_3 \rho_{k,m} = 0$, so that the underlying relation $>$ between solutions is circuit-free: namely going from S_d to S_o when $S_d > S_o$ strictly increases $(-\alpha_1 \beta_1 + \alpha_3 \beta_3, v)$ for $>_{\text{lex}}$.

Moreover the exclusion between $P(j, k, l, m, v)$ and $P(l, m, j, k, 1 - v)$ is satisfied, which means that whenever there is a solution with a disobedient free variable, it suffices to flip the value of this variable so that it becomes obedient. We investigated the best ratio between α_1 and α_3 by numerical experiments.

3.2 Summary of Results

As one can see in table 1, our method yields in all models a slight improvement on the bounds obtained by former methods. Note that for some models there is a range of values for α which give the same upper bound.

In the model where signs as well as occurrences are balanced, the method of NPS+imbalance is of course the same as the method of NPS, whereas our method is somewhat better than the method of NPS.

The bound we obtain in the standard model is 4.500; this is not better than the bound of 4.490 obtained by Díaz *et al.* in [11]. Their calculation adds 2 ingredients to the method of [12]: typicality of clauses and elimination of pure literals. These 2 ingredients might be combined to our approach to improve on the 4.490, but this would involve too complicated calculations with respect to the expected improvement. However in models where signs are balanced it is irrelevant to eliminate pure literals.

Table 1. Summary of our results

model	standard	almost balanced signs	almost balanced occurrences	almost balanced signs and occurrences
all solutions	5.040	3.858	5.046	3.783
NPS $v > 0$	4.552	3.521	4.662	3.548
NPS+imbalance $(\rho_{j,l} + \rho_{k,m}, v) > (0,0)$	4.506	3.514	4.628	3.548
our method $(\alpha\rho_{j,l} + \rho_{k,m}, v) > (0,0)$	4.500	3.509	4.623	3.546
our α	$\alpha = 2.00$	$1.01 \leq \alpha \leq 1.16$	$2.01 \leq \alpha \leq 2.24$	$\alpha \geq 1.01$

4 The First Moment Method

4.1 Types of Variables

We split the set of variables into several sets and subsets of variables. In order to be able to match the original random 3-CNF model of formulas where all literals are drawn independently, we should consider p and q to range in $\mathbf{N}$. For convenience of our forthcoming maximization, we only take into account bounded values of p and q. So we are going to consider 2 kinds of variables, according to their numbers of occurrences. We follow the notations of [11]. We denote by M some integer whose value will be determined according to the required accuracy of the calculations; in practice we shall take $M = 21$. M enables us to define 2 kinds of variables:

1. the set of light variables, that is variables which indices are in the set $\mathcal{L} = \{(p, q) \in \mathbf{N}^2, p \leq M \wedge q \leq M \wedge d_{p,q} > 0\}$; they are the most important variables since almost all variables are light in the models we consider; we call $\delta_{p,q}$ the proportion of light variables having p positive occurrences, q negative occurrences, and assigned 1. As a further refinement, we call $\pi_{i,j,k,l,m,v}$ the proportion of variables of type (i, j, k, l, m, v) whose corresponding weight $\omega_{i,j,k,l,m,v}$ is non zero, and omit the other ones because we shall need all active $\pi_{i,j,k,l,m,v}$ to be non zero. To connect $\pi_{i,j,k,l,m,v}$'s with $\delta_{p,q}$'s we introduce the following set of tuples of integers: $Q_{p,q} = \{(i, j, k, l, m) \subset \mathbf{N}^5, i + j + k = p \wedge l + m = q\}$; thus we have

$$\sum_{(i,j,k,l,m)\in Q_{p,q}} \pi_{i,j,k,l,m,1} = \delta_{p,q} \; ; \tag{2}$$

$$\sum_{(i,j,k,l,m)\in Q_{q,p}} \pi_{i,j,k,l,m,0} = d_{p,q} - \delta_{p,q} \; . \tag{3}$$

Note that equality 3 involves $Q_{q,p}$ whereas equality 2 involves $Q_{p,q}$.

2. the set of heavy variables, that is all other variables; their indices are thus in the set $\mathcal{H} = \{(p,q) \in \mathbf{N}^2, p > M \vee q > M \vee d_{p,q} = 0\}$; we weaken the notion of satisfiability by considering that heavy variables are always satisfied, regardless of their signs and values. Doing so is harmless for the validity of the first moment method because we can only increase the number of solutions. In other words we are going to consider heavy variables as undistinguishable members of a tote bag. We call τ the global scaled number of heavy variables: $\tau = \sum_{(p,q)\in\mathcal{H}} d_{p,q}$.

We also need to distinguish some types of occurrences of heavy variables. We call H the global scaled number of occurrences of heavy variables: $H = \sum_{(p,q)\in\mathcal{H}} (p+q)\, d_{p,q} = 3c - \sum_{(p,q)\in\mathcal{L}} (p+q)\, d_{p,q}$. According to the types of clauses where occurrences appear, H is divided into H_t's, where H_t is the scaled number of occurrences of heavy variables in clauses of type t.

We are now ready to write down the expression of the first moment of X, the weight of all solutions.

4.2   Expression of the First Moment and Its Constraints

We recall that all occurrences of literals are drawn according to the distribution $d_{p,q}$ (see section 2). Thus the sample space we consider consists in the $(3cn)!$ permutations of labelled occurrences of literals, and our parameters are n, c, $d_{t,p}$, τ, H and $\omega_{i,j,k,l,m,v}$'s (although we must carefully choose the weights $\omega_{i,j,k,l,m,v}$, as explained below in section 4.4).

All other quantities: β_t, H_t, $\delta_{t,p}$ and $\pi_{i,j,k,l,m,v}$ are variables, and the first moment of X can be split up into a big sum over all variables of the product of the following factors depending on variables: number of assignments, weight of an assignment and probability for an assignment to be a solution.

1. number of assignments: each variable is assigned 0 or 1: $2^{\tau n} \prod_{(p,q)\in\mathcal{L}} \binom{d_{p,q}n}{\delta_{p,q}n}$;
2. weight of an assignment: $\prod_{(p,q)\in\mathcal{L}} \prod_{\substack{(i,j,k,l,m)\in Q_{p,q} \\ v\in\{0,1\}}} \omega_{i,j,k,l,m,v}^{\pi_{i,j,k,l,m,v} n}$;
3. probability for an assignment to be a solution: quotient of the number of satisfied configurations by the total number of configurations:
   (a) number of satisfied configurations: a configuration can be seen as a set of bins filled with occurrences of literals:
      i. each of the $3cn$ bins is first given a truth value: there are $\binom{cn}{\beta_1 cn, \beta_2 cn, \beta_3 cn} 3^{(\beta_1+\beta_2)cn}$ possibilities, and the following constraint appears:

$$\beta_1 + \beta_2 + \beta_3 = 1 \ . \tag{4}$$

   ii. each light literal is given a tuple (i,j,k,l,m) consistently with $d_{p,q}$ and $\delta_{p,q}$. This gives a series of constraints:

$$\sum_{(i,j,k,l,m)\in Q_{p,q}} \pi_{i,j,k,l,m,1} + \sum_{(i,j,k,l,m)\in Q_{q,p}} \pi_{i,j,k,l,m,0} = d_{p,q} \ . \tag{5}$$

Note that $\delta_{p,q} = \sum_{(i,j,k,l,m)\in Q_{p,q}} \pi_{i,j,k,l,m,1}$. Thus, given a family $(\pi_{i,j,k,l,m,v})$, there are

$$\prod_{(p,q)\in\mathcal{L}} \binom{\delta_{p,q}n}{\cdots \pi_{i,j,k,l,m,1}n \cdots}_{(i,j,k,l,m)\in Q_{p,q}}$$
$$\cdot \prod_{(p,q)\in\mathcal{L}} \binom{(d_{p,q}-\delta_{p,q})n}{\cdots \pi_{i,j,k,l,m,0}n \cdots}_{(i,j,k,l,m)\in Q_{q,p}}$$

possible allocations. Moreover the following constraints appear, so that all occurrences of literals can fit into the destined types of clauses:

$$\sum_{\substack{(p,q)\in\mathcal{L}\\(i,j,k,l,m)\in Q_{p,q}\\v\in\{0,1\}}} i\pi_{i,j,k,l,m,v} + H_1 = \beta_1 c \; ; \tag{6}$$

$$\sum_{\substack{(p,q)\in\mathcal{L}\\(i,j,k,l,m)\in Q_{p,q}\\v\in\{0,1\}}} j\pi_{i,j,k,l,m,v} + H_2 = 2\beta_2 c \; ; \tag{7}$$

$$\sum_{\substack{(p,q)\in\mathcal{L}\\(i,j,k,l,m)\in Q_{p,q}\\v\in\{0,1\}}} k\pi_{i,j,k,l,m,v} + H_3 = 3\beta_3 c \; ; \tag{8}$$

$$\sum_{\substack{(p,q)\in\mathcal{L}\\(i,j,k,l,m)\in Q_{p,q}\\v\in\{0,1\}}} l\pi_{i,j,k,l,m,v} = 2\beta_1 c \; ; \tag{9}$$

$$\sum_{\substack{(p,q)\in\mathcal{L}\\(i,j,k,l,m)\in Q_{p,q}\\v\in\{0,1\}}} m\pi_{i,j,k,l,m,v} = \beta_2 c \; . \tag{10}$$

iii. all occurrences of light variables are allocated to the 5 regions:
$\prod_{\substack{(p,q)\in\mathcal{L}\\(i,j,k,l,m)\in Q_{p,q}\\v\in\{0,1\}}} \left(\binom{i+j+k}{i,j,k}\binom{l+m}{l,m}\right)^{\pi_{i,j,k,l,m,v}n}$ allocations are possible;

iv. all occurrences of heavy variables are allocated to the 3 satisfied regions, which yields $\binom{Hn}{H_1n,H_2n,H_3n}$ possible allocations; and we must add the following constraint:

$$H_1 + H_2 + H_3 = H \; . \tag{11}$$

v. all permutations of occurrences of literals are possible inside the 5 regions: their number is $(\beta_1 cn)! \, (2\beta_2 cn)! \, (3\beta_3 cn)! \, (2\beta_1 cn)! \, (\beta_2 cn)!$;

(b) total number of configurations: the occurrences of literals can be in any order: $(3cn)!$ permutations are possible.

We denote by $\mathcal{P}$ the set of all families ζ of non negative numbers

$$\left(\left(\pi_{i,j,k,l,m,v} \right)_{\substack{(p,q)\in\mathcal{L}\\(i,j,k,l,m)\in Q_{p,q}\\v\in\{0,1\}}} , (H_1, H_2, H_3), (\beta_1, \beta_2, \beta_3) \right) \tag{12}$$

satisfying the above constraints; note that $\mathcal{P}$ is convex (by linearity of constraints). We denote by $\mathcal{I}(n)$ the intersection of $\mathcal{P}$ with the multiples of $\frac{1}{n}$; we get the following expression of the first moment: $EX = \sum_{\zeta\in\mathcal{I}(n)} T(n)$ where

$$T(n) = 2^{\tau n} \binom{Hn}{H_1 n, H_2 n, H_3 n} \binom{cn}{\beta_1 cn, \beta_2 cn, \beta_3 cn} 3^{(\beta_1+\beta_2)cn}$$
$$\cdot \frac{(\beta_1 cn)! \, (2\beta_2 cn)! \, (3\beta_3 cn)! \, (2\beta_1 cn)! \, (\beta_2 cn)!}{(3cn)!}$$
$$\cdot \prod_{(p,q)\in\mathcal{L}} \binom{d_{p,q}n}{\delta_{p,q}n} \prod_{(p,q)\in\mathcal{L}} \binom{\delta_{p,q}n}{\ldots \pi_{i,j,k,l,m,1}n \ldots}_{(i,j,k,l,m)\in Q_{p,q}}$$
$$\cdot \prod_{(p,q)\in\mathcal{L}} \binom{(d_{p,q}-\delta_{p,q})n}{\ldots \pi_{i,j,k,l,m,0}n \ldots}_{(i,j,k,l,m)\in Q_{q,p}}$$
$$\cdot \prod_{\substack{(p,q)\in\mathcal{L}\\(i,j,k,l,m)\in Q_{p,q}\\v\in\{0,1\}}} \left(\omega_{i,j,k,l,m,v} \binom{i+j+k}{i,j,k}\binom{l+m}{l,m} \right)^{\pi_{i,j,k,l,m,v}n} . \tag{13}$$

We get rid of all factorials thanks to the following Stirling's inequalities due to Batir [14]: $\left(\frac{k}{e}\right)^k \sqrt{2\pi\left(k+\frac{1}{6}\right)} < k! < \left(\frac{k}{e}\right)^k \sqrt{2\pi\left(k+\left(\frac{e^2}{2\pi}-1\right)\right)}$.

The boundedness of the set $\mathcal{L}$ of light variables (and thus the boundedness of the sets $Q_{p,q}$) allows to write that $T(n) \leq \mathrm{poly}_1(n) F^n$ where

$$F = 2^\tau \frac{H^H}{H_1^{H_1} H_2^{H_2} H_3^{H_3}} \left(\frac{1}{3} (2\beta_1)^{\beta_1} (2\beta_2)^{\beta_2} (3\beta_3)^{\beta_3} \right)^{2c}$$
$$\prod_{(p,q)\in\mathcal{L}} d_{p,q}^{d_{p,q}} \prod_{\substack{(p,q)\in\mathcal{L}\\(i,j,k,l,m)\in Q_{p,q}\\v\in\{0,1\}}} \left(\omega_{i,j,k,l,m,v} \frac{\binom{i+j+k}{i,j,k}\binom{l+m}{l,m}}{\pi_{i,j,k,l,m,v}} \right)^{\pi_{i,j,k,l,m,v}} . \tag{14}$$

Once again, by the lightness property, $\mathcal{I}(n)$ consists of a bounded number of variables, each of which can take at most $n+1$ values (as a multiple of $\frac{1}{n}$ ranging between 0 and 1). It follows that the size of $\mathcal{I}(n)$ is bounded by a polynomial $\mathrm{poly}_2(n)$. And since $\mathcal{I}(n) \subseteq \mathcal{P}$, we have $EX \leq \mathrm{poly}_2(n) \, \mathrm{poly}_1(n) \, (\max_{\zeta\in\mathcal{P}} F)^n$.

4.3   Maximization of ln F

This is the technical part of our work. We mainly use the same techniques as [11]. For lack of space here we only give the sketch of our proof, but the details are available in [15].

1. In order to maximize ln F under our constraints, we use the standard Lagrange multipliers technique. This is appendix A of [15]. The following equations come from the Lagrange derivations and are important for our study:

$$\pi_{i,j,k,l,m,1} = \omega_{i,j,k,l,m,1} \binom{i+j+k}{i,j,k} \binom{l+m}{l,m} r_{i+j+k,l+m} x_1^{2i} x_2^j y_1^l y_2^{2m} \quad (15)$$

$$\pi_{i,j,k,l,m,0} = \omega_{i,j,k,l,m,0} \binom{i+j+k}{i,j,k} \binom{l+m}{l,m} r_{l+m,i+j+k} x_1^{2i} x_2^j y_1^l y_2^{2m} \quad (16)$$

x_1, x_2, y_1 and y_2 are Lagrange multipliers, that is positive numbers; moreover $r_{p,q}$ is defined as follows:

$$r_{p,q} = \frac{d_{p,q}}{A_{p,q}} \; ; \quad\quad\quad\quad\quad\quad\quad\quad\quad\quad (17)$$

$$A_{p,q} = \sum_{(i,j,k,l,m)\in Q_{p,q}} \omega_{i,j,k,l,m,1} \binom{p}{i,j,k} \binom{q}{l,m} x_1^{2i} x_2^j y_1^l y_2^{2m}$$

$$+ \sum_{(i,j,k,l,m)\in Q_{q,p}} \omega_{i,j,k,l,m,0} \binom{q}{i,j,k} \binom{p}{l,m} x_1^{2i} x_2^j y_1^l y_2^{2m} \; . \quad (18)$$

2. In order to justify the use of this technique we must show that the function ln F does not maximize on the boundary of the polytope of constraints; to do so we show that starting at a boundary point there is always a "good" direction inside the polytope which makes ln F greater. This is appendix B of [15].
3. Finally we must ensure that the solution we found by the Lagrange multiplier technique is indeed a global maximum; to do so we make a sweep over different values of the parameters β_t; indeed when these β_t are fixed the function ln F is strictly concave relative to the remaining variables, thus easier to maximize. This is appendix C of [15].

4.4   Minimization of Global Weight

Let us see how one can minimize F (or equivalently ln F) by a good choice of the weights. The following reasoning is not rigorous; we only aim at giving some hints to explain the choice of the weights we made in section 3.

Remember that F is given by equation 14. We want to minimize ln F by tuning the weights $\omega_{0,j,k,l,m,v}$, so we are going to differentiate ln F with respect to an individual $\omega_{0,j,k,l,m,1}$. Of course due to the constraints every variable depend on $\omega_{0,j,k,l,m,1}$ in the process of maximizing ln F under these constraints. But we consider that the variations on all variables are negligible except for $\pi_{0,j,k,l,m,1}$ (because of equation 15) and $\pi_{0,l,m,j,k,0}$ (because of equations 16 and 1), so we can write:

$$\frac{\partial(\ln F)}{\partial\omega_{0,j,k,l,m,1}} \simeq \frac{\partial(\ln F)}{\partial\pi_{0,j,k,l,m,1}}\frac{\partial\pi_{0,j,k,l,m,1}}{\partial\omega_{0,j,k,l,m,1}} + \frac{\partial(\ln F)}{\partial\pi_{0,l,m,j,k,0}}\frac{\partial\pi_{0,l,m,j,k,0}}{\partial\omega_{0,j,k,l,m,1}} . \quad (19)$$

Using equations 15, 16 and 1 we find that:

$$\frac{\partial(\ln F)}{\partial\omega_{0,j,k,l,m,1}} \simeq -\binom{j+k}{j,k}\binom{l+m}{l,m}r_{j+k,l+m}x_2^j y_1^l y_2^{2m} \ln\left(r_{j+k,l+m}x_2^j y_1^l y_2^{2m}\right)$$
$$+\binom{j+k}{j,k}\binom{l+m}{l,m}r_{j+k,l+m}x_2^l y_1^j y_2^{2k} \ln\left(r_{j+k,l+m}x_2^l y_1^j y_2^{2k}\right) \quad (20)$$

Now due to equations 17 and 18 and numerical experiments we make the following approximations: $r_{j+k,l+m}x_2^j y_1^l y_2^{2m} \ll 1$ and $r_{j+k,l+m}x_2^l y_1^j y_2^{2k} \ll 1$. As the function $x \mapsto x\ln(ax)$ is strictly decreasing between 0 and $\frac{1}{ea}$, we can infer the following property: $\frac{\partial(\ln F)}{\partial\omega_{0,j,k,l,m,1}} > 0$ iff $x_2^l y_1^j y_2^{2k} < x_2^j y_1^l y_2^{2m}$, i.e. $\left(\frac{y_1}{x_2}\right)^{j-l}\left(y_2^2\right)^{k-m} < 1$.

Now let us consider we are at the minimum point of $\ln F$. If $\frac{\partial\ln(F)}{\partial\omega_{0,j,k,l,m,1}} \neq 0$, then $\omega_{0,j,k,l,m,1}$ must be at the boundary, i.e. 0 or 1.

$\frac{\partial(\ln F)}{\partial\omega_{0,j,k,l,m,1}} > 0$ iff $\alpha_1\rho_{j,l} + \alpha_3\rho_{k,m} < 0$, where $\alpha_1 = \ln\frac{y_1}{x_2}$ and $\alpha_3 = \ln\left(y_2^2\right)$. Thus:

1. if $\alpha_1\rho_{j,l} + \alpha_3\rho_{k,m} < 0$, then $\omega_{0,j,k,l,m,1} = 0$;
2. if $\alpha_1\rho_{j,l} + \alpha_3\rho_{k,m} > 0$, then $\omega_{0,j,k,l,m,1} = 1$;
3. if $\alpha_1\rho_{j,l} + \alpha_3\rho_{k,m} = 0$, nothing can be said about $\omega_{0,j,k,l,m,1}$.

What about $\omega_{0,j,k,l,m,0}$?

1. if $\alpha_1\rho_{j,l} + \alpha_3\rho_{k,m} < 0$, then $\alpha_1\rho_{l,j} + \alpha_3\rho_{m,k} > 0$, thus $\omega_{0,l,m,j,k,1} = 1$, so $\omega_{0,j,k,l,m,0} = 0$;
2. if $\alpha_1\rho_{j,l} + \alpha_3\rho_{k,m} > 0$, then by the same argument, $\omega_{0,j,k,l,m,0} = 1$;
3. if $\alpha_1\rho_{j,l} + \alpha_3\rho_{k,m} = 0$, nothing can be said about $\omega_{0,j,k,l,m,0}$.

5   Conclusion

We hope that the new track we opened will help gain some more insight and some more decimals in the quest of the 3-SAT threshold. In particular note that we required the relation $>$ between solutions to be circuit-free although this might not be necessary; indeed we only used the fact that this relation had at least one minimal element. The same remark holds for the constraints we put onto the weights of two neighboring solutions as introduced in equation 1, since this might be too strong. Thus there may be better orientations or weighting schemes than ours.

References

1. Mitchell, D., Selman, B., Levesque, H.: Hard and easy distributions of SAT problems. In: Proceedings of the National Conference on Artificial Intelligence, pp. 459–459. AAAI, Menlo Park (1992)
2. Boufkhad, Y., Dubois, O., Interian, Y., Selman, B.: Regular Random k-{SAT}: Properties of Balanced Formulas. J. Autom. Reasoning 35(1-3), 181–200 (2005)
3. Friedgut, E., Bourgain, J.: Sharp thresholds of graph properties, and the k-sat problem. Journal of the American Mathematical Society 12(4), 1017–1054 (1999)
4. Achlioptas, D., Peres, Y.: The Threshold for Random k-SAT is 2^k ln2 - O(k). JAMS: Journal of the American Mathematical Society 17, 947–973 (2004)
5. Franco, J., Paull, M.: Probabilistic analysis of the davis putnam procedure for solving the satisfiability problem. Discrete Appl. Math. 5, 77–87 (1983)
6. Maftouhi, A.E., de la Vega, W.F.: On random 3-sat. Combinatorics, Probability & Computing 4, 189–195 (1995)
7. Kamath, A., Motwani, R., Palem, K., Spirakis, P.: Tail bounds for occupancy and the satisfiability threshold conjecture. Random Structures and Algorithms 7(1), 59–80 (1995)
8. Dubois, O., Boufkhad, Y.: A General Upper Bound for the Satisfiability Threshold of Random r-{SAT} Formulae. J. Algorithms 24(2), 395–420 (1997)
9. Kirousis, L.M., Kranakis, E., Krizanc, D., Stamatiou, Y.C.: Approximating the unsatisfiability threshold of random formulas. Random Structures and Algorithms 12(3), 253–269 (1998)
10. Dubois, O., Boufkhad, Y., Mandler, J.: Typical random 3-SAT formulae and the satisfiability threshold. In: Proceedings of the eleventh annual ACM-SIAM symposium on Discrete algorithms, pp. 126–127. Society for Industrial and Applied Mathematics (2000)
11. Díaz, J., Kirousis, L., Mitsche, D., Pérez-Giménez, X.: On the satisfiability threshold of formulas with three literals per clause. Theoretical Computer Science 410(30-32), 2920–2934 (2009)
12. Dubois, O., Boufkhad, Y., Mandler, J.: Typical random 3-SAT formulae and the satisfiability threshold. Technical report, ECCC (2003)
13. Ardila, F., Maneva, E.N.: Pruning processes and a new characterization of convex geometries. Discrete Mathematics 309(10), 3083–3091 (2009)
14. Batir, N.: Inequalities for the gamma function. Archiv der Mathematik 91(6), 554–563 (2008)
15. Hugel, T., Boufkhad, Y.: Non Uniform Selection of Solutions for Upper Bounding the 3-SAT Threshold. Technical report, CoRR abs/1002.1636 (2010)

Symmetry and Satisfiability: An Update

Hadi Katebi, Karem A. Sakallah, and Igor L. Markov

EECS Department, University of Michigan
{hadik,karem,imarkov}@umich.edu

Abstract. The past few years have seen significant progress in algorithms and heuristics for both SAT and symmetry detection. Additionally, the thesis that some of SAT's intractability can be explained by the presence of symmetry, and that it can be addressed by the introduction of symmetry-breaking constraints, was tested, albeit only to a rather limited extent. In this paper we explore further connections between symmetry and satisfiability and demonstrate the existence of intractable SAT instances that exhibit few or no symmetries. Specifically, we describe a highly scalable symmetry detection algorithm based on a decision tree that combines elements of group-theoretic computation and SAT-inspired backtracking search, and provide results of its application on the SAT 2009 competition benchmarks. For SAT instances with significant symmetry we also compare SAT runtimes with and without the addition of symmetry-breaking constraints.

1 Introduction

Over the past several years a fruitful interplay developed between the algorithms for graph automorphism and those of CNF satisfiability. The initial trigger was the black-box use of the **nauty** graph automorphism and canonical labeling package [12,11] to detect the symmetries in CNF formulas. This was accomplished by encoding a CNF formula as a colored graph [5,6,3] that was processed by **nauty** to produce an irredundant set of generators for the graph's automorphism group, and hence the formula's symmetries. These symmetries were subsequently used to augment the original formula with symmetry-breaking predicates that preclude a SAT solver from redundant search in symmetric portions of the solution space. It quickly became apparent, however, that the graphs of typical CNF formulas were too large (hundreds of thousands to millions of vertices) and unwieldy for **nauty** which was more geared towards small dense graphs (hundreds of vertices). The obvious remedy, changing the data structure for storing graphs from an incidence matrix to a linked list, yielded the **saucy** system which demonstrated the viability of graph automorphism detection on very large sparse graphs [7]. Unlike **nauty**, which also solved the canonical labeling problem, **saucy** was limited to just finding an irredundant set of symmetry generators. Canonical labeling seeks to assign a unique signature to a graph that captures its structure and is invariant under all possible labelings of its vertices. The **bliss** tool [10] adopted, and improved upon, **saucy**'s sparse data structures and solved both

O. Strichman and S. Szeider (Eds.): SAT 2010, LNCS 6175, pp. 113–127, 2010.

the symmetry detection and canonical labeling problems for both small dense and large sparse graphs. Close analysis of the search trees used in **nauty** and **bliss** revealed that they were primarily designed to solve the canonical labeling problem, and that symmetry generators were detected "along the way." Both tools employed sophisticated group-theoretic pruning heuristics to narrow the search for the canonical labeling of an input graph. The detection of symmetries benefited from these pruning rules, but also helped prune the "canonical labeling" tree since labelings that are related by a symmetry (i.e., a permutation of graph vertices that preserve the graph's edge relation) yield the same signature.

The next version of the **saucy** tool [8] introduced a major algorithmic change that delinked the search for symmetries from the search for a canonical labeling. This yielded a remarkable 1000-fold improvement in run time for many large sparse graphs with sparse symmetry generators, i.e., generators that "move" only a tiny fraction of the graph's vertices. This change also made the search for symmetries resemble, at least superficially, the search for satisfying assignments by a SAT solver. In this paper we further explore the connection between symmetry detection and satisfiability to better understand and improve symmetry detection algorithms. We present the **saucy** 2.1 algorithm and highlight its key feature, namely the organization of its search for symmetries along lines similar to those of CNF satisfiability. We also present and analyze the results of applying **saucy** 2.1 on the entire suite of SAT 2009 competition benchmarks. Finally, we examine the effect of static symmetry breaking on the most challenging benchmarks in this suite.

2   Preliminaries

We assume familiarity with basic notions from group theory, including such concepts as subgroups, cosets, group generators, group action, orbit partition, etc. Most of these concepts can be found in standard textbooks on abstract algebra, e.g. [9]. We will mainly focus on the automorphism group of a colored graph, i.e., the group of vertex permutations that preserve the graph's edge relation. We assume an n-vertex graph whose vertices are labeled with the integers $\{0, 1, \cdots, n-1\}$. For the rest of the paper, we will use V to denote this set. Permutations of V are bijections from V to V and are combined by functional composition. We will use γ and η to refer to permutations and employ both tabular and cycle notation to express them. The identity permutation will be denoted as ι. When clear from context $\gamma\eta$ will mean $\gamma \circ \eta$ where $\circ$ denotes functional composition. Finally, we will denote the symmetric group on the m-element set T as $S_m(T)$. The order of $S_m(T)$ is $m!$.

An *ordered partition* $\pi = [W_1|W_2|\cdots|W_m]$ of V is an ordered list of nonempty pair-wise disjoint subsets of V whose union is V. The subsets W_i are referred to as *cells* of the partition. Ordered partition π is *unit* if $m = 1$ (i.e., $W_1 = V$) and *discrete* if $m = n$ (i.e., $|W_i| = 1$ for $i = 1, \cdots, n$). An *ordered partition pair* Π is specified as

$$\Pi = \begin{bmatrix} \pi_T \\ \pi_B \end{bmatrix} = \begin{bmatrix} T_1 \,|T_2\,|\cdots|T_m \\ B_1\,|B_2\,|\cdots|B_k \end{bmatrix}$$

with π_T and π_B referred to, respectively, as the top and bottom ordered partitions of Π. An ordered partition pair (OPP for short) Π is *isomorphic* if $m = k$ and $|T_i| = |B_i|$ for $i = 1, \cdots, m$; otherwise it is *non-isomorphic*. In other words, an OPP is isomorphic if its top and bottom partitions have the same number of cells, and corresponding cells have the same cardinality. An isomorphic OPP is *matching* if its corresponding non-singleton cells are *identical*. We will refer to an OPP as discrete (resp. unit) if its top and bottom partitions are discrete (resp. unit).

3 Implicit Representation of Permutation Sets

OPPs play a central role in the **saucy** symmetry detection algorithm we describe in this paper since they provide a compact implicit representation of sets of permutations. Specifically, a discrete OPP represents a single permutation, whereas a unit OPP represents all $n!$ permutations of V. In general, an isomorphic OPP

$$\Pi = \begin{bmatrix} T_1 & T_2 & \cdots & T_m \\ B_1 & B_2 & \cdots & B_m \end{bmatrix} \tag{1}$$

represents $\prod_{1 \le i \le n} |T_i|!$ permutations. On the other hand, note that it is not possible to obtain well-defined mappings between the top and bottom partitions of a non-isomorphic OPP. Thus, non-isomorphic OPPs conveniently serve as empty sets of permutations.

Example 1. Here are several example OPPs and the permutation sets they encode.

- Discrete OPP: $\begin{bmatrix} 2 & 0 & 1 \\ 1 & 2 & 0 \end{bmatrix} = \{(0\ 2\ 1)\}$
- Unit OPP: $\begin{bmatrix} 0,1,2 \\ 0,1,2 \end{bmatrix} = \{\iota, (0\ 1), (0\ 2), (1\ 2), (0\ 1\ 2), (0\ 2\ 1)\}$
- Isomorphic OPP: $\begin{bmatrix} 2 & 0,1 \\ 1 & 2,0 \end{bmatrix} = \{(1\ 2), (0\ 2\ 1)\}$
- Matching OPP: $\begin{bmatrix} 1 & 0,2,4 & 3 \\ 3 & 0,2,4 & 1 \end{bmatrix} = (1\ 3) \circ S_3 (\{0,2,4\})$
- Non-isomorphic OPPs: $\begin{bmatrix} 0,2| 1 \\ 1| 2,0 \end{bmatrix} = \emptyset, \begin{bmatrix} 2|0|1 \\ 1|2,0 \end{bmatrix} = \emptyset$

4 Basic Enumeration of the Permutation Search Space

OPPs play a role similar to partial variable assignments in CNF-SAT solvers. Recall that a partial variable assignment on n Boolean variables can be encoded by an n-element array whose ith element indicates the value of the ith variable:

0, 1, or X for *unassigned*. A *complete* assignment is one in which all variables
have been assigned a binary value; otherwise the assignment is *partial* and corresponds to a set of complete assignments that can be enumerated by considering
all possible 0, 1 combinations of the unassigned variables. A backtracking SAT
solver extends a given partial assignment by choosing an unassigned variable and
assigning to it one of the two binary values. This is referred to as a *decision* step
and SAT solvers use a variety of decision heuristics to determine which variable
to assign next and what value to assign to it. SAT solvers also employ *propagation* to avoid making decisions on variables whose values are implied (forced) by
prior decisions. Finally, SAT solvers backtrack from "conflicts", i.e. assignments
that cause the formula being checked to become unsatisfied.

As described earlier, a non-discrete OPP can be viewed as a representation
of a set of permutations. The basic skeleton of a permutation enumeration algorithm can thus be patterned after a backtracking SAT algorithm that finds *all*
satisfying assignments to a given CNF formula. An OPP is extended by:

- choosing a non-singleton cell (the *target* cell) from the top partition,
- choosing a vertex from the target cell (the *target* vertex), and
- mapping the target vertex to a vertex from the corresponding cell of the
  bottom partition.

The mapping step is accomplished by splitting the target cell so that the target
vertex is in a cell of its own. The corresponding cell of the bottom partition is
split similarly, placing the vertex to which the target vertex is mapped in a new
singleton cell. Symbolically, given the isomorphic OPP in (1) assume that the
ith cell is the target cell and let $j \in T_i$ be the target vertex. Mapping j to $k \in B_i$
refines the m-cell OPP Π to the following $(m + 1)$-cell OPP Π':

$$\Pi' = \left[\begin{array}{c|c|c|c|c|c|c} T_1' & T_2' & \cdots & T_i' & T_{i+1}' & \cdots & T_{m+1}' \\ B_1' & B_2' & \cdots & B_i' & B_{i+1}' & \cdots & B_{m+1}' \end{array} \right]$$

where

$$\begin{array}{lll} T_l' = T_l & B_l' = B_l & l = 1, \cdots, i - 1 \\ T_i' = T_i - \{j\} & B_i' = B_i - \{k\} & \\ T_{i+1}' = \{j\} & B_{i+1}' = \{k\} & \\ T_l' = T_{l-1} & B_l' = B_{l-1} & l = i + 2, \cdots, m + 1 \end{array}$$

To illustrate, consider the search tree in Figure 1(a) which enumerates all permutations of $V = \{0, 1, 2\}$ and checks which are valid symmetries of the indicated
3-vertex 2-edge graph. Each node of the search tree corresponds to an OPP which
is the root of a subtree obtained by mapping a target vertex in all possible ways.
For example, the unit OPP at the root of the search tree is extended into a 3-way
branch by mapping target vertex 1 to 0, 1, and 2. It is important to point out
that the choice of target vertex at each tree node and the order in which each
of its possible mappings are processed does not affect the final set of permutations produced at the leaves of the search tree. It does, however, alter the order
in which these permutations are produced. Note that valid automorphisms can
be viewed as satisfying assignments whereas invalid ones are analogous to SAT

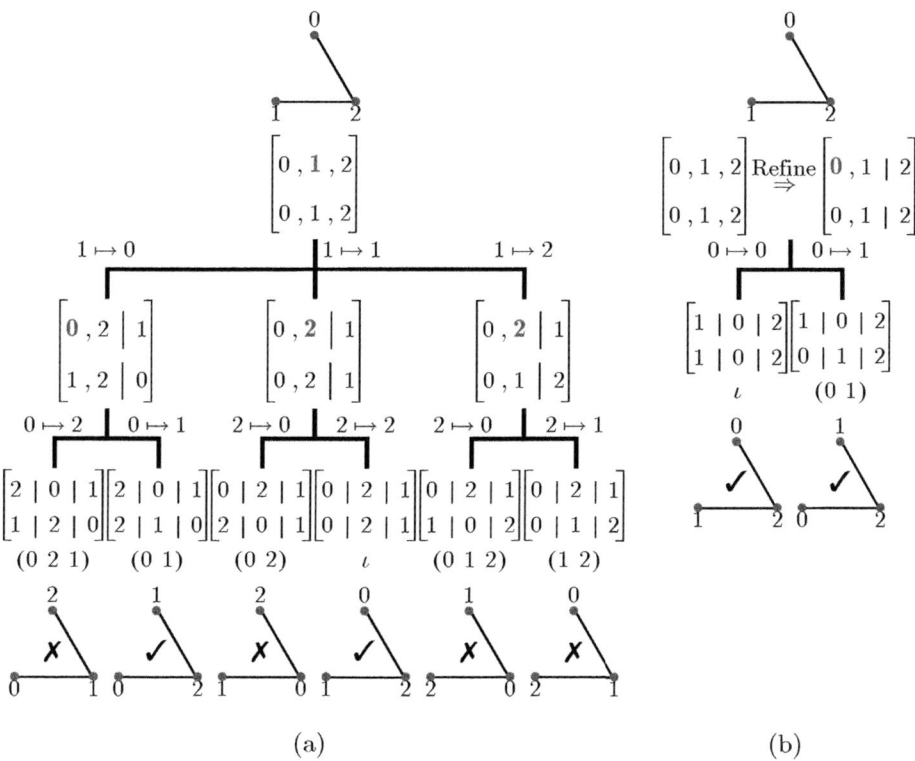

Fig. 1. Search trees for the automorphisms of a 3-vertex "line" graph. The target vertex ("decision variable") at each tree node is highlighted. (a) without partition refinement. (b) with partition refinement.

conflicts. The permutation search tree can be pruned significantly by performing *partition refinement* [1,7,12] before selecting and branching on a target vertex. This is analogous to Boolean constraint propagation in the SAT space and is standard in all algorithms for graph automorphism and canonical labeling. In the present context, partition refinement is applied *simultaneously* to the top and bottom partitions of the current OPP. This is illustrated in Figure 1(b) where vertex 2 is split from vertices 0 and 1 because it has a different degree.

As in SAT search, partition refinement is invoked after each decision assignment to determine the consequences of that decision. In some cases, this allows for the early detection of conflicts, i.e., concluding that the subtree rooted at the current tree node does not contain valid permutations. To illustrate, consider the 7-vertex graph in Figure 2 and assume that the decision to map vertex 0 to vertex 4 has just been made. This decision triggers partition refinement which causes the top and bottom partitions of the OPP to refine *non-isomorphically* proving that there are no automorphisms of this graph that map vertex 0 to vertex 4.

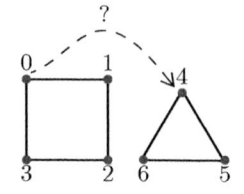

$$\left[\begin{array}{c|c} 1,2,3,4,5,6 & 0 \\ 0,1,2,3,5,6 & 4 \end{array}\right] \Rightarrow \left[\begin{array}{c|c|c} 2,4,5,6 & 1,3 & 0 \\ 0,1,2,3 & 5,6 & 4 \end{array}\right] \Rightarrow \left[\begin{array}{c|c|c|c} 4,5,6 & 2 & 1,3 & 0 \\ 0,1,2,3 & & 5,6 & 4 \end{array}\right]$$

Fig. 2. Example of non-isomorphic refinement. Attempting to map vertex 0 to vertex 4 causes the top and bottom partitions to split non-isomorphically into 4 and 3 cells, respectively.

5   Group-Theoretic Pruning

There are two primary pruning mechanisms anchored in group theory: *coset* pruning and *orbit* pruning. Both are routinely employed by symmetry detection and canonical labeling algorithms. The choice of OPPs to encode permutation sets introduces further opportunities to prune the search space as we show later in this section. To understand how coset and orbit pruning are employed in the search for a set of irredundant group generators requires the introduction of a few more group-theoretic concepts.

Let G be the automorphism group of our graph. The *action* of G on the graph vertices V is a map $* : G \times V \rightarrow V$ such that a) $\iota i = i$ for all $i \in V$, and b) $(\gamma \eta)(i) = \gamma(\eta i)$ for all $i \in V$ and all $\gamma, \eta \in G$. This group action induces an equivalence relation $\sim$ on the vertex set such that $i \sim j$ if and only if there exists $\gamma \in G$ with $\gamma i = j$. The resulting equivalence partition is referred to as the *orbit* partition and will be denoted by $\widehat{\pi}$. The orbit of $i \in V$ under G is the cell in $\widehat{\pi}$ that contains i and is conventionally written as Gi.

Let G_i denote the subgroup of G that "fixes" i, i.e., $G_i = \{\gamma \in G | \gamma i = i\}$. This is referred to as the *stabilizer* subgroup of i. The (left) coset of G_i in G containing η is defined as the set $\{\eta \gamma | \gamma \in G_i\}$. Note how this definition implies that *any* coset element can generate the entire coset by composing that element with the elements of G_i. The set of (left) cosets of G_i partitions G into equal-sized subsets. Now assume that Z is a set of irredundant generators for G_i. A set of generators for the parent group G can be obtained by augmenting Z with a *single* representative from each coset of G_i. This set may, however, contain redundant generators that must be eliminated with the aid of the orbit partition.

To place these pruning mechanisms in the context of the permutation search tree, consider a tree node that represents a group G and assume that the subtree under G is expanded by mapping vertex i to vertices $i, i_1, i_2, \cdots, i_k$ in that order (see Figure 3). As above, the permutation subset corresponding to mapping i to itself is G_i, the stabilizer subgroup of i. The other subsets will be denoted by $H_{i \rightarrow i_j}$ and correspond to those permutations that, among other things, map i

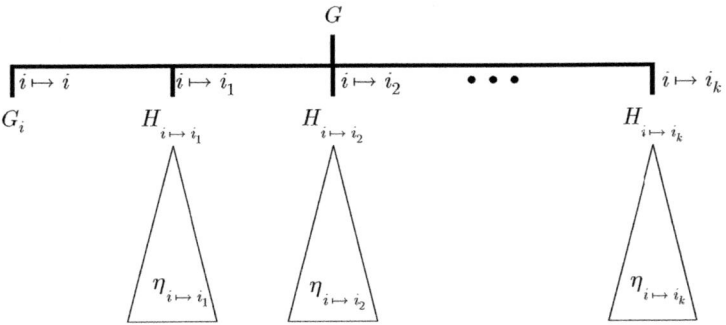

Fig. 3. Structure of the permutation search tree

to i_j. To find a set of irredundant generators for G we must now "solve" up to k independent problems where problem i_j seeks to determine whether the set of permutations $H_{i \mapsto i_j}$ is a coset of G_i. This is accomplished by searching $H_{i \mapsto i_j}$ for a *single permutation* that "satisfies" the graph edge relation, i.e., a permutation that is an automorphism of the graph. If no such permutation exists, then $H_{i \mapsto i_j}$ is "unsatisfiable", i.e., it is not a coset of G_i. This problem is remarkably similar, structurally, to the problem of finding a satisfying assignment to a CNF formula or proving that no such assignment exists.

Let permutation $\eta_{i \mapsto i_j}$ denote the "solution" to problem i_j. Clearly, $\eta_{i \mapsto i_j}$ serves as a coset representative for $H_{i \mapsto i_j}$ and can be added to the set of generators for G. Additionally, vertices i and i_j must now be in the same orbit. Thus, if the orbit of i_j contains vertex i_l with $l > j$, then problem i_l can be skipped since its corresponding coset must necessarily contain redundant generators.

A key pruning mechanism that is enabled by the OPP encoding of permutation sets is the quick discovery of candidate coset representatives. This occurs when the OPP at a given tree node is *matching*. For example, the matching OPP

$$\begin{bmatrix} 1 & 0,2 & 4,6,7 & 3 & 5 \\ 3 & 0,2 & 4,6,7 & 5 & 1 \end{bmatrix}$$

encodes the permutation set:

$$(1\ 3\ 5) \circ S_2\left(\{0,2\}\right) \circ S_3\left(\{4,6,7\}\right)$$

which clearly include the permutation $(1\ 3\ 5)$. If this permutation is found to be a symmetry of the graph, we can terminate the search in this coset and return this permutation as the coset representative. Significantly, if this permutation is found not to be a symmetry of the graph, then we can also terminate the search in this subtree since all other permutations in this subset are composed with this permutation! For large graphs, this pruning mechanism leads to a drastic reduction in the size of the search tree and a commensurate reduction in run time.

Finally, it is interesting to note that in addition to finding a set of irredundant generators for G, symmetry detection algorithms can also compute the order of G using the orbit-stabilizer and Lagrange theorems [9]: $|G| = |G_i| \cdot |Gi|$.

6   The Algorithm

The symmetry detection algorithm is basically a depth-first traversal of the permutation search tree. To enable coset and orbit pruning, the left-most tree path must correspond to a sequence of subgroup stabilizers ending in the identity (a so-called *subgroup decomposition*). In other words, "decisions" along this path must map each selected target vertex to itself. This requirement does not apply to decisions in other parts of the tree. The tree is pruned by systematic application of the four pruning rules elaborated earlier, namely:

- **Coset pruning** which terminates the search in a coset subtree as soon as a coset representative is found.
- **Orbit pruning** which avoids searching the subtree of coset $H_{i \mapsto j}$ if j is already in the orbit of i.
- **Matching OPP pruning** which can identify a candidate permutation at a tree node without the need to explore the subtree rooted at that node.
- **Non-isomorphic OPP pruning** which indicates that there are no permutations in the subtree rooted at that node which are symmetries of the graph.

It is important to note that coset and orbit pruning are, in some sense, *intrinsic* and can (should?) be viewed as part of the "specification" of the automorphism problem. In other words, any graph automorphism algorithm must return a set of irredundant generators, and thus, must employ coset and orbit pruning. The two other pruning rules, based on the OPP encoding of permutation sets, represent algorithmic enhancements that assist in eliminating unnecessary search.

This algorithm has been implemented in the **saucy** 2.1 symmetry detection tool. A trace of the algorithm illustrating all four pruning mechanisms is shown in Figure 4.

7   Experimental Evaluation

We ran symmetry detection using **saucy** 2.1 on the complete set of 1183 SAT 2009 competition benchmarks and checked satisfiability with symmetry-breaking on the 47 most difficult ones (see below). Experiments were conducted on a SUN workstation equipped with a 3GHz Intel Dual-Core CPU, a 6MB cache and an 8GB RAM, running the 64-bit version of Redhat Linux. The run time results are shown in Figure 5. With a time-out of 500 seconds, **saucy** finished on all but 18 benchmarks from the crafted category belonging to three families: connum (6 instances), equilarge (3 instances), and mod2-rand3bip (9 instances). By varying the branching heuristics, **saucy** was able to quickly

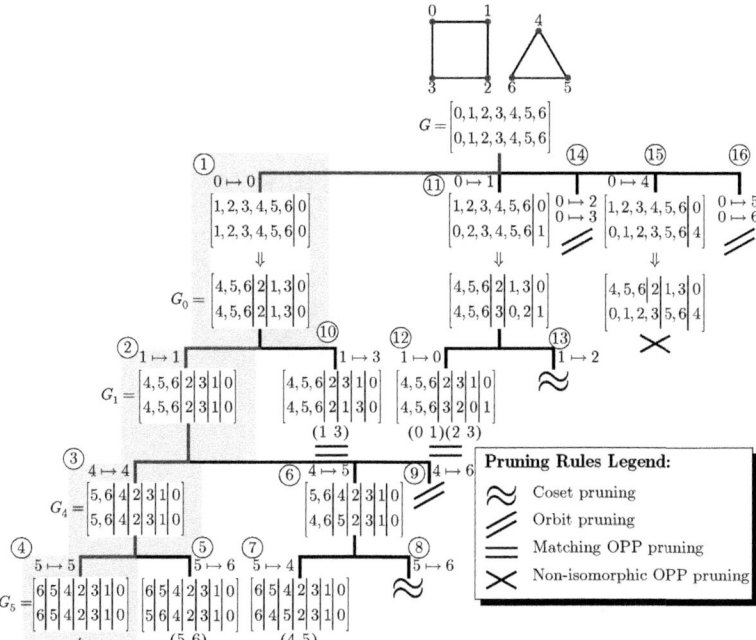

Initialization: $\widehat{\pi} = \{0|\ 1|\ 2|\ 3|\ 4|\ 5|\ 6\}$, $Z = \emptyset$.

1. Fix vertex 0 and refine
2. Fix vertex 1
3. Fix vertex 4
4. Fix vertex 5; $G_5 = \{\iota\}$
5. Search for representative of coset $H_{5\mapsto6}$;
   $Z = \{(5\ 6)\}$; $\widehat{\pi} = \{0|\ 1|\ 2|\ 3|\ 4|\ 5,6\}$; $|G_4| = |G_5| \cdot |G5| = 1 \cdot 2 = 2$
6. Search for representative of coset $H_{4\mapsto5}$
7. Found representative of coset $H_{4\mapsto5}$;
   $Z = \{(5\ 6), (4\ 5)\}$; $\widehat{\pi} = \{0|\ 1|\ 2|\ 3|\ 4,5,6\}$
8. **Coset pruning**: no need to explore since we have already found a coset representative for $H_{4\mapsto5}$
9. **Orbit pruning**: no need to explore since 6 is already in the orbit of 4.
   $|G_1| = |G_4| \cdot |G4| = 2 \cdot 3 = 6$
10. Search for representative of coset $H_{1\mapsto3}$;
    Matching OPP pruning: found representative of coset $H_{1\mapsto3}$.
    $Z = \{(5\ 6), (4\ 5), (1\ 3)\}$; $\widehat{\pi} = \{0|\ 2|1,\ 3|\ 4,5,6\}$; $|G_0| = |G_1| \cdot |G1| = 6 \cdot 2 = 12$
11. Search for representative of coset $H_{0\mapsto1}$
12. **Matching OPP pruning**: found representative of coset $H_{0\mapsto1}$.
    $Z = \{(5\ 6), (4\ 5), (1\ 3), (0\ 1)(2\ 3)\}$; $\widehat{\pi} = \{0,1,2,3|\ 4,5,6\}$
13. **Coset pruning**: no need to explore since we have already found a coset representative for $H_{0\mapsto1}$
14. **Orbit pruning**: no need to explore since 2 and 3 are already in the orbit of 0.
15. **Non-isomorphic OPP pruning**: 0 cannot map to 4.
16. **Orbit pruning**: no need to explore since 5 and 6 are already in the orbit of 4.
    $|G| = |G_0| \cdot |G0| = 12 \cdot 4 = 48$

Fig. 4. Search tree for graph automorphisms of the "square and triangle" graph and relevant computations at each node. The shaded region corresponds to subgroup decomposition.

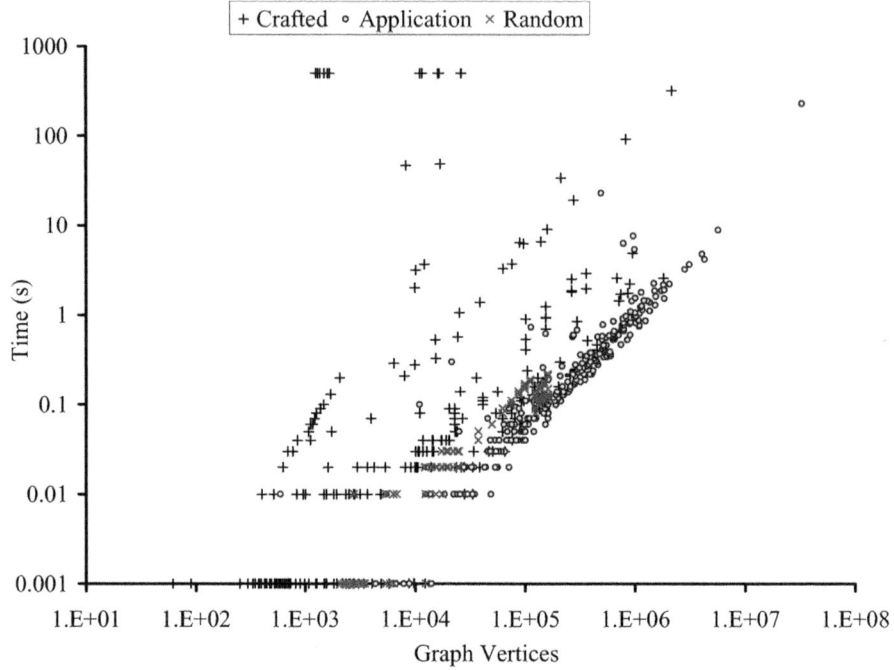

Fig. 5. saucy 2.1 run time, in seconds, as a function of graph size for the SAT 2009 competition benchmarks. A time-out of 500 seconds was applied.

solve the six connum instances (in less than 5 seconds each) but still failed to process the remaining twelve even with a much larger time-out limit. In general, instances from the crafted category were more challenging for **saucy** than similarly-sized instances from the random or application suites. The largest benchmark post-cbmc-zfcp-2.8-u2-noholes, an application instance with about 11 million variables and 33 million clauses, was modeled by a graph with over 32 million vertices and required about 231 seconds to process. As the figure shows, there is a weak trend towards larger run times for larger graphs. However, run time seems to also depend on other attributes of a graph besides its absolute size (number of vertices.) In any case, **saucy** is extremely fast, finishing in less than one second on 93% (1101) of all benchmarks.

The "amount" of symmetry present (order of the automorphism group) in each benchmark is shown in Figure 6. In total, only 323 benchmarks exhibited non-trivial symmetries, and the order of the largest automorphism group (for benchmark hsat_vc11813) was an astronomical $5.091978 \times 10^{142761}$. The figure only lists those benchmarks whose automorphism group has an order between 2 (meaning one non-trivial symmetry) and 10^{60} (a total of 293 out of 323.) Of the 610 benchmarks in the random category, 606 had no symmetry at all, and the remaining four had just one symmetry. In the application category, **saucy** reported the presence of symmetry in about 50% of the benchmarks (144 out of

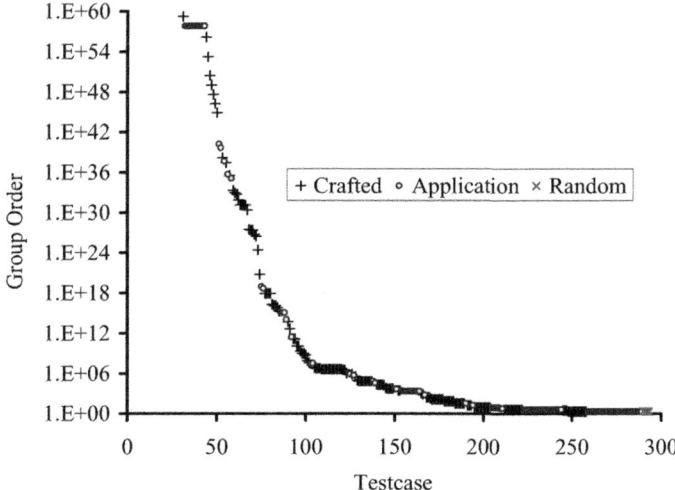

Fig. 6. saucy 2.1 group order for the SAT 2009 competition benchmarks

292), and it found symmetry in about two-thirds (175 out of 263) of the crafted benchmarks which it was able to process within the time-out limit.

Figure 7 shows the relation between the order of the automorphism group and the number of generators returned by **saucy** for the 293 benchmarks. Symmetry detection algorithms, including **saucy**, guarantee to produce no more than $n-1$ generators for an n-vertex graph. The number of reported generators in these results is significantly less than $n-1$. This, however, is not inconsistent with the well-known fact that the number of (irredundant) generators is exponentially smaller than the order of the corresponding symmetry group.

To evaluate the effectiveness of static symmetry breaking, we applied **shatter** [2] to 47 "difficult" benchmarks. These included 13 application and 34 crafted benchmarks that had significant symmetry and either could not be solved by any of the SAT 2009 competition solvers (38 benchmarks), or required at least 1000 seconds to be solved (9 benchmarks). The **shatter** flow consists of running **saucy** on a CNF instance to obtain its symmetry generators, followed by the creation of CNF symmetry-breaking predicates (SBPs) using the encoding in [4], and finally passing the original instance augmented with the SBPs to a SAT solver. Figures 8(a) and 8(b) depict the increase in instance size (variables and clauses) for each of these benchmarks due to the addition of the SBPs. For 29 of the benchmarks, the number of added SBP clauses was quite insignificant (less than 4%). The additional clauses for the remaining 18 benchmarks ranged from 25% to 133% of the original number. The number of variables increased by less than 1% for 23 benchmarks and by 9% to an order of magnitude for the remaining 24 benchmarks.

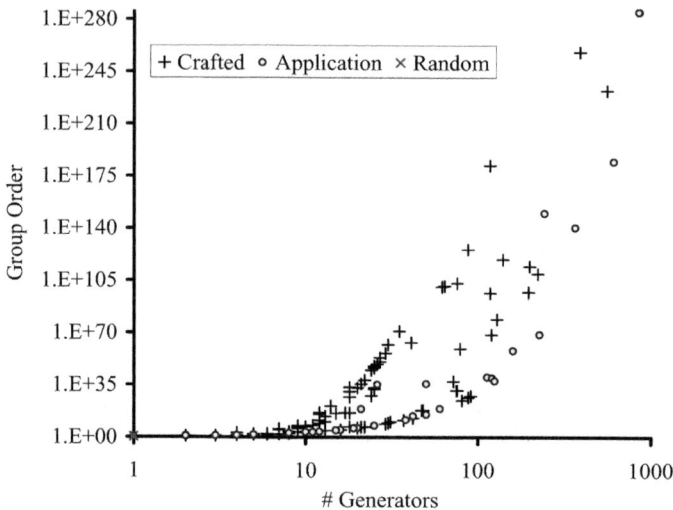

Fig. 7. saucy 2.1 group order as a function of the number of group generators for the SAT 2009 competition benchmarks

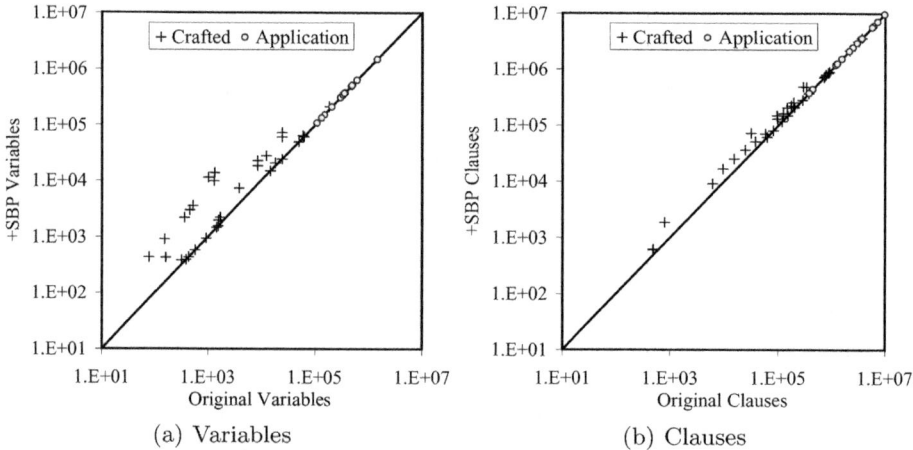

(a) Variables

(b) Clauses

Fig. 8. Number of variables and clauses before and after the addition of SBPs

To obtain meaningful statistical data, we used a script that re-orders the variables and clauses in a CNF instance using a random seed[1] to create twenty different versions of each benchmark: ten for the original and ten for the SBP-augmented benchmark. We then applied the best solver for a given benchmark, based on the 2009 competition results, to these twenty versions. The run time

---

[1] We obtained the reorder.c script and a seed generator from Laurent Simon. The script was originally written by Edward Hirsh and later modified by Simon to handle large benchmarks.

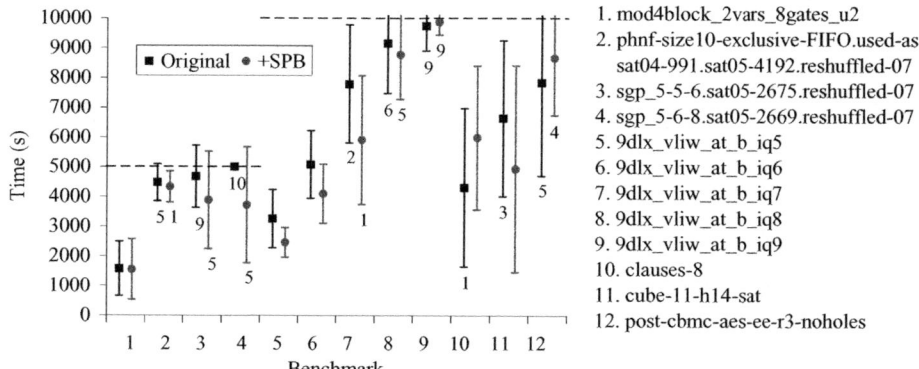

Fig. 9. SAT solver run time results before and after the addition of SBPs. The first 4 benchmarks are from the crafted category with a time-out limit of 5,000 seconds, whereas the last 8 are from the application category with a time-out limit of 10,000 seconds. minisat 2.0 was used for benchmarks 10 and 11 and glucose 1.0 was used for the others. The data for each benchmark (original or with SBPs) show the mean and standard deviation over the ten runs, including the runs that timed out. Since time-outs skew the statistics, the number of runs that timed out is indicated below the "error" bars (absence of a number indicates that all ten runs finished within the time-out limit.).

results comparing search times with and without the addition of SBPs for 12 of the 47 benchmarks are shown in Fig. 9; experiments on the remaining 35 benchmarks were still running at press time. For this limited subset, the SBP-augmented versions generally led to fewer time-outs and, in all but three cases, were solved faster than the original versions. Four of these benchmarks (2, 4, 8, and 9) which were reported to be unsolvable within the time-out limits of the competition, were solved with the addition of SBPs. Interestingly, though, benchmarks 2, 8, and 9 were solved on our experimental machine even without the addition of SBPs. These anomalies are possibly due to the use of different machines with varying configurations in the SAT 2009 competition and merely point out that we must be careful not to draw incorrect conclusions from empirical data.

8   Conclusions

It has been conjectured that symmetries in CNF formulas contribute to the intractability of SAT. The availability of extremely-efficient scalable symmetry detection algorithms, such as **saucy** 2.1, has enabled the testing of this hypothesis on very large CNF formulas. The question, however, remains open. Many intractable CNF instances (e.g., random instances) possess no or little symmetry. Those that possess significant symmetry may or may not benefit from *static* symmetry breaking for a number of possible reasons. For example, the generators produced by a symmetry detection algorithm may not be the most suitable for symmetry breaking. Better branching heuristics while searching the permutation space might yield more useful generators for SAT solving. A more promising direction is the integration of symmetry detection within the SAT solver itself [13]. The raw speed of modern symmetry detectors like **saucy** suggests that they can be invoked during the SAT search with minimal overhead. And unlike static symmetry breaking, dynamic symmetry detection does not require the addition of large SBPs, and can uncover hidden/conditional symmetries adaptively. We plan to pursue this in our future research.

Acknowledgments

This work was funded in part by NSF award number 0705103. We gratefully acknowledge Timothy Lane's help at various stages of this project and offer our sincere thanks to Allen Van Gelder for his careful reading of, and thoughtful comments on, the manuscript.

References

1. Aho, A.V., Hopcroft, J.E., Ullman, J.D.: The Design and Analysis of Computer Algorithms. Addison-Wesley, Reading (1974)
2. Aloul, F.A., Markov, I.L., Sakallah, K.A.: Shatter: Efficient symmetry-breaking for boolean satisfiability. In: Proc. 40th IEEE/ACM Design Automation Conference (DAC), Anaheim, California, pp. 836–839 (2003)
3. Aloul, F.A., Ramani, A., Markov, I.L., Sakallah, K.A.: Solving difficult instances of boolean satisfiability in the presence of symmetry. IEEE Transactions on Computer-Aided Design of Integrated Circuits and Systems 22(9), 1117–1137 (2003)
4. Aloul, F.A., Sakallah, K.A., MarkovEfficient, I.L.: symmetry breaking for boolean satisfiability. In: Proc. 18th International Joint Conference on Artificial Intelligence (IJCAI 2003), Acapulco, Mexico, pp. 271–282 (2003)
5. Crawford, J.: A theoretical analysis of reasoning by symmetry in first-order logic (extended abstract). In: AAAI 1992 Workshop on Tractable Reasoning, San Jose, CA, pp. 17–22 (1992)
6. Crawford, J., Ginsberg, M., Luks, E., Roy, A.: Symmetry-breaking predicates for search problems. In: Principles of Knowledge Representation and Reasoning (KR 1996), pp. 148–159 (1996)
7. Darga, P.T., Liffiton, M.H., Sakallah, K.A., Markov, I.L.: Exploiting structure in symmetry detection for CNF. In: Proc. 41st IEEE/ACM Design Automation Conference (DAC), San Diego, California, pp. 530–534 (2004)
8. Darga, P.T., Sakallah, K.A., Markov, I.L.: Faster symmetry discovery using sparsity of symmetries. In: Proc. 45th IEEE/ACM Design Automation Conference (DAC), Anaheim, California, pp. 149–154 (2008)

9. Fraleigh, J.B.: A First Course in Abstract Algebra, 6th edn. Addison Wesley Longman, Reading (2000)
10. Junttila, T., Kaski, P.: Engineering an efficient canonical labeling tool for large and sparse graphs. In: Ninth Workshop on Algorithm Engineering and Experiments (ALENEX 2007), New Orleans, LA (2007)
11. McKay, B.D.: Nauty user's guide (version 2.2),
    http://cs.anu.edu.au/~bdm/nauty/nug.pdf
12. McKay, B.D.: Practical graph isomorphism. Congressus Numerantium 30, 45–87 (1981)
13. Schaafsma, B., Heule, M.J., Maaren, H.: Dynamic symmetry breaking by simulating Zykov contraction. In: Kullmann, O. (ed.) SAT 2009. LNCS, vol. 5584, pp. 223–236. Springer, Heidelberg (2009)

A Non-prenex, Non-clausal QBF Solver with Game-State Learning

William Klieber, Samir Sapra, Sicun Gao, and Edmund Clarke*

Computer Science Department
Carnegie Mellon University
Pittsburgh, Pennsylvania

Abstract. We describe a DPLL-based solver for the problem of quantified boolean formulas (QBF) in non-prenex, non-CNF form. We make two contributions. First, we reformulate clause/cube learning, extending it to non-prenex instances. We call the resulting technique *game-state learning*. Second, we introduce a propagation technique using *ghost literals* that exploits the structure of a non-CNF instance in a manner that is symmetric between the universal and existential variables. Experimental results on the QBFLIB benchmarks indicate our approach outperforms other state-of-the-art solvers on certain benchmark families, including the `tipfixpoint` and `tipdiam` families of model checking problems.

Keywords: QBF, DPLL, non-clausal, non-prenex, clause learning.

1 Introduction

Many problems in formal verification (among other areas) are naturally expressed in the language of QBF. Traditionally, QBF solvers have used conjunctive normal form (CNF). Although CNF works well for SAT solvers, it hinders the work of QBF solvers by impeding the ability to detect and learn from satisfying assignments. In fact, a family of problems that are trivially satisfiable in negation-normal form (NNF) were experimentally found to require exponential time (in the problem size) for existing CNF solvers [18].

Various techniques have been proposed for avoiding the drawbacks of a CNF encoding. Zhang et al. have investigated dual CNF-DNF representations in which a boolean formula is transformed into a combination of an equi-satisfiable CNF formula and an equi-tautological DNF [18]. Sabharwal et al. have developed a QBF modeling approach based a game-theoretic view of QBF [14]. Ansotegui

---

* This research was sponsored by the GSRC under contract no. 1041377 (Princeton University), National Science Foundation under contracts no. CCF0429120, no. CNS0926181, no. CCF0541245, and no. CNS0931985, Semiconductor Research Corporation under contract no. 2005TJ1366, General Motors under contract no. GM-CMUCRLNV301, Air Force (Vanderbilt University) under contract no. 18727S3, International Collaboration for Advanced Security Technology of the National Science Council, Taiwan, under contract no. 1010717, and the Office of Naval Research under award no. N000141010188.

O. Strichman and S. Szeider (Eds.): SAT 2010, LNCS 6175, pp. 128–142, 2010.
© Springer-Verlag Berlin Heidelberg 2010

et al. have investigated the use of *indicator variables* [1]. These approaches all help to alleviate the problems of a pure CNF encoding, but we argue that a fully non-clausal approach can lead to even greater improvements, especially for instances produced from deeply-nested circuits.

In addition to combined CNF-DNF techniques, fully non-clausal techniques have recently been investigated. A prenex circuit-based DPLL solver with "don't care" reasoning and clause/cube learning has been developed by Goultiaeva et al. [8]. A non-prenex NNF-based DPLL solver with dependency-directed (non-chronological) backtracking, but without learning, was developed by Egly, Seidl, and Woltran [4]. Non-clausal techniques using symbolic quantifier expansion (rather than DPLL) have been developed by Lonsing and Biere [10] and by Pigorsch and Scholl [13]. Giunchiglia et al. have developed a technique for mini-scoping quantifiers (pushing quantifiers inward so as to minimize their scope) [7]. Non-clausal representations have also been investigated in the context of SAT solvers [9,16,5].

Most existing DPLL-based QBF solvers perform clause/cube learning. However, traditional clause/cube learning was designed for prenex QBF instances, and it is not optimal for (or even directly applicable to) non-prenex QBF instances. We reformulate clause/cube learning and extend it to the non-prenex case. Additionally, we develop a new propagation technique using *ghost literals*. Experimental results indicate that our approach can beat other state-of-the-art solvers on fixed-point computation instances of the type found in the `tipfixpoint` benchmark family.

2   Preliminaries

We consider non-prenex QBF formulas in negation-normal form[1], as described by the following abstract grammar:

$$\phi ::= e_i \mid \neg e_i \mid u_i \mid \neg u_i \mid \phi \vee ... \vee \phi \mid \phi \wedge ... \wedge \phi \mid \exists e_i \, \phi \mid \forall u_i \, \phi$$

We label each conjunction and disjunction with a *gate variable* of the form g_i, as illustrated in Figure 1. The conjunction/disjunction labelled g_i, together with its quantifier prefix (if any), is labelled with the primed gate variable g_i', as illustrated in Figure 1. As indicated in the abstract grammar, each labelled conjunction/disjunction may have any number of conjuncts/disjuncts.

The term "gate variable" arises from the circuit representation of a propositional formula, in which a gate variable labels a logic gate.

Let "*InFmla*" denote the formula that the QBF solver is given as input. We impose the following restriction on *InFmla*: Every variable in *InFmla* must be quantified exactly once, and no variable may occur free (i.e., outside the scope of its quantifier). The variables that occur in *InFmla* are said to be *input variables*.

---

[1] Our solver does not require the use of strict NNF. Subformulas containing no quantifiers can be represented in circuit form.

$$\exists e_{10} \left[\left[\underbrace{\exists e_{11} \, \forall u_{21} \, \overbrace{(e_{10} \wedge e_{11} \wedge u_{21})}^{g_1}}_{g_1'} \right] \wedge \left[\underbrace{\forall u_{22} \, \exists e_{30} \, \overbrace{(e_{10} \wedge u_{22} \wedge e_{30})}^{g_2}}_{g_2'} \right] \right]$$

Fig. 1. Example QBF instance with gate labels

We represent an assignment π by the set of literals assigned `true` by π. For example, the assignment $\{e_1, \neg u_2\}$ assigns e_1 `true` and assigns u_2 `false`, while leaving all other variables unassigned. We write "$\pi(\ell)$" to denote the value (`true`, `false`, or `undef`) that π assigns to ℓ, as defined as follows: $\pi(\ell) = $ `true` if $\ell \in \pi$, $\pi(\ell) = $ `false` if $\neg\ell \in \pi$, and $\pi(\ell) = $ `undef` otherwise. For any variable x, we treat $\neg\neg x$ as equivalent to x. An assignment may not include both a variable and its negation. An *input assignment* is an assignment in which every assigned variable is an input variable (as opposed to a gate variable).

Definition 1 (Reduction). The *reduction* of a formula f under an input assignment π, denoted by "$f|\pi$", is constructed from f as follows: For each variable x which is assigned a value by π, we delete the quantifier of x and replace each occurrence of x with its assigned value. For example, if $\pi = \{e_1\}$, then $[\exists e_1. \forall u_2. (e_1 \wedge u_2)]|\pi = [\forall u_2. (\text{true} \wedge u_2)]$. Formally:

$$\ell|\pi = \begin{cases} \pi(\ell) & \text{if } \pi(\ell) \neq \text{undef} \\ \ell & \text{if } \pi(\ell) = \text{undef} \end{cases} \qquad (\exists x.f)|\pi = \begin{cases} f|\pi & \text{if } \pi(x) \neq \text{undef} \\ \exists x.(f|\pi) & \text{if } \pi(x) = \text{undef} \end{cases}$$

$$(f_1 \wedge \ldots \wedge f_n)|\pi = (f_1|\pi) \wedge \ldots \wedge (f_n|\pi)$$

$$(f_1 \vee \ldots \vee f_n)|\pi = (f_1|\pi) \vee \ldots \vee (f_n|\pi) \qquad (\forall x.f)|\pi = \begin{cases} f|\pi & \text{if } \pi(x) \neq \text{undef} \\ \forall x.(f|\pi) & \text{if } \pi(x) = \text{undef} \end{cases}$$

Given two input literals x and y, we say that x is *upstream* of y iff the scope of the quantifier of x contains the quantifier of y. We say that a gate literal g is *upstream* of an input literal y iff every variable that occurs in the subformula g is upstream of y.

2.1   QBF as a Two-Player Game

It is helpful to view QBF as a game between two players, Player E and Player U. We make the following formal definitions:

- The existentially quantified variables are *owned* by Player E.
- The universally quantified variables are *owned* by Player U.

Informally, the game formulation goes as follows. Throughout the course of the game, the two players assign values to the variables that they own. The order in which the players assign variables is the quantification order of the variables. On each turn of the game, the owner of the outermost-quantified unassigned

variable assigns it a value. The goal of Player E is to make *InFmla* true, and
the goal of Player U is to make *InFmla* false. For non-prenex instances, we
say that each quantifier-prefixed subformula (e.g., g_1' and g_2' in Figure 1) is a
subgame. It may happen that two or more variables are quantified outermost;
e.g., in Figure 1 on page 130, after e_{10} is assigned a value, both e_{11} and u_{22} are
quantified outermost. In this case, two subgames have become independent of
each other; they may be played in parallel or in series.

Definition 2 (Winning under an assignment). Player U *wins* a formula f
under π iff $f|\pi$ is false. Player E *wins* a formula f under π iff $f|\pi$ is true. (See
Definition 1 for the meaning of $f|\pi$.) (It would be more proper to say "has a
winning strategy for" instead of "wins", but for brevity, we'll say simply "wins".)

For example, in Figure 1, Player U wins g_2' under the empty assignment, and
Player E wins g_2' under $\{e_{10} : \texttt{true}, u_{22} : \texttt{true}\}$.

Proposition 1. Player E wins $[\exists x\ \phi]$ under π if he wins ϕ under either $\pi \cup \{x\}$
or $\pi \cup \{\neg x\}$. Player U wins $[\forall x\ \phi]$ under π if he wins ϕ under either $\pi \cup \{x\}$ or
$\pi \cup \{\neg x\}$.

3 Symbolic Game States

In this section, introduce *game-state learning*, a reformulation of clause/cube
learning. For prenex instances, the game-state formulation is isomorphic to
clause/cube learning; the differences are merely cosmetic. However, the game-
state formulation is more convenient to extend to the non-prenex case.

   To motivate the notation of game-state learning, we start by reviewing certain
aspects of clause learning. Suppose the input formula *InFmla* is a prenex CNF
QBF whose first clause is $(e_1 \vee e_3 \vee u_4 \vee e_5)$. Under an assignment π, if all the
literals in the clause are false, then clearly $InFmla|\pi$ is false. Moreover, if, under
π, all the clause's existential literals are assigned false and none of the clause's
universal literals are assigned true (i.e., they may either be assigned false or be
unassigned), then $InFmla|\pi$ is false, since the universal player can win by making
all the universal literals in the clause false.

   As shown in [19], when the QBF clause learning algorithm is applied to

$$\exists e_1 \exists e_3 \forall u_4 \exists e_5 \exists e_7. \, (e_1 \vee e_3 \vee u_4 \vee e_5) \wedge (e_1 \vee \neg e_3 \vee \neg u_4 \vee e_7) \wedge \ldots$$

it can yield the tautological learned clause $(e_1 \vee u_4 \vee \neg u_4 \vee e_5 \vee e_7)$. Although
counter-intuitive, this learned clause can be interpreted in the same way as a
non-tautological clause: Under an assignment π, if all the clause's existential
literals are assigned false and none of the clause's universal literals are assigned
true, then $InFmla|\pi$ is false.

   Learned cubes are similar: Under an assignment π, if all the cube's universal
literals are assigned true and none of the cube's existential literals are assigned
false, then $InFmla|\pi$ is true. With game-state learning, we explicitly separate the
"must be true" literals from the "may be either true or unassigned" literals. (For

non-prenex instances, the division is more complicated than just existential-vs-universal.) Instead of writing a cube $(e_1 \lor u_2 \lor \neg e_3)$, we will write a game-state sequent $\langle \{u_2\}, \{e_1, \neg e_3\} \rangle \models (\text{E wins } InFmla)$.

Definition 3. A *symbolic game state* is a tuple $\langle L^{\text{now}}, L^{\text{fut}} \rangle$, where L^{now} is a set of literals and L^{fut} is a set of input literals. $\langle L^{\text{now}}, L^{\text{fut}} \rangle$ symbolically represents (or *matches*) exactly those input assignments under which:

1. every literal in L^{now} reduces to `true`, and
2. no literal in L^{fut} is assigned `false` — i.e., for every literal ℓ in L^{fut}, either ℓ is already true or ℓ has not yet been assigned a value (and therefore may become true in the future).

For example, consider again the QBF instance in Figure 1 on page 130. The assignment $\{\neg e_{10}\}$ matches both $\langle \{\neg g_1'\}, \varnothing \rangle$ and $\langle \{\neg g_1'\}, \{u_{21}, \neg u_{21}\} \rangle$ (because $\neg e_{10}$ implies $\neg g_1'$), but not $\langle \{\neg g_1'\}, \{e_{10}\} \rangle$. No assignment matches $\langle \{\neg e_{10}\}, \{e_{10}\} \rangle$.

Definition 4 (Winning under a game state). We say that player P *wins* a formula f under a game state GS, written "$GS \models (P \text{ wins } f)$", iff P wins f under all assignments that match GS. Additionally, we say that P *loses* f under GS, written "$GS \models (P \text{ loses } f)$", iff the opponent of P wins f under GS.

For example, for the QBF instance in Figure 1:

- Neither player wins g_1' under the game state $\langle \varnothing, \varnothing \rangle$, because Player U loses under the matching assignment $\{e_{10}, e_{11}, u_{21}\}$ and Player E loses under the matching assignment $\{\neg e_{10}\}$.
- Player U wins g_1' under $\langle \varnothing, \{\neg u_{21}\} \rangle$. For example, under the assignment $\pi = \{e_{11}\}$, $g_1'|\pi$ is $[\forall u_{21} (e_{10} \land \texttt{true} \land u_{21})]$, which evaluates to `false`.
- Player E wins g_1' under $\langle \{u_{21}\}, \{e_{10}, e_{11}\} \rangle$.

In our solver, instead of learning clauses or cubes, we maintain a game-state database with *sequents* of the form $GS \models (P \text{ wins } g_i')$. It turns out that whenever we learn a new game-state sequent for a prenex instance, the literals owned by the winner all go in L^{fut}, and the literals owned by the loser and the gate literals go in L^{now}. The relationship between learned game-state sequents and learned clauses/cubes (for prenex instances) is as follows. $\langle L^{\text{now}}, L^{\text{fut}} \rangle \models (\text{U wins } InFmla)$ is equivalent to the learned clause $[\neg \ell_1 \lor ... \lor \neg \ell_n]$ where $\{\ell_1, ..., \ell_n\} = L^{\text{now}} \cup L^{\text{fut}}$ (where L^{now} contains the loser/gate literals and L^{fut} contains the winner literals). This equivalence is easily verified using the interpretation of learned clauses developed on the previous page. Likewise, $\langle L^{\text{now}}, L^{\text{fut}} \rangle \models (\text{E wins } InFmla)$ is equivalent to the learned cube $[\ell_1 \land ... \land \ell_n]$ where $\{\ell_1, ..., \ell_n\} = L^{\text{now}} \cup L^{\text{fut}}$.

Proposition 2. If $\langle L^{\text{now}} \cup \{\ell\}, L^{\text{fut}} \rangle \models (P \text{ wins } f)$, and ℓ is owned by Player P and the quantifier of ℓ is inside f, then $\langle L^{\text{now}}, L^{\text{fut}} \cup \{\ell\} \rangle \models (P \text{ wins } f)$, provided that $\neg \ell \notin L^{\text{fut}}$.

For example, consider the QBF instance $\forall u_1. \exists e_2. (u_1 \oplus e_2)$, where "$u_1 \oplus e_2$" means "$(u_1 \land \neg e_2) \lor (\neg u_1 \land e_2)$". If Player E wins under $\langle \{u_1, \neg e_2\}, \varnothing \rangle$, then Proposition 2 tells us that Player E wins under $\langle \{u_1\}, \{\neg e_2\} \rangle$.

4   Algorithm

An overview of the top-level solver algorithm is provided in Figure 2. Initially, the current assignment *CurAsgn* is empty. For non-prenex instances, we may temporarily target in on a subgame of the input formula *InFmla* and ignore the rest; the subgame being targetted is recorded in the *TargFmla* global variable. On each iteration of the main loop, we first test to see if we know who wins *TargFmla* under the current assignment. There are two cases:

- If the winner of *TargFmla* is unknown, then we call DecideLit, which picks an unassigned input variable (from the first available quantifier block in the prefix of *TargFmla*) and assigns it a value in *CurAsgn*. If there are no more unassigned variables in the quantifier prefix of the current *TargFmla*, then we pick a new *TargFmla* from among the unassigned immediate subformulas of *TargFmla* and try again. After adding a new literal to *CurAsgn*, we call Propagate to perform boolean constraint propagation (BCP).
- If the winner is known, then we call LearnNewGS to learn a new game-state sequent, adding it to the database. If the new game-state sequent reveals that *InFmla* evaluates to a value v under the empty assignment, then we return v as our final answer. Otherwise, we backtrack. We follow the well-known non-chronological backtracking technique, with the addition that we must also undo changes to *TargFmla* as appropriate. (That is, if we backtrack to the beginning of the kth decision level, then we must restore *TargFmla* to the value that it held at the beginning of the kth decision level. For this purpose, we maintain an array UndoTarg that maps each decision level to the value of *TargFmla* to be restored.) After backtracking, the newly-learned game-state sequent will force a literal, so we call Propagate

```
func Solve() {
    CurAsgn = ∅;
    TargFmla = InFmla;
    while (true) {
        if (the winner of TargFmla under CurAsgn is unknown) {
            DecideLit();   // Picks new TargFmla if necessary.
            Propagate();
        } else {
            GS = LearnNewGS();
            if (TargFmla == InFmla and ∅ matches GS) return winner;
            Backtrack to the earliest point at which GS will force a literal;
            Propagate();
        }
    }
}
```

Fig. 2. Overview of top-level solver algorithm

to perform BCP. (Is a literal forced even when we leave a subgame b by restoring an old value of *TargFmla* during backtracking? Yes; ghosts of b are forced, as per case 1(b) in Section 4.3.)

4.1   Ghost Literals

Goultiaeva et al. [8] introduce a powerful propagation technique for QBF that significantly improves on existing QBF solvers on a variety of benchmarks. With their technique, if the solver notices that a gate literal g must be true in order for the existential player to win, then g becomes forced. However, this technique is asymmetric between the existential and universal players. A gate literal g is forced if it is needed for the existential player to win, but not if it is needed for the universal player to win. We adapt this technique so that the universal variables benefit from the same propagation technique as do the existential variables and so that the learning procedure for satisfying assignments is just as powerful as for falsifying assignments.

In a prenex solver, for each gate variable g, we would introduce two *ghost* variables, $g\langle U \rangle$ for Player U and $g\langle E \rangle$ for Player E. A ghost literal $g\langle P \rangle$ would be forced whenever we detect that Player P cannot win unless g is made true.

For our non-prenex solver, we need to consider subgames (quantifier-prefixed subformulas, such as g_1' and g_2' in Figure 1). We introduce ghost variables of the form $g\langle U, b \rangle$ and $g\langle E, b \rangle$ where b is a subgame which contains g as a subformula. A ghost literal $g\langle P, b \rangle$ becomes forced when we detect that Player P cannot win subgame b without g being true. For example, consider the below QBF instance (where g_1 is some propositional formula involving e_1, u_2, and e_3):

$$\exists e_1 \, \forall u_2 \, \exists e_3 \, \forall u_4. \, [[\underbrace{\forall u_5. \, g_1 \lor u_5}_{g_2'}] \land u_4] \lor [\underbrace{\forall u_6. \, \neg g_1 \lor u_6}_{g_3'}]$$

Under the empty assignment, $g_1\langle E, g_2' \rangle$ is forced (because Player E cannot win g_2' under $\varnothing$ unless g_1 is true) and likewise $\neg g_1\langle E, g_3' \rangle$ is forced.

In order to simplify the propagation and learning procedures, we allow game states to contain ghost literals. A game state with a ghost literal is said to *match* the same input assignments as if the game state contained the corresponding non-ghost gate literal; e.g., $\langle L^{\text{now}} \cup \{g\langle P, b \rangle\}, L^{\text{fut}} \rangle$ matches the same input assignments as $\langle L^{\text{now}} \cup \{g\}, L^{\text{fut}} \rangle$.

4.2   Initialization of Game-State Database

In CNF-based QBF solvers, the existential player owns the gate variables[2], and there are clauses (generated from the Tseitin transformation [17]) that ensure that the existential player loses if he assigns a value to a gate variable that turns out to be inconsistent with the inputs to the gate. For example, if $g = e_1 \land e_2$, then Player E would lose if he assigns $g = \texttt{true}$ and $e_1 = \texttt{false}$.

---

[2] For CNF solvers, gate variables are introduced when formulas are converted to CNF via the Tseitin transformation [17]; these gate variables are existentially quantified.

In our solver, instead of generating clauses via the Tseitin transformation, we generate game-state sequents. In a prenex solver, we would generate game-state sequents that ensure that a player P loses if he assigns a ghost gate variable a value inconsistent with the gate's inputs. In our non-prenex solver, for each subgame b, we generate game-state sequents that ensure that a player P loses subgame b if he assigns a ghost gate variable $g\langle P, b\rangle$ a value inconsistent with the gate's inputs. For example, if $g = e_1 \wedge e_2$ and subformula g appears in a subgame b, then Player E would lose b if he assigns $g\langle E, b\rangle = \mathtt{true}$ and $e_1 = \mathtt{false}$. We construct such game-state sequents as follows. For every gate literal g, if g labels a formula $\ell_1 \wedge ... \wedge \ell_n$ (or $\neg g$ labels a formula $\neg\ell_1 \vee ... \vee \neg\ell_n$), we add the following game-state sequents for each player $P \in \{E, U\}$ and each quantifier-prefixed formula b which contains g as a subformula:

- $\langle\{\ell_1, ..., \ell_n, \neg g\}, \varnothing\rangle \models (P \text{ loses } b)$
- $\langle\{\neg\ell_i, g\}, \varnothing\rangle \models (P \text{ loses } b)$ for every $i \in \{1, ..., n\}$

For example, if $g_3 = \neg e_1 \vee \neg u_2$ and g_3 is a subformula of a subgame g_7', then we add game-state sequents $\langle\{e_1, u_2, g_3\}, \varnothing\rangle \models (E \text{ loses } g_7')$, $\langle\{\neg e_1, \neg g_3\}, \varnothing\rangle \models$ (E loses g_7'), and $\langle\{\neg u_2, \neg g_3\}, \varnothing\rangle \models (E \text{ loses } g_7')$, among others.

After adding the game-state sequents to the database, we normalize them as follows. Consider a game-state sequent of the form $\langle L^{\text{now}}, L^{\text{fut}}\rangle \models (P \text{ loses } b)$. First, we use Proposition 2 (on page 132) to move input literals owned by the winning player from L^{now} to L^{fut}. Second, we replace each gate literal g in L^{now} with the ghost literal $g\langle P, b\rangle$. For example, consider a game-state sequent $\langle\{e_1, u_2, g_3\}, \varnothing\rangle \models$ (E loses g_7'). We move u_2 using Proposition 2 (assuming that the quantifier of u_2 is within the formula g_7') and replace g_3 with $g_3\langle E, g_7'\rangle$, yielding $\langle\{e_1, g_3\langle E, g_7'\rangle\}, \{u_2\}\rangle \models$ (E loses g_7').

Recall that a ghost literal $g\langle P, b\rangle$ should become forced when g must be true in order for P to win b. Thus, for every quantifier-prefixed subformula b, the ghost literals $\neg b\langle U, b\rangle$ and $b\langle E, b\rangle$ should be forced. To ensure that the propagation procedure in Section 4.3 forces these literals, we add the following game-state sequents for every gate variable b that labels a quantifier-prefixed formula:

- $\langle\{b\langle U, b\rangle\}, \varnothing\rangle \models (U \text{ loses } b)$ (to force $\neg b\langle U, b\rangle$)
- $\langle\{\neg b\langle E, b\rangle\}, \varnothing\rangle \models (E \text{ loses } b)$ (to force $b\langle E, b\rangle$)

4.3   Propagation and Forced Literals

CurAsgn may contain forced ghost literals, so in general we can't say *CurAsgn* is a match for a game-state in the sense of Definition 3, because *CurAsgn* is not necessarily an input assignment. Instead, let us say that *CurAsgn* is a *ghost match* for a game-state sequent $\langle L^{\text{now}}, L^{\text{fut}}\rangle \models (P \text{ loses } b)$ iff every literal in L^{now} is assigned true by *CurAsgn* and no literal in L^{fut} is assigned false by *CurAsgn*.

During the `Propagate` procedure, conceptually we examine each learned game-state sequent GS of the form $\langle L^{\text{now}}, L^{\text{fut}}\rangle \models (P \text{ loses } b)$ in which none of the literals in $L^{\text{now}} \cup L^{\text{fut}}$ are assigned `false` and b is a subformula of *TargFmla*. There are three cases:

1. If all literals in L^{now} are true, then *CurAsgn* is a ghost match for *GS*, so P loses b under the current assignment.[3] There are two subcases to consider:

   (a) If $b = \textit{TargFmla}$, then we know who wins *TargFmla* under the current assignment, so we stop propagation and return to the `Solve` procedure.

   (b) If $b \neq \textit{TargFmla}$, then for all subgames s that contain b, the ghost variables $b\langle \text{E}, s\rangle$ and $b\langle \text{U}, s\rangle$ are forced to be false (if $P{=}\text{E}$) or true (if $P{=}\text{U}$).

2. If there is exactly one unassigned literal ℓ_U in L^{now}, then $\neg\ell_U$ is forced if:

   (1) ℓ_U is owned by P or is a ghost literal of the form $g\langle P, b\rangle$, and

   (2) ℓ_U is upstream of all unassigned literals in L^{fut}, and

   (3) ℓ_U does not appear outside subgame b if ℓ_U is an input literal

   (so that forcing $\neg\ell_U$ can't cause P to lose a different subgame).

   For example, consider again the QBF instance in Figure 1 on page 130. The game-state sequent $\langle\{u_{22}, \neg g_2\langle\text{U}, g_2'\rangle\}, \{e_{10}, e_{30}\}\rangle \models (\text{U loses } g_2')$ will force $\neg u_{22}$ if $\textit{CurAsgn} = \{\neg g_2\langle\text{U}, g_2'\rangle, e_{10}\}$. However, $\neg u_{22}$ will not be forced if $\textit{CurAsgn} = \{\neg g_2\langle\text{U}, g_2'\rangle, e_{30}\}$, since e_{10} is upstream of u_{22}, and thus Player U can delay assigning a value to u_{22} until E has assigned a value for e_{10}.

3. If more than one literal in L^{now} is unassigned, then *GS* doesn't force a literal.

When a game-state sequent *GS* forces a literal ℓ, we set `antecedent`$[\ell] = GS$.

Watched Literals. We use a straightforward adaptation of the watched-literals rule [11,6]. For each game-state sequent $\langle L^{\text{now}}, L^{\text{fut}}\rangle \models (P \text{ wins } g)$, we watch two literals in L^{now} and one literal in L^{fut}.

Optimized Implementation of Ghost Literals. If a subformula g occurs in a subgame b, and b itself occurs in a larger subgame s, then we say that this occurrence of g is an *indirect* occurrence in s. For example, in Figure 1, e_{10} occurs directly in g_1' and g_2' but occurs only indirectly in *InFmla*.

If a subformula g occurs directly in only a single subgame b, then we only need to explicitly record only two ghost variables, $g\langle\text{U}, b\rangle$ and $g\langle\text{E}, b\rangle$. For any other quantified formula s that contains g as a subformula,

$$\text{we infer } \underbrace{g\langle P, s\rangle \in \textit{CurAsgn}}_{(P \text{ needs } g \text{ to win } s)} \text{ iff } \underbrace{g\langle P, b\rangle \in \textit{CurAsgn}}_{(P \text{ needs } g \text{ to win } b)} \text{ and } \underbrace{b\langle P, s\rangle \in \textit{CurAsgn}}_{(P \text{ needs } b \text{ to win } s)}$$

since the only way g can influence the value of s is via b. If a subformula g occurs directly in multiple subgames, then we must record two ghost variables (existential and universal) for each subgame in which it directly occurs.

---

[3] Let $\textit{CurAsgn}_I = \{\ell \mid \ell \in \textit{CurAsgn} \text{ and } \ell \text{ is an input literal}\}$. If all literals in L^{now} are input literals, then $\textit{CurAsgn}_I$ matches *GS*, because all literals in L^{now} are assigned true by $\textit{CurAsgn}_I$ and no literals in L^{fut} are assigned false by $\textit{CurAsgn}_I$. If there are ghost literals in L^{now}, then P is still doomed to lose b, because P needs the corresponding gate literals to be true in order to win, but if these gate literals become true, then $\textit{CurAsgn}_I$ will match *GS* and P loses under *GS*.

4.4   Learning New Game States

As shown in Figure 2 on page 133, when it becomes known which player wins *TargFmla* under the current assignment, we call LearnNewGS to learn a new game-state sequent. The only way for it to become known who wins *TargFmla* under *CurAsgn* is for *CurAsgn* to become a ghost match for a game-state sequent in the database (see case 1(a) in Section 4.3). Thus, when we enter LearnNewGS, the current assignment is a ghost match for some game state.

```
func LearnNewGS() {
    GS = GetMatchingGS().copy();
    do { ℓ = (most recently forced literal in GS not owned by winner);
         if (ℓ is quantified outside TargFmla) break;
         Discharge(GS, ℓ);
    } until (GS.now.IsEmpty() || HasGoodUIP(GS));
    return GS;
}

func Discharge(GS, ℓ) {
    GS.now.remove(ℓ);
    GS.now = (GS.now ∪ (antecedent[ℓ].now - {¬ℓ}));
    GS.fut = (GS.fut ∪  antecedent[ℓ].fut);
}
```

Fig. 3. Overview of Learning Algorithm

The procedure for learning a new game-state sequent is shown in Figure 3. We first make a copy of the existing game state that is a ghost match for the current assignment. We then remove the most recently forced literal in L^{now} (not owned by the winner) by *discharging* it via its antecedent, as detailed in Figure 3. We continue to discharge until the L^{now} slot either is empty or has a *good* unique implication point (UIP), as determined by the criteria from [20][4], or until we hit a literal quantified outside *TargFmla*.

For prenex instances, the procedure for discharging a forced literal is similar to *resolution* in clause learning: If $[x_1 \vee ... \vee x_n \vee \ell]$ and $[\neg \ell \vee y_1 \vee ... \vee y_m]$ are true, then $[x_1 \vee ... \vee x_n \vee y_1 \vee ... \vee y_m]$ is also true. The basic argument for the soundness of the discharge method goes as follows. Let $\langle L_A^{now} \cup \{\ell\}, L_A^{fut} \rangle \models (P$ wins $f)$ be GS, and let $\langle L_B^{now} \cup \{\neg\ell\}, L_B^{fut} \rangle \models (P$ wins $h)$ be the antecedent of ℓ. Discharging ℓ via

---

[4] Specifically, an input literal ℓ (owned by the loser) in $\langle L^{now}, L^{fut} \rangle$ is a *good* UIP if (1) the decision variable of ℓ's decision level belongs to the losing player, (2) every literal in $(L^{now} \setminus \{\ell\})$ belongs to an earlier decision level than ℓ, and (3) every literal in L^{fut} that is upstream of ℓ belongs to a decision level earlier than that of ℓ.

Example. Consider the QBF below.

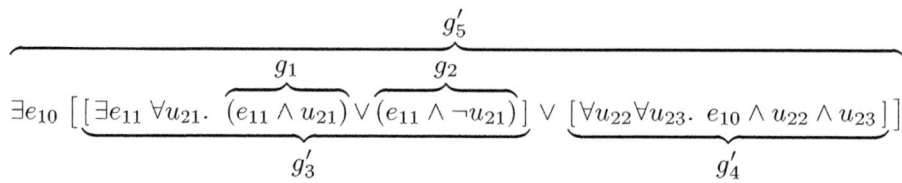

Fig. 4. Example non-prenex QBF instance

1. The initial assignment includes $g\langle \mathrm{E}, g'\rangle$ and $\neg g\langle \mathrm{U}, g'\rangle$ for $g \in \{g_3, g_4, g_5\}$.
2. $\langle\{g_1\langle \mathrm{U}, g_3'\rangle, \neg g_3\langle \mathrm{U}, g_3'\rangle\}, \varnothing\rangle \models (\mathrm{E}\text{ wins }g_3')$ forces $\neg g_1\langle \mathrm{U}, g_3'\rangle$.
3. $\langle\{g_2\langle \mathrm{U}, g_3'\rangle, \neg g_3\langle \mathrm{U}, g_3'\rangle\}, \varnothing\rangle \models (\mathrm{E}\text{ wins }g_3')$ forces $\neg g_2\langle \mathrm{U}, g_3'\rangle$.
4. Player E decides to assign $e_{10} = \mathbf{true}$.
5. All the variables in the outermost quantifier prefix are now assigned, so we must pick a subformula to investigate. We pick g_3' as the new target subformula.
6. Player E decides to assign $e_{11} = \mathbf{true}$.
7. $\langle\{u_{21}, \neg g_1\langle \mathrm{U}, g_3'\rangle\}, \{e_{11}\}\rangle \models (\mathrm{E}\text{ wins }g_3')$ forces $\neg u_{21}$.
8. $\langle\{\neg u_{21}, \neg g_2\langle \mathrm{U}, g_3'\rangle\}, \{e_{11}\}\rangle \models (\mathrm{E}\text{ wins }g_3')$ is a (ghost) match for the current assignment. Since g_3' is the current *TargFmla*, we learn a game state. We discharge $\neg u_{21}$, then $\neg g_2\langle \mathrm{U}, g_3'\rangle$, then $\neg g_1\langle \mathrm{U}, g_3'\rangle$, and finally $\neg g_3\langle \mathrm{U}, g_3'\rangle$, yielding the new game-state sequent $\langle\varnothing, \{e_{11}\}\rangle \models (\mathrm{E}\text{ wins }g_3')$.
9. We now backtrack, removing e_{11} and e_{10} from the current assignment and reverting *TargFmla* to *InFmla*.
10. Having backtracked, our newly learned game-state sequent now forces $g_3\langle \mathrm{U}, g_5'\rangle$.
11. $\langle\{g_3\langle \mathrm{U}, g_5'\rangle, \neg g_5\langle \mathrm{U}, g_5'\rangle\}, \varnothing\rangle \models (\mathrm{E}\text{ wins }\mathit{InFmla})$ matches current assignment.
12. We learn the new game-state sequent $\langle\varnothing, \{e_{11}\}\rangle \models (\mathrm{E}\text{ wins }\mathit{InFmla})$.
13. The empty assignment matches this new game-state, so our final answer is that $\mathit{InFmla} = \mathbf{true}$.

its antecedent yields $\langle L_A^{\mathrm{now}}\cup L_B^{\mathrm{now}},\ L_A^{\mathrm{fut}}\cup L_B^{\mathrm{fut}}\rangle \models (P\text{ wins }f)$. To simplify matters, let us assume that ℓ is upstream of every literal in L_B^{fut}, so that ℓ is forced under any assignment that matches $\langle L_B^{\mathrm{now}}, L_B^{\mathrm{fut}}\rangle$. Since P wins f under any assignment that matches $\langle L_A^{\mathrm{now}} \cup \{\ell\}, L_A^{\mathrm{fut}}\rangle$, we conclude that if an assignment π matches both $\langle L_B^{\mathrm{now}}, L_B^{\mathrm{fut}}\rangle$ and $\langle L_A^{\mathrm{now}}, L_A^{\mathrm{fut}}\rangle$ (i.e., if π matches $\langle L_A^{\mathrm{now}}\cup L_B^{\mathrm{now}}, L_A^{\mathrm{fut}}\cup L_B^{\mathrm{fut}}\rangle$) then ℓ is forced and P wins f.

5   Experimental Results

We implemented the ideas in this paper in a solver which we call *GhostQ*. In our experimental results, GhostQ always did at least as well as CirQit and it outperformed Qube on the k, tipdiam, and tipfixpoint families.

Table 1. Comparison between GhostQ and CirQit

Family	inst.	GhostQ		CirQit	
Seidl	*150*	**150**	(1606 s)	147	(2281 s)
assertion	*120*	**12**	(141 s)	3	(1 s)
consistency	*10*	0	(0 s)	0	(0 s)
counter	*45*	**40**	(370 s)	39	(1315 s)
dme	*11*	**11**	(13 s)	10	(15 s)
possibility	*120*	**14**	(274 s)	10	(1707 s)
ring	*20*	**18**	(28 s)	15	(60 s)
semaphore	*16*	**16**	(4 s)	16	(7 s)
Total	*492*	261	(2435 s)	240	(5389 s)

Table 2. Comparison between GhostQ and Qube

Family	inst.	GhostQ		Qube	
bbox-01x	*450*	171	(133 s)	**341**	(1192 s)
bbox_design	*28*	19	(256 s)	**28**	(15 s)
bmc	*132*	43	(266 s)	**49**	(239 s)
k	*61*	**42**	(355 s)	13	(55 s)
s	*10*	10	(1 s)	10	(5 s)
tipdiam	*85*	**72**	(143 s)	60	(235 s)
tipfixpoint	*196*	**165**	(503 s)	100	(543 s)
sort_net	*53*	0	(0 s)	**19**	(176 s)
all other	*121*	9	(38 s)	**23**	(227 s)
Total	*1136*	531	(1695 s)	643	(2687 s)

We ran GhostQ on the non-CNF instances from QBFLIB on 2.66 GHz machine with a timeout of 300 seconds. For comparison we show the results for CirQit published in [8] (which were conducted on a 2.8 GHz machine with a timeout of 1200 seconds). (CirQit is not publicly available.) As shown in Table 1, GhostQ performs better CirQit on every benchmark family except consistency. The ring and semaphore families consist of prenex instances. The other families are non-prenex, so our solver took advantage of its ability to perform non-prenex

Table 3. Comparison between GhostQ and Non-DPLL Solvers

Family	inst.	Timeout 60 s			Timeout 600 s	
		GhostQ	Quantor	sKizzo	GhostQ	AIGsolve
bbox-01x	*450*	171	130	166	178	173
bbox_design	*28*	19	0	0	22	23
bmc	*132*	43	106	83	51	30
k	*61*	42	37	47	51	56
s	*10*	10	8	8	10	10
tipdiam	*85*	72	23	35	72	77
tipfixpoint	*196*	165	8	25	170	133
sort_net	*53*	0	27	1	0	0
all other	*121*	9	49	31	17	35
Total	*1136*	531	388	396	571	537

In Tables 1–2, we give the number of instances solved and the time needed to solve them. (Times shown do not include time spent trying to solve instances where the solver timed out.) In Table 3, we give the number of instances solved.

game-state learning. During testing of our solver, it was noted that non-prenex learning was especially helpful on the dme family.[5]

We compared GhostQ to the state-of-the-art solvers Qube 6.6 [7], Quantor 3.0 [3], and sKizzo 0.8.2 [2]. We ran these solvers on the QBFLIB QBFEVAL 2007 benchmarks [12] on a 2.66 GHz machine, with a time limit of 60 seconds and a memory limit of 1 GB. The results are shown in Tables 2 and 3. We also show the results for AIGsolve published in [13], but these numbers are not directly comparable because they were obtained on a different machine and with a timeout of 600 seconds.

For the CNF benchmarks, we wrote a script to reverse-engineer the QDIMACS file to circuit form and convert it to our solver's input format. (This is similar to the technique in [13], but we also looked for "if-then-else" gates of the form $g = (x ? y : z)$.) Of the four other solvers shown in Tables 2 and 3, Qube is the only other DPLL-based solver, so it is most similar to our solver. Our experimental results show that GhostQ does better than Qube on the tipdiam and tipfixpoint families (which concern diameter and fixpoint calculations for model checking problems on the TIP benchmarks) and on the k family.

The use of ghost literals can help GhostQ in two ways: (1) By treating the gate literals specially instead of treating them as belonging to the existential player, we can more readily detect satisfactions and we can learn more powerful cubes; (2) By using universal ghost literals, we have a more powerful propagation procedure for the universal input literals. (We did not perform unprenexing on

---

[5] The dme family instances were originally given in prenex form, but we pushed the quantifiers inward as a preprocessing step. The unprenexing time was about 0.8 seconds per instance and is included in our solver's total time shown in the table.

any of the originally-CNF benchmarks, so our use of game-state learning doesn't improve performance here.) To further investigate, we turned off downward propagation of universal ghost literals; on most families the effect was negligible, but on `tipfixpoint` we solved only 149 instances instead of 165.

6 Conclusion

In this paper, we have made two contributions. First, we have introduced the concept of *symbolic game states* and used this concept to reformulate clause/cube learning and extend it to the non-prenex case. Using game states, we have also been able to reformulate the techniques for conflict/satisfaction analysis, BCP, and non-chronological backtracking. In all cases, we give a unified presentation which is applicable to both the existential and universal players, instead of using separate terminology and notation for the two players. Further, game states are 'well-behaved' theoretically, in that we no longer need learn and store tautological clauses (or contradictory cubes). Our second contribution is introducing the concept of *ghost literals*, allowing us to improve upon the propagation technique introduced in [8] by eliminating the asymmetry between the players so that the technique can reduce the search space for both the universal and existential players (instead of only the existential player). Experiments show that our techniques work particularly well on certain benchmarks related to formal verification. For future work, it may be worthwhile to investigate whether the ideas of dynamic partitioning [15] can be extended to allow dynamic unprenexing.

References

1. Ansótegui, C., Gomes, C.P., Selman, B.: The Achilles' Heel of QBF. In: AAAI (2005)
2. Benedetti, M.: Evaluating QBFs via Symbolic Skolemization. In: Baader, F., Voronkov, A. (eds.) LPAR 2004. LNCS (LNAI), vol. 3452, pp. 285–300. Springer, Heidelberg (2005)
3. Biere, A.: Resolve and Expand. In: Hoos, H., Mitchell, D.G. (eds.) SAT 2004. LNCS, vol. 3542, pp. 59–70. Springer, Heidelberg (2005)
4. Egly, U., Seidl, M., Woltran, S.: A Solver for QBFs in Nonprenex Form. In: ECAI (2006)
5. Ganai, M.K., Ashar, P., Gupta, A., Zhang, L., Malik, S.: Combining strengths of circuit-based and CNF-based algorithms for a high-performance SAT solver. In: DAC (2002)
6. Gent, I.P., Giunchiglia, E., Narizzano, M., Rowley, A.G.D., Tacchella, A.: Watched Data Structures for QBF Solvers. In: Giunchiglia, E., Tacchella, A. (eds.) SAT 2003. LNCS, vol. 2919, pp. 25–36. Springer, Heidelberg (2004)
7. Giunchiglia, E., Narizzano, M., Tacchella, A.: Quantifier structure in search based procedures for QBFs. In: DATE 2006 (2006)
8. Goultiaeva, A., Iverson, V., Bacchus, F.: Beyond CNF: A Circuit-Based QBF Solver. In: Kullmann, O. (ed.) SAT 2009. LNCS, vol. 5584, pp. 412–426. Springer, Heidelberg (2009)

9. Jain, H., Bartzis, C., Clarke, E.: Satisfiability Checking of Non-clausal Formulas Using General Matings. In: Biere, A., Gomes, C.P. (eds.) SAT 2006. LNCS, vol. 4121, pp. 75–89. Springer, Heidelberg (2006)

10. Lonsing, F., Biere, A.: Nenofex: Expanding NNF for QBF solving. In: Kleine Büning, H., Zhao, X. (eds.) SAT 2008. LNCS, vol. 4996, pp. 196–210. Springer, Heidelberg (2008)

11. Moskewicz, M.W., Madigan, C.F., Zhao, Y., Zhang, L., Malik, S.: Chaff: Engineering an Efficient SAT Solver. In: DAC 2001 (2001)

12. Narizzano, M., Pulina, L., Tacchella, A.: QBFEVAL, http://www.qbfeval.org/

13. Pigorsch, F., Scholl, C.: Exploiting structure in an AIG based QBF solver. In: DATE 2009 (2009)

14. Sabharwal, A., Ansótegui, C., Gomes, C.P., Hart, J.W., Selman, B.: QBF Modeling: Exploiting Player Symmetry for Simplicity and Efficiency. In: Biere, A., Gomes, C.P. (eds.) SAT 2006. LNCS, vol. 4121, pp. 382–395. Springer, Heidelberg (2006)

15. Samulowitz, H., Bacchus, F.: Dynamically Partitioning for Solving QBF. In: Marques-Silva, J., Sakallah, K.A. (eds.) SAT 2007. LNCS, vol. 4501, pp. 215–229. Springer, Heidelberg (2007)

16. Thiffault, C., Bacchus, F., Walsh, T.: Solving Non-clausal Formulas with DPLL Search. In: Wallace, M. (ed.) CP 2004. LNCS, vol. 3258, pp. 663–678. Springer, Heidelberg (2004)

17. Tseitin, G.S.: On the complexity of derivations in the propositional calculus. In: Slisenko, A.O. (ed.) Studies in Constructive Mathematics and Mathematical Logic, Part-II (1968)

18. Zhang, L.: Solving QBF by Combining Conjunctive and Disjunctive Normal Forms. In: AAAI 2006 (2006)

19. Zhang, L., Malik, S.: Conflict Driven Learning in a Quantified Boolean Satisfiability Solver. In: ICCAD 2002 (2002)

20. Zhang, L., Malik, S.: Towards a Symmetric Treatment of Satisfaction and Conflicts in Quantified Boolean Formula Evaluation. In: Van Hentenryck, P. (ed.) CP 2002. LNCS, vol. 2470, p. 200. Springer, Heidelberg (2002)

SAT Solving with Reference Points[*]

Stephan Kottler

Wilhelm–Schickard–Institute, University of Tübingen, Germany

Abstract. Many state-of-the-art SAT solvers use the VSIDS heuristic to make branching decisions based on the activity of variables or literals. In combination with rapid restarts and phase saving this yields a powerful decision heuristic in practice. However, there are approaches that motivate more in-depth reasoning to guide the search of the SAT solver. But more reasoning often requires more information and comes along with more complex data structures. This may sometimes even cause strong concepts to be inapplicable in practice.

In this paper we present a suitable data structure for the DMRP approach to overcome the problem above. Moreover, we show how DMRP can be combined with CDCL solving to be competitive to state-of-the-art solvers and to even improve on some families of industrial instances.

1 Introduction

Research in satisfiability checking (SAT) has managed to bridge the gap between theory and practice in many aspects. There are several kinds of real-world problems that are actually tackled by modelling those problems as SAT instances like hardware and software verification [21,10], planning [11], automotive product configuration [13] and haplotype inference in bioinformatics [15] (cf. [16]).

In the domain of SAT solving there are different schemes and even more variants of these schemes to decide whether there exists a satisfying assignment to the variables of a Boolean formula in CNF or if a formula cannot be satisfied by any assignment. Both experiments and applications show that there is no perfect SAT solving approach that is suited for all different categories and families of problem instances. However, conflict-driven solving has proven itself to be very successful on a wide range of benchmarks. In this paper we study the quite new DMRP algorithm (decision making with a reference point) [8,9] from a practical point of view. Moreover, a hybrid approach that combines DMRP and CDCL solving is presented which is also motivated by experimental evaluations.

The paper is organised as follows: In section 2 we sketch related work in the domain of CDCL and DMRP solving. Section 3 examines the DMRP approach from a practical point of view and we introduce a new implementation for this approach. In section 4 we motivate the combination of CDCL and DMRP to a new hybrid approach. In section 5 some experimental results are presented.

---

[*] This work was supported by DFG-SPP 1307, project "Structure-based Algorithm Engineering for SAT-Solving".

O. Strichman and S. Szeider (Eds.): SAT 2010, LNCS 6175, pp. 143–157, 2010.

2   Related Work

In this chapter two different SAT solving approaches are sketched. The state-of-the-art conflict-driven solving and the quite recent DMRP approach that operates on complete assignments. By $\mathcal{V}(F)$ resp. $\Gamma(F)$ we state the set of variables resp. clauses of a formula F (we omit F if evident). A clause consists of literals l_i that are variables v or their negations $\bar{v}$. The polarity of a literal is *true* or *false* respectively. $var(l)$ indicates the variable of literal l.

Conflict-Driven Solving. Conflict-driven solving with clause learning (CDCL) is a leading approach and is especially but not only successful for industrial problems. It is based on the GRASP algorithm [17] which extends the original DPLL branch-and-bound procedure [5,4] by the idea of learning from conflicting assignments. Moreover, conflicts are analysed to jump over parts of the search space that would cause further conflicts. There are several improvements to the original algorithm like the *two-watched-literal* data structure and the VSIDS (variable state independent decaying sum) variable selection heuristic [18]. In recent years further improvements have been achieved by developing different restart strategies like the concept of rapid and adaptive restarts [3,2] and so-called Luby restarts [14]. In combination with phase-saving [19] frequent restarts constitute a strong concept especially for industrial SAT instances.

Decision Making with a Reference Point. DMRP is a new SAT solving approach that was proposed by Goldberg in [8,9]. Even though DMRP uses Boolean constraint propagation (BCP) with backtracking and learning from conflicting assignments it is not a simple variant of CDCL. In difference to CDCL solvers DMRP additionally holds a complete assignment (a so-called reference point). The algorithm aims for modifying the current reference point $\mathcal{P}$ to $\mathcal{P}'$ in order to satisfy a clause under consideration. Furthermore, it is crucial that all clauses being satisfied by $\mathcal{P}$ remain satisfied by the modified reference point $\mathcal{P}'$.

Algorithm 1 gives an overview of the DMRP approach, though this notation varies in some ways from the original notation in [9]. One invocation of the DMRP subsolver (line 7 of Algorithm 1) takes a clause and a reference point as arguments. It may either compute a modification to the reference point or it may learn the empty clause or else it times out. The latter case causes the surrounding algorithm to call the DMRP subsolver with another unsatisfied clause.

3   A Closer Look at DMRP

Taking the set of clauses that are not satisfied by a current assignment as basis for branching decisions requires the solver to know this set of clauses. This could be realized analogously to how it is implemented in many local search approaches [20,7] where the solver keeps track of clauses that change their state from 'satisfied' to 'not satisfied' and vice versa whenever the value of a variable changes. However, for any variable v this implies the solver to know all clauses where v

---

Algorithm 1. Sketch of the DMRP approach

Require Formula F in CNF with $\mathcal{V}, \Gamma$ the set of variables and clauses, a reference point $\mathcal{P}$ and any two timeout criteria T_1, T_2

Function solveDMRP$(F, \mathcal{P}, T_1, T_2)$

    $\mathcal{M} \leftarrow \{C \in \Gamma(F) \mid C \text{ not satisfied by } \mathcal{P}\}$;

    while $\neg T_1$ **do**

        if $\mathcal{M} = \emptyset$ **then return** 'Satisfiable';

6        $C \leftarrow$ remove any clause from $\mathcal{M}$;

7        $res \leftarrow$ dmrpTryModifyPoint$(F \setminus \mathcal{M}, C, \mathcal{P}, T_2)$;

        if $res =$ *'Unsatisfiable'* **then return** 'Unsatisfiable';

        else if $res =$ *'Timeout'* **then** $\mathcal{M} \leftarrow \mathcal{M} \cup \{C\}$;

        else

            $\mathcal{P} \leftarrow$ modify$(\mathcal{P}, res)$;   /* adapt ref. point */

            $\mathcal{M} \leftarrow \{C \in \Gamma \mid C \text{ not satisfied by } \mathcal{P}\}$;

Require (Sub)formula F', a clause C that shall be satisfied by modification of the current point $\mathcal{P}$, and a timeout criteria T

14 **Function** dmrpTryModifyPoint$(F', C, \mathcal{P}, T)$

    $\mathcal{D} \leftarrow \{C\}$ $\mathcal{P}_t \leftarrow \mathcal{P}$;

    while $\neg T$ **do**

        if $\mathcal{D} = \emptyset$ **then return** $\mathcal{P}_t$;   /* found valid modification of $\mathcal{P}$ */

        $C \leftarrow$ choose any clause from $\mathcal{D}$;

        $l \leftarrow l \in C \mid \mathcal{P}_t \setminus \{\bar{l}\} \cup \{l\}$ satisfies maximal number of clauses in $\mathcal{D}$;

        $< res, \mathcal{P}_t > \leftarrow$ boolean-constraint-propagation$(F', l := true, \mathcal{P}_t)$;

        while $res =$ *'Conflict'* **do**

22            $lemma \leftarrow$ analyze-conflict(F', res) ;

            if $lemma = \emptyset$ **then return** 'Unsatisfiable';

24            $\mathcal{P}_t \leftarrow$ backtrack-reset-point$(F', lemma)$;

25            $< res, \mathcal{P}_t > \leftarrow$ learn-and-propagate$(F', lemma, \mathcal{P}_t)$;

        $\mathcal{D} \leftarrow \{C \in \Gamma(F') \mid C \text{ not satisfied by } \mathcal{P}_t\}$;

    return 'Timeout'

---

resp. $\bar{v}$ occurs in. Since the *two watched literals* scheme was introduced [18] most CDCL based solvers do not maintain complete *occurrence-lists* of variables.

In this section we present a data structure that allows for a fast computation of the most frequently required information in the DMRP approach by simultaneously avoiding the maintenance of complete occurrence-lists.

3.1 Different States of Variables

In CDCL solvers each variable v can actually have three values: $val(v) \in \{true, false, unknown\}$. In general, any variable whose value is known has either been chosen as decision variable or its value was implied by BCP. To undo decisions and their implications both types of assignments (decisions and implications) are placed on a stack (often called trail) in the order they are assigned [6,3].

In the DMRP algorithm we introduce two different kinds of values expressed by the functions *pval* and *tval*: The DMRP algorithm maintains a reference point $\mathcal{P}$ which is an assignment to all the variables in the formula. Hence, for any variable v in the formula the reference point $\mathcal{P}$ either contains v or its negation $\bar{v}$. For a variable v we refer to its value in $\mathcal{P}$ by $pval(v) \in \{true, false\}$. The second kind of value $tval(v)$ is used to state a temporary modification of $pval(v)$. The default of $tval(v) \in \{true, false, ref\}$ is ref which indicates that the corresponding variable is not affected by the current temporary modification of $\mathcal{P}$ and hence the value given by $pval(v)$ is valid. During the search for a modification of $\mathcal{P}$ to $\mathcal{P}'$ (line 14 of Algorithm 1) that reduces the set of unsatisfied clauses $\mathcal{M}$ to $\mathcal{M}' \subset \mathcal{M}$ the temporary value $tval(v) \neq ref$ hides $pval(v)$ for any variable v. For any literal l with polarity b the function $pval(l)$ (resp. $tval(l)$) is true iff $pval(var(l)) = b$ (resp. $tval(var(l)) = b$) and it is $tval(l) = ref$ iff $tval(var(l)) = ref$.

3.2 Clauses Satisfied by the Reference Point

In addition to standard SAT solving the algorithm has to maintain a reference point $\mathcal{P}$. Obviously, if all clauses Γ are satisfied by $\mathcal{P}$ the algorithm has found a model for the formula. Hence, for the remaining section we assume the set of clauses $\mathcal{M}$ that contains all clauses not satisfied by $\mathcal{P}$ to be non-empty.

After any initialisation of $\mathcal{P}$ the set $\mathcal{M}$ can be computed by simply traversing Γ. However, whilst the algorithm tries to modify $\mathcal{P}$ in order to satisfy more clauses of $\mathcal{M}$ we have to keep track of those clauses in $\Gamma \setminus \mathcal{M}$ that become temporarily unsatisfied by a temporarily modified reference point. These clauses are put onto a stack $\mathcal{D}$ which is described further below. The first matter is how to compute the clauses that become unsatisfied by a modification of the reference point.

Similar to the concept of watched literals [18] for each clause C in $\Gamma \setminus \mathcal{M}$ we choose one literal $l \in \mathcal{P}$ to take on responsibility for C regarding its satisfiability by the current reference point $\mathcal{P}$. By definition for any clause in $\Gamma \setminus \mathcal{M}$ at least one such literal $l \in C$ has to exist with $pval(l) = true$. We say a literal l is responsible for a set of clauses $R(l)$. Whenever the value of a variable $v := var(l)$ changes from $tval(v) = ref$ to $\neg pval(v)$ all clauses in $R(l)$ have to be traversed. For each clause $C \in R(l)$ a new literal from the current (modified) reference point has to be found that takes on responsibility for C.

Note that - in addition to the responsibilities regarding the reference point - there are also two literals per clause that watch this clause in the sense of the usual two-watched-literal scheme [18]. This is necessary to notice whenever a temporary modification ($tval$) generates a unit clause or completely unsatisfies a clause. Let the set of clauses that are watched by a literal l be $W(l)$. We examine this in more detail now.

Whenever the value of a variable is changed whilst searching for a modified reference point $\mathcal{P}'$ ($tval$ is changed) we have to take care of $W(l)$ and $R(l)$ of the corresponding literal l that became false under $tval$. When examining $W(l)$ the usual three cases may happen for any affected clause C that is watched by l and any other literal $l_w \in C$. For these cases only the values of $tval$ act a part:

W.1 Another literal $l_j \in \{C \setminus l_w\}$ with $tval(l_j) \neq false$ can watch C.

W.2 There is no other literal in $\{C \setminus l_w\}$ that is not false. Hence $tval(l_w)$ has to be set to $true$ to satisfy C.

W.3 If in the second case above $tval(l_w)$ is already set to $false$ a conflicting assignment is generated and the algorithm jumps back to resolve the conflict.

The following update is done after the list $W(l)$ was examined successfully:

R.1 If $tval(l)$ equals $pval(l)$ nothing has to be done.

R.2 $tval(l)$ differs from $pval(l)$ and another literal in C can be found to take responsibility for C. This might be any literal $l_j \in C$ for which it is $tval(l_j) = ref$ and $pval(l_j) = true$. In that case C is removed from $R(l)$ and put into $R(l_j)$. Or we might find a literal with $tval(l_j) = true$. In that case C remains in $R(l)$ since $tval(l_j)$ was obviously assigned before the current modification of l in the reference point.

R.3 $tval(l)$ differs from $pval(l)$ but no other literal $\in C$ satisfies C under the current temporary point. In that case C is put on the stack $\mathcal{D}$ that keeps track of all clauses that are not satisfied by the current temporary reference point. Note that since $W(l)$ was examined first there are at least two literals $l_i, l_j \in C$ for which $tval(l_i) = tval(l_j) = ref$ and $pval(l_i) = pval(l_j) = false$. If this did not hold one of the cases W.2, W.3 or R.2 would apply.

Note that this implementation (sketched in Algorithm 2) allows for backtracking without any updates of the sets $R(l)$ of any literal l. The responsibility list[1] $R(l)$ only has to be examined when $tval$ of a variable changes from ref to $true$ or $false$ not for the opposite case. Moreover, the data structure is sound in the sense that no clause that becomes unsatisfied by $\mathcal{P}$ will be missed.

3.3   Keeping Track of Temporarily Unsatisfied Clauses

While trying to modify a reference point $\mathcal{P}$ to $\mathcal{P}'$ to reduce the set $\mathcal{M}$ of clauses that are unsatisfied by $\mathcal{P}$ to $\mathcal{M}' \subset \mathcal{M}$ a data structure $\mathcal{D}$ is used to store those clauses that are unsatisfied by any temporary reference point $\mathcal{P}_t$. In the subsection above we described when clauses are added to $\mathcal{D}$. The data structure $\mathcal{D}$ has to meet three main demands:

- Clauses that are not satisfied by the current point $\mathcal{P}_t$ have to be found in reasonable time without having to traverse the clause's literals at each lookup in the data structure.
- Backjumping over parts of the temporary modification (due to a conflict - see case W.3) has to be very fast with least possible overhead to update $\mathcal{D}$.

---

[1] For any literal l the list $R(l)$ is only meaningful if $pval(l) = true$. To save memory, responsibility lists can be associated with variables in practice.

Algorithm 2. Update data structure when the value of a variable changed

---

Require A variable $v \in \mathcal{V}$ where $tval(v)$ was changed from ref to
$b \in \{true, false\}$. Let l_c be the literal of v with $tval(l_c) = false$.
Function onChangeOfVariableTVal(v)

> **forall** $C \in W(l_c)$ **do**
>> $l_w \leftarrow$ other watcher of C $(l_w \neq l_c)$;
>> **if** $tval(l_w) = true$ **then** continue;
>> **if** $\exists\, l_n \in C :\ tval(l_n) \neq false \wedge l_n \neq l_w$ **then**
>>> $W(l_c) \leftarrow W(l_c) \setminus \{C\};\quad W(l_n) \leftarrow W(l_n) \cup \{C\}$
>>
>> **else if** $tval(l_w) = false$ **then return** 'conflict';
>> **else** $tval(l_w) \leftarrow true$; ;   /* usual two watched literal scheme */
>
> **if** $tval(v) = pval(v)$ **then return** 'ok' ;   /* R.1 */
> **forall** $C \in R(\overline{l_c})$ **do**
>> **if** $\exists\, l_0 \in C :\ tval(l_0) = ref\ \wedge pval(l_0) = true$ **then**
>>> $R(\overline{l_c}) \leftarrow R(\overline{l_c}) \setminus \{C\}; R(l_0) \leftarrow R(l_0) \cup \{C\}$;   /* R.2.1 */
>>
>> **else if** $\exists\, l_0 \in C : tval(l_0) = true$ **then** continue ;   /* R.2.2 */
>> **else** push C at $\mathcal{D}$;   /* R.3 */

- When a temporarily unsatisfied clause C is chosen from $\mathcal{D}$ by the decision procedure the set $\Lambda_C = \{l \in C\ :\ tval(l) = ref\}$ contains those literals whose value in $\mathcal{P}_t$ may possibly be modified to satisfy C (as stated in case R.3 above it is $|\Lambda_C| \geq 2$). Our data structure has to support finding that literal of Λ_C which satisfies the most clauses in $\mathcal{D}$.

Fast Backjumping. The main data structure is depicted in Figure 1. Since the second issue above is fundamental we use a stack to realise $\mathcal{D}$ which has one entry pL for each decision level. Each entry itself basically points to a set of clauses L that became unsatisfied by the current point $\mathcal{P}_t$ at this level. In addition each L has a flag that indicates if the referred clauses still have to be considered to belong to $\mathcal{D}$. This allows for very fast backjumping: For each level we jump back the according flag in L is set to false, the set of clauses in L is deleted and pL is popped from the stack. This means a negligible overhead compared to backjumping in CDCL solving. Note, that an entry pL which is removed from the stack does not destroy the corresponding set L which allows other data still to refer to L. These invalid references may be updated lazily later on. Another important advantage of this implementation will become evident further below.

Finding clauses not satisfied by $\mathcal{P}_t$. To find those clauses in $\mathcal{D}$ that still have to be satisfied by further modifications of the current point $\mathcal{P}_t$ the clause sets L that are referred by entries pL of the stack have to be traversed. Let this procedure be called $findUnsat(\mathcal{D})$. We do not remove any satisfied clause C from any set L that is still flagged to be valid since this would require to put such clauses back into L whenever the satisfying modification to C is undone. Instead, we additionally cache a literal for each clause in L as a kind of representative.

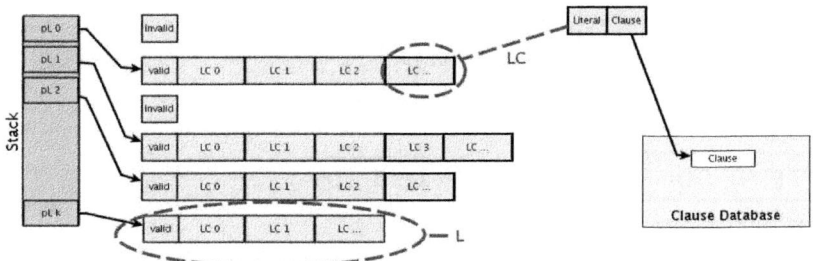

Fig. 1. Basic data structure to represent $\mathcal{D}$

Thus any entry in L (besides the flag) is an element LC which itself consists of a clause C and one representative literal $l \in C$. When a clause C in a set L is found to be satisfied by a temporary point $\mathcal{P}_t$ the representative literal l is set to that one which actually satisfies this clause $(tval(l) = true)$.

When searching for unsatisfied clauses in $\mathcal{D}$ we first check the state of the representative literal before the entire clause is checked. On the one hand this guarantees that a satisfied clause is not checked twice unless a modification makes this necessary. On the other hand significant changes of $\mathcal{P}_t$ to the satisfiability state of any clause are not missed. The latter issue would require extra maintenance if only Boolean flags were used to mark satisfied clauses. To find clauses in $\mathcal{D}$ that are still unsatisfied by the current point $\mathcal{P}_t$ we traverse the stack from its top[2] to its bottom. This prefers the most recently added clauses. The first clause that is found to be unsatisfied by $\mathcal{P}_t$ is taken as basis for the next branching decision.

Computation of the MakeCount of variables. Given a clause C^* that is unsatisfied under the current point $\mathcal{P}_t$ the algorithm has to find that literal of $\Lambda_{C^*} = \{l \in C^* : tval(l) = ref\}$ which satisfies most clauses in $\mathcal{D}$ (or optionally $\mathcal{D} \cup \mathcal{M}$) when its value in $\mathcal{P}_t$ is changed. To compute this so-called *MakeCount* of a variable we use another data structure that interplays with the above one. This data structure allows for lazy computation of the MakeCount of a variable and is basically organised as follows:

Each variable v that is not yet affected by the temporary modification of the reference point $\mathcal{P}_t$ $(tval(v) = ref)$ is associated with a list Ω_v of elements M. Each element represents a clause in $\mathcal{D}$ that can be satisfied by flipping the current value of v in the point $\mathcal{P}_t$. Due to the laziness of the data structure it might be that an element M is out of date. More precisely (see Figure 2) each element M (representing a clause C) in a list of variable v consists of two fields:

The first field references the set L of clauses in which C is contained. The second field is an index into L that indicates the particular clause C (i.e. the according element LC) that can be satisfied by flipping the value of v in $\mathcal{P}_t$.

---

[2] An index into $\mathcal{D}$ can be cached such that search only starts from the top of $\mathcal{D}$ if a conflict occurred at the previous decision.

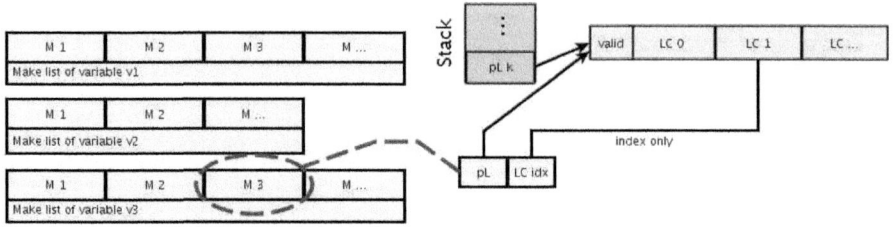

Fig. 2. Computation of the *MakeCount* of variables

Whenever case $R.3$ from above applies for a clause C a pointer to C and any representing literal are wrapped into an object LC. This data is appended to the set of clauses L which is referenced from the topmost entry on the stack $\mathcal{D}$. At this point we also add an entry M to the Make-lists (Ω_v) of all variables v of $\Lambda_C = \{l \in C : tval(l) = ref\}$. We take a closer look at the different cases when computing the valid MakeCount from the possibly out-of-date information:

- It might be that an element M refers to a clause that has already been removed from $\mathcal{D}$. In that case the flag of the structure L referenced by M has been invalidated during backtracking. Hence, this case can be realised immediately and M can also be deleted from the list.
- We can assume now that the clause C indirectly referenced by M is still contained in $\mathcal{D}$. Let us assume for now that C is already satisfied by a further modification of the point $\mathcal{P}_t$. Recall, that what we actually get from M is a reference to LC - the clause C and a representing literal l of C. We can distinguish between two cases:
  - C might have been already considered by the procedure *findUnsat(D)* to find unsatisfied clauses in $\mathcal{D}$ as described above. In that case *findUnsat()* has changed the representative literal l such that by checking the value $(tval)$ of l in $\mathcal{P}_t$ we know that C is satisfied and we are done.
  - If C was not considered by *findUnsat(D)* yet, the satisfiability state of the clause has to be computed by checking its literals. Given that C is satisfiable under $\mathcal{P}_t$ a literal that satisfies C will be found and will be made the representative literal for this clause in LC. This allows for fast detection of the satisfiability of C later on and will relieve *findUnsat(D)* from checking all literals of C. The representing literal guarantees that for each temporary point $\mathcal{P}_t$ there is at most one traversal through all literals of a clause to recognise that this clause is satisfied by $\mathcal{P}_t$.
- In case C is not satisfied by the current point $\mathcal{P}_t$ this is recognised by a check of all literals in C.

The realisation of the set $\mathcal{D}$ and the data structure to compute MakeCounts of variables follows the concept of lazy data structures and avoids to store complete occurrence lists for literals. MakeCount lists do not require any update operation on backjumping. Even though indices in M become undefined when the referred set L is cleared during backjumping, this is not problematic since an index is only used after L is asserted to be still valid by its flag.

The size of a list Ω_v of a variable v gives an upper bound $\widehat{\Omega_v}$ on the valid MakeCount of v. Hence, to determine the variable with maximum MakeCount of a clause, the variables are traversed in descending order regarding $\widehat{\Omega}$. This allows for early termination when $\widehat{\Omega}$ becomes smaller than the actual valid maximum MakeCount.

3.4   Learning

Two aspects that have to be considered for the realisation of the DMRP approach are related to learning (Algorithm 1, line 22) as mentioned in case W.3:

Whenever a unit clause $C = (l)$ is learned the algorithm jumps back to decision level 0, assigns l to be true and propagates all implications of this assignment. This also requires a modification of the current reference point $\mathcal{P}$ to $\mathcal{P}'$ with a difference to previously described modifications: The set of clauses $\mathcal{M}'$ unsatisfied by $\mathcal{P}'$ does not necessarily have to be a subset of $\mathcal{M}$ – the set of clauses unsatisfied by the previous reference point $\mathcal{P}$.

Secondly, for any learned lemma that is generated when a conflict is analysed the data structure has to be updated properly. We use the following property.

Property: Any lemma generated by the function *analyze-conflict* in line 22 of Algorithm 1 contains at least one literal l with $l \in \mathcal{P}$ ($pval(l) = true$) regarding the current valid reference point $\mathcal{P}$. In other words: No generated lemma extends[3] the current set $\mathcal{M}$.

Proof: We prove this property by the construction of learned lemmata. The surrounding function *dmrpTryModifyPoint* (lines 7, 14 of Algorithm 1) only considers clauses in $\Gamma \setminus \mathcal{M} \cup C$ where a modification of $\mathcal{P}$ is wanted that additionally satisfies C. Since C is always the base for the decision at the first decision level any temporary point $\mathcal{P}_t \neq \mathcal{P}$ will always satisfy C during one execution of *dmrpTryModifyPoint*. Hence, C can never be an assign-reason to an assignment of a variable, since assign-reasons are clauses that become unit during the search. Thus, all assign-reasons are clauses from the set $\Gamma \setminus \mathcal{M}$ that, by definition, are satisfied by at least one literal $\in \mathcal{P}$.

Running into a conflict means that for a clause C_0 (conflicting clause) all literals are set to false. The lemma λ_* is generated by recursively resolving out variables (that were no decisions) from the conflicting clause by using the according assign-reasons. C_0 can be seen as the first version (λ_0) of the generated lemma λ_*. Given that $C_0 \in \Gamma \setminus \mathcal{M}$ one of the literals of λ_0 is in $\mathcal{P}$. Let $l*$ be one literal $\in \lambda_i$ with $l* \in \mathcal{P}$ ($pval(l*) = true$). If any literal $l' \neq l*$ is resolved out from λ_i the resolvent λ_{i+1} still contains literal $l*$. If on the other hand $l*$ is resolved out by the use of its assign reason C^*, clause C^* has to contain literal $\overline{l*}$. Since $C^* \in \Gamma \setminus \mathcal{M}$ it also has to contain a literal $l^* \in \mathcal{P}$. Moreover, with $l* \in \mathcal{P}$ it is $l^* \neq \overline{l*}$ and the new resolvent λ_{i+1} contains literal $l^* \in \mathcal{P}$. By induction the final lemma contains at least one literal that is in $\mathcal{P}$.   □

---

[3] Note the difference that generated lemmata always extend the formula.

For any generated lemma λ_* we chose that literal $l \in \mathcal{P} \cap \lambda_*$ which was assigned at the highest decision level d (most recently). By the above property at least one such literal l has to exist. Literal l takes on responsibility for λ_*: $R(l) \leftarrow R(l) \cup \{\lambda_*\}$. The functions in lines 24 and 25 of Algorithm 1 determine a new point $\mathcal{P}_t$. If $l \notin \mathcal{P}_t$ the lemma λ_* is also appended to the list L that is referred by the stack $\mathcal{D}$ for decision level d and considered for the MakeCounts as described in the previous section 3.3. These two actions guarantee a proper update of the entire data structure and no more special treatments are needed.

4   Combining DMRP and CDCL to a Hybrid Solver

In Algorithm 1 we assume the initial reference point to be given from outside. In the original paper [9] reference points are chosen at random and are then tried to modify by a call to function *solveDMRP*. In case no result can be computed within a certain amount of time (i.e. number of conflicts) *solveDMRP* will be invoked with a new initial point. This is similar to local search restarts but with the difference that the DMRP algorithm itself carefully reasons on how to modify a reference point. However, the choices of initial points are crucial for the algorithm as presented in section 5.

As mentioned in section 2 CDCL solvers perform restarts quite frequently. At a restart activity values of variables or literals are kept and also a subset of the learned clauses is carried along for the next start. However, the current partial assignment (all literals in the trail) is almost completely rejected, even though phase saving keeps some information. This motivates a hybrid approach that reasonably alternates the CDCL and the DMRP algorithms. The DMRP approach offers a suitable possibility to take a closer look at the drawback of a partial assignment before it is rejected. It may focus on the not yet satisfied clauses.

Our recent implementation that is shown in Algorithm 3 combines both approaches by the use of the Luby et al. restart strategy [14] which proved itself successful in both theory and practice. The Luby strategy assumes that the algorithm does not have any external information and does not know when it is best to perform a restart. In that case the available computation time is shared almost equally among different restart strategies [14]. The function *maxConflict-Count* in Algorithm 3 returns the number of conflicts for the next run due to the Luby strategy. That is the product of a constant factor f and the next number of the sequence $(1, 1, 2, 1, 1, 2, 4, 1, 1, 2, 4, 8, 1, \ldots)$ (see [14] for details).

The function *chooseAlgo* decides on which algorithm to use for the next run. On average we achieved the best results when running the DMRP algorithm exactly for the smallest conflict limit (when $cl = f$).

Since the DMRP algorithm requires a reference point i.e. an assignment to all variables the last partial assignment of the CDCL solver has to be extended to a complete assignment (*extendPartialAssignmToRefPoint*). This is done by continuing the previous CDCL search with the last partial assignment. However, within this execution only binary clauses are considered during search until all

Algorithm 3. The hybrid approach

Require Formula F in CNF with $\mathcal{V}, \Gamma$ the set of variables and clauses
Function solveHybrid(F)

> $last \leftarrow$ 'CDCL' $res \leftarrow$ 'Unknown' ;
> **while** $res =$ '*Unknown*' **do**
>> $cl \leftarrow$ maxConflictCount() ;   /* Use Luby strategy */
>> $algo \leftarrow$ chooseAlgo(cl) ;   /* Apply CDCL or DMRP ? */
>> **if** $algo =$ '*DMRP*' **then**
>>> **if** $last =$ '*CDCL*' **then**
>>>> $< res, \mathcal{P} > \leftarrow$ extendPartialAssignmToRefPoint() ;
>>>> **if** $res =$ '*Unsatisfiable*' **then return** res;
>>>
>>> $res \leftarrow$ solveDMRP($F, \mathcal{P}, cl, cl$);
>>
>> **else** $res \leftarrow$ solveCDCL(F, cl);
>> $last \leftarrow algo$;
>
> **return** res;

variables are assigned a value. This assignment constitutes the initial reference point for the DMRP algorithm. In this phase the solver may also realise that the formula is unsatisfiable. For the case the partial assignment is empty (at algorithm start) this function simply computes a reference point that satisfies all binary clauses. Taking care of binary clauses at first is also motivated by the work in [22] and [1] where the idea to primarily focus on binary clauses has also improved solving for some families of instances. This also guarantees an additional invariant for our data structure that a binary clause can neither be contained in the set $\mathcal{M}$ nor in the delta stack $\mathcal{D}$ (resp. its elements).

Some Adaptions for the Hybrid Approach

In addition to standard CDCL solving each clause of the formula is assigned an activity value initialised to zero at the beginning. Whenever a clause is involved in a conflict (i.e. it is used for resolution during the generation of a lemma) its activity value is increased. In some solvers (for instance [6]) this technique is common for learned clauses to clear the clause database of inactive learned clauses periodically. Our hybrid solver maintains an activity value for every clause.

This activity value of a clause is taken into account when the next clause from set $\mathcal{M}$ has to be chosen (line 6 of Algorithm 1) to be handled by the function *dmrpTryModifyPoint*. We always choose the clause with the highest activity value for the next attempt to modify the current reference point. However, if the call to *dmrpTryModifyPoint* times out or for two subsequent calls to *solveDMRP* the next clause with the second highest activity value is chosen.

In contrast to the original DMRP algorithm the conflict limit (timeout) for the function *solveDMRP* depends on the success of its subroutine *dmrpTryModifyPoint* in line 7 of Algorithm 1. If the current reference point could be improved the initial conflict limit is reset.

The solver also differs in the computation of the MakeCount of a variable. For the MakeCount one can count only the clauses currently in $\mathcal{D}$ to get the most local improvement or all clauses in $\mathcal{D} \cup \mathcal{M}$ can be considered to make decisions more globally. For variables that have the same MakeCount ties can be broken in favour of different issues which is explained in more detail in the next section.

5   Experiments and Evaluation

For the evaluation presented in this section we have run our solver for all industrial (application) instances of the SAT competitions resp. SAT Races of the years 2006 - 2009 that add up to 564 non-trivial[4] instances. Each instance is preprocessed in advance and the timeout for the solvers was set to 1200 seconds. As a reference and also to check results we have run our CDCL solver using the Luby restart strategy (without DMRP) and MiniSAT 2.0 [6].

Figure 3 shows the results of different configurations of the hybrid approach. Furthermore, there are results that show performance of a pure DMRP solver. The presented configurations differ in the following issues that are related to decision making: As mentioned above the MakeCount may consider all clauses in $\mathcal{D} \cup \mathcal{M}$ (global) or only clauses in $\mathcal{D}$ (local). If two variables have the same MakeCount ties are broken in favour of the variable v that:

(Act) has the highest activity value similar to the VSIDS heuristic [18].
(BC) has the smallest set $R(v)$. This can be seen as a simple approximation of the *BreakCount* of the variable. In difference to the MakeCount the BreakCount of a variable v states the number of clauses that become unsatisfied by a flip of the value of variable v.
(DC) was chosen least often for DMRP decisions. This avoids flipping always the same variables back and forth in different calls to *solveDMRP*.

The left plot of Figure 3 shows clearly that pure DMRP solving could not compete with CDCL solving. Both pure DMRP configurations (global and local MakeCount) solve around 224 of 564 instances within 1200 seconds. Initial reference points are always chosen at random. Timeouts for the analysis of one reference point (one call to *solveDMRP*) are changed according to the Luby sequence. Modifying the strategy on how to choose initial reference points showed quite some impact. Our assumption was that DMRP requires a better guidance on where to start search and how to choose its initial reference points. That motivates our hybrid approach where DMRP gets its initial reference points indirectly from the CDCL solver. As the plots show this clearly improves the performance of the solver.

A previous version of our hybrid approach [12] has taken part in the SAT competition 2009. That version mainly differs from the presented one regarding the restart strategy and the choice of when to perform DMRP resp. CDCL solving. It also implemented a more extensive solving of particular subsets of clauses which

---

[4] Instances that are not solved by preprocessing.

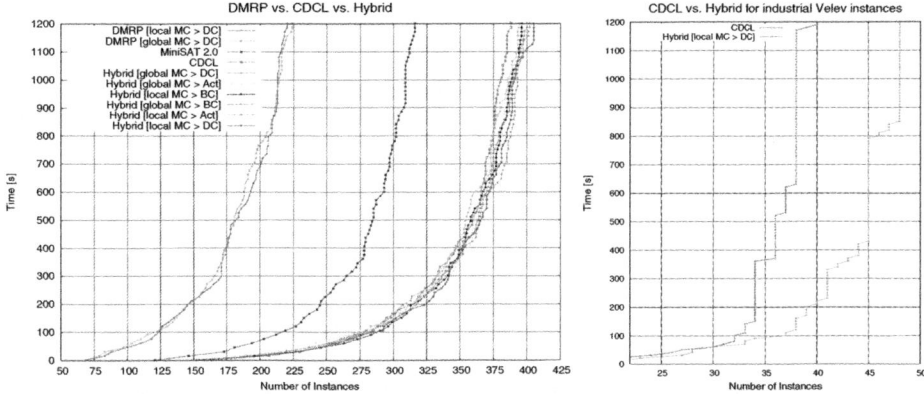

Fig. 3. The left plot compares DMRP, CDCL and our hybrid approach on 564 industrial benchmarks. The right plot compares CDCL and the hybrid approach on 51 instances from hardware verification. A point (x, y) states that x instances were solved within y sec. by that solver. Legends are ordered regarding the number of solved instances after 1200 seconds. Using local (resp. global) MakeCount and smaller decision count (resp. Activity or BreakCount) to break ties is indicated by [local MC > DC].

is only done for binary clauses in this improved approach. However, the results indicate that the older version did not utilize the DMRP approach in a sufficient way. Compared to MiniSAT 2.0 our hybrid approach also performs much better. Admittedly, this is not only due to the hybridisation with DMRP. This version of MiniSAT does neither use the Luby restart strategy nor phase saving. However, the hybrid approach also clearly outperforms our CDCL implementation with Luby restarts and phase saving.

The hybrid configuration where the MakeCount is computed locally outperformes the other configurations. It is interesting to notice that using the activity of variables to break ties does not achieve the best results. It turns out that it is better to prefer variables that were flipped least often at the current call of *solveDMRP*.

The right plot of Figure 3 compares pure CDCL with the hybrid approach on the 51 "Velev" instances of last years SAT competitions. For these instances that stem from the domain of hardware verification the hybrid approach clearly outperforms pure CDCL by solving 8 more instances.

Even though the hybrid implementation beats our pure CDCL solver on the entire benchmark set it turns out that for the most instances solved by the hybrid solver the answer was given by the CDCL part. Only about 6% were finally solved by the DMRP subsolver. Moreover, the improvement due to the hybridisation was mainly for unsatisfiable instances (17 more unsat results).

Our conjecture about this phenomenon is that DMRP generates some important lemmata: When the CDCL solver reaches the (current) maximum number of conflicts it delivers work to the DMRP solver. DMRP starts with an extension R of the last partial assignment P of the CDCL solver and hence focuses on a nearby part of the search space. When analysing this part it purposely examines clauses $\mathcal{M}$ that are not satisfied by R. In CDCL these clauses in $\mathcal{M}$ could likely become conflicting clauses if decisions were made similar to the values in R. Up to a certain point phase saving would do this after a normal CDCL restart. However, DMRP immediately considers clauses in $\mathcal{M}$ for search and resolution.

6   Conclusion

In this paper we have presented a data structure to implement the DMRP approach in an efficient way. Similar to the two-watched-literals scheme we choose one literal for each clause. The literal takes on responsibility so that a clause which is satisfied by a reference point is also satisfied by a modification of the point. Moreover, we present a way how to determine that variable which satisfies the most previously unsatisfied clauses when its value is flipped (MakeCount). Based on this implementation we motivate a hybrid SAT solver that combines CDCL and DMRP solving. Experiments have shown that our hybrid approach is competetive to the highly optimised state-of-the-art CDCL solvers and that the maintenance of complete assignments may definitely turn to account.

References

1. Bacchus, F.: Exploring the computational tradeoff of more reasoning and less searching. In: SAT 2002, pp. 7–16 (2002)
2. Biere, A.: Adaptive restart strategies for conflict driven SAT solvers. In: Kleine Büning, H., Zhao, X. (eds.) SAT 2008. LNCS, vol. 4996, pp. 28–33. Springer, Heidelberg (2008)
3. Biere, A.: Picosat essentials. JSAT 4, 75–97 (2008)
4. Davis, M., Logemann, G., Loveland, D.: A machine program for theorem-proving. ACM Commun. 5(7), 394–397 (1962)
5. Davis, M., Putnam, H.: A computing procedure for quantification theory. J. ACM 7(3), 201–215 (1960)
6. Eén, N., Sörensson, N.: An extensible SAT-solver. In: Giunchiglia, E., Tacchella, A. (eds.) SAT 2003. LNCS, vol. 2919, pp. 502–518. Springer, Heidelberg (2004)
7. Fukunaga, A.S.: Efficient Implementations of SAT Local Search. In: SAT (2004)
8. Goldberg, E.: Determinization of resolution by an algorithm operating on complete assignments. In: Biere, A., Gomes, C.P. (eds.) SAT 2006. LNCS, vol. 4121, pp. 90–95. Springer, Heidelberg (2006)
9. Goldberg, E.: A decision-making procedure for resolution-based SAT-solvers. In: Kleine Büning, H., Zhao, X. (eds.) SAT 2008. LNCS, vol. 4996, pp. 119–132. Springer, Heidelberg (2008)
10. Ivancic, F., Yang, Z., Ganai, M., Gupta, A., Ashar, P.: Efficient SAT-based bounded model checking for software verification. Theoretical Computer Science 404(3) (2008)
11. Kautz, H.A., Selman, B.: Planning as satisfiability. In: Proceedings of the Tenth European Conference on Artificial Intelligence ECAI 1992, pp. 359–363 (1992)
12. Kottler, S.: Solver descriptions for the SAT competition (2009), satcompetition.org

13. Küchlin, W., Sinz, C.: Proving consistency assertions for automotive product data management. J. Automated Reasoning 24(1-2), 145–163 (2000)
14. Luby, M., Sinclair, A., Zuckerman, D.: Optimal speedup of las vegas algorithms. In: ISTCS, pp. 128–133 (1993)
15. Lynce, I., Marques-Silva, J.: SAT in bioinformatics: Making the case with haplotype inference. In: Biere, A., Gomes, C.P. (eds.) SAT 2006. LNCS, vol. 4121, pp. 136–141. Springer, Heidelberg (2006)
16. Marques-Silva, J.P.: Practical Applications of Boolean Satisfiability. In: Workshop on Discrete Event Systems, WODES 2008 (2008)
17. Marques-Silva, J.P., Sakallah, K.A.: Grasp: A search algorithm for propositional satisfiability. IEEE Trans. Comput. 48(5), 506–521 (1999)
18. Moskewicz, M.W., Madigan, C.F., Zhao, Y., Zhang, L., Malik, S.: Chaff: engineering an efficient SAT solver. In: DAC (2001)
19. Pipatsrisawat, K., Darwiche, A.: A lightweight component caching scheme for satisfiability solvers. In: Marques-Silva, J., Sakallah, K.A. (eds.) SAT 2007. LNCS, vol. 4501, pp. 294–299. Springer, Heidelberg (2007)
20. Selman, B., Levesque, H., Mitchell, D.: A new method for solving hard satisfiability problems. In: Tenth National Conference on Artificial Intelligence (1992)
21. Velev, M.N.: Using rewriting rules and positive equality to formally verify wide-issue out-of-order microprocessors with a reorder buffer. In: DATE 2002 (2002)
22. Zheng, L., Stuckey, P.J.: Improving SAT using 2SAT. In: Proceedings of the 25th Australasian Computer Science Conference, pp. 331–340. E (2002)

Integrating Dependency Schemes in Search-Based QBF Solvers

Florian Lonsing and Armin Biere

Institute for Formal Models and Verification
Johannes Kepler University, Linz, Austria
http://fmv.jku.at/

Abstract. Many search-based QBF solvers implementing the DPLL algorithm for QBF (QDPLL) process formulae in prenex conjunctive normal form (PCNF). The quantifier prefix of PCNFs often results in strong variable dependencies which can influence solver performance negatively. A common approach to overcome this problem is to reconstruct quantifier structure e.g. by quantifier trees. Dependency schemes are a generalization of quantifier trees in the sense that more general dependency graphs can be obtained. So far, dependency graphs have not been applied in QBF solving. In this work we consider the problem of efficiently integrating dependency graphs in QDPLL. Thereby we generalize related work on integrating quantifier trees. By analyzing the core parts of QDPLL, we report on modifications necessary to profit from general dependency graphs. In comprehensive experiments we show that QDPLL using a particular dependency graph, despite of increased overhead, outperforms classical QDPLL relying on quantifier prefixes of PCNFs.

1 Introduction

The satisfiability problem of *quantified boolean formulae (QBF)* is the canonical PSPACE-complete decision problem. QBF often allows many practically relevant-problems from the domains of model checking or automated planning to be encoded succinctly. As propositional logic (SAT), which is widely applied for modelling NP-complete problems in practice, QBF requires efficient and scalable decision procedures to be accepted for practical application.

Many QBF solvers process formulae in *prenex conjunctive normal form (PCNF)*, hence QBF encodings of problems have to be converted into PCNF first. Such conversion often comes with a loss of structural properties of the original formula. This can influence solver performance negatively.

Structure can be partially recovered to tackle this problem. A special case in this respect is the analysis of quantifier structure in QBFs, either before [8,16] or after [2] conversion to PCNF. Such approaches allow a QBF solver to overcome the restrictions of linear quantifier prefixes in PCNFs to some extent. This applies to search- and elimination-based solvers, e.g. [4,7,12,19,20,28,32].

Exploiting tree-shaped quantifier structure is well-known and has been applied in different contexts. This can be achieved either by reconstructing *quantifier*

O. Strichman and S. Szeider (Eds.): SAT 2010, LNCS 6175, pp. 158–171, 2010.
© Springer-Verlag Berlin Heidelberg 2010

trees from PCNFs [2], which is closely related to minimizing quantifier scopes by miniscoping [1], or by analyzing tree structure present in non-PCNF formulae as e.g. in [8,16]. The latter corresponds to directly considering the parse tree of a formula and can be integrated in non-PCNF solvers such as [9,18,28].

Dependency schemes [30] based on [4,5], which are relations over variables, can be regarded as a generalization of tree-shaped quantifier structure. Given a dependency scheme D, a variable x is associated with all the variables y that "depend" on x with respect to D. Informally, if y depends on x, i.e. $y \in D(x)$, then the result obtained from assigning y before x in a search-based solver may not be sound in general. Quantifier prefixes of PCNFs as well as quantifier trees fit into that framework since dependency schemes can be obtained from the prefix or the tree, respectively. Sophisticated dependency schemes were introduced in [30], all of which can be computed efficiently by syntactically analyzing PCNFs.

1.1   From Quantifier Trees to Dependency Graphs

A well-known drawback when reconstructing quantifier trees in PCNFs is non-determinism [2,8,9,16]. This is related to preferring some variable over another, which can result in different trees and hence in different sets of dependencies.

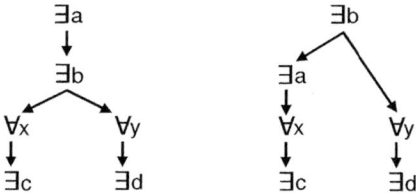

Fig. 1. Quantifier trees for the PCNF $\exists a, b \forall x, y \exists c, d.$ $(a \lor b) \land (a \lor x \lor c) \land (b \lor c) \land (b \lor y \lor d)$. Minimizing the scope of $\exists a$ in the left tree yields the tree on the right. See also Ex. 1.

Example 1. Consider the PCNF $\exists a, b \forall x, y \exists c, d.$ $(a \lor b) \land (a \lor x \lor c) \land (b \lor c) \land (b \lor y \lor d)$. Minimizing the scopes of $\exists c, \exists d, \forall x$ and $\forall y$ is deterministic and yields $\exists a, b.(a \lor b) \land (\forall x \exists c.(a \lor x \lor c) \land (b \lor c)) \land (\forall y \exists d.(b \lor y \lor d))$. Now there is the non-deterministic choice of whether to first minimize $\exists a$ and then $\exists b$ (right tree in Fig. 1) or vice versa (left tree in Fig. 1). Note that the left tree induces a dependency between a and y which is not the case in the right tree. Further, the left tree can be transformed into the tree on the right by first swapping $\exists a$ and $\exists b$ according to the rule $\exists a \exists b.\phi \equiv \exists b \exists a.\phi$ and then minimizing $\exists a$.

In addition to the problem described in Ex. 1, analyzing tree-shaped quantifier structure in general is not optimal among syntactic methods for structure analysis. This applies to reconstructed quantifier trees as well as considering tree-shaped structure present in non-PCNFs. For example, the *standard dependency scheme* D^{std} [30] is superior to tree-based approaches since it is deterministic and yields less dependencies.[1] This was pointed out in Ex. 2 in [24]. D^{std} can be efficiently constructed by analyzing connections between variables over clauses.

For those reasons and given the drawbacks of tree-shaped quantifier structure, we suggest to apply the more general concept of dependency schemes for analyzing quantifier structure in PCNFs. Similar to quantifier trees, which have already been implemented in QBF solvers, we apply *directed acyclic dependency graphs (DAGs)* in QBF solving. This generalizes quantifier trees. Dependency DAGs can be obtained from dependency schemes such as the ones introduced in [30]. When integrating dependency DAGs in QBF solvers, the drawbacks of tree-shaped quantifier structure as pointed out above can be overcome.

The core parts of our work presented here are as follows. We focus on search-based QBF solvers for PCNF which implement the *DPLL algorithm for QBF* [7] *(QDPLL)* with learning like [12,32,33]. By considering the main parts of QDPLL such as boolean constraint propagation, decision making or learning, we show how to integrate dependency DAGs into QDPLL in order to profit from dependency schemes in practice (Sec. 3 and 4). This analysis is closely related to [16] which aims at exploiting tree-based quantifier structure in QDPLL. Our work generalizes observations made in [16] to *arbitrary* dependency schemes. Further we address implementation-related issues indispensable for practical efficiency of dependency DAGs. Although we focus on PCNF and QDPLL, our results are, just as quantifier trees, relevant for any QBF solver.

We provide a comprehensive experimental evaluation (Sec. 5) of dependency DAGs in practice. For this purpose we have implemented QDPLL with learning in a new QBF solver DepQBF [23] which tightly integrates dependency DAGs. We analyze the costs of moving from relatively simple structures like linear quantifier prefixes of PCNFs or trees to more general dependency DAGs. This is closely related to practical applicability. Finally, we evaluate dynamic effects on QDPLL when using dependency DAGs for different dependency schemes.

In DepQBF we implemented a common framework for dependency DAGs which can represent linear quantifier prefixes and trees as well, thus enabling us to compare these approaches. Apart from that, we have implemented D^{std} as suggested in [30]. The remarks on how to profit from dependency schemes in QDPLL (Sec. 4) are general and hold independently from our implementation.

We show in experiments (Sec. 5) that, despite increased overhead, QDPLL with a DAG representation of D^{std} outperforms QDPLL relying on quantifier prefixes and trees. Our results indicate the potential of using dependency schemes in QDPLL in terms of more powerful rules for detecting unit literals and learning.

2 Preliminaries

For a set of propositional variables V, a *literal* is either a variable $x \in V$ or its negation $\neg x$ where $v(x) = x$ and $v(\neg x) = x$ denotes the variable of a literal. A *clause* (*cube*) is a disjunction (conjunction) over literals. A propositional formula

---

[1] Note that here we ignore variable polarities in dependency analysis. Otherwise, quantifier trees would have to be compared to the polarity-aware triangle scheme $D^{\triangle}$ [30].

is in *conjunctive normal form (CNF)* if it consists of a conjunction over clauses. A quantified boolean formula (QBF) $F = S_1 \ldots S_n. \phi$ in *prenex conjunctive normal form (PCNF)* consists of a propositional formula ϕ in CNF over a set of variables V and a *quantifier prefix* $S_1 \ldots S_n$. The quantifier prefix is a linearly ordered set of *scopes* S_i forming a partition on V. A scope S_i is *existential* ($q(S_i) = \exists$) if it is associated with an existential quantifier and *universal* ($q(S_i) = \forall$) otherwise. For scopes S_i and S_{i+1}, $q(S_i) \neq q(S_{i+1})$ for $1 \leq i < n$. The set of existential and universal variables is denoted by $V_\exists = \bigcup S_i$ for $q(S_i) = \exists$ and $V_\forall = \bigcup S_i$ for $q(S_i) = \forall$. For a literal x with $v(x) \in S_i$, $q(x) = q(S_i)$ is the *type* of x. For a clause (cube) C and $Q \in \{\forall, \exists\}$, $L_Q(C) := \{l \in C \mid q(l) = Q\}$. For literals l, k with $v(l) \in S_i$ and $v(k) \in S_j$, $l \leq k$ if, and only if $i \leq j$ for $1 \leq i, j \leq n$.

3   Dependency Schemes in Theory

Due to space limitations, we introduce dependency schemes only informally and refer to the original definition in [30]. As we focus on QDPLL with learning [12,22,32,33] (see Sec. 4.1), we confine the theoretical framework in that respect.

Definition 1. *For a PCNF F, a* dependency scheme *is a relation $D \subseteq ((V_\exists \times V_\forall) \cup (V_\forall \times V_\exists))$ with the following property when applied in QDPLL: for variables x and y with $y \notin D(x)$, the result of QDPLL when assigning y before x will be* sound. *The* inverse *of D is $\overline{D} := \{(y, x) \mid (x, y) \in D\}$.*

Def. 1 is related to the semantical evaluation of a QBF by QDPLL and corresponds to *cumulative* dependency schemes as defined in [30], which guarantees soundness of assigning y before x if $y \notin D(x)$. It is based on independence ($y \notin D(x)$), rather than dependence ($y \in D(x)$). In practice, independence between x and y with respect to D allows y to be assigned earlier. Consequently, if $y \in D(x)$ then the result of QDPLL when assigning y before x as a decision variable is *not* sound in general[2]. At the same time it is *not always* unsound. This is due to different amounts of independence identified by different dependency schemes. For a PCNF there could be dependency schemes D and D' such that $y \in D'(x)$ but $y \notin D(x)$. Hence dependency schemes can be compared according to the amount independence.

Definition 2. *For a PCNF F and dependency schemes D and D', D is* less restrictive *if, and only if $|D| \subseteq |D'|$.*

Example 2. For the QBF from Ex. 1, let D^{triv} be the trivial dependency scheme obtained from the prefix of F: $y \in D^{\mathrm{triv}}(x)$ if, and only if, $q(x) \neq q(y)$ and $x \leq y$. Let D^{tree} be obtained from the left tree in Fig. 1: $y \in D^{\mathrm{tree}}(x)$ if, and only if $q(x) \neq q(y)$ and y is a successor of x in the tree. Then D^{tree} is less restrictive than D^{triv} since $D^{\mathrm{tree}} \subseteq D^{\mathrm{triv}}$. For example, $d \in D^{\mathrm{triv}}(x)$ but $d \notin D^{\mathrm{tree}}(x)$.

A dependency scheme induces a partial order on the set of variables V which can be represented as a directed acyclic graph (DAG) over V.

---

[2] Assignments by unit and pure literals [7] are always sound independently from D.

Definition 3. *Given a dependency scheme D, the* dependency graph *for D is a DAG G(D) with vertices V and edges* $E := \{(x, y) \mid y \in D(x)\}$.

4   Dependency Schemes in QDPLL

Many implementations of QDPLL rely on the quantifier prefixes of PCNFs, which corresponds to D^{triv} as defined in Ex. 2. In this section we analyze the core parts of QDPLL. We point out how to modify those parts in order to profit from less restrictive dependency schemes other than D^{triv} in QDPLL. In our analysis, we generalize observations from using quantifier trees in QDPLL [16]. Our results are *independent* from a particular dependency scheme (Def. 1).

In the following, let D be an arbitrary dependency scheme for a PCNF. For literals x, y, we write $x \prec y$ if $v(y) \in D(v(x))$. G denotes the dependency graph for D. D is integrated into QDPLL by means of G, which is used to check if $x \prec y$. This corresponds to checking if there is an edge (x, y) in G. However, D has $\mathcal{O}(V^2)$ elements and storing all edges of G can be prohibitive. Instead, transitive edges are discarded and variables are merged into equivalence classes. Checking $x \prec y$ is done by checking successor relation between x and y in G. As shown (Sec. 5), these optimizations are indispensable for efficiency in practice.

4.1   QDPLL with Learning

We briefly introduce QDPLL with *conflict-driven clause and solution-driven cube learning* based on [33]. For a PCNF $S_1 \ldots S_n. \phi$, an additional disjunction ψ over learnt cubes is stored: $S_1 \ldots S_n. (\phi \vee \psi)$, also called *augmented CNF*. Fig. 2 shows a high-level view. Clauses and cubes (*constraints*) are derived by clause resolutions [6] and cube resolutions [13,22,33] (*constraint resolutions*). Cube "resolution" is actually consensus. Different from [32,33] we do *not* consider to learn constraints containing complementary literals and rather follow the algorithms from [15]. Further details can also be found in e.g. [13,22].

The core of algorithm qdpll in Fig. 2 is propagation of *implications* (unit and pure literals) which is carried out in bcp until saturation. If neither a conflicting (*conflict*), nor a satisfying assignment (*solution*) was found, i.e. the formula state is undefined under the current assignment, then a variable x is assigned as next *decision* in select_dec_var. Decisions are numbered ascendingly by *decision levels* $dl(x)$, starting at 1. Having assigned x as decision, all implications y are propagated by bcp again, where $dl(y) := dl(x)$.

Otherwise the solver has either derived a conflict or a solution. This situation corresponds to a leaf in the search tree enumerated by QDPLL. For conflicts the formula contains an *empty clause* (see Def. 5) returned by get_initial_reason in analyze_leaf. By means of successive clause resolutions, the backtrack level and a learnt clause (called *asserting clause*, see Def. 8) containing a *forced literal* are computed which is unit at the backtrack level (also called *asserting level*). We assume that qdpll learns asserting clauses only. The current clause R is resolved (constraint_res) with the *antecedent clause* (get_antecedent) of an existential unit literal (get_pivot) in R. The antecedent clause is the clause where that literal became unit. If R is asserting then resolution stops (stop_res).

For handling solutions, get_initial_reason either returns a *satisfied learnt cube* (see Def. 5) already present in the cube set ψ of the formula or a new one generated from the current assignment. Dually to clauses, an asserting cube is learnt by cube resolutions using antecedent cubes of universal unit literals.

After backtracking and unassigning variables (backtrack), the forced literal is assigned as unit at the backtrack level and the learnt constraint is added to the formula. Again bcp propagates all implications. If an empty clause or satisfied cube is derived by resolutions then qdpll terminates (btlevel == INVALID).

```
State qdpll ()
  while (true)
    State s = bcp ();          DecLevel analyze_leaf (State s)
    if (s == UNDEF)              R = get_initial_reason (s);
      // Make decision.          // s == UNSAT: 'R' is empty clause.
      v = select_dec_var ();     // s == SAT: 'R' is sat. cube...
      assign_dec_var (v);        // ..or new cube from assignment.
    else                         while (!stop_res (R))
      // Conflict or solution.     p = get_pivot (R);
      // s == UNSAT or s == SAT.   A = get_antecedent (p);
      btlevel = analyze_leaf (s);  R = constraint_res (R, p, A);
      if (btlevel == INVALID)    add_to_formula (R);
        return s;                assign_forced_lit (R);
      else                       return get_asserting_level (R);
        backtrack (btlevel);
```

Fig. 2. Pseudo-code of QDPLL with conflict-driven clause and solution-driven cube learning [13,22,33]. Code blocks are indicated by indentation level. See also Sec. 4.1.

In the following, we generalize unit literals and learning (Def. 4, 6, 8) to arbitrary dependency schemes. Soundness is explained by reordering the quantifiers in the prefix of a PCNF F based on D by Def. 1 to obtain an equivalent PCNF F' [30]. This is possible, as Def. 1 corresponds to cumulative schemes [30]. Then original versions of Def. 4, 6, 8 in the context of prefixes (i.e. D^{triv} in Def. 2) apply to F'. Finally, F' can be converted back to F by reordering.

4.2   Unit Literal Detection

Unit literals were introduced in [7] for clauses and extended to cubes in [13,33]. The original definition is based on quantifier prefixes of PCNFs, i.e. on D^{triv} as defined in Ex. 2, and can be generalized to arbitrary dependency schemes.

Definition 4. *A clause (cube) C is* unit *if, and only if no $l \in C$ is assigned true (false), exactly one $l_e \in L_\exists(C)$ $(l_u \in L_\forall(C))$ is unassigned, and for all unassigned $l_u \in L_\forall(C)$ $(l_e \in L_\exists(C))$: $l_u \not\prec l_e$ $(l_e \not\prec l_u)$.*

Analogously, Def. 4 generalizes the definition based on quantifier trees from [16]. If a clause (cube) C is unit according to Def. 4 then l_e (l_u) can be assigned as a unit literal (bcp in Fig. 2). Detecting unit literals involves checking dependencies. Using a two-literal watching scheme based on [10,26], this can be achieved lazily as follows. In each clause two unassigned literals l_1 and l_2 are watched such that either $q(l_1) = q(l_2) = \exists$ or $q(l_1) = \forall$, $q(l_2) = \exists$ and $l_1 \prec l_2$. If watched literals are updated during BCP then condition $l_1 \prec l_2$ needs to be checked in the latter case only. Literal watching in cubes can be handled dually.

4.3   Constraint Learning

In QDPLL as shown in Fig. 2, new constraints are added to the formula whenever a conflicting or satisfying assignment was found. These constraints are derived by successive resolutions, each potentially eliminating literals from the resolvent.

Definition 5. *A clause (cube) C is* empty (satisfied) *if, and only if no $l \in C$ is assigned true (false), and all $l \in L_\exists(C)$ ($l \in L_\forall(C)$) are assigned false (true).*

Definition 6. *Universal reduction (existential reduction) eliminates from a clause (cube) C all $l_u \in L_\forall(C)$ ($l_e \in L_\exists(C)$) for which there is no $l_e \in L_\exists(C)$ ($l_u \in L_\forall(C)$) with $l_u \prec l_e$ ($l_e \prec l_u$).*

Definition 7. *For universally-reduced clauses (existentially-reduced cubes) C_1 and C_2 with $v \in C_1$ and $\neg v \in C_2$ for a variable v, let $C := (C_1 \cup C_2) \setminus \{v, \neg v\}$. If C does not contain complementary literals then let C' be the result of applying universal (existential) reduction to C; C' is the* resolvent *of C_1 and C_2 on v.*

Soundness of universal reduction as part of clause resolution for QBF was proved in [6]. Existential reduction for cube resolution was applied in [13,33]. Def. 6 generalizes the reduction rules from D^{triv} to arbitrary dependency schemes. In [16] such generalization was given for quantifier trees.

Among several learning strategies which add and remove learnt constraints according to particular quality measures [13,22,33], QDPLL as shown in Fig. 2 learns exactly one constraint for each conflict or solution. The learnt constraint is asserting, i.e. it is unit at the level QDPLL backtracks to, and hence will trigger a unit literal to be assigned at the backtrack level. Resolution continues until the current resolvent is asserting. This is controlled by a stop criterion.

Definition 8. *Let R denote the clause (cube) derived after some resolution steps in the learning process. For $Q := \exists$ ($Q := \forall$), let $d := max(\{dl(l) \mid l \in L_Q(R)\})$. Then R is* asserting at level *$a := max(\{dl(l) \mid l \in R \wedge dl(l) < d\})$ if, and only if*

1. *the decision variable at level d is existential (universal).*
2. *there is exactly one $l \in L_\exists(R)$ ($l \in L_\forall(R)$) with $dl(l) = d$*
3. *for all $l_u \in L_\forall(R)$ ($l_e \in L_\exists(R)$) where $l_u \prec l$ ($l_e \prec l$): l_u (l_e) must be assigned false (true) with $dl(l_u) < d$ ($dl(l_e) < d$).*

Def. 8 generalizes the stop criteria for generating asserting constraints given in [15,33] from D^{triv} to arbitrary dependency schemes. This affects condition 3 in Def. 8 only, where dependency has to be checked. In practice, this check is deferred as far as possible by checking conditions 1 and 2 before condition 3.

4.4   Decision Making

The quantifier prefix of PCNFs restricts the freedom of QDPLL to select decision variables, as variables must be assigned "from left to right" according to the prefix (i.e. D^{triv}). In the context of dependency schemes (see Def. 1), a variable y may be assigned as decision as soon as all variables in $\overline{D}(y)$ have been assigned.

Definition 9. *A variable y is* enabled *in QDPLL if, and only if all variables in $\overline{D}(y)$ are assigned. Otherwise, y is* disabled. *A variable is a* (decision) candidate *if, and only if it is unassigned and enabled.*

Example 3. For the PCNF $\exists a \forall x, y \exists b.\ \phi$, $\overline{D^{\mathrm{triv}}}(a) = \emptyset$, $\overline{D^{\mathrm{triv}}}(x) = \overline{D^{\mathrm{triv}}}(y) = \{a\}$ and $\overline{D^{\mathrm{triv}}}(b) = \{x, y\}$. Variable a is always enabled, b is enabled as soon as both x and y are assigned and if a is assigned then both x and y are enabled.

Following from Def. 1 and 9, assigning disabled variables as decisions is not sound in general. Using less restrictive dependency schemes (see Def. 2) than e.g. D^{triv} allows more freedom to select candidates in QDPLL because $\overline{D}$ is smaller and hence variables become enabled earlier.

One candidate is heuristically selected as next decision by select_dec_var in Fig. 2. In practice, it is prohibitive to maintain the exact candidate set explicitly. First, this set is needed precisely in select_dec_var and not e.g. in bcp. Further, not every assignment enables, not every backtrack disables new variables.

Based on these observations, we apply the dependency graph G and maintain the set of decision candidates (DC) incrementally as follows. Before QDPLL starts, $DC := \{x \in V \mid \overline{D}(x) = \emptyset\}$, i.e. DC corresponds to the roots of G. Each time a decision is made (i.e. each time select_dec_var in Fig. 2 is called), DC is updated by taking into account the effects of assignments $l_1, \ldots, l_k$ made since the previous decision only. Each l_i in $l_1, \ldots, l_k$ is processed one after the other. The assignment l_i possibly enables some, not necessarily all variables in $D(v(l_i))$. There might be other variables $x \neq v(l_i)$ with $D(x) = D(v(l_i))$. If any such x is still unassigned at the time l_i is processed then l_i will *not* enable any variable in $D(v(l_i))$. This observation can be exploited by constructing G as a graph [24] over equivalence classes $[x]$ of variables: $x \approx y \Leftrightarrow D(x) = D(y)$ for $x, y \in V$. Assuming $l_i \in [x]$, *no* variable will be enabled unless *all* variables in $[x]$ are assigned. Only if this is the case, set $D(v(l_i))$ is inspected by traversing successors of $[x]$ in G and new candidates are added to DC. If successor $[y]$ with $D(y) \subseteq D(v(l_i))$ is visited, then it is checked if $[y]$ is fully assigned.

When backtracking in backtrack, assignments l_i in $l_1, \ldots, l_k$ made between the backtrack level and the current decision level are cleared one after the other, which possibly disables variables in $D(v(l_i))$. Assuming $l_i \in [x]$, this can happen only if *all* variables in $[x]$ are assigned at the time l_i is cleared. Only if this is the case, set $D(v(l_i))$ is inspected and disabled variables are removed from DC.

Maintaining DC as described is independent from any decision heuristic for QBF and therefore can be integrated in any implementation of select_dec_var. Furthermore, this approach generalizes quantifier watching [10] from quantifier prefixes to arbitrary dependency schemes.

5   Experimental Results

We have implemented QDPLL with dependency schemes as described in Sec. 4 in our PCNF-based solver DepQBF [23], which also participated in QBFEVAL'10 [11]. It differs from other search-based solvers mainly in a tight integration of dependency schemes. Apart from that, approaches implemented comprise watched data structures for detection of unit and pure literals [10,14,26], conflict-driven clause and solution-driven cube learning [13,22,32,33], assignment caching [29], activity heuristics based on VSIDS [26] and partial restarts based on [3].

As pointed out in Sec. 4, dependency graphs G in DepQBF are represented as compact graphs over equivalence classes of variables. The data structure evolved from previous work in [24]. Although originally being tailored to the standard dependency scheme D^{std}, we also implemented dependency graphs for D^{triv} and D^{tree} (see Ex. 2) within the *same* framework. This enables us to directly compare QDPLL using those three schemes without changing any other part of the solver. To build one out of possibly many non-deterministic dependency graphs (i.e. trees) for D^{tree}, we adapted the approach from [2] to our framework.

Table 1. Performance comparison of DepQBF with quantifier prefixes (D^{triv}), quantifier trees (D^{tree}) and the standard dependency scheme (D^{std}), which is less restrictive than the other two. Average run times are given in seconds. Benchmarks include all structured formulae from *QBFEVAL'07*, *QBFEVAL'08* and from set *Herbstritt* [11]. The three versions of DepQBF do *not* apply preprocessing and differ only in the integrated dependency schemes, all other parts are *exactly* the same. For external reference, statistics of PCNF-based QuBE6.6 [12] with and without preprocessing (*QuBE6.6-np*) are listed. We did not add other solvers as we focus on evaluating QDPLL with dependency schemes and, given the results of QBF competitions [11], QuBE6.x is the state-of-the-art QDPLL-based solver.

QBFEVAL'08 (3326 formulae)					
	D^{triv}	D^{tree}	D^{std}	QuBE6.6-np	QuBE6.6
solved	1223	1221	1252	1106	2277
time	579.94	580.64	572.31	608.97	302.49
QBFEVAL'07 (1136 formulae)					
solved	533	548	567	458	734
time	497.12	484.69	469.97	549.29	348.05
Herbstritt (478 formulae)					
solved	321	357	357	296	395
time	316.06	248.20	248.07	357.52	173.53

Table 2. Comparing combinations of DepQBF with quantifier prefixes (D^{triv}), quantifier trees (D^{tree}) and the standard dependency scheme (D^{std}). Only formulae solved by *both* solvers ($\cap$) were considered. E.g. in section "$D^{\text{triv}} \cap D^{\text{std}}$", the left column reports statistics for D^{triv}, the right one for D^{std}. Average values are given for run time in seconds, ratio of implications among all assignments, number of backtracks, ratio of satisfied learnt cubes among all identified solutions and size (i.e. number of literals) of learnt constraints. See also Sec. 4.1 for terminology.

	$D^{\text{triv}} \cap D^{\text{tree}}$		$D^{\text{triv}} \cap D^{\text{std}}$		$D^{\text{tree}} \cap D^{\text{std}}$	
	QBFEVAL'08 (solved only)					
solved	1172		1196		1206	
time	**23.15**	26.68	**23.73**	25.93	25.63	**22.37**
implied/assigned	90.4%	**90.7%**	88.6%	**90.5%**	90.9%	**92.1%**
backtracks	32431	**27938**	34323	**31085**	**25106**	26136
sat. cubes/sol.	1.8%	**2.9%**	1.8%	**2.6%**	**3.6%**	3.1%
learnt constr. size	157	**99**	150	**96**	102	**95**
	QBFEVAL'07 (solved only)					
solved	501		513		537	
time	**31.22**	34.46	32.76	**32.66**	33.31	**28.33**
implied/assigned	89.0%	**89.2%**	87.7%	**89.5%**	89.9%	**91.9%**
backtracks	35131	**22334**	39906	**26362**	**21945**	22323
sat. cubes/sol.	4.0%	**10.0%**	4.0%	**9.5%**	**10.8%**	9.9%
learnt constr. size	150	**101**	134	**113**	100	**96**
	Herbstritt (solved only)					
solved	312		308		348	
time	26.86	**19.28**	24.41	**19.28**	**20.46**	20.83
implied/assigned	96.6%	**97.4%**	96.2%	**97.4%**	97.4%	97.4%
backtracks	26565	**1329**	26733	**1482**	**1615**	1800
sat. cubes/sol.	0%	0%	0%	0%	0%	0%
learnt constr. size	**174**	306	**173**	323	**407**	410

Tab. 1 shows a comparison[3] of DepQBF with D^{triv}, D^{tree} and D^{std} on structured formulae from previous QBF competitions [11]. Dependency checking as needed in Def. 4, 6 and 8 was optimized in D^{triv}: for $x, y \in V$, $x \prec y$ if, and only if $x < y$, which can be checked in constant time. This is impossible for arbitrary schemes where $x \prec y$ if, and only if $q(x) \neq q(y)$ and y is a successor of x in G. Despite that additional overhead, QDPLL with D^{std} is best on *QBFEVAL'07* and *QBFEVAL'08* and is slightly faster than D^{tree} on set *Herbstritt*. There is a large performance gap to QuBE6.6 which, different from DepQBF, uses preprocessing. However, any version of DepQBF outperforms QuBE6.6 when preprocessing is disabled. Note that, in our terminology, QuBE6.6 uses D^{triv}.

A more detailed comparison of all three versions of DepQBF considering the intersection of solved formulae is shown in Tab. 2. D^{triv} is slightly faster on the

---

[3] Setup for *all* experiments reported: Ubuntu Linux 9.04, Intel® Q9550@2.83 GHz, 3 GB/900 sec. mem and time limit. Data: http://fmv.jku.at/papers/sat10qbf.7z

QBFEVAL sets. On the other hand, D^{triv} yields more backtracks than D^{tree} and D^{std} on all sets. On set *Herbstritt*, the difference in this respect is a factor of up to 20. D^{tree} and D^{std}, both being less restrictive than D^{triv}, produce smaller learnt constraints on the *QBFEVAL* sets. Furthermore, D^{std} triggers more implications on all sets and D^{triv} fewer satisfied learnt cubes. These effects can be attributed to more powerful rules for unit detection and constraint reduction (Def. 4, 6).

The results from Tab. 2 indicate that moving from D^{triv} to more sophisticated dependency DAGs incurs run time overhead (except on set *Herbstritt*), but also allows QDPLL to produce shorter runs in terms of backtracks. As mentioned above, checking if $x \prec y$, which is required in unit literal detection and constraint learning, is not a constant-time operation in general dependency DAGs. Instead, G must be inspected. However, QDPLL still seems to profit from using less restrictive dependency schemes such as D^{tree} and D^{std}, as indicated in Tab. 1.

In order to assess both the costs and benefits of integrating dependency DAGs in QDPLL in more detail, we carried out the following experiment. In addition to the dependency DAG which is used for dependency checking and constraint reduction in QDPLL, called primary DAG G_1, another dependency DAG, the secondary DAG G_2, is maintained *independently* and *in parallel* for statistical computations. The idea is to compare the effects of using different DAGs *dynamically*, i.e. during a solver run. This setup allows to compute more fine-grain

Table 3. Comparing costs and benefits of different dependency schemes in DepQBF (all benchmarks, time out 900 sec.). The solver maintains *two* dependency DAGs G_1 (primary) and G_2 (secondary) in parallel. E.g. in section "$D^{\text{triv}} \ltimes D^{\text{std}}$", G_1 is obtained from D^{triv} (left column), G_2 from D^{std} (right column). Note that columns "D^{std}" in "$D^{\text{std}} \ltimes D^{\text{triv}}$" and "$D^{\text{std}} \ltimes D^{\text{tree}}$" are incomparable since G_2 influences run time, i.e. "$D^{\text{std}} \ltimes D^{\text{triv}}$" and "$D^{\text{std}} \ltimes D^{\text{tree}}$" may run at different speeds. Numbers of decision candidates (DC, see Sec. 4.4, Def. 9) when using different DAGs are compared. Each time before decision making, the number of DC under the current assignment is computed. Row "DC/d" shows the total sum of DC over the total number of decisions in the benchmark set after max. 900 sec. run time. Average costs are listed for (un)assigning an l_i as defined in Sec. 4.4 for updating DC (DC-*updt.*), dependency checks ($\prec$) as needed in unit detection (Def. 4) and for the stop criterion (Def. 8), and constraint reduction (*C-red.*) per resolution. The latter are irrelevant for G_2 ("-").

	$D^{\text{triv}} \ltimes D^{\text{std}}$		$D^{\text{std}} \ltimes D^{\text{triv}}$		$D^{\text{tree}} \ltimes D^{\text{std}}$		$D^{\text{std}} \ltimes D^{\text{tree}}$	
QBFEVAL'08 (3326 formulae)								
DC/d	13801.0	13801.6	11409.7	11409.0	8932.5	8933.0	15625.6	15625.3
DC-updt.	3.23	3.16	3.30	3.43	3.38	3.37	3.30	3.36
$\prec$	1	-	6.21	-	7.15	-	6.26	-
C-red.	1.18	-	535.62	-	538.30	-	540.94	-
Herbstritt (478 formulae)								
DC/d	21.3	26.55	20.14	20.13	20.67	20.67	20.16	20.16
Pan (384 formulae) ∪ Sorting-Networks (84 formulae)								
DC/d	75.81	93.87	117.50	109.66	86.89	86.90	120.03	119.98

statistics than overall run time or number of backtracks, as listed in Tab. 1 and 2. During a run of QDPLL, it is interesting to compare the numbers of decision candidates (DC) with respect to G_1 and G_2 under the current assignment. These numbers are computed each time before a decision is made and reflect the degree of freedom resulting from less restrictive dependency schemes (see Sec. 4.4). E.g. we expect D^{std} to allow more candidates than D^{triv} and D^{tree}. Apart from that, we want to measure average costs of dependency checking and candidate maintenance for DAGs resulting from different dependency schemes.

Tab. 3 shows results of the experiments described above. For G_1 and G_2, we compared D^{std} to D^{triv} and D^{tree} , where all four combinations were run to even out biased solver behaviour. Due to limited computational resources, we did not compare D^{triv} to D^{tree} and omitted *QBFEVAL'07*. As indicated for sets *QBFEVAL'08* and *Herbstritt*, the difference in DC statistics is very small in general, sometimes less than 1 candidate on average per decision. However, it seems that this is already enough for QDPLL with D^{std} to outperform D^{triv} and D^{tree} by Tab. 1. Further, DC statistics are also family-dependent, as shown by the results for sets *Pan* and *Sorting-Networks* in Tab. 3.

Cost statistics in Tab. 3 (rows "*DC-updt.*", "$\prec$", "*C-red.*") are correlated to the number of variables that have to be visited (i.e. pointer dereferences in our implementation) when inspecting a dependency DAG. Average costs for dependency checking and (un)assigning variables for updating DC before decisions or during backtracking are small. This is due to the class-based approaches described in Sec. 4. On the other hand, costs of constraint reduction are very high for D^{tree} and D^{std}. These effects are closely related to implementation. When using D^{triv}, all constraints C can be kept sorted according to scope order, which allows efficient reduction. This was implemented in DepQBF with D^{triv} and is reflected by low costs in Tab. 3. In general, such an approach is not possible and we rather reduce constraints based on classes in the dependency DAG for D^{tree} and D^{std}. Classes are collected for all literals in C before reduction, where the size of C (particularly for cubes) can be large. The statistics in Tab. 3 also include that effort. Instead of collecting from scratch, the set of classes could also be maintained incrementally for all constraints, which is currently not implemented in DepQBF. However, despite that overhead in D^{tree} and D^{std}, overall performance by Tab. 1 is still better than with D^{triv} .

6   Conclusion

Structure analysis of formulae can improve QBF solvers considerably. A common approach is the analysis of quantifier structure in PCNFs by quantifier trees. Dependency schemes generalize trees and allow to overcome related drawbacks.

In this work, we considered the problem of efficiently integrating dependency DAGs into search-based QBF solvers (QDPLL) for PCNFs. Dependency DAGs result from dependency schemes and, just as trees, represent quantifier structure. By analyzing core parts of QDPLL, we have pointed out how to profit from DAGs. Thereby we generalized related work on quantifier trees in QDPLL. The results of our analysis are independent from a particular dependency scheme. Further, quantifier DAGs are relevant for QBF solvers of any kind.

Our experiments demonstrate that a careful implementation of QDPLL integrating the standard dependency scheme D^{std} outperforms classical approaches based on quantifier prefixes and trees. Despite increased overhead, our results indicate the potential of using less restrictive dependency schemes in QDPLL, which is supported by DepQBF's performance in QBFEVAL'10 [11]. More powerful unit literal detection and constraint reduction produce more implications and shorter learnt constraints. However, we also argue that the effects to a large extent differ with respect to problem domains and QBF encodings.

As future work we want to extend our implementation to arbitrary dependency schemes. Particularly, the *triangle dependency scheme* seems to be promising since it is provably less restrictive than D^{std} [27,30].

Finally, we want to thank Paolo Marin and Enrico Giunchiglia for providing us with a version of QuBE6.6 without preprocessing.

References

1. Ayari, A., Basin, D.A.: QUBOS: Deciding Quantified Boolean Logic Using Propositional Satisfiability Solvers. In: Aagaard, M., O'Leary, J.W. (eds.) FMCAD 2002. LNCS, vol. 2517, pp. 187–201. Springer, Heidelberg (2002)
2. Benedetti, M.: Quantifier Trees for QBFs. In: Bacchus, F., Walsh, T. (eds.) SAT 2005. LNCS, vol. 3569, pp. 378–385. Springer, Heidelberg (2005)
3. Bhalla, A., Lynce, I., de Sousa, J.T., Marques-Silva, J.: Heuristic-Based Backtracking Relaxation for Propositional Satisfiability. Journal of Automated Reasoning (JAR) 35(1-3), 3–24 (2005)
4. Biere, A.: Resolve and Expand. In: Hoos, H.H., Mitchell, D.G. (eds.) SAT 2004. LNCS, vol. 3542, pp. 59–70. Springer, Heidelberg (2005)
5. Bubeck, U., Büning, H.K.: Bounded Universal Expansion for Preprocessing QBF. In: Marques-Silva, Sakallah (eds.) [25], pp. 244–257.
6. Kleine Büning, H., Karpinski, M., Flögel, A.: Resolution for Quantified Boolean Formulas. Inf. Comput. 117(1), 12–18 (1995)
7. Cadoli, M., Giovanardi, A., Schaerf, M.: An Algorithm to Evaluate Quantified Boolean Formulae. In: AAAI/IAAI, pp. 262–267 (1998)
8. Egly, U., Seidl, M., Tompits, H., Woltran, S., Zolda, M.: Comparing Different Prenexing Strategies for Quantified Boolean Formulas. In: Giunchiglia, Tacchella (eds.) [17], pp. 214–228
9. Egly, U., Seidl, M., Woltran, S.: A Solver for QBFs in Nonprenex Form. In: Brewka, G., Coradeschi, S., Perini, A., Traverso, P. (eds.) ECAI. FAIA, vol. 141, pp. 477–481. IOS Press, Amsterdam (2006)
10. Gent, I.P., Giunchiglia, E., Narizzano, M., Rowley, A.G.D., Tacchella, A.: Watched Data Structures for QBF Solvers. In: Giunchiglia, Tacchella (eds.) [17], pp. 25–36 (2003)
11. Giunchiglia, E., Narizzano, M., Tacchella, A.: Quantified Boolean Formulas Satisfiability Library, QBFLIB (2001), http://www.qbflib.org
12. Giunchiglia, E., Narizzano, M., Tacchella, A.: QUBE: A System for Deciding Quantified Boolean Formulas Satisfiability. In: Goré, R.P., Leitsch, A., Nipkow, T. (eds.) IJCAR 2001. LNCS (LNAI), vol. 2083, pp. 364–369. Springer, Heidelberg (2001)

13. Giunchiglia, E., Narizzano, M., Tacchella, A.: Learning for Quantified Boolean Logic Satisfiability. In: AAAI/IAAI, pp. 649–654 (2002)
14. Giunchiglia, E., Narizzano, M., Tacchella, A.: Monotone Literals and Learning in QBF Reasoning. In: Wallace (ed.) [31], pp. 260–273 (2004)
15. Giunchiglia, E., Narizzano, M., Tacchella, A.: Clause/Term Resolution and Learning in the Evaluation of Quantified Boolean Formulas. J. Artif. Intell. Res. (JAIR) 26, 371–416 (2006)
16. Giunchiglia, E., Narizzano, M., Tacchella, A.: Quantifier Structure in Search-Based Procedures for QBFs. TCAD 26(3), 497–507 (2007)
17. Giunchiglia, E., Tacchella, A. (eds.): SAT 2003. LNCS, vol. 2919. Springer, Heidelberg (2004)
18. Goultiaeva, A., Iverson, V., Bacchus, F.: Beyond CNF: A Circuit-Based QBF Solver. In: Kullmann (ed.) [21], pp. 412–426
19. Pan, G., Vardi, M.Y.: Symbolic Decision Procedures for QBF. In: Wallace (ed.) [31], pp. 453–467
20. Jussila, T., Biere, A., Sinz, C., Kröning, D., Wintersteiger, C.M.: A First Step Towards a Unified Proof Checker for QBF. In: Marques-Silva, Sakallah (eds.) [25], pp. 201–214
21. Kullmann, O. (ed.): SAT 2009. LNCS, vol. 5584. Springer, Heidelberg (2009)
22. Letz, R.: Lemma and Model Caching in Decision Procedures for Quantified Boolean Formulas. In: Egly, U., Fermüller, C.G. (eds.) TABLEAUX 2002. LNCS (LNAI), vol. 2381, pp. 160–175. Springer, Heidelberg (2002)
23. Lonsing, F.: DepQBF 0.1 Source Code (2010), http://fmv.jku.at/depqbf/
24. Lonsing, F., Biere, A.: A Compact Representation for Syntactic Dependencies in QBFs. In: Kullmann (ed.) [21], pp. 398–411
25. Marques-Silva, J., Sakallah, K.A. (eds.): SAT 2007. LNCS, vol. 4501. Springer, Heidelberg (2007)
26. Moskewicz, M.W., Madigan, C.F., Zhao, Y., Zhang, L., Malik, S.: Chaff: Engineering an Efficient SAT Solver. In: DAC 2001, pp. 530–535. ACM, New York (2001)
27. Samer, M.: Variable Dependencies of Quantified CSPs. In: Cervesato, I., Veith, H., Voronkov, A. (eds.) LPAR 2008. LNCS (LNAI), vol. 5330, pp. 512–527. Springer, Heidelberg (2008)
28. Pigorsch, F., Scholl, C.: Exploiting structure in an AIG based QBF solver. In: DATE, pp. 1596–1601. IEEE, Los Alamitos (2009)
29. Pipatsrisawat, K., Darwiche, A.: A Lightweight Component Caching Scheme for Satisfiability Solvers. In: Marques-Silva, Sakallah (eds.) [25], pp. 294–299
30. Samer, M., Szeider, S.: Backdoor Sets of Quantified Boolean Formulas. Journal of Automated Reasoning (JAR) 42(1), 77–97 (2009)
31. Wallace, M. (ed.): CP 2004. LNCS, vol. 3258. Springer, Heidelberg (2004)
32. Zhang, L., Malik, S.: Conflict Driven Learning in a Quantified Boolean Satisfiability Solver. In: Pileggi, L.T., Kuehlmann, A. (eds.) ICCAD, pp. 442–449. ACM, New York (2002)
33. Zhang, L., Malik, S.: Towards a Symmetric Treatment of Satisfaction and Conflicts in Quantified Boolean Formula Evaluation. In: Van Hentenryck, P. (ed.) CP 2002. LNCS, vol. 2470, pp. 200–215. Springer, Heidelberg (2002)

An Exact Algorithm for the Boolean Connectivity Problem for k-CNF

Kazuhisa Makino[1], Suguru Tamaki[2], and Masaki Yamamoto[3]

[1] Graduate School of Information Science and Technology, University of Tokyo
makino@mist.i.u-tokyo.ac.jp
[2] Graduate School of Informatics, Kyoto University
tamak@kuis.kyoto-u.ac.jp
[3] Dept. of Mathematical Sciences, School of Science, Tokai University
yamamoto@tokai-u.jp

Abstract. We present an exact algorithm for a PSPACE-complete problem, denoted by CONNkSAT, which asks if the solution space for a given k-CNF formula is connected on the n-dimensional hypercube. The problem is known to be PSPACE-complete for $k \geq 3$, and polynomial solvable for $k \leq 2$ [6]. We show that CONNkSAT for $k \geq 3$ is solvable in time $O((2 - \epsilon_k)^n)$ for some constant $\epsilon_k > 0$, where ϵ_k depends only on k, but not on n. This result is considered to be interesting due to the following fact shown by [5]: QBF-3-SAT, which is a typical PSPACE-complete problem, is not solvable in time $O((2 - \epsilon)^n)$ for any constant $\epsilon > 0$, provided that the SAT problem (with no restriction to the clause length) is not solvable in time $O((2 - \epsilon)^n)$ for any constant $\epsilon > 0$.

1 Introduction

There are so many NP-hard problems around the world, which are considered to be intractable. To deal with those intractable problems, efficient algorithms with good approximation ratio or working well on average, have been proposed. Another approach to dealing with intractable problems is to develop algorithms that exactly solve the problems, so-called *exact algorithms*, where exact algorithms usually run in super-polynomial time, but exponentially faster than trivial ones. See [20,17] for surveys on this topic. A number of exact algorithms for typical NP-hard problems have been proposed, and novel techniques for bounding the running time have been found: E.g., [1,7,2] for the traveling salesman problem, [4,9,3] for graph partitioning problems such as the graph coloring problem, and [14,13,16,8,15] for the satisfiability problem.

Viewing this approach in terms of computational complexity, we are concerned with the following question: Given an NP-hard problem of solution length n or witness length n, (for example, n denotes the number of vertices of a graph for the traveling salesman problem, or the number of variables of a formula for the satisfiability problem) is there an exact algorithm for the problem in time $O(2^n)$, or $O((2-\epsilon)^n)$ for some constant $\epsilon > 0$? Here, we assume that the length of instances with solution length n or witness length n is bounded by a polynomial

O. Strichman and S. Szeider (Eds.): SAT 2010, LNCS 6175, pp. 172–180, 2010.
© Springer-Verlag Berlin Heidelberg 2010

in n. Moreover, as usual, we omit the polynomial factor in the O-notation when concerning with an upper bound of exponential-time.

The oldest result for this kind of questions is for the traveling salesman problem by Bellman [1] and by Held and Karp [7]. Given an undirected graph $G = (V, E)$ and a length function $\ell : E \to R_+$, the problem asks for finding a shortest Hamilton cycle. It is easy to see that the problem is solvable in time $O(n!)$. However, it is indeed not so easy to see that it is solvable in time $O(2^n)$. These two papers [1,7] gave an affirmative answer to this question. There are several results that give such an affirmative answer: for example, [4,9] for the k-coloring problem showed that it is solvable in time $O(2^n)$ for any k (not necessarily constant) while it is trivially solvable in time $O(k^n)$, and [19] for the maximum satisfiability problem where the clause length of an instance is at most two showed that it is solvable in time $O(1.731^n)$ while it is trivially solvable in time $O(2^n)$.

One of the most notable questions of this kind, which are still open, is for the satisfiability problem (SAT). This problem asks if there is a satisfying assignment for a given conjunctive normal form (CNF) formula φ with *no* restriction to the clause length. It is clear that the problem is solvable in time $O(2^n)$. However, it is still open whether it is solvable in time $O((2 - \epsilon)^n)$ for some constant $\epsilon > 0$. Another well-known open question is to ask whether the traveling salesman problem is solvable in time $O((2 - \epsilon)^n)$ for some constant $\epsilon > 0$.

While developing exact algorithms for NP-complete problems and their optimization problems, we *rarely* see exact algorithms for decision problems in complexity classes beyond NP, such as the second and higher levels of PH, PSPACE, EXP, etc. There is one exceptional problem as far as we know: the quantified Boolean formula (QBF) problem, that is a typical PSPACE-complete problem, even if given Boolean formulas are restricted to 3-CNF. Williams [18] proposed an exact algorithm for this problem. However, he analyzed the running time with respect to *the number of clauses*, but not the number of variables. (Apart from decision problems, there are several problems solvable in time $O((2 - \epsilon)^n)$, e.g., #k-SAT problem, which is #P-complete for $k \geq 2$. For this problem, we easily obtain an $O((2 - \epsilon))^n$-time exact algorithm, by using a simple backtracking algorithm for k-SAT. The best upper bound for #3-SAT, for example, can be found in [10].)

In this paper, we show that the following PSPACE-complete problem, denoted by CONNkSAT, is solvable in time $O((2 - \epsilon)^n)$ for n variables: given a k-CNF formula φ over n variables, decide whether the solution space of φ is connected on the n-dimensional hypercube. (See the next section for the precise definition.) This problem was proposed by Gopalan, Kolaitis, Maneva, and Papadimitriou to investigate connectivity properties on Boolean formulas. It is known that the problem is PSPACE-complete for $k \geq 3$, while it is in P for $k \leq 2$ [6]. Moreover, it is known to be coNP-complete, if given formulas are restricted to Horn 3-CNF [12]. We show that CONNkSAT for $k \geq 3$ is solvable in time $O((2 - \epsilon_k)^n)$ for some constant $\epsilon_k > 0$, where ϵ_k depends only on k, but not on n. It seems to be the first nontrivial result that gives an $O((2 - \epsilon)^n)$-time algorithm for a certain

PSPACE-complete problem in terms of the number of *variables*. Furthermore, this result is considered to be interesting because Calabro, Impagliazzo, and Paturi [5] recently showed the following fact on Π_2-3-SAT: this problem, which is a typical Π_2^P-complete problem, is the QBF problem over 3-CNF formulas, where the quantifier starts with $\forall$, and the number of changes between two types of consecutive quantifiers is at most one. They showed that Π_2-3-SAT is *not* solvable in time $O((2 - \epsilon)^n)$ for any constant $\epsilon > 0$, provided that the SAT problem (with no restriction to the clause length) is not solvable in time $O((2-\epsilon)^n)$ for any constant $\epsilon > 0$. It means that the (general) QBF over 3-CNF formulas, which is a typical PSPACE-complete problem, is not solvable in time $O((2 - \epsilon)^n)$ for any constant $\epsilon > 0$ under the same assumption.

2   Preliminaries

In this paper, we deal with k-CNF formulas, where the length of each clause of a formula is at most k. Let $X = \{x_1, \ldots, x_n\}$ be a set of Boolean variables. An *assignment* to X is an element of $\{0, 1\}^n$. A *partial* assignment to X is an element of $\{0, 1, *\}^n$, where we regard a variable assigned $*$ as *unassigned*. We alternately express partial assignments by pulling out coordinates assigned 0 or 1, e.g., a partial assignment $(x_1 = 1, x_2 = *, x_3 = 0, x_4 = *, \ldots, x_n = *) \in \{0, 1, *\}^n$ is denoted by $(x_1 = 1, x_3 = 0)$. For two assignments $t_1, t_2 \in \{0, 1\}^n$ to X, the *Hamming distance* d between t_1 and t_2 is $d(t_1, t_2) = |\{i \in [n] : t_1(i) \neq t_2(i)\}|$. We extend this notion to partial assignments as follows[1]: for two partial assignments $t_1, t_2 \in \{0, 1, *\}^n$, the *Hamming distance* d between t_1 and t_2 is

$$d(t_1, t_2) \stackrel{\text{def}}{=} \left| \left\{ i \in [n] : \begin{array}{l} t_1(i) \neq *, \ t_2(i) \neq *, \ \text{and} \\ t_1(i) \neq t_2(i) \end{array} \right\} \right|.$$

Given a partial assignment t, we simplify φ in the standard way, that is, eliminating any clause from φ if a literal of the clause is assigned 1 under t, and eliminating any literal from φ if the literal is assigned 0 under t. The resulting formula is denoted by $\varphi|_t$. For later use, we present a typical algorithm for k-SAT, denoted by simple-sat, in Fig. 1 below.

Proposition 1. *Given a k-CNF formula φ, the running time of* simple-sat(φ) *is $O(c_k^n)$ for some constant $c_k < 2$ depending only on k.*

This is the historically first non-trivial exact algorithm for k-SAT proposed by Monien and Speckenmeyer [11]. They showed that the running time is $O(\alpha_k^n)$ for k-SAT, where α_k satisfies $\alpha_k = 2 - 1/\alpha_k^{k-1}$. (For example, $\alpha_3 = 1.618$.)

We slightly modify this algorithm for our purpose. First, we omit the second "return YES" from the algorithm, that is, we just run simple-sat$(\varphi|_t)$ for each partial assignment $t \in S$. Second, we therefore omit the second "return NO" from the algorithm. Note that Proposition 1 also holds for this modified algorithm.

---

[1] It might be better to give it another term since the extension is no longer "distance": it does not satisfy the triangle inequality.

```
simple-sat(φ) // φ is a k-CNF formula
```

> if $\emptyset \in \varphi$ (i.e., $\varphi \notin$ SAT), return NO
> if $\varphi = \{\}$ (i.e., $\varphi \in$ SAT), return YES
>
> Choose a clause $(\ell_1 \vee \cdots \vee \ell_{k'}) \in \varphi$ arbitrarily $(k' \leq k)$
> Let $S = \{(\ell_1 = 0, \ldots, \ell_{i-1} = 0, \ell_i = 1) : 1 \leq i \leq k'\} \subset \{0, 1, *\}^{k'}$
> for each partial assignment $t \in S$
>   if simple-sat$(\varphi|_t)$ returns YES, then return YES
> end-for-each
>
> return NO

Fig. 1. A simple backtracking algorithm for k-SAT

However, the base constant of the running time is worse than what Monien and Speckenmeyer [11] gave: we only obtain the running time of $O(\beta_k^n)$, where β_k is the largest real number $x > 0$ that satisfies $x^k - x^{k-1} - \cdots - x^2 - x - 1 = 0$. (For example, $\beta_3 = 1.840$.) This modification comes from our strategies for solving CONNkSAT: we enumerate *all* satisfying partial assignments. In what follows, we call this modified algorithm simple-sat.

Given a k-CNF formula, a *binary decision diagram* is constructed by the execution of simple-sat(φ). It is viewed as a rooted *binary* tree shown in Fig. 2: we only depict one part of the binary tree, where a recursive call of simple-sat$(\varphi|_t)$ with $t = (l_1 = 0, \ldots, l_{k'-1} = 0, l_{k'} = 1)$ for some $k' \leq k$ is executed. In such a binary tree, each non-leaf vertex represents a variable, and each edge is labelled with 0 or 1. Alternatively, in such a representation, every vertex can be viewed as a partial assignment. The *depth* of a vertex v in a binary tree is the number of ancestors of v.

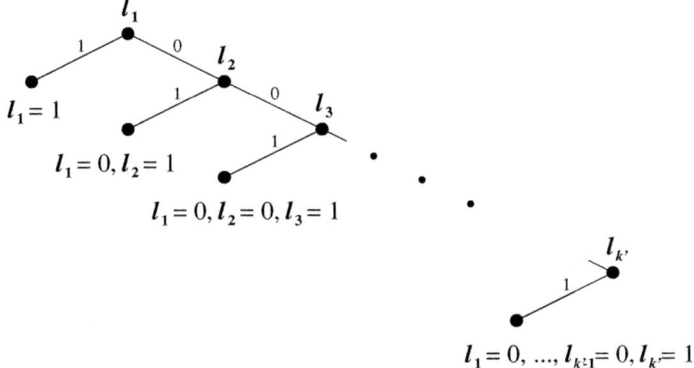

Fig. 2. A binary tree

Given a k-CNF formula, let T_φ be the rooted binary tree obtained by running `simple-sat`(φ). Let SAT_φ be the set of leaves of T_φ that satisfy φ. We alternatively view SAT_φ as the set of partial satisfying assignments. For simplicity, we assume that every leaf of T_φ corresponds to a partial satisfying assignment so that SAT_φ is exactly the set of leaves of T_φ. This is because such a tree is constructed by erasing from T_φ all sub-trees every leaf of which are not satisfying assignments. Moreover, this construction is done in time $O(\beta_k^n)$, where $\beta_k \leq 2 - \epsilon_k$ for our target bound $O((2 - \epsilon_k)^n)$. Then, we note the following two facts about SAT_φ.

Note 1. For any pair of distinct vertices $u, v \in \mathrm{SAT}_\varphi$, $d(u, v) \geq 1$.

Note 2. The vertex set SAT_φ is a partition of the set of all satisfying assignments of φ, that is, for any satisfying assignment $t \in \{0, 1\}^n$ of φ, there is a *unique* vertex $v \in \mathrm{SAT}_\varphi$ (i.e., v is a satisfying *partial* assignment) such that $d(t, v) = 0$.

Given a k-CNF formula φ over n variables, let SAT_φ be as above, and let H_φ be the graph induced from the n dimensional hypercube by SAT_φ. The *solution space induced by* $S \subset SAT_\varphi$ is the graph induced from H_φ by S, that is, by $\{t \in \{0, 1\}^n : \exists s \in S[d(s, t) = 0]\}$. We here note the following two facts, which are easily shown.

Note 3. The solution space induced by a single vertex of SAT_φ is connected.

Note 4. Let v_1, v_2 be distinct vertices of SAT_φ. Suppose that $d(v_1, v_2) = 1$. Then, the solution space induced by $\{v_1, v_2\}$ is connected.

Given a k-CNF formula φ over n variables, the connectivity problem which we study, denoted by *CONNkSAT*, is to ask if the graph H_φ is connected.

Theorem 1 (Gopalan et al. [6]). *CONNkSAT is PSPACE-complete for $k \geq 3$. On the other hand, CONNkSAT is in P for $k \leq 2$.*

3   An Exact Algorithm for CONNkSAT

We present an exact algorithm for CONNkSAT, and show the running time is $O((2 - \epsilon_k)^n)$ for some constant $\epsilon_k > 0$. The algorithm, denoted by `conn-sat`(φ) given a k-CNF formula φ, is shown in Fig. 3, where β_k is the constant specified in the preliminary section.

The idea of this algorithm is to enumerate all satisfying partial assignments, and to construct a graph over those assignments such that there is an edge between two satisfying partial assignments if and only if the Hamming distance between them is *exactly* one. (Recall Note 1 that $d(u, v) \geq 1$ for any pair of distinct vertices $u, v \in \mathrm{SAT}_\varphi$.) Then, we can easily check the connectivity of the graph. The crucial point is to bound the number of edges of the graph. We will show that it is at most $O((2 - \epsilon)^n)$ for some constant $\epsilon > 0$.

conn-sat(φ) // parameter α is a real number that satisfies $(2\beta_k)^{\alpha n} = \beta_k^n$

> Run simple-sat(φ)
> Let T_φ and SAT_φ be as defined above
> Let $V_\varphi = V(T_\varphi)$ be the set of vertices of T_φ,
>
> Let $E_\varphi = \emptyset$
> Construct an undirected graph $G_{\mathrm{SAT}} = (\mathrm{SAT}_\varphi, E_\varphi)$ as follows:
> (1) for each pair of vertices $u, v \in \mathrm{SAT}_\varphi$ with depth at most $(1 - \alpha)n$
>   if $d(u, v) = 1$, then add (u, v) to E_φ
> (2) for each $u \in \mathrm{SAT}_\varphi$ with depth more than $(1 - \alpha)n$
>   Visit $v \in V_\varphi$ in the depth-first search manner
>   starting from the root of T_φ that
>   if $d(u, v) \geq 2$, then do not visit vertices below v any

longer

>   else if $v \in \mathrm{SAT}_\varphi$ and $v \neq u$, then add (u, v) to E_φ
>
>   if $G_{\mathrm{SAT}} = (\mathrm{SAT}_\varphi, E_\varphi)$ is connected, output YES, else output

NO

Fig. 3. An exact algorithm for CONNkSAT

Lemma 1. *Given a k-CNF formula φ, let $G_{SAT} = (SAT_\varphi, E_\varphi)$ be the final G_{SAT} obtained by constructing E_φ. Let $v_1, v_2 \in SAT_\varphi$ be distinct vertices. Then,*

$$d(v_1, v_2) = 1 \iff (v_1, v_2) \in E_\varphi.$$

Proof. Note first that $d(v_1, v_2) \geq 1$, which comes from Note 1. It is easy to see that $(u, v) \in E_\varphi$ implies $d(u, v) = 1$ since our algorithm adds an edge (u, v) to E_φ only if $d(u, v) = 1$.

Suppose that $d(u, v) = 1$. We see that it means our algorithm adds an edge (u, v) to E_φ because of the following observation: if $d(u, v) = 1$, then $d(u, w) \leq 1$ for any ancestor w of v.   □

Lemma 2. *Given a k-CNF formula φ, let $G_{SAT} = (SAT_\varphi, E_\varphi)$ be the final G_{SAT} obtained by constructing E_φ. Then,*

$$\varphi \in CONNkSAT \iff G_{SAT} \text{ is connected.}$$

Proof. We first consider the case of $|\mathrm{SAT}_\varphi| \leq 1$. In this case, it is obvious that $G_{\mathrm{SAT}} = (\mathrm{SAT}_\varphi, E_\varphi)$ is connected. Moreover, $\varphi \in \mathrm{CONN}k\mathrm{SAT}$ holds because of Note 3. Thus, this lemma holds for $|\mathrm{SAT}_\varphi| \leq 1$.

Next, we assume that $|\mathrm{SAT}_\varphi| \geq 2$. Suppose that G_{SAT} is connected. We will show that any pair of two satisfying assignments of φ is connected on H_φ. Let $t_1, t_2 \in \{0, 1\}^n$ be distinct satisfying assignments of φ. Let $v_1 \in \mathrm{SAT}_\varphi$ (resp. $v_2 \in \mathrm{SAT}_\varphi$) be a vertex of G_{SAT} corresponding to t_1 (resp. t_2), that is, $d(v_1, t_1) = 0$ (resp. $d(v_2, t_2) = 0$). From Note 2, there is such a vertex which is unique. We may assume $v_1 \neq v_2$ since otherwise it is the same as the case of

$|SAT_\varphi| \leq 1$. Since G_{SAT} is connected, there is a path between v_1 and v_2 (on G_{SAT}). Consider any pair of adjacent vertices on the path, say, $u_1, u_2 \in SAT_\varphi$, From the previous lemma, $d(u_1, u_2) = 1$ since $(u_1, u_2) \in E_\varphi$. Moreover, from Note 4, the solution space of H_φ induced by $\{u_1, u_2\}$ is connected. Applying this argument repeatedly to every pair of adjacent vertices on the path, we see that the solution space of H_φ induced by all vertices on the path is connected, and hence t_1 and t_2 are connected on H_φ. This holds for any pair of two satisfying assignments of φ. Thus, we conclude $\varphi \in CONNkSAT$.

Suppose that $\varphi \in CONNkSAT$. We will show that any pair of two vertices of SAT_φ is connected on G_{SAT}. Let $v_1, v_2 \in SAT_\varphi$ be distinct vertices of G_{SAT}. Let t_1 (resp. t_2) be an arbitrary satisfying assignment of φ such that $d(t_1, v_1) = 0$ (resp. $d(t_2, v_2) = 0$). Since $\varphi \in CONNkSAT$, there exists a path $t_1 = a_0 \to a_1 \to \cdots \to a_\ell = t_2$ on H_φ. Consider any pair of a_i and a_{i+1}. There are two cases: (1) there is a vertex $u \in SAT_\varphi$ such that $d(a_i, u) = d(a_{i+1}, u) = 0$, and (2) there are distinct vertices $u_1, u_2 \in SAT_\varphi$ such that $d(a_i, u_1) = d(a_{i+1}, u_2) = 0$. Consider the second case. (We do not need to care for the first case.) Since $d(a_i, a_{i+1}) = 1$, we must have $d(u_1, u_2) = 1$. (We do not have $d(u_1, u_2) = 0$ since $u_1 \neq u_2$.) From the previous lemma, it means $(u_1, u_2) \in E_\varphi$. Applying this argument repeatedly to every pair of adjacent vertices on the path, we see that v_1 and v_2 are connected on G_{SAT}. This holds for any pair of two vertices of SAT_φ. Thus, we conclude that G_φ is connected. $\qquad\square$

From this lemma, we conclude that the output of conn-sat(φ) is correct for any φ. It remains to show the upper bound on the running time of conn-sat(φ).

Lemma 3. *The running time of conn-sat(φ) is $O((2 - \epsilon_k)^n)$ for some constant $\epsilon_k > 0$ depending only on k.*

Proof. Given a k-CNF formula φ, let T_φ be the rooted binary tree obtained by running simple-sat(φ). Let $G_{SAT} = (SAT_\varphi, E_\varphi)$ be the final G_{SAT} obtained by constructing E_φ. Note here that the running time of constructing T_φ is $O(\beta_k^n)$, where β_k is the constant specified in the preliminary section. For showing the worst-case running time, it suffices to estimate an upper bound of $|E_\varphi|$. For any $\alpha : 0 \leq \alpha \leq 1$, let

$$U \stackrel{\text{def}}{=} \{u \in SAT_\varphi : depth(u) \leq (1 - \alpha)n\},$$
$$W \stackrel{\text{def}}{=} \{w \in SAT_\varphi : depth(w) > (1 - \alpha)n\},$$

where $depth(u)$ is the depth of u in T_φ. Then,

$$|E_\varphi| = |E_1| + |E_2|, \text{ where } \begin{cases} E_1 = E_\varphi \cap (U \times U), \\ E_2 = (E_\varphi \cap (U \times W)) \cup (E_\varphi \cap (W \times W)). \end{cases}$$

Claim. For any $\alpha : 0 \leq \alpha \leq 1$,

1. $|E_1| \leq \left(\beta_k^{(1-\alpha)n}\right)^2 \quad \left(= \beta_k^{2(1-\alpha)n}\right)$,
2. $|E_2| \leq \sum_{0 \leq t \leq \alpha n} \beta_k^{n-t}\left((n-t)2^t\right) \quad \left(\leq n^2 \cdot 2^{\alpha n} \cdot \beta_k^{(1-\alpha)n}\right)$.

Proof. The first inequality comes from the fact that the number of vertices of T_φ with depth at most $(1 - \alpha)n$ is at most $\beta_k^{(1-\alpha)n}$.

Fix t with $0 \le t \le \alpha n$ arbitrarily. Consider an arbitrary vertex $w \in W$ with depth $n - t$. We will estimate the possible number of edges $(w, v) \in E_\varphi$ where $v \in \mathrm{SAT}_\varphi$. Let r_i be the ancestor of w at depth i $(0 \le i < n - t)$. Let r_i' be the child vertex of r_i that is not an ancestor of w. Let $T_{w,i}$ be the sub-tree of T_φ rooted at r_i'. Then, the number of assignments (not necessarily satisfying ones) $a \in \{0, 1\}^n$ such that $d(r_i', a) = 0$ and $d(w, a) \le 1$ is exactly 2^t since the number of variables assigned $*$ under w is t. Let $A \subset \{0, 1\}^n$ be the set of those assignments. Then, the number of leaves v of $T_{w,i}$ such that $d(w, v) \le 1$ is at most 2^t since each assignment $a \in A$ corresponds to a unique leaf v if a is a satisfying assignment. Thus, for any $w \in W$ with depth $n - t$, the number of leaves v such that $d(w, v) \le 1$ is at most $(n - t)2^t$. Since the number of vertices $w \in W$ with depth $n - t$ is at most β_k^{n-t}, the second inequality holds. $\qquad \square$

From this claim, we have $|E_\varphi| \le \beta_k^{2(1-\alpha)n} + n^2 2^{\alpha n} \beta_k^{(1-\alpha)n}$ for any $0 \le \alpha \le 1$. By fixing α to a constant satisfying $\beta_k^{2(1-\alpha)n} = 2^{\alpha n} \beta_k^{(1-\alpha)n}$, which is equivalent to $(2\beta_k)^{\alpha n} = \beta_k^n$, we obtain an upper bound on $|E_\varphi|$ as follows:

$$|E_\varphi| \le 2 \cdot \mathrm{poly}(n) \cdot 2^{\alpha n} \beta_k^{(1-\alpha)n}.$$

We see that the formula on the right-hand side is $O((2 - \epsilon_k)^n)$ for some constant $\epsilon_k > 0$ (depending only on k) since β_k is a constant less than 2. $\qquad \square$

From Lemma 2, we see that our algorithm solves CONNkSAT. From Lemma 3, we see that our algorithm runs in time $O((2 - \epsilon_k)^n)$ for some constant $\epsilon_k > 0$. Therefore, we obtain the following theorem:

Theorem 2. *The problem CONNkSAT is solvable in time $O((2 - \epsilon_k)^n)$ for some constant $\epsilon_k > 0$ depending only on k. (For example, it is $O(1.914^n)$ for $k = 3$.)*

4 Conclusion

We have presented an $O(2 - \epsilon_k)^n$-time exact algorithm for CONNkSAT. One of our future work is to improve the analysis of the running time of our algorithm, and to obtain the upper bound $O(\beta_k^n)$ which is same as the running time of simple-sat: our bound is slightly worse than $O(\beta_k^n)$. Instead of doing that, we may be able to reduce the running time just by replacing simple-sat with a more sophisticated backtrack-type algorithm A that satisfies the following: all leaves of the rooted binary tree constructed by A constitute a partition of all satisfying assignments. However, we encounter the same problem as above: we cannot derive the running time as much as that of A from our analysis.

References

1. Bellman, R.: Dynamic programming treatment of the travelling salesman problem. Journal of the ACM 9(1), 61–63 (1962)
2. Björklund, A., Husfeldt, T., Kaski, P., Koivisto, M.: The travelling salesman problem in bounded degree graphs. In: Aceto, L., Damgård, I., Goldberg, L.A., Halldórsson, M.M., Ingólfsdóttir, A., Walukiewicz, I. (eds.) ICALP 2008, Part I. LNCS, vol. 5125, pp. 198–209. Springer, Heidelberg (2008)
3. Björklund, A., Husfeldt, T., Kaski, P., Koivisto, M.: Fourier meets möbius: fast subset convolution. In: Proc. STOC 2007, pp. 67–74 (2007)
4. Björklund, A., Husfeldt, T.: Inclusion-exclusion algorithms for counting set partitions. In: Proc. FOCS 2006, pp. 575–582 (2006)
5. Calabro, C.: The exponential complexity of satisfiability problems, Ph.D. thesis, University of California, San Diego (2009)
6. Gopalan, P., Kolaitis, P.G., Maneva, E.N., Papadimitriou, C.H.: The connectivity of Boolean satisfiability: Computational and structural dichotomies. SIAM J. Comput. 38(6), 2330–2355 (2009)
7. Held, M., Karp, R.M.: The traveling-salesman problem and minimum spanning trees. Operations Res. 18, 1138–1162 (1970)
8. Iwama, K., Tamaki, S.: Improved upper bounds for 3-SAT. In: Proc. SODA 2004, pp. 328–329 (2004)
9. Koivisto, M.: An $O(2^n)$ algorithm for graph coloring and other partitioning problems via inclusion-exclusion. In: Proc. FOCS 2006, pp. 583–590 (2006)
10. Kutzkov, K.: New upper bound for the #3-SAT problem. Inf. Process. Lett. 105(1), 1–5 (2007)
11. Monien, B., Speckenmeyer, E.: Solving satisfiability in less than 2^n steps. Discrete Applied Mathematics 10, 287–295 (1985)
12. Makino, K., Tamaki, S., Yamamoto, M.: On the Boolean connectivity problem for Horn relations. In: Marques-Silva, J., Sakallah, K.A. (eds.) SAT 2007. LNCS, vol. 4501, pp. 187–200. Springer, Heidelberg (2007)
13. Paturi, R., Pudlák, P., Saks, M.E., Zane, F.: An improved exponential-time algorithm for k-SAT. J. ACM 52(3), 337–364 (2005)
14. Paturi, R., Pudlák, P., Zane, F.: Satisfiability coding lemma. Chicago J. Theor. Comput. Sci. (1999)
15. Rolf, D.: Improved bound for the PPSZ/Schöning-algorithm for 3-SAT. JSAT 1, 111–122 (2006)
16. Schöning, U.: A probabilistic algorithm for k-SAT based on limited local search and restart. Algorithmica 32(4), 615–623 (2002)
17. Schöning, U.: Algorithmics in exponential time. In: Diekert, V., Durand, B. (eds.) STACS 2005. LNCS, vol. 3404, pp. 36–43. Springer, Heidelberg (2005)
18. Williams, R.: Algorithms for quantified Boolean formulas. In: Proc. SODA 2002, pp. 299–307 (2002)
19. Williams, R.: A new algorithm for optimal constraint satisfaction and its implications. In: Díaz, J., Karhumäki, J., Lepistö, A., Sannella, D. (eds.) ICALP 2004. LNCS, vol. 3142, pp. 1227–1237. Springer, Heidelberg (2004)
20. Woeginger, G.J.: Exact algorithms for NP-hard problems: a survey. In: Jünger, M., Reinelt, G., Rinaldi, G. (eds.) Combinatorial Optimization - Eureka, You Shrink! LNCS, vol. 2570, pp. 185–207. Springer, Heidelberg (2003)

Improving Unsatisfiability-Based Algorithms for Boolean Optimization

Vasco Manquinho, Ruben Martins, and Inês Lynce

IST/INESC-ID, Technical University of Lisbon, Portugal
{vmm,ruben,ines}@sat.inesc-id.pt

Abstract. Recently, several unsatisfiability-based algorithms have been proposed for Maximum Satisfiability (MaxSAT) and other Boolean Optimization problems. These algorithms are based on being able to iteratively identify and relax unsatisfiable sub-formulas with the use of fast Boolean satisfiability solvers. It has been shown that this approach is very effective for several classes of instances, but it can perform poorly on others for which classical Boolean optimization algorithms find it easy to solve. This paper proposes the use of Pseudo-Boolean Optimization (PBO) solvers as a preprocessor for unsatisfiability-based algorithms in order to increase its robustness. Moreover, the use of constraint branching, a well-known technique from Integer Linear Programming, is also proposed into the unsatisfiability-based framework. Experimental results show that the integration of these features in an unsatisfiability-based algorithm results in an improved and more effective solver for Boolean optimization problems.

1 Introduction

The success of Propositional Satisfiability (SAT) solvers has increased the interest in several generalizations of SAT, namely in Boolean optimization problems. As a result, several techniques first proposed for SAT algorithms have been extended for Pseudo-Boolean Optimization (PBO), Maximum Satisfiability (MaxSAT) and the more general problem of Weighted Boolean Optimization (WBO). Moreover, the acknowledgment of the strong relation between all these problems has led to the development of new algorithms based on the translation between these Boolean formalisms [8,12,15].

Algorithms based on the identification of unsatisfiable sub-formulas have also been developed and are now able to tackle all these Boolean optimization problems. The first proposal of unsatisfiability-based algorithm [13] was restricted to MaxSAT and partial MaxSAT problems. However, recent work has been done on improving this algorithmic solution [21] and generalizing it for weighted MaxSAT, PBO and WBO [2,19].

The proposal in this paper is for a further integration of procedures in an unique Boolean optimization framework. Hence, it is proposed the encoding into PBO and the use of a PBO solver as a preprocessing step for finding a tight upper bound on the optimal solution before applying an unsatisfiability-based

O. Strichman and S. Szeider (Eds.): SAT 2010, LNCS 6175, pp. 181–193, 2010.

algorithm. Moreover, the use of constraint branching, a well-known technique initially presented for (Mixed) Integer Linear Programming, can also be integrated with success in an unsatisfiability-based algorithm for Boolean optimization.

The paper is organized as follows: in section 2 several formalisms used for Boolean optimization are introduced, namely Weighted Boolean Optimization (WBO), pseudo-Boolean Optimization (PBO) and the Maximum Satisfiability (MaxSAT) problem and its variants. Furthermore, several relations between these formalisms are reviewed, as well as the most common algorithmic solutions. In section 3, it is proposed the use of pseudo-Boolean solvers as a preprocessing step for an unsatisfiability-based algorithm. Next, in section 4, it is shown how to integrate constraint branching into an unsatisfiability-based solver for WBO. Finally, experimental results are presented in section 5 and the paper concludes in section 6.

2   Preliminaries

In this section several Boolean optimization problems are defined, starting with the more general Weighted Boolean Optimization problem. Next, translations between several formalisms are reviewed and the most common algorithmic solutions are briefly described. The approach based on the identification of unsatisfiable sub-formulas is presented in more detail since it will be extensively referred in the remaining of the paper.

2.1   Weighted Boolean Optimization

Weighted Boolean Optimization (WBO) is a natural extension of other Boolean problems, such as Maximum Satisfiability (MaxSAT) and Pseudo-Boolean Optimization (PBO). In WBO, constraints can be any linear inequality with integer coefficients (also known as pseudo-Boolean constraints) defined over a set of Boolean variables. In general, one can define a pseudo-Boolean constraint as follows:

$$\sum_{j \in N} a_j \, l_j \geq b \tag{1}$$

where a_j and b are positive integers and l_j is a propositional literal that either denotes a variable x_j or its complement $\bar{x}_j$. It is well-known that all other types of linear constraints with Boolean variables can be easily translated into this one [7]. Notice that propositional clauses are a particular case of pseudo-Boolean constraints where all coefficients a_j and the right-hand side b are equal to 1. If all a_j are equal to 1 and $b > 1$, then the constraint is called a cardinality constraint.

A WBO formula φ is defined as the conjunction of two pseudo-Boolean formulas φ_h and φ_s, where φ_h contains the *hard* constraints and φ_s contains the *soft* constraints. Moreover, each soft constraint ω_i has an associated positive weight c_i that represents the cost of not satisfying constraint ω_i. The WBO problem can be defined as finding an assignment to problem variables that satisfies all hard constraints in φ_h and minimizes the total weight of unsatisfied soft constraints in φ_s.

Example 1. Consider the following example of a WBO formula:

$$\varphi_h = \{\ x_1 + x_2 + x_3 \geq 2, \quad 2\bar{x}_1 + \bar{x}_2 + x_3 \geq 2\}$$
$$\varphi_s = \{\ (x_1 + \bar{x}_2 \geq 1, 2), \quad (\bar{x}_1 + \bar{x}_3 \geq 1, 3)\} \tag{2}$$

In this example, there are only two possible assignments that satisfy all hard constraints in φ_h. These assignments are $x_1 = x_3 = 1, x_2 = 0$ and $x_1 = 0, x_2 = x_3 = 1$. However, for each of these assignments, at least one soft constraint in φ_s is made unsatisfied. Therefore, the solution to the WBO instance would be $x_1 = 0, x_2 = x_3 = 1$ since it is the assignment that minimizes the total cost of unsatisfied soft constraints while satisfying all hard constraints.

2.2 Relating with MaxSAT and Pseudo-Boolean Optimization

One should note that WBO is a direct generalization of the Maximum Satisfiability (MaxSAT) problem and variants. The MaxSAT problem can be defined as finding an assignment that minimizes the number of unsatisfied clauses in a given CNF formula φ. Hence, a WBO instance where $\varphi_h = \emptyset$ and φ_s contains only propositional clauses with weight 1 is in fact a MaxSAT instance.

The *partial* MaxSAT problem differs from MaxSAT since there is a set of clauses declared as hard and a set of clauses declared as soft. The objective in partial MaxSAT is to find an assignment such that all hard clauses are satisfied while minimizing the number of unsatisfied soft clauses. Again, a WBO formula where all constraints in φ_h and φ_s are propositional clauses and all soft clauses have weight 1 is a partial MaxSAT instance. Finally, there are also variants of MaxSAT and partial MaxSAT with weights greater than 1 which are respectively known as *weighted* MaxSAT and *partial weighted* MaxSAT. Clearly, the resulting instances are also specific cases of WBO instances.

Another well-known Boolean optimization formalism is Pseudo-Boolean Optimization (PBO), also known as 0-1 Integer Linear Programming (0-1 ILP). The PBO problem can be defined as finding an assignment to the Boolean variables such that a set of pseudo-Boolean constraints is satisfied and the value of a linear cost function is minimized. Formally, it is possible to define PBO as follows:

$$\begin{aligned}
\text{minimize} \quad & \sum_{j \in N} c_j\, x_j \\
\text{subject to} \quad & \sum_{j \in N} a_{ij}\, l_j \geq b_i, \\
& l_j \in \{x_j, \bar{x}_j\}, x_j \in \{0, 1\}, a_{ij}, b_i, c_j \in \mathbf{N}_0^+
\end{aligned} \tag{3}$$

It is also possible to encode a PBO instance into the WBO formalism. The constraint set of the PBO instance can be directly mapped into the set of hard constraints φ_h of the resulting WBO instance, while the objective function is mapped using soft constraints. For each term $c_j x_j$ in the objective function, a new soft constraint $\bar{x}_j \geq 1$ is added to φ_s with weight c_j. The optimal solution to the resulting WBO instance will also be an optimal solution to the original PBO instance [19].

2.3   Algorithmic Solutions

For each of these Boolean formalisms (WBO, MaxSAT and PBO), there is a
wide variety of algorithmic solutions. One classical approach is the use of a
branch and bound algorithm where an *upper bound* on the value of the objective
function is updated whenever a better solution is found and *lower bounds* on
the value of the objective function are estimated considering a set of variable
assignments. Whenever the lower bound value is higher or equal to the upper
bound, the search procedure can safely backtrack since it is guaranteed that the
current best solution cannot be improved by extending the current set of variable
assignments. Several MaxSAT and PBO algorithms follow this approach using
different lower bounding procedures [16,17,3,14,18].

Algorithm 1. Unsatisfiability-based Algorithm for MaxSAT and partial
MaxSAT

FuMalikAlg(φ)

1   **while true**
2   **do** (st, φ_C) $\leftarrow$ SAT(φ)
3   **if** st = **UNSAT**
4   **then** $V_R \leftarrow \emptyset$
5   **for each** $\omega \in \varphi_C \wedge$ soft(ω)
6   **do** r is a new relaxation variable
7                           $\omega_R \leftarrow \omega \cup \{r\}$
8                           $\varphi \leftarrow \varphi \setminus \{\omega\} \cup \{\omega_R\}$
9                           $V_R \leftarrow V_R \cup \{r\}$
10   **if** $V_R = \emptyset$
11   **then return UNSAT**
12   **else** $\varphi \leftarrow \varphi \cup \text{CNF}(\sum_{r \in V_R} r = 1)$
13                      $\triangleright$ Additional clauses for Equals1 constraint are marked as hard clauses
14   **else return** Satisfiable assignment to φ

Another approach used in PBO solvers is to perform a linear search on the
value of the objective function by iterating on the possible upper bound values [7]. Whenever a new solution to the problem constraints is found, the upper
bound value is updated and a new constraint is added such that all solutions
with an higher value are discarded. Several state of the art PBO solvers use
this approach such as Pueblo [26], minisat+ [12], among others [8,1]. These
solvers rely on the generalization of the most effective techniques already used
in SAT solvers, such as Boolean Constraint Propagation, conflict-based learning
and conflict-directed backtracking [18,10].

There are other successful solvers that perform conversions of one Boolean formalism to another and subsequently use a specific solver on the new formalism.
For instance, PBO solver minisat+ [12] converts all pseudo-Boolean constraints
to propositional clauses and uses a SAT solver to find an assignment that satisfies the problem constraints; SAT4J MAXSAT [8] converts MaxSAT instances into

---

Algorithm 2. Unsatisfiability-based Weighted Boolean Optimization algorithm

---

WBO(φ)

```
 1  while true
 2      do (st, φ_C) ← PB(φ)
 3          if st = UNSAT
 4              then  min_c ← ∞
 5                  for each ω ∈ φ_C
 6                      do if soft(ω) and cost(ω) < min_c
 7                          then min_c ← cost(ω)
 8                  V_R ← ∅
 9                  for each ω ∈ φ_C ∧  soft(ω)
10                      do r is a new relaxation variable and ω = ∑ a_j l_j ≥ b
11                          V_R ← V_R ∪ {r}
12                          ω_R ← (b r + ∑ a_j l_j ≥ b)
13                          cost(ω_R) ← min_c
14                          if cost(ω) > min_c
15                              then φ ← φ ∪ {ω_R}
16                                  cost(ω) ← cost(ω) − min_c
17                              else φ ← φ\{ω} ∪ {ω_R}
18                  if V_R = ∅
19                      then return UNSAT
20                      else φ_W ← φ_W ∪ {∑_{r∈V_R} r = 1}
21          else  return Satisfiable assignment to φ
```

---

a PBO instance; `Toolbar` [15] converts MaxSAT instances into a weighted constraint network and uses a Constraint Satisfaction Problem (CSP) solver, among other solvers [23,24].

A recent approach initially proposed by Fu and Malik [13] for MaxSAT and partial MaxSAT problems is based on the iterated use of a SAT solver to identify unsatisfiable sub-formulas. Algorithm 1 presents the pseudo-code for the original Fu and Malik's proposal. Consider that φ is the Boolean working formula where constraints are marked as either soft or hard. At each iteration, a SAT solver is used and its output is a pair (st, φ_C) where st denotes the resulting status of the solver (satisfiable or unsatisfiable) and φ_C contains the unsatisfiable sub-formula provided by the SAT solver if φ is unsatisfiable. In this latter case, for each soft constraint in φ_C, a new relaxation variable is added. Moreover, φ is changed to encode that exactly one of the new relaxation variables can be assigned value 1 (Equals1 constraint in line 12) and the algorithm continues to the next iteration. Otherwise, if φ is satisfiable, the SAT solver was able to find an assignment which is an optimal solution to the original MaxSAT or partial MaxSAT problem [13].

Different algorithms have been proposed for MaxSAT and partial MaxSAT based on this approach. For instance, effective encodings for the Equals1 constraint have been proposed with better results [21,20] than the pairwise encoding of the original algorithm [13]. Moreover, different strategies have been used regarding the total number of relaxation variables needed [20,2].

Finally, the unsatisfiability-based approach has also been extended for weighted and partial weighted MaxSAT [2,19] and generalized to WBO [19] by using a pseudo-Boolean solver instead of a SAT solver. Algorithm 2 presents the pseudo-code for the WBO solver and one can clearly notice that it follows the same structure as Algorithm 1. However, in this case, φ is now a WBO formula, i.e. constraints can be any type of pseudo-Boolean constraints and a positive cost is associated with each soft constraint. One difference from Algorithm 1 is in lines 4-7 where min_c denotes the cost associated to the unsatisfiable sub-formula φ_C, defined as the minimum cost of soft constraints in φ_C. Moreover, if the weight of a soft constraint in φ_C is larger than min_c, then the relaxation also differs since the original constraint is kept, but with a smaller weight as shown in lines 9-17. Finally, notice that the Equals1 constraint in line 20 does not need to be encoded into CNF, since a pseudo-Boolean solver is used instead of a SAT solver.

3   Improving Unsatisfiability-Based Algorithms

As shown previously, unsatisfiability-based algorithms are able to tackle several Boolean optimization problems. These algorithms work by making a linear search on the lower bounds of the optimal solution value. However, it has been shown that in some cases, it is preferable to search on the upper bounds of the optimal solution.

In this section, we propose to translate Weighted Boolean Optimization (WBO) to the more specific Pseudo-Boolean Optimization (PBO) problem before applying an unsatisfiability-based algorithm. This approach has two main goals: (i) to apply simplification techniques that are used as preprocessing procedures in PBO and (ii) to find a tight upper bound on the optimal solution. Afterwards, the problem is again translated into WBO and solved using an unsatisfiability-based algorithm.

3.1   Pseudo-Boolean Optimization as Preprocessing

We start by reviewing the translation from WBO formulas into PBO. Clearly, hard constraints φ_h can be directly mapped as constraints into the resulting PBO formula. However, for soft constraints in φ_s, additional variables are needed. Each soft constraint of the form $\sum a_j \, l_j \geq b$, is mapped into a new PBO constraint $b \, r + \sum a_j l_j \geq b$, where r is a new relaxation variable. The objective function will be to minimize the weighted sum of the relaxation variables. The coefficient of variable r in the objective function is the weight of the original constraint associated with variable r.

Example 2. Consider the following WBO formula:

$$\varphi_h = \{\, x_1 + x_2 + x_3 \geq 2, \quad 2\bar{x}_1 + \bar{x}_2 + x_3 \geq 2, \quad x_1 + x_4 \geq 1\}$$
$$\varphi_s = \{\, (x_1 + \bar{x}_2 \geq 1, 2), \quad (\bar{x}_1 + \bar{x}_3 \geq 1, 3), \quad (\bar{x}_4 \geq 1, 4)\} \tag{4}$$

The resulting PBO instance would be:

$$\begin{array}{ll}
\text{minimize} & 2r_1 + 3r_2 + 4r_3 \\
\text{subject to} & x_1 + x_2 + x_3 \geq 2 \\
& 2\bar{x}_1 + \bar{x}_2 + x_3 \geq 2 \\
& x_1 + x_4 \geq 1 \\
& r_1 + x_1 + \bar{x}_2 \geq 1 \\
& r_2 + \bar{x}_1 + \bar{x}_3 \geq 1 \\
& r_3 + \bar{x}_4 \geq 1
\end{array} \tag{5}$$

Notice that in this example variable r_3 is not necessary in the resulting PBO instance. Since $\bar{x}_4 \geq 1$ is a unit clause in (4), one can remove this constraint and just add x_4 with weight 4 to the objective function, which would result in minimizing $2r_1 + 3r_2 + 4x_4$. This is an important simplification, as many industrial instances have unit clauses as soft constraints [4].

After translating the WBO formula to PBO, two steps are applied:

1. Simplification techniques are used in the PBO formula;
2. The PBO formula is solved using tight limits.

In the first step we use a generalization of Hypre [5] for pseudo-Boolean formulas. As a result, literal equivalence detection and hyper-binary resolution are used to eliminate variables from the formula. In fact, besides these techniques, other preprocessing procedures could have also been used, such as clause and variable subsumption, among others [22].

After the first step, a search procedure is carried out using a pseudo-Boolean solver and making the classical linear search on the upper bound of the optimal solution [7,1,26]. However, our use of the pseudo-Boolean solver is limited to 10% of the time limit given to solve the formula. Given a tight time limit, the pseudo-Boolean solver will not find the optimal solution in most cases. Therefore, if the solver is unable to prove optimality, the problem instance is encoded back to WBO (as described in section 2.2) and solved using an unsatisfiability-based algorithm. Moreover, learned conflict constraints by the PBO solver can be kept as hard constraints on the WBO instance, thus pruning the search space. Additionally, remember that the unsatisfiability-based algorithm will make a search on the lower bound of the optimal solutions, but in this case it will be already limited on the upper bound side. We note that searching on both the upper and the lower bound on the value of the objective function is not new [21], but to the best of our knowledge, the presented approach is novel.

Although the objective is to find a tight upper bound, it is possible that the PBO solver proves the optimality of the found upper bound. In that case, the optimal solution to the original problem has been found without having to make the search on the lower bound value. However, even if the solver is unable to prove optimality, small clauses learned by the pseudo-Boolean solver are kept in the WBO formula as hard clauses, further constraining the search space.

4   Using Constraint Branching

One of the main problems of using unsatisfiability-based algorithms for WBO is that after a given number of iterations, the number of relaxation variables can be much larger than the initial number of problem variables [21]. This might occur even when using a pseudo-Boolean solver where the encoding of the Equals1 constraint to CNF is not necessary. Furthermore, when solving a formula with several Equals1 constraints, setting a single variable to 0 or 1 may cause a dramatic difference on the number of propagations that results from this assignment.

Remember that in each iteration of Algorithm 2, a new Equals1 constraint is added (line 20), thus constraining that only one of the new relaxation variables can be assigned value 1. Consider that, at any given iteration, k new relaxation variables are added. As a result, a new Equals1 constraint is added as follows:

$$\sum_{i=1}^{k} r_i = 1 \tag{6}$$

Notice that, by assigning one variable r_i with value 1, all other variables $r_j \neq r_i$ (with $1 \leq j \leq k$) must be assigned value 0. However, if r_i is assigned value 0, no propagation occurs due to (6). As a result, assigning a value to any of these variables tends to produce very different search trees, in particular for large values of k. Therefore, if the solver assigns one single variable that appears in these constraints early in the search tree, that assignment might be too strong or too weak depending on the chosen value. This problem has already been observed in (Mixed) Integer Linear Programming problems [6] and one way to balance the search tree is to use constraint branching [25].

Constraint branching is a well-known technique used in specific cases of (Mixed) Integer Linear Programming in which the formula to be solved is split into two sub-problems such that new constraints are added to each branch. In our case, we would like to take advantage of the Equals1 constraints in order to assign large sets (hundreds or even thousands) of variables in a single step. Therefore, it is proposed the use of a branching step due on Equals1 constraints and integrate it into an unsatisfiability-based algorithm.

By using constraint branching on an Equals1 constraint, instead of assigning just one variable, half of the k variables in (6) are assigned value 0. Without loss of generality, assume that variables r_1 to $r_{k/2}$ are assigned value 0. This is done by adding the following constraint to the working formula φ:

$$\omega_{c1} : \sum_{i=1}^{k/2} r_i = 0 \tag{7}$$

This means that the variable to be assigned value 1 is one of $r_{k/2+1}$ to r_k. If the formula $\varphi \cup \{\omega_{c1}\}$ is not satisfiable, then it is possible to infer that one of the variables from r_1 to $r_{k/2}$ must be assigned value 1, while all others from $r_{k/2+1}$ to r_k must be assigned value 0. Hence, if $\varphi \cup \{\omega_{c1}\}$ is unsatisfiable, the following constraint can be safely inferred:

$$\omega_{c2} : \sum_{i=k/2+1}^{k} r_i = 0 \tag{8}$$

Algorithm 3. Using Constraint Branching in Unsatisfiability-based WBO Algorithm

---

COMPUTE_CORE(φ)

1   ▷ Compute an unsatisfiable sub-formula from φ
2   **if** (no large Equals1 constraint exist in φ)
3   **then** (st, φ_C) ← PB(φ)
4   **return** (st, φ_C)
5   **else** Select a large Equals1 constraint ω from φ
6       $k = \text{size}(\omega)$
7       $\omega_{c1} : \sum_{i=1}^{k/2} r_i = 0$
8   (st, φ_{C1}) ← COMPUTE_CORE($\varphi \cup \{\omega_{c1}\}$)
9   **if** (st = **SAT** $\vee$ $\omega_{c1} \notin \varphi_{C1}$)
10   **then return** (st, φ_{C1})
11   **else** $\omega_{c2} : \sum_{i=k/2+1}^{k} r_i = 0$
12   (st, φ_{C2}) ← COMPUTE_CORE($\varphi \cup \{\omega_{c2}\}$)
13   **if** (st = **SAT** $\vee$ $\omega_{c2} \notin \varphi_{C2}$)
14   **then return** (st, φ_{C2})
15   **else return** (st, $\varphi_{C1} \cup \varphi_{C2}$)

Algorithm 3 illustrates the use of constraint branching in the computation of unsatisfiable sub-formulas. This procedure can replace the call for the pseudo-Boolean solver in line 2 of Algorithm 2. In Algorithm 3 we start by selecting a *large*[1] Equals1 constraint in order to maximize the number of variables to be assigned due to ω_{c1}. Notice that φ_{C1} denotes an unsatisfiable sub-formula from $\varphi \cup \{\omega_{c1}\}$. If φ_{C1} does not include ω_{c1}, then φ_{C1} is also an unsatisfiable sub-formula from φ and the procedure returns. The same applies to φ_{C2} when it does not include ω_{c2}. Otherwise, if both φ_{C1} and φ_{C2} include the respective added constraints, then an unsatisfiable sub-formula for φ is $\varphi_{C1} \cup \varphi_{C2}$.

Finally, it should be noticed that, in practice, this technique is applied parsimoniously. It was observed that if we were to make constraint branching on *all* large Equals1 constraints, then the resulting unsatisfiable sub-formula would usually be much larger than a single call to the pseudo-Boolean solver. This occurs, since the search space is explored differently in each sub-problem and the set union of both unsatisfiable sub-formulas results in a larger unsatisfiable

---

[1] An Equals1 constraint with than 100 relaxation variables is considered large in the context of our solver.

sub-formula for the main problem. Hence, before making a constraint branching step, the solver is called with a limited number of conflicts (approx. 30,000). Afterwards, if the solver has been unable to produce an unsatisfiable sub-formula, a constraint branching step is applied.

5   Experimental Results

In order to evaluate the techniques proposed in the paper, solver wbo was modified to include pseudo-Boolean optimization techniques described in section 3, as well as the use of constraint branching, presented in section 4. The new version of solver wbo is 1.2, while version 1.0 is the one submitted to the last MaxSAT evaluation [4].

For the experimental evaluation, the industrial benchmark sets of the partial MaxSAT problem (a specific case of WBO) were selected. Besides wbo, we also run other solvers among the most effective for these benchmark sets, namely MSUncore [21,19], SAT4J MaxSAT [8] and pm2 [2]. Experiments were run on a set of Intel Xeon 5160 servers (3.0GHZ, 1333Mhz, 3GB) running Red Hat Enterprise Linux WS 4. For each instance, the CPU time limit was 1800 seconds.

Table 1. Solved Instances for Industrial Partial MaxSAT

Benchmark set	#I	MSUncore	SAT4J (MS)	pm2	wbo1.0	wbo1.2
bcp-fir	59	49	10	58	40	47
bcp-hipp-yRa1	176	139	140	166	144	137
bcp-msp	148	121	95	93	26	95
bcp-mtg	215	173	196	215	181	207
bcp-syn	74	32	21	39	34	33
CircuitTraceCompaction	4	0	4	4	0	4
HaplotypeAssembly	6	5	0	5	5	5
pbo-mqc	256	119	250	217	131	210
pbo-routing	15	15	13	15	15	15
PROTEIN_INS	12	0	2	3	1	2
Total	965	553	731	815	577	755

Table 1 shows the number of solved instances by each solver for all benchmark sets. The improvements from version 1.0 to version 1.2 of wbo are clear. The overall number of solved instances is vastly improved as it now solves more instances than MSUncore and SAT4J MaxSAT. Nevertheless, version 1.2 of wbo is not as effective as the current version of pm2. However, wbo has an additional overhead since it is a more general solver able to tackle any WBO problem instance, whereas pm2 is specific for partial MaxSAT and it cannot handle formulas with weights. Furthermore, pm2 needs to use the encoding of cardinality constraints to CNF that depends on the number of iterations, but since the number of iterations for most instances is not large, the respective CNF encoding should tend to produce manageable CNF formulas. Finally, wbo is built on top of minisat 2.0 [11], while pm2 is built on the more effective PicoSAT solver [9].

The improvements of wbo are due to different reasons for the several benchmark sets. For example, improvements in bcp-fir are due to the use of constraint branching technique, while in bcp-msp several instances are trivially solved by the use of a PBO solver at preprocessing. Preprocessing techniques from PBO are also extensively applied in the CircuitTraceCompaction where the initial formula can be significantly reduced. Overall, the integration of all these techniques into an unsatisfiability-based algorithm improve its performance and robustness for several sets of industrial instances.

Observe that it was chosen not to present results from other industrial categories from the MaxSAT evaluation for two main reasons: (i) version 1.0 of the wbo solver was already able to solve all instances from the partial weighted MaxSAT problem and (ii) the proposed techniques do not apply on the industrial MaxSAT instances without hard constraints for which wbo was already one of the best performing solvers [4]. Note that PBO preprocessing techniques can only be applied when literal implications can be extracted from the formula and that does not occur for those benchmark sets. Furthermore, most of industrial MaxSAT instances are solved after finding a single unsatisfiable sub-formula. Hence, there are not enough iterations to apply constraint branching and overall results from version 1.0 and 1.2 for solver wbo would be the same for these sets of instances.

6   Conclusions

This paper proposes to extend an unsatisfiability-based algorithm for Weighted Boolean Optimization, by first encoding the problem into pseudo-Boolean Optimization such that powerful inference preprocessing techniques can be used. Furthermore, the pseudo-Boolean solver can also be used to learn hard constraints and deal with problem instances that are trivially solved using a linear search on the upper bound value of the solution. Moreover, the paper also shows how to selectively apply constraint branching in the unsatisfiability-based framework.

Preliminary experimental results show that these techniques significantly improve the performance of our unsatisfiability-based algorithm when solving industrial instances of the partial MaxSAT problem (a special case of Weighted Boolean Optimization). As a result, our solver is now competitive with dedicated algorithms for the partial MaxSAT problem.

The success obtained on solving these problem instances with the integration of techniques from Pseudo-Boolean Optimization and constraint branching, first proposed for (Mixed) Integer Linear Programming, provide a strong stimulus for further integration of several Boolean optimization techniques into an unique framework.

Acknowledgement. This work was partially supported by FCT grant PTDC-/EIA/76572/2006 and FCT (INESC-ID multiannual funding) through the PID-DAC Program funds.

References

1. Aloul, F., Ramani, A., Markov, I., Sakallah, K.A.: Generic ILP versus specialized 0-1 ILP: An update. In: International Conference on Computer-Aided Design, pp. 450–457 (2002)
2. Ansótegui, C., Bonet, M., Levy, J.: Solving (Weighted) Partial MaxSAT through Satisfiability Testing. In: Kullmann, O. (ed.) SAT 2009. LNCS, vol. 5584, pp. 427–440. Springer, Heidelberg (2009)
3. Argelich, J., Li, C.M., Manyà, F.: An improved exact solver for partial max-sat. In: Proceedings of the International Conference on Nonconvex Programming: Local and Global Approaches (NCP-2007), pp. 230–231 (2007)
4. Argelich, J., Li, C.M., Manyà, F., Planes, J.: Fourth Max-SAT evaluation (2009), http://www.maxsat.udl.cat/09/
5. Bacchus, F., Winter, J.: Effective preprocessing with hyper-resolution and equality reduction. In: Giunchiglia, E., Tacchella, A. (eds.) SAT 2003. LNCS, vol. 2919, pp. 183–192. Springer, Heidelberg (2004)
6. Barnhart, C., Johnson, E., Nemhauser, G., Savelsbergh, M., Vance, P.: Branch-and-price: Column generation for solving huge integer programs. Operations Research 46(3), 316–329 (1998)
7. Barth, P.: A Davis-Putnam Enumeration Algorithm for Linear Pseudo-Boolean Optimization. Technical Report MPI-I-95-2-003, Max Plank Institute for Computer Science (1995)
8. Berre, D.L.: SAT4J library, http://www.sat4j.org
9. Biere, A.: PicoSAT essentials. Journal on Satisfiability, Boolean Modeling and Computation 2, 75–97 (2008)
10. Chai, D., Kuehlmann, A.: A fast pseudo-Boolean constraint solver. In: Design Automation Conference, pp. 830–835 (2003)
11. Eén, N., Sörensson, N.: Minisat 2.0 sat solver, http://minisat.se/MiniSat.html
12. Eén, N., Sörensson, N.: Translating pseudo-Boolean constraints into SAT. Journal on Satisfiability, Boolean Modeling and Computation 2 (2006)
13. Fu, Z., Malik, S.: On solving the partial MAX-SAT problem. In: Biere, A., Gomes, C.P. (eds.) SAT 2006. LNCS, vol. 4121, pp. 252–265. Springer, Heidelberg (2006)
14. Heras, F., Larrosa, J., Oliveras, A.: MiniMaxSAT: An efficient weighted Max-SAT solver. Journal of Artificial Intelligence Research 31, 1–32 (2008)
15. Larrosa, J., Heras, F., de Givry, S.: A logical approach to efficient Max-SAT solving. Artificial Intelligence 172(2-3), 204–233 (2008)
16. Li, C.M., Manyà, F., Planes, J.: New inference rules for Max-SAT. Journal of Artificial Intelligence Research 30, 321–359 (2007)
17. Lin, H., Su, K.: Exploiting inference rules to compute lower bounds for MAX-SAT solving. In: International Joint Conference on Artificial Intelligence, pp. 2334–2339 (2007)
18. Manquinho, V., Marques-Silva, J.: Search pruning techniques in SAT-based branch-and-bound algorithms for the binate covering problem. IEEE Transactions on Computer-Aided Design 21(5), 505–516 (2002)
19. Manquinho, V., Marques-Silva, J., Planes, J.: Algorithms for weighted boolean optimization. In: Kullmann, O. (ed.) SAT 2009. LNCS, vol. 5584, pp. 495–508. Springer, Heidelberg (2009)
20. Marques-Silva, J., Manquinho, V.: Towards more effective unsatisfiability-based maximum satisfiability algorithms. In: Kleine Büning, H., Zhao, X. (eds.) SAT 2008. LNCS, vol. 4996, pp. 225–230. Springer, Heidelberg (2008)

21. Marques-Silva, J., Planes, J.: Algorithms for maximum satisfiability using unsatisfiable cores. In: Design, Automation and Testing in Europe Conference, March 2008, pp. 408–413 (2008)
22. Martins, R., Lynce, I., Manquinho, V.: Preprocessing in pseudo-boolean optimization: An experimental evaluation. In: Eighth International Workshop on Constraint Modelling and Reformulation (2009)
23. Pipatsrisawat, K., Palyan, A., Chavira, M., Choi, A., Darwiche, A.: Solving weighted Max-SAT problems in a reduced search space: A performance analysis. Journal on Satisfiability Boolean Modeling and Computation 4, 191–217 (2008)
24. Ramírez, M., Geffner, H.: Structural relaxations by variable renaming and their compilation for solving MinCostSAT. In: Bessière, C. (ed.) CP 2007. LNCS, vol. 4741, pp. 605–619. Springer, Heidelberg (2007)
25. Ryan, D., Foster, B.: An integer programming approach to scheduling. In: Computer Scheduling of Public Transport, pp. 269–280 (1981)
26. Sheini, H., Sakallah, K.: Pueblo: A Modern Pseudo-Boolean SAT Solver. In: Design, Automation and Testing in Europe Conference, pp. 684–685 (March 2005)

Encoding Techniques, Craig Interpolants and Bounded Model Checking for Incomplete Designs

Christian Miller, Stefan Kupferschmid, Matthew Lewis, and Bernd Becker

Albert-Ludwigs-Universität, Freiburg, Germany
{millerc,skupfers,lewis,becker}@informatik.uni-freiburg.de

Abstract. This paper focuses on bounded invariant checking for partially specified circuits – designs containing so-called blackboxes – using the well known 01X- and QBF-encoding techniques. For detecting counterexamples, modeling the behavior of a blackbox using 01X-encoding is fast, but rather coarse as it limits what problems can be verified. We introduce the idea of 01X-hardness, mainly the classification of problems for which this encoding technique does not provide any useful information about the existence of a counterexample. Furthermore, we provide a proof for 01X-hardness based on Craig interpolation, and show how the information contained within the Craig interpolant or unsat-core can be used to determine heuristically which blackbox outputs to model in a more precise way. We then compare 01X, QBF and multiple hybrid modeling methods. Finally, our total workflow along with multiple state-of-the-art QBF-solvers are shown to perform well on a range of industrial blackbox circuit problems.

Keywords: BMC, blackbox, SAT, QBF, Craig interpolation, unsat-core.

1 Introduction

Recently, Bounded Model Checking (BMC) has become an important method for finding errors in sequential circuits [1,2]. BMC accomplishes this by iteratively unfolding a circuit k times for $k = 0, 1, \ldots$, adding the negated property, and then finally converting the BMC instance into a SAT formula so that a SAT-solver can be used. If the SAT-solver finds the k-th problem instance satisfiable, a path of length k violating the property has been found. In this paper we focus on BMC for *incomplete* designs, meaning that certain parts of the circuit (combined into a so-called blackbox) are not specified. The interest on verifying incomplete designs is becoming popular as larger system-on-chip (SoC) designs, that contain multiple blackbox IP cores, become more prevalent. Blackboxes can also add a layer of abstraction if a design is too large to verify in its entirety. Additionally, blackboxes allow us to start the verification process earlier in the design stages of a chip when certain components are only partially completed.

In these cases we want to answer the question of *unrealizability*, that is, is there a path of length k violating the property regardless of the implementation of the blackbox. If so, the property is unrealizable. For example, a processor with a blackbox covering the ALU is shown in Figure 1. Since the behavior of blackbox outputs is unknown we need to model them in an adequate way. One option is to use 01X-logic. This approach applies the value X to all blackbox outputs, and then encodes the circuit as done e.g.

O. Strichman and S. Szeider (Eds.): SAT 2010, LNCS 6175, pp. 194–208, 2010.
© Springer-Verlag Berlin Heidelberg 2010

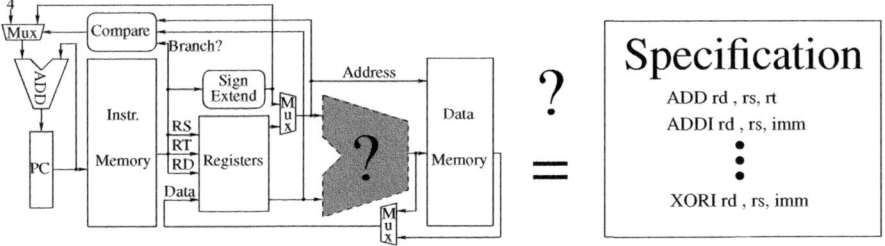

Fig. 1. Example MIPS type processor with blackboxes

in [8] by Jain et al. This again yields a propositional SAT formula, and a modern SAT-solver can be used. Counterexamples found by this approach are independent of the blackbox's behavior. However, using 01X-logic may be to coarse to prove the unrealizability of the property when the counterexamples depend on the blackbox's behavior. Therefore, we must sometimes model the blackbox outputs in a more precise way by universally quantifying them. This results in a quantified boolean formula (QBF).

QBF formulas are hard (PSPACE-Complete), and in this work we introduce improvements with respect to encoding a blackbox BMC problem using a combination of 01X-logic and QBF. Further, we present a method based on Craig interpolation to prove that using 01X-logic is too coarse to provide any information about the existence of a counterexample (problems of this type will be called *01X-hard*). For such problems, we show that it is usually not the case that *all* blackbox outputs of the incomplete design have to be encoded using QBF to obtain a counterexample. We then introduce two techniques for obtaining which blackboxes need to be modeled using QBF. The first is based on exploiting the computed Craig interpolants, and the second method uses the clauses from the unsatisfiable core to illuminate the problematic blackboxes. Our work, which incorporates all this into one tool, allows us to automatically combine the advantages of both 01X- and QBF-encodings so that we can verify more problems.

The paper is structured as follows. In Section 2 we introduce the concepts and related work for BMC of incomplete designs, Z-modeling using 01X-logic, Z_i-modeling using QBF as well as the combination of these modeling techniques. Section 3 then introduces our tool, optimizations, definitions for 01X-hard and 01X-easy, as well as new ideas for heuristically combining Z- and Z_i-modeling. Results and analysis of multiple industrial circuits are given in Section 4, and Section 5 concludes the paper.

2   Bounded Model Checking for Incomplete Designs

Standard BMC has been shown to be able to refute invariants on industrial sequential circuits [1,2]. Starting with the initial state of the circuit, BMC iteratively unfolds the system k times with $k = 0, 1, \ldots$ and checks in every iteration whether a counterexample for the given invariant exists or not. The algorithm stops, if a counterexample is found or a predefined unfolding limit is reached. Let I_0 characterize the initial state, $T_{i,i+1}$ the transition relation, which is *true*, if there is a transition from a state at time

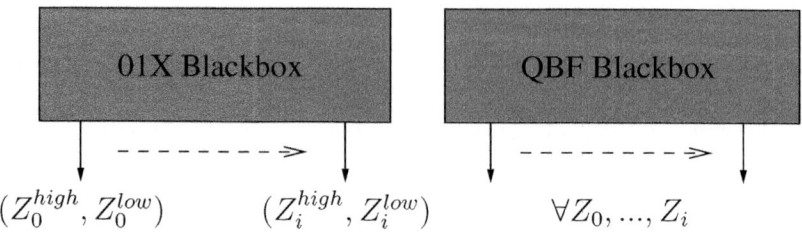

Fig. 2. Example modeling of blackbox outputs (01X vs. Z_i)

step i to a state at time step $i + 1$. Let P_k represent the invariant (property) at depth k. Then, the BMC formula (1) is satisfied iff there exists a counterexample of length k that violates the property.

$$I_0 \wedge T_{0,1} \wedge \ldots \wedge T_{k-1,k} \wedge \neg P_k \tag{1}$$

In this paper we focus on bounded invariant checking of *incomplete* designs. An example of this is shown in Figure 1. In this pipelined processor, the ALU is not yet completed. To deal with this, we replace the ALU with a *blackbox*. The simplest way to model a blackbox is to assign *one* extra value denoting the unknown behavior to each blackbox output. With accordance to [18] we call this Z-modeling. For encoding the BMC problem this way we can make use of 01X-logic, an extension to propositional logic by a third logical value X, denoting an unknown state. This means that all signal lines in a circuit are now encoded using two bits instead of one. This allows each bus in the circuit to take on one of the three possible logic values (0,1, or X). For the blackbox outputs this is shown in the left part of Figure 2.

Encoding all blackbox outputs using 01X-logic is only sufficient for finding counterexamples that are independent of the blackbox's behavior. However, these problem can be encoded as a propositional SAT formula at the expense of additional variables and clauses. To encode all gates and buses using 01X-logic and Jain encoding [8], we use the following transformation functions:

$$AND_{01X}((g^{high}, g^{low}), (f^{high}, f^{low})) := (g^{high} \cdot f^{high}, g^{low} + f^{low})$$
$$OR_{01X}((g^{high}, g^{low}), (f^{high}, f^{low})) := (g^{high} + f^{high}, g^{low} \cdot f^{low})$$
$$NOT_{01X}((g^{high}, g^{low})) := (g^{low}, g^{high})$$

These transformation functions define the functional relationship of each gate using 01X-logic and our two bit encoding. This allows us to define the functionality of the AND_{01X} gate when one or both of its inputs are in the X (the unknown) state. For example, "0 AND_{01X} X" is always 0. However, "1 AND_{01X} X" results in X because the output of the AND_{01X} gate depends on the X input. Using this encoding, we can now convert the circuit and property into CNF form so that a high performance SAT-solver can be used.

Once a problem is encoded, the SAT-solver tries to find a counterexample that violates the property P_k, or prove that no such counterexample exists. Some counterexamples for an invariant can depend on the behavior of the blackbox. Unfortunately, these types of counterexample cannot be found using Z-modeling. The reason for this is that the X values assigned to the blackbox outputs propagate to the relevant signals checked by the invariant. Finding out if blackbox outputs cause this issue motivates our definition of 01X-hardness in Section 3.2.

To compute counterexamples of incomplete designs which depend on the blackbox's behavior, a more precise modeling method is needed. Similar to [18] we call this technique Z_i-modeling, and it is shown on the right side of Figure 2. This technique introduces one universally quantified variable for each blackbox output resulting in a QBF formula. QBF extends propositional formulas by allowing variables to be either universally ($\forall$) or existentially ($\exists$) quantified. Using the QBF formulation, there is no longer a need for three valued logic and the X state, as we can check all possible logic values of the blackbox outputs (0 and 1) using universal quantifiers.

Since we are interested in one input sequence, such that for every blackbox behavior the invariant is violated, we build the quantifier prefix stepwise as follows: let $x_0, \ldots, x_n$ be the primary inputs and $Z_0, \ldots, Z_m$ the blackbox outputs of the design. A second index denotes the unfolding depth of the variable (e.g. $Z_{3,2}$ is blackbox output Z_3 at unfolding depth 2). Further let H_i be the set of additional Tseitin-variables needed to encode the circuit at unfolding depth i. We end up in the following quantifier prefix (referred to as $pref_1$ and presented in [6]):

$$\underbrace{\exists x_{0,0}, \ldots, x_{n,0} \forall Z_{0,0}, \ldots, Z_{m,0} \exists H_0}_{\text{depth } 0} \cdots \underbrace{\exists x_{0,k}, \ldots, x_{n,k} \forall Z_{0,k}, \ldots, Z_{m,k} \exists H_k}_{\text{depth } k} \text{ CNF}$$

$$(pref_1)$$

By combining both these methods we now can encode a combination of Z- and Z_i-modeled blackbox outputs. This allows use to reduce the number of universal variables in the problem with the aim of making the resulting QBF problems easier to solve. Concerning the combined Z/Z_i-modeling in [6], and unlike Figure 2, they used two variables for each Z_i-modeled blackbox output, one universally and one existentially quantified. This resulted in two quantifier alternations per Z_i-modeled blackbox output. This allowed them to keep transformation functions constant for both Z- and Z_i-modeled blackboxes. However, the resulting QBF formula was very complex, and could contain thousands of quantifier alternations. In Section 3 we introduce new transformation functions that remove these unneeded quantifier alternation and variables.

Finally, other work similar to [6] examined the BMC problem of incomplete designs in the context of BDD based model checking [12,13]. However, most previous work used randomly placed blackboxes, and random selection of which type of model (Z or Z_i) each blackbox output should use. In this work, we place blackboxes for specific circuit components (e.g. adders, multiplier units, control units, ...), and not just random gates. Moreover, we introduce heuristics to automatically find which blackbox outputs must be modelled using Z_i and QBF, and which can remain using 01X-logic.

3 Workflow

Our blackbox BMC tool is called Bounce. Bounce supports multiple 01X, QBF, and hybrid encoding modes, as well as multiple QBF-solvers. The basic workflow of our tool is shown in Figure 3. This workflow consist of three major stages: a BMC problem encoder and a SAT-solver with Craig interpolation and unsat-core support; a heuristic search based component that in the case of 01X-hardness, finds the reasons for this; and thirdly, a hybrid Z/Z_i BMC problem encoder and QBF-solver.

In the first stage of the workflow, a behavioral level VHDL or Verilog circuit description is taken, and blackboxes are inserted for the components that are not fully specified. Then a small VHDL wrapper is added to the circuit so the reset (and/or other) functionality can be controlled. This allows us to initialize a circuit into a predefined state which is sometimes required. For instance, some circuits are only guaranteed to operate correctly if initialized properly. Once this completed, the circuit is then compiled with Synopsys Design Compiler (Version B-2008.09) and linked to a gate library containing only basic one and two input logic gates and storage elements.

The resulting gate level HDL code is then used to generate the BMC equation from Section 2. This is then sent to our multi-threaded SAT-solver MiraXT [9] which includes additional support for Craig interpolation and unsat-core production. If the solver for a specific unrolling returns *true*, the resulting variable assignment is our counterexample and the problem is classified as 01X-easy. Otherwise, if the solver continually returns *false* for each unrolling, then at some point a fixed-point should be reached. When a fixed-point is found, we say the problem is 01X-hard, and will show in Section 3.2 that this means no satisfiable solution using 01X-encoding will be possible.

For 01X-hard instances, we then can use the Craig interpolants (or alternatively the unsat-core of the problem at the fixed-point depth), to find the reasons why the instance is unsatisfiable. This is done by tracing the Craig interpolants backwards to the blackbox outputs, or scanning the unsat-core for blackbox related variables. The strategies Bounce uses for this are discussed in Sections 3.3 and 3.4. Note, however, that these

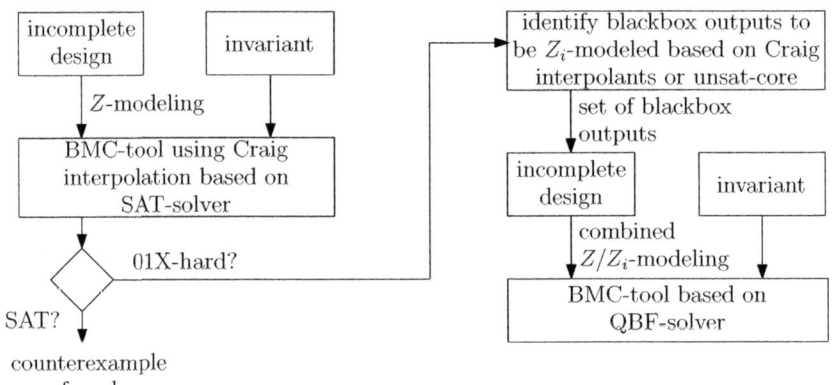

Fig. 3. Workflow

strategies are not guaranteed to find all necessary blackbox outputs (see Section 3.5) but on practical problems they seem to perform well enough.

Finally, once a list of blackbox outputs that need to be modeled precisely is identified, we can then use our optimized techniques (discussed in Section 3.1) to produce hybrid problems containing both Z- and Z_i-modeled blackboxes, hopefully resulting in an easier to solve QBF formula (due to less universally quantified variables). If the QBF-solver returns *true*, we know that the invariant will always be unsatisfied within the current design, irrelevant of the implementation of the blackbox (i.e. the circuit cannot be fixed by adding functionality to the blackbox). Otherwise, when a QBF-solver continually returns *false* for every unrolling of the problem, we cannot prove anything about the current invariant, unless we are able to prove the maximum depth of the circuit, which in practice can be infeasible.

3.1 Optimizing Blackbox Bounded Model Checking

The 01X transformation scheme maps the three logical values 0, 1 and X to the binary tuples (0,1), (1,0) and (0,0), respectively. However, in [5,6,7] all signals were encoded using 01X-logic. This resulted in a doubling of the number of variables in SAT formula that transformation functions would produce. We have extended this mapping so that signals not associated with the blackbox require only the regular one bit encoding. To combine 01X- with 01-logic, we introduce the new transformation functions:

$$AND_{01X}((g^{high}, g^{low}), f) := (g^{high} \cdot f, g^{low} + \neg f)$$
$$OR_{01X}((g^{high}, g^{low}), f) := (g^{high} + f, g^{low} \cdot \neg f)$$

Note, that in contrast to [6] we allow arbitrary sequential behavior of every blackbox (i.e. a blackbox output can produce different output values for the same input values at different time steps). Hence we do not care about the blackbox inputs. Furthermore, with respect to Z/Z_i-modeling in [6], and covered in Section 2, each Z_i-modeled blackbox output introduced two variables and quantifier alternations. However, using our new transformation functions this is no longer required, and we can encode all Z_i-modeled blackbox outputs with only one bit as shown on the right side of Figure 2.

When encoding the problems as QBF, we have an additional optimization that can reduce quantifier alternations even further. In $pref_1$ equation from Section 2, the inputs in each step can 'react' to the values of the blackbox outputs from the previous steps. However, in total this quantifier prefix yields at most $2 \cdot (k + 1)$ quantifier alternations when unfolding k times. This number can be lowered to 2 when restricted to uniform counterexamples [14], having one block of the existential quantified inputs of all unfolding depths followed by all universally quantified blackbox outputs and then all existentially quantified Tseitin-variables.

$$\underbrace{\exists x_{0,0}, \ldots, x_{n,k}}_{\text{primary inputs depth } 0 \ldots k} \quad \underbrace{\forall Z_{0,0}, \ldots, Z_{m,k}}_{\text{blackbox outputs depth } 0 \ldots k} \quad \underbrace{\exists H_0, \ldots, H_k}_{\text{Tseitin depth } 0 \ldots k} \quad \text{CNF} \qquad (pref_2)$$

This formulation implies the first one, since every uniform counterexample is also a counterexample resulting from the first formulation, but we avoid the number of quantifier alternations increasing with the unfolding depth. However, this formulation is not as exact, and may not be able to verify as many problems or invariants.

3.2   Craig Interpolation and Proving 01X-Hardness

In this section we will introduce Craig interpolants [3] and how interpolation can lift a classical BMC procedure to a complete model checking technique to prove invariants [11]. Later we describe how we benefit from this procedure.

Theorem 1 (Craig). *Given two propositional formulas A and B with the property that $A \wedge B$ is unsatisfiable, then there exists a Craig interpolant C for A and B. This Craig interpolant has the following properties:*

- *C contains only variables which occur in A and B (AB-common variables).*
- *$\models A \Rightarrow C$*
- *$\models C \Rightarrow \neg B$*

Craig interpolants in BMC are used as an over-approximated forward image of reachable states in a transition system. If the computed over-approximated forward image reaches a fixed-point, that is no new states are reachable, and the given invariant still holds, no counterexample is possible for any unrolling depth. Let I_k be the initial state, P_k the invariant to disprove and $T_{i,i+1}$ the transition relation from a state at time step i to a state at time step $i+1$[1]. After showing that $I_0 \wedge \neg P_0$ is unsatisfiable (that is initially the property is not violated), the procedure first solves the BMC formula $\Phi = A \wedge B$, where $A := R_0 \wedge T_{0,1}$, $B := \neg P_1$ and initially $R_0 := I_0$. If Φ is unsatisfiable then a Craig interpolant C_1 for the formulas A and B is computed[2]. By $A \Rightarrow C_1$, the interpolant C_1 is an over-approximation of the states reachable in one step from R_0. If this over-approximation shifted to the zeroth instantiation of the variables (as described by C_0) is a subset of the so far reachable states, that is $C_0 \Rightarrow R_0$, then further transitions can only lead to states already characterized by R_0. As a consequence, the target states are unreachable and the verification procedure terminates. Otherwise, we expand the set of reachable states such that it also covers the reachable states given by the shifted interpolant, that is $R_0 := R_0 \vee C_0$. Then, the procedure is iterated until the above termination criterion holds. For a more detailed account, confer [11].

As we are focusing on BMC problems for hardware designs, the only variables occurring in both formulas A and B are the latch variables. Now we describe how this procedure can be used to classify incomplete designs. This not only prevents the 01X BMC tool from endlessly returning *false*, but also provides us with a proof that we must use a more precise, but harder to solve QBF formulation.

Proving 01X-Hardness. For a pure 01X-encoded BMC problem, unsatisfiability for every unfolding depth has two possible reasons. First, as demonstrated earlier, Z-modeling all blackbox outputs may be too coarse. Second, there may exist no counterexample. However, in both of these cases 01X-encoding is not suitable to get to a result. This motivates the following definition of *01X-hardness*.

---

[1] In the following a lower index denotes the timed instantiation of a boolean formula.
[2] Note, that C_1 only contain *AB*-common variables.

Definition 1 (01X-hardness). *An incomplete design combined with an invariant is 01X-hard iff the pure 01X-encoded BMC problem is unsatisfiable for all unfoldings.*

Example. Figure 4 shows a 01X-hard incomplete design with two storage elements q_0 and q_1, two primary inputs x_0 and x_1 and one blackbox with two outputs Z_0 and Z_1. For this design we want to disprove the invariant $AG(\neg q_0 \vee \neg q_1)$ stating that this circuit never reaches a state where both storage elements are *true* at the same time.

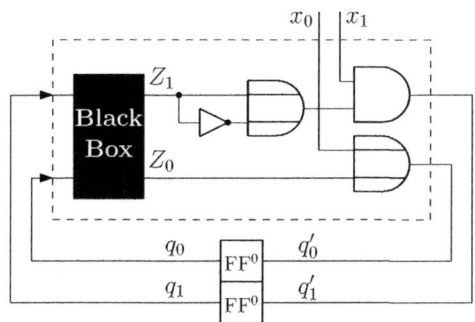

Fig. 4. 01X-hard incomplete design

Using Z-modeling for blackbox output Z_1, one can see that the assigned X value propagates to the storage element q_1 independent of the value x_1. Obviously this fact holds for every unroll depth.

In order to apply BMC with Craig interpolation to this example, the entire circuit is Jain encoded. Figure 5 illustrates the first unfolding $I_0 \wedge T_{0,1} \wedge \neg P_1$. The gates responsible for the unsatisfiability (unsat-core) are highlighted in gray. Applying the construction rules for Craig interpolants to this unsat-core yields the formula $C_1^0 = \neg(q'^h_1)$. This formula represents an over-approximation of the reachable states after one transition step. Using this set of states as new initial states, the Craig interpolant C_1^1 derived from $C_0^0 \wedge T_{0,1} \wedge \neg P_1$ is equivalent to the one computed before and thus a fixed-point is reached and the 01X-hardness is proven.

Here, the Craig interpolant $C_1^0 = \neg(q'^h_1)$ forces the storage element $q'_1 = (q'^h_1, q'^l_1)$ to be either $0_{01X} = (0,1)$ or $X_{01X} = (0,0)$. In 01X-logic, X represents both 0 and 1 simultaneously, meaning it is not possible to determine under which circumstances the values 0 and 1 appear. This can only be determined using Z_i-modeling. In our example, when Z_1 is Z_i-modeled, the output of the OR-gate is a constant 1 for all values of Z_1. The resulting QBF formula is then satisfiable after one transition step. Furthermore, this shows it is sufficient to model only the first blackbox output using QBF.

Now, we are in the advantageous situation to gather information about the reason for the unsatisfiability of every unrolling depth. This information is located in the Craig interpolant and also in the underlying unsat-core. How this information can be exploited to refine our encoding, will be discussed in the next sections.

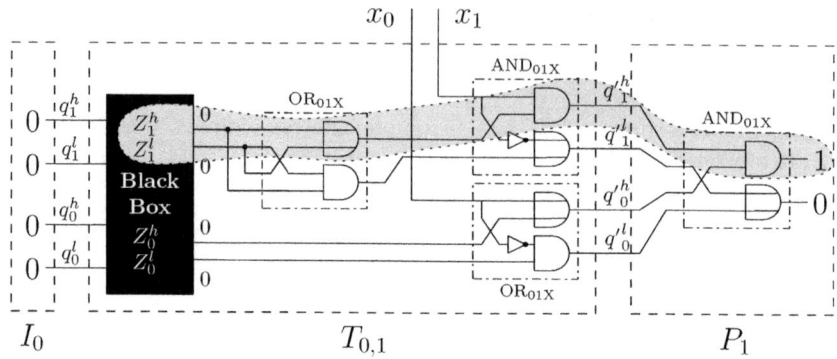

Fig. 5. Jain encoded incomplete design unfolded one time

3.3 Exploiting Craig Interpolants

For a given incomplete design and an invariant we apply the unbounded model checking procedure described in the previous section with all blackbox outputs Z-modeled and thus 01X-encoded. Once a fixed-point is reached the last Craig interpolant is analyzed.

This interpolant only contains latch variables[3], which are collected in a set L. All blackbox outputs affecting these latches are one reason for the 01X-hardness. Starting from each latch in L, a recursive backward traversal on the incomplete design is performed. If a blackbox output is reached, this blackbox output is marked to be Z_i-modeled. When we reach a latch which is not yet an element of L, we add it to L and continue the backward traversal until a fixed-point in L is reached. With this algorithm we identify at least all blackbox outputs having influence on the latches in L.

Coming back to our example in Figure 5, traversing backward from $L = \{q'^h_1\}$ only blackbox output Z_1 is marked to be Z_i-modeled, which is sufficient to prove the unrealizability applying the combined modeling technique presented in Section 3.1.

3.4 Utilizing Unsatisfiable Cores

Instead of exploiting the Craig interpolant as described above we also can make use of the unsat-core that the SAT-solver produces. After proving 01X-hardness we determine the unsat-core of the BMC problem of the unrolling depth where the fixed-point was found (it is clear that this problem must be unsat as the 01X-hardness was already proven). To determine the unsat-core, the solver's conflict analysis routine is modified so that each new conflict clause contains a link to the original problems clauses it was derived from. Simply put, this means the unsat-core contains all the clauses needed by the underlying SAT-solver to derive the empty clause. Among these clauses we detect every literal that represents a Jain-encoded blackbox output. All these outputs will now be marked to be Z_i-modeled.

In our example the clauses derived from the gray highlighted gates build the unsat-core and thus Z_1 is immediately detected to be Z_i-modeled. With this approach we

---

[3] Due to the 01X-encoding these latches are still Jain encoded.

only identify blackbox outputs which directly have an influence on the unsatisfiability, whereas the backward traversal of the method based on Craig interpolants also can identify blackbox outputs which may not effect any relevant latch.

3.5   Challenges and Limitations

For a 01X-hard system there is not necessarily only one single proof or unsat-core. This fact can result in situations, where not all necessary Z_i-modeled blackbox outputs are identified in order to find a counterexample. If we know the structure of the circuit in detail this problem can be avoided by decomposing the circuit and applying our method for each component. Since in general this is not a trivial task, a further approach is to collect the latch variables of *all* computed interpolants during the proof, not only those involved in the successful fixed-point-check. This is motivated by the heuristic search of the underlying SAT-solver computing the Craig interpolants. However, it will be shown in Section 4 that our unsat-cores seem to be sufficient on our industrial benchmarks to prove unrealizability.

Further, it is not guaranteed that a fixed-point will be found. If so, we can not prove the 01X-hardness of the system, but we can continue our procedure by aborting at a given timeout or depth, and analyzing the Craig interpolants or unsat-cores computed so far. After several unrollings, this technique would hopefully allow us to still identify the required blackbox outputs. However, as will be shown in Section 4, both of these limitations did not present problems in any of the circuits that were tested.

4   Experimental Results

To evaluate our methods we used industrial opensource FPGA verified designs from opencores.org [15] and some instances of the VLIW ALU benchmark suite presented in [14]. This allows us to test our ideas on implementable circuits designs that are very similar to the ones used in industry. In our current design flow, the blackboxes were inserted into the behavioral VHDL source code of each design allowing us to select and replace specific entities (i.e. entire multipliers, shifters, dividers) from each circuit. The same method was used to insert errors into some circuits. A description of each benchmarks circuit is given below. To compile the behavioral VHDL source code of each design, Synopsys Design Compiler Version B-2008.09 was used along with a minimized gate library containing only one and two input basic logic gates and latches. All benchmarks were run on a dual AMD Opteron(tm) 250 processor machine (2.4GHz) with 4GB of RAM, running the 2.6.24 SMP enabled Linux kernel. Lastly, for all benchmarks a timeout (TO) value of 3600 seconds was used.

- Plasma - A simple opensource pipelined 32-bit RISC microprocessor supporting all MIPS Version I user mode instructions. The multiplier and shifter inside the ALU were replaced with blackboxes, and the boot loader was simplified. Proper functionality and reliability of the boot loader are being verified. The circuit contains 16,603 gates and 2,463 latches.

- FPU - An IEEE-754 compliant pipelined double precision floating point unit that supports four operations (+, -, *, /) and multiple rounding modes and exceptions. VHDL and gate level faults were inserted into the FPU. The multiplication and division units were replaced with blackboxes and some of the functionality of the remaining units was tested. The remaining circuit contains 21,280 gates and 2,701 latches.
- UART - A configurable, pipelined serial transmitter/receiver UART pair. Here, by defining the UART controller as a blackbox, we verify that all possible 8 bit inputs to the UART will be transmitted and received properly at different bit rates. The circuit contains 555 gates and 70 latches.
- QALU - A SoC design containing a configurable VLIW ALU and Timer. The VLIW ALU consists of four separate functional units (1xLogic, 2xArithmetic, 1xLoad/Store) and multiple working registers. The 12 bit Timer has set, reset, exception, and overflow functionality. Blackboxes replaced different arithmetic units in the two ALU sizes and some of the parts of the Timer. The functionality of the remaining Timer and ALU logic unit which contains gate level design errors are being verified. QALU_32 contains 31,866 gates, 538 latches, and QALU_64 contains 35,496 gates and 1,054 latches.

The results for the first stage of our tool are presented in Table 1. Here, the first column is the benchmark name, followed by the found counterexample or fixed-point depth and time (CE/FP Depth and Time). If the benchmark is 01X-hard, the number of detected blackbox outputs for both the Craig interpolant and unsat-core methods are given (# Detected Craig and Unsat). The first fact that is apparent from this table is that almost half of the properties we check are 01X-easy, meaning they can be solved using only a SAT-solver. Secondly, if problems are 01X-hard, only a small fraction of blackbox outputs are selected to be modeled using QBF. Overall, unsat-core method seems to outperform Craig interpolation. Finally, for all these benchmarks the first phase of our tools does not require a significant amount of time even though this circuits are quite

Table 1. Initial 01X results

Bench.	CE/FP Depth	Time	01X Hard	# Detected Craig	Unsat	Bench.	CE/FP Depth	Time	01X Hard	# Detected Craig	Unsat
FPU-ec01	27	67.39	no	—	—	UART-ec02	245	74.44	no	—	—
FPU-ec02	28	70.36	no	—	—	UART-ec03	475	354.3	no	—	—
FPU-ee01	27	75.16	no	—	—	UART-hc01	8	0.29	yes	01/16	01/16
FPU-hc01	33	54.78	yes	3/141	6/141	UART-hc02	47	3.93	yes	08/16	08/16
FPU-hc02	33	58.87	yes	3/141	6/141	UART-hc03	4	0.12	yes	01/16	01/16
FPU-he01	32	57.14	yes	139/141	6/141	UART-hc04	133	39.88	yes	08/16	08/16
FPU-he02	49	433.52	yes	141/141	34/141	UART-hc05	15	0.76	yes	01/16	01/16
Plasma-ec01	15	8.86	no	—	—	UART-hc06	121	23.07	yes	08/16	08/16
Plasma-ee01	10	6.32	no	—	—	UART-hc07	4	0.14	yes	01/16	01/16
Plasma hc01	24	22.82	yes	64/64	12/64	qualu32-e	4	1.37	no	—	—
Plasma-hc03	46	55.09	yes	64/64	12/64	qualu64-e	4	1.45	no	—	—
Plasma-he01	28	26.11	yes	64/64	10/64	qualu32b-h	4	2.46	yes	2/66	2/66
UART-ec01	126	19.93	no	—	—	qualu64b-h	2	1.93	yes	2/386	2/386

Table 2. $pref_1$ encoding results

Bench.	CE Depth	AIGsolve			QMiraXT			QuBE		
		Craig	Unsat	All Z_i	Craig	Unsat	All Z_i	Craig	Unsat	All Z_i
FPU-hc01	27	33.41	74.44	52.88	14.10	8.00	7.93	44.5	44.88	42.13
FPU-hc02	28	54.99	35.77	64.79	13.79	11.12	10.69	54.61	53.98	49.41
FPU-he01	27	64.96	62.15	57.96	16.13	9.01	8.76	TO	143.83	TO
FPU-he02	27	TO	TO	TO	TO	74.03	TO	TO	TO	TO
Plasma-hc01	10	0.03	0.05	0.05	0.06	0.03	0.03	0.04	0.03	0.04
Plasma-hc03	15	0.09	0.05	0.07	0.08	0.07	0.09	0.05	0.04	0.03
Plasma-he01	10	0.06	0.04	0.04	0.09	0.03	0.05	0.05	0.04	0.05
UART-hc01	126	0.70	0.70	1.13	0.82	0.91	0.90	1.27	1.31	1.53
UART-hc02	126	2.12	2.21	2.12	0.78	0.86	0.88	1.80	1.78	1.76
UART-hc03	245	2.49	2.52	4.85	2.06	2.3	2.33	4.85	4.80	5.49
UART-hc04	245	13.42	13.38	13.38	2.1	2.29	2.42	5.78	6.05	6.30
UART-hc05	475	8.61	8.69	25.75	5.86	5.9	5.87	19.02	19.04	22.78
UART-hc06	475	177.98	178.07	177.74	6.18	6.12	6.17	23.58	23.57	23.93
UART-hc07	1,860	156.97	157.56	1,103.54	66.86	68.14	66.37	400.66	419.00	582.14
qualu32b-h	9	105.28	105.58	TO	TO	TO	TO	TO	TO	258.60
qualu64b-h	9	576.94	575.45	TO	TO	TO	TO	TO	TO	718.29
Total Solved		15	15	13	13	14	13	12	13	14
Total Time		1,198.05	1,216.66	1,504.3	128.91	188.81	112.49	556.21	718.35	1,712.48

complex. For example, a circuit with 30,000 gates could require well over 100,000 clauses for the transition relation $T_{i,i+1}$ described in Section 2 alone. In fact, some of the benchmarks contain over a million variables. Also, as an important side note, the inclusion of blackboxes in the FPU allows an extra level of abstraction that makes this verification possible. If the FPU contained the pipelined multiplier and divider, we are no longer able to verify the functionality of adder or subtractor as the entire circuit is too complex. This highlights another application of blackbox modeling and our tool.

For the remaining 01X-hard problems we ran the second phase of our tool using the two different QBF prefix introduced as $pref_1$ and $pref_2$ in Section 2 and 3. The results for each type of prefix are shown in Tables 2 and 3. In both tables, we test three different QBF-solvers, mainly: AIGsolve [16]; QMiraXT [10]; and QuBE [4]. QuBE was chosen as it was the only solver from the 2007 and 2008 QBF Competitions [17] that offered good performance on the blackbox benchmark set. AIGsolve and QMiraXT are newer solvers that have also been shown to perform well on blackbox benchmarks. For each of these solvers, we compared the performance using the Craig interpolation and unsat-core blackbox detection techniques, and the case where all blackboxes were modeled using Z_i. Finally the counterexample depth for each circuit is given (CE Depth).

In Table 2, when we are using the more complex, but also more accurate QBF prefix, AIGsolve seems to perform the best, followed by QMiraXT and QuBE. Note, that on the problems that QMiraXT was able to solve, it is generally the fastest. Also, with the exception of QuBE, the unsat-core technique performs best, and the All Z_i case performs the worst with respect to time and problems solved. The reason for the reverse performance trends with QuBE remain unclear, especially considering it as a DPLL based QBF-solver like QMiraXT. More interesting, is that this table shows that large

Table 3. $pref_2$ encoding results

Bench.	CE Depth	AIGsolve Craig	AIGsolve Unsat	AIGsolve All Z_i	QMiraXT Craig	QMiraXT Unsat	QMiraXT All Z_i	QuBE Craig	QuBE Unsat	QuBE All Z_i
FPU-hc01	27	75.91	34.04	54.48	8.26	8.61	7.88	47.27	47.19	42.41
FPU-hc02	28	58.81	35.74	65.06	11.49	11.15	27.07	55.68	58.41	50.87
FPU-he01	27	62.61	55.66	59.92	9.19	9.15	17.01	TO	119.65	TO
FPU-he02	27	TO	TO	TO	113.46	TO	116.30	TO	TO	TO
Plasma-hc01	10	0.05	0.03	0.04	0.06	0.04	0.07	0.04	0.03	0.06
Plasma-hc03	15	0.04	0.04	0.06	0.07	0.07	0.06	0.06	0.03	0.06
Plasma-he01	10	0.05	0.02	0.05	0.05	0.06	0.08	0.05	0.02	0.04
UART-hc01	126	0.68	0.67	1.15	0.95	0.81	1.00	1.30	1.37	1.52
UART-hc02	126	2.19	2.07	2.17	0.88	0.9	0.77	1.64	1.79	1.62
UART-hc03	245	2.57	2.43	4.81	2.7	2.48	2.53	5.04	4.89	5.51
UART-hc04	245	13.38	13.23	13.34	2.58	2.66	2.81	5.59	5.75	5.60
UART-hc05	475	8.76	8.84	25.60	7.09	6.92	7.45	18.54	18.64	22.13
UART-hc06	475	177.83	177.72	177.9	7.14	7.25	7.04	22.99	22.90	22.87
UART-hc07	1,860	161.98	163.47	1,078.84	TO	TO	TO	411.46	412.71	565.46
qualu32b-h	9	108.77	108.85	97.55	18.77	18.69	4.41	TO	TO	28.20
qualu64b-h	9	604.72	606.06	242.69	134.6	134.55	9.27	TO	TO	85.20
Total Solved		15	15	15	15	14	15	12	13	14
Total Time		1,278.35	1,208.87	1,823.66	317.29	203.34	203.75	569.66	693.38	831.55

benchmarks, as well as really deep (i.e. high depth) benchmarks can be solved with current QBF-solvers in a reasonable amount of time.

Table 3 then shows the results for our more compact encoding style. For all these benchmarks the number of quantifier alternations is restricted to 2, unlike the results in Table 2 where some benchmarks contain thousands of alternations. The new simpler prefix seems to be more effective overall. This is shown by the fact that in Table 3 there are only 16 unsolved instances, where as in Table 2 there are 22. This is especially true for QMiraXT and AIGsolve as in almost all cases $pref_2$ performs better. QuBE benefits from this as well, but only in the *All* Z_i case where the total solve time is cut in half. Finally, even though $pref_2$ is less accurate than $pref_1$, it did not effect the results as in all cases $pref_2$ was still exact enough to solve all the benchmarks.

Lastly, when considering the larger picture, the use of blackbox verification techniques is now feasible. In Table 1 we showed that using 01X-logic alone is good enough to verify many properties. Furthermore, this table and the results in Tables 2 and 3 show that our blackbox detection procedures perform well, with our optimized encoding in Table 3 performing better. Finally, we showed that modern QBF-solvers perform well on a range of benchmarks, showing that they are finally ready for industrial uses.

5 Conclusion

In this work, we presented our tool Bounce that can automatically and efficiently verify a wide array of blackbox BMC problems. We introduced a novel and efficient encoding for the BMC problem of incomplete designs when combining Z- and Z_i-modeling techniques. We also provided a procedure for proving the 01X-hardness of an incomplete design with the help of Craig interpolation. Furthermore, we showed how the Craig interpolants and unsat-cores of a problem can be used for heuristically determining which blackbox outputs should be Z_i-modeled in order to find a counterexample. Moreover, we showed that all these techniques and our tool perform well on a large set of industrial problems. Lastly, we compared our tools performance when connected to three different state-of-the-art QBF-solvers.

Our results show that blackbox verification is becoming feasible, and that it can be used for the testing of designs with incomplete information. These include early stage prototypes where every component is not yet developed, or SoC designs where the exact functionality of a component is unknown. Furthermore, the ability of blackbox methods to verify designs that are to complex if the entire design is considered also poses multiple interesting applications. Additionally, we are currently looking at ways to make the blackboxes gray by introducing the ability to give each blackbox certain properties, opening up many more opportunities for our tool in the future.

References

1. Biere, A., Cimatti, A., Clarke, E., Zhu, Y.: Symbolic Model Checking without BDDs. In: Cleaveland, W.R. (ed.) TACAS 1999. LNCS, vol. 1579, pp. 193–207. Springer, Heidelberg (1999)
2. Clarke, E., Biere, A., Raimi, R., Zhu, Y.: Bounded Model Checking Using Satisfiability Solving. Formal Methods in System Design 19, 7–34 (2001)
3. Craig, W.: Linear Reasoning: A New Form of the Herbrand-Gentzen Theorem. Journal of Symbolic Logic 22(3), 250–268 (1957)
4. Giunchiglia, E., Narizzano, M., Tacchella, A.: Clause/Term Resolution and Learning in the Evaluation of Quantified Boolean Formulas. Journal of Artificial Intelligence Research (JAIR) 26, 371–416 (2006)
5. Herbstritt, M., Becker, B.: On SAT-based Bounded Invariant Checking of Blackbox Designs. In: Microprocessor Test and Verification Workshop (MTV), pp. 23–28 (2005)
6. Herbstritt, M., Becker, B.: On Combining 01X-Logic and QBF. In: Moreno Díaz, R., Pichler, F., Quesada Arencibia, A. (eds.) EUROCAST 2007. LNCS, vol. 4739, pp. 531–538. Springer, Heidelberg (2007)
7. Herbstritt, M., Becker, B., Scholl, C.: Advanced SAT-Techniques for Bounded Model Checking of Blackbox Designs. In: Microprocessor Test and Verification (MTV), pp. 37–44 (2006)
8. Jain, A., Boppana, V., Mukherjee, R., Jain, J., Fujita, M., Hsiao, M.: Testing, Verification, and Diagnosis in the Presence of Unknowns. In: IEEE VLSI Test Symposium (VTS), pp. 263–269 (2000)
9. Lewis, M., Schubert, T., Becker, B.: Multithreaded SAT Solving. In: 12th Asia and South Pacific Design Automation Conference, pp. 926–931 (2007)
10. Lewis, M., Schubert, T., Becker, B.: QMiraXT – A Multithreaded QBF Solver. In: Methoden und Beschreibungssprachen zur Modellierung und Verifikation von Schaltungen und Systemen (2009)
11. McMillan, K.L.: Interpolation and SAT-Based Model Checking. In: Hunt Jr., W.A., Somenzi, F. (eds.) CAV 2003. LNCS, vol. 2725, pp. 1–13. Springer, Heidelberg (2003)

12. Nopper, T., Scholl, C.: Approximate Symbolic Model Checking for Incomplete Designs. In: Formal Methods in Computer-Aided Design, pp. 290–305 (2004)
13. Nopper, T., Scholl, C.: Flexible Modeling of Unknowns in Model Checking for Incomplete Designs. In: 8. GI/ITG/GMM Workshop Methoden und Beschreibungssprachen zur Modellierung und Verifikation von Schaltungen und Systemen (2005)
14. Nopper, T., Scholl, C., Becker, B.: Computation of Minimal Counterexamples by Using Black Box Techniques and Symbolic Methods. In: IEEE Int'l Conf. on Computer-Aided Design, pp. 273–280 (2007)
15. OpenCores, http://www.opencores.org
16. Pigorsch, F., Scholl, C.: Exploiting Structure in an AIG Based QBF Solver. In: Conf. on Design, Automation and Test in Europe (DATE), April 2009, pp. 1596–1601 (2009)
17. QBF Solver Evaluation, http://www.qbflib.org/index_eval.php
18. Scholl, C., Becker, B.: Checking Equivalence for Partial Implementations. In: Design Automation Conf., pp. 238–243 (2000)

Statistical Methodology for Comparison of SAT Solvers[*]

Mladen Nikolić

Faculty of Mathematics, University of Belgrade,
Belgrade, Studentski Trg 16, Serbia
nikolic@matf.bg.ac.rs

Abstract. Evaluating improvements to modern SAT solvers and comparison of two arbitrary solvers is a challenging and important task. Relative performance of two solvers is usually assessed by running them on a set of SAT instances and comparing the number of solved instances and their running time in a straightforward manner. In this paper we point to shortcomings of this approach and advocate more reliable, statistically founded methodologies that could discriminate better between good and bad ideas. We present one such methodology and illustrate its application.

1 Introduction

Many SAT solvers have been developed and various improvements to them have been proposed over the years, especially in the domain of heuristic components. Solver comparisons as a method for detecting good ideas are widely recognized in the SAT community. This is the main purpose of competitions of SAT solvers.[1] Their importance is growing, especially because of the significant number of new ideas and solvers that appear each year. Nevertheless, main responsibility for evaluation of new ideas is on the researchers themselves.

In order to assess the quality of a proposed modification, one usually runs a modified and the base version of the solver on some set of SAT instances. The solver that solves more instances, or the same number of instances in less time is considered to be better. This approach can be flawed because solving times of instances can significantly vary depending only on trivial properties of the formula like ordering of clauses and literals, or on random seeds used, which can lead to different experimental results by chance.

We performed experiments to investigate this claim. Four solvers were chosen from the MiniSAT hack track of the SAT 2009 competition — the first, the last, the baseline and one of the medium performance according to the results of the track.[2] We used two benchmark sets. The first consisted of 292 industrial instances used at the MiniSAT hack track and the second of 300 graph coloring

---

[*] This work was partially supported by Serbian Ministry of Science grant 144030.
[1] http://www.satcompetition.org/,http://baldur.iti.uka.de/sat-race-2008/
[2] http://www.cril.univ-artois.fr/SAT09/results/ranking.php?idev=25

O. Strichman and S. Szeider (Eds.): SAT 2010, LNCS 6175, pp. 209–222, 2010.

Table 1. Number of solved instances for "lucky" and "unlucky" case of each solver

Solver	Industrial			Graph coloring	
	Lucky	Original	Unlucky	Lucky	Unlucky
MiniSAT 09z	161	142	111	180	157
minisat_cumr r	156	139	107	190	180
minisat2	141	121	93	200	183
MiniSat2hack	144	121	93	200	183

instances from the SAT 2002 competition. Each solver was run on 50 shuffled variants of each benchmark (obtained by reordering the clauses, literals in each clause, and renaming the variables) with cutoff time of 1200 seconds.

First we checked how much the number of solved formulae can vary. A solver was "lucky" if for each formula it was given the shuffled variant that it solves in the shortest time. The solver was "unlucky" if for each formula it was given the shuffled variant that it solves in greatest time (unsolved if such variant exists). For each benchmark set and each solver, results for both the "lucky" and the "unlucky" case are presented in Table 1. For industrial formulae, the number of formulae solved in their original form is also given. The graph coloring instances were already shuffled, so we don't give such information for them. One can see from the table that the variation of the number of solved formulae can be large.

Second, we investigated the effect of this variation on solver comparison. We checked that for each two solvers, on the industrial instances it is possible to suitably select shuffled variants of each instance to make one benchmark set on which the first solver is better than the second, and another on which the second is better than the first (in this case, both solvers are run on the same shuffled variant of each formula). However, the probability of such event should be also taken into the consideration. For each pair of solvers we performed 10000 simulated pairwise comparisons with shuffled variants chosen on random for each formula in order to estimate the probabilities of each solver in the comparison being the winner. For most of the pairs, changing the outcome of the comparison turned out to be very unlikely. However, when comparing MiniSAT 09z and minisat cumr r on industrial instances the odds of winning are 92% to 8%, when comparing minisat2 and minisat2hack on industrial instances the odds are 6% to 94%, and when comparing minisat2 and minisat2hack on graph coloring instances the odds are 74% to 26%. It is interesting to notice that on industrial instances, the solver that appears to be the best, can be beaten in practice as a result of chance. Also, ordering of minisat2 and minisat2hack would be different from the one obtained at the competition in most of the cases.

Sometimes the use of shuffling is disputed. Its use is not essential for the methodology that will be proposed. The purpose of shuffling is to make a solver choose different paths of the solving process on different runs, and thus obtain information about its runtime distribution. Such an effect could also be achieved without shuffling by changing the random seed the solver uses, and we certainly don't prefer some random seeds to the others. We also performed the similar experiment with random seeds instead of shuffling. The "lucky version" of MiniSAT solved 144 instances, and the "unlucky" one solved 96, which is close to the results obtained by shuffling. Note that the use of randomization is a common practice in modern SAT solvers.

In addition to the problem just discussed, there is a problem of drawing conclusions from the available experimental results. Sometimes, the results are presented by tables showing that the new SAT solver is performing better than the base one on some subsets of instances, and worse on the others, without clear conclusion about the overall effect. Also, SAT solver comparisons are concluded without discussion if the observed differences could be obtained by chance or are a consequence of a genuine effect.

The goal of this work is the formulation of statistically founded methodology of SAT solver comparison that would *i*) eliminate chance effects from the results, *ii*) give an answer if there is a positive (or negative) overall effect of the proposed modification to SAT solver performance, and *iii*) give an information of statistical significance of that effect. Such a methodology would enable more reliable discrimination between good and bad ideas, enabling the community to focus on the more promising ones.

There are several issues that have to be addressed in devising such methodology. The first is a presence of censored data. If the formula is not solved in a given cutoff time, it is only known that it needs more time to be solved, but not how much exactly. The second is a need to compare runtime distributions instead of single solving times that are unreliable. The third issue is finding a way to combine conclusions for different formulae to derive an overall conclusion.

The methodology we propose was conceived for detection of improvements over some base solver, but it can be used without limitation to comparison of two arbitrary solvers. Also, it will be shown how it can be extended for ranking of several solvers. This methodology *is not* concerned with selection of benchmarks. One should choose the benchmarks representative for the problems of interest.

The rest of the paper is organized as follows. In Sect. 2, a brief information on relevant concepts and techniques is given. The proposed methodology is described in Sect. 3 and the experimental results are given in Sect. 4. In Sect. 5, related work is discussed. In Sect. 6 final conclusions are drawn and some directions of possible further work are pointed to. In the appendix, a proof of the theorem from Sect. 3 is given.

2   Preliminaries

In this section we describe concepts and techniques important for understanding the proposed methodology and introduce needed notation.

2.1   Distributions of Solver Running Times

It is well known that solving times for a propositional formula can vary substantially from one solver run to another if the solver uses some random decisions during its work. Also, solving times can change substantially if a syntactical representation of the formula is changed. Distributions of these solving times have been a subject of intensive study [GSCK00, FRV97], resulting in important theoretical insights and understanding of randomized restarts. A runtime distribution of a solver on some instance bears much more information about solver performance than a single run, but it is considerably more expensive to obtain.

2.2   Statistical Hypothesis Testing and the Notion of the Effect Size

Statistical hypothesis testing is concerned with determining if a proposed hypothesis about some populations hold, based on sample data from those populations. The test is performed by trying to reject the *null hypothesis* H_0. H_0 is usually a statement of "no effect" claiming that the effect considered is not present in the populations.

In order to test if H_0 holds, one computes a value t of some test statistic T (depending on the purpose and formulation of the test) with a known probability distribution. The probability of obtaining the computed or more extreme value of the statistic, assuming that H_0 is true, is called a p value. If the p value is less than some predetermined threshold α (usually 0.05), the observed event is considered to be too improbable to be observed if H_0 holds, and the hypothesis H_0 is rejected. Such a result is said to be *statistically significant at the level* α. Otherwise $(p > \alpha)$, one cannot reject the hypothesis H_0.

The smaller the p value, the greater the confidence that the observed effect is not obtained by chance. Nevertheless, a small p value is not enough to conclude that the effect is large, because it depends both on the size of the effect and the sample size. Even if the effect is statistically highly significant, it can still be too small to be of any practical importance. In order to measure the magnitude of the underlying effect, an *effect size* has to be computed. There are several standard effect size statistics [Ros91, GK05]. One, commonly used when comparing two samples, is a point biserial correlation (often referred to as Pearson's r) [Ros91].

2.3   Point Biserial Correlation

Point biserial correlation ρ_{pb} between two random variables is the correlation between their outcomes and an *indicator variable* with value 1 for outcomes of the first random variable, and value -1 for the outcomes of the other. Its sample estimate r_{pb} is calculated by the formula:

$$r_{pb} = \frac{\sum_{i=1}^{N}(X_i - \overline{X})(Y_i - \overline{Y})}{\sqrt{\sum_{i=1}^{N}(X_i - \overline{X})^2}\sqrt{\sum_{i=1}^{N}(Y_i - \overline{Y})^2}}$$

where X_i denote observations from both samples, and Y_i are indicator variables. $\overline{X}$ and $\overline{Y}$ are the means of X_i and Y_i. N is the total number of observations. Quantities ρ_{pb} and r_{pb} have values ranging from -1 to $+1$. Absolute values closer to 1 mean that the distributions of random variables exhibit better separation. Values near 0 indicate great overlapping between distributions.

If there is no information about the distribution of the data, the data are often transformed by ranking — each observation in either sample is replaced by its rank in the sorted sample. If there are tied (equal) observations, each of them is assigned the average rank of the ranks that would be attributed to them. The point biserial correlation calculated on ranked data has different properties to the original statistic and is an instance of the Spearman correlation coefficient [DKS51, DM61].

The estimate r_{pb} is asymptotically normally distributed with the mean ρ_{pb}. The variance of r_{pb} is not easy to determine if the ranking is used and if the distribution of the data is not normal except for the case $\rho_{pb} = 0$ [DKS51, DM61]. Nevertheless, it can be estimated by methods like bootstrapping or jackknife [Efr79, ES81]. The variance of r_{pb} is strongly dependent on value of ρ_{pb}, and r_{pb} is usually used in statistical tests only after the Fisher's variance stabilizing transformation $z(x) = arctanh(x)$ is applied [Hot53]. Also, the transformed variable is much closer to normal distribution than the original one. It has the mean $z(\rho_{pb})$ and its variance can be estimated by $var(r_{pb})(1 - r_{pb}^2)^{-2}$.

In order to interpret the magnitude of r_{pb}, one can follow commonly accepted recommendations by Cohen [Coh88] — effects with $|r_{pb}|$ in the intervals [0,0.1), [0.1,0.3), [0.3,0.5), and [0.5,1], are considered respectively, negligible, small, medium, and large. However, note that these are not strict rules, but rather, reasonable guidelines.

2.4   Accounting for Censored Data

By *censored data* we mean data known to be greater than some threshold value, but of unknown exact value. One well-known test for comparison of two samples which include censored data is the Gehan test [Geh65]. The statistic used in this test can be formulated as follows [Man67]. The *pooled sample* is the sample that includes elements of both samples that are compared. Note that the repetitions of elements are possible. Let U_i be the number of observations in the pooled sample than which the i-th observation in the pooled sample is strictly greater minus the number than which it is strictly less. In the case of unique censoring time, censored observations are treated as equal and greater than all the uncensored observations.[3] Then Gehan statistic is defined by

$$W_G = \frac{1}{|A_1||A_2|} \sum_{i \in A_1} U_i$$

where A_j is a set of indices in the pooled sample of the observations from the j-th sample ($j = 1, 2$). As shown by Gehan [Geh65], using the theory of U statistics [Hoe48, Leh51], Gehan statistic is a consistent estimate of $\omega = P(X > Y) - P(X < Y)$. It is asymptotically normally distributed with the mean ω. The variance of W_G is easy to calculate if $\omega = 0$. In other cases bootstrapping or jackknife estimates can be used [Efr79, ES81]. As in case of r_{pb}, the variance depends on ω, diminishing as ω approaches extreme values -1 or 1.

---

[3] In the case of varying censoring times, more sophisticated statistics might be used.

3   The Methodology

An overall idea of the proposed methodology for comparing two solvers is simple. For each SAT instance from some benchmark set one should calculate suitably defined difference of performance of two solvers on that instance. If the performances of two solvers are approximately the same for the benchmark set, then the differences on considered instances should mainly cancel out, and the average of the differences couldn't be too large. Note that the concept of runtime distribution is important for our methodology, but in formulation of the methodology we leave the sampling mechanism unspecified. The methodology will be applicable regardless of that choice. First, we outline the methodology, and then, discuss its various aspects.

3.1   The Outline of the Proposed Methodology

Let random variable τ^j represent runtimes of the solver S_j ($j = 1, 2$) on SAT instance F. Since solving times can be too large for practical evaluation, a cutoff time T is used, and thus distributions of random variables τ^j are truncated to the right at the point T. The difference of SAT solver performances should be defined by some function $\delta(\tau^1, \tau^2)$ measuring the suitably chosen difference between distributions of these variables. Since the random variables themselves are not available, inferences about them are made using samples of runtimes. The value of the function δ should be approximated by a difference d between samples. The differences δ_i of random variables corresponding to formulae F_i can be averaged to obtain a value $\bar{\delta}$ which measures the overall difference between solvers on given corpus of formulae. Sample estimate of $\bar{\delta}$, the average of d_i values, will be denoted $\bar{d}$. Distribution of the average of $\bar{d}$ under the hypothesis $\bar{\delta} = 0$ will be denoted by Θ.

The methodology is outlined in Fig. 1. It can be considered as a statistical test with the null hypothesis that there is no overall effect — H_0: $\bar{\delta} = 0$.

Obviously, in order to use this methodology, its various aspects must be discussed. The most important ones are the choice of the function d, estimation of distribution Θ, and interpretation of the magnitude of $\bar{d}$. We will propose some choices for each of these aspects.

3.2   Choosing Function d

The role of function d is to quantify the difference in performance of two solvers on one instance based on samples of corresponding solving times. For that we use effect size measures for difference between two samples. Three possible effect size measures will be introduced, and their relations will be analyzed.

Probably the most intuitive indicator of two solvers performing equally on some instance F would be that the probability that the first solver solves the instance in more time than the second solver is equal to the probability that the second solver solves the instance in more time than the first solver. More

- INPUT: Solvers S_1 and S_2, and the set of benchmark instances
- OUTPUT: Information if one solver is better than the other and estimate of the effect size
- Choose the level of statistical significance α ($\alpha < 1$)
- For each formula F_i from corpus $\mathcal{F}$ consisting of M instances:
    - Take a sample T_i^j of size N of random variable τ_i^j ($j = 1, 2$)
    - Calculate the difference $d_i = d(T_i^1, T_i^2)$ between obtained solving times
- Calculate the average $\overline{d}$ of values d_i
- Estimate Θ — the distribution of $\overline{d}$ under the null hypothesis
- Calculate the p value for $\overline{d}$ according to the distribution Θ
- If $p \leq \alpha$
    - Declare the first solver to be better if $\overline{d} < 0$
    - Declare the second solver to be better if $\overline{d} > 0$
    - Report $\overline{d}$ as the estimate of the magnitude of the difference between performances of two solvers
- otherwise, declare the difference insignificant

Fig. 1. Outline of the proposed methodology

formally
$$P(\tau^1 > \tau^2) = P(\tau^1 < \tau^2)$$

or equivalently
$$\omega = P(\tau^1 > \tau^2) - P(\tau^1 < \tau^2) = 0$$

where τ^j is a random variable representing solving times of the solver S_j on instance F. These two probabilities need not sum to 1 in case that censored data are present. In that case

$$\pi = \frac{1 - \omega}{2} = P(\tau^1 < \tau^2) + \frac{1}{2}P(\tau^1 = \tau^2)$$

which is a quite intuitive measure that combines the evidence of one solver performing better than the other with the uncertainty that appears if both solvers haven't solved the same benchmarks. Namely, the case $\tau^1 = \tau^2$ is possible only for censored observations since, practically, all uncensored solving times differ even slightly if measured with enough precision. The value π is a known effect size measure [GK05]. Recall that ω is estimated by W_G and π is estimated by $(1 - W_G)/2$. Drawback of using ω or π is a lack of variance stabilizing transformation like the one available for the point biserial correlation (see Sect. 2).

Point biserial correlation ρ_{pb} is a commonly used and well understood effect size measure (as described in Sect. 2). It is estimated by r_{pb}. Since there is no information about distribution of the data, estimate should be calculated on ranked data (see Sect. 2). Technical advantage of using this measure is availability of Fisher's transformation which stabilizes the variance and makes the distribution closer to normal. This makes determining statistical significance much more reliable. On the other hand, it is not obvious if this measure makes sense with censored data. Also, without prior experience with this measure, one might feel uncomfortable interpreting its magnitude.

To establish a relation between estimates of technically more suitable ρ_{pb}, and more intuitive ω and π, we present the following theorem, showing that all three can be used interchangeably (the proof is given in the appendix). For observations X_i of a random variable X, by S_X^2 we denote $\sum(X_i - \overline{X})^2$ where $\overline{X}$ is an average of observations X_i.

Theorem 1. *Let T^1 and T^2 be two samples of two random variables τ^1 and τ^2. Let X_i be the i-th element in the sorted pooled sample, R_i its rank in that sample, Y_i the corresponding indicator variable, and r_{pb} the sample point biserial correlation between R_i and Y_i. Then, if there are no ties in uncensored data and the censoring time is unique, the following relation holds*

$$W_G = r_{pb}S_R S_Y /|T^1||T^2| \tag{1}$$

Additionally, if $|T^1|/|T^2|$ approaches finite positive constant when $|T^1| \to \infty$,

$$var(W_G) \to var(r_{pb})S_R^2 S_Y^2 /|T^1|^2|T^2|^2 \tag{2}$$

also holds when $|T^1| \to \infty$.

Note that the assumptions of the theorem are fulfilled in the context of SAT solving. As already noticed, the assumption of no ties is quite realistic for uncensored data. The assumption of unique censoring time is standard in SAT solving. The last assumption is trivially satisfied as one can always use samples of equal size. This theorem allows us to use either of the proposed effect size measures for function d since one can be easily calculated from the other. Since p value depends on the value of the test statistic and its variance, the second relation ensures that p value estimates are practically the same for large samples regardless which of the proposed measures is used.

For our primary effect size measure, we take point biserial correlation due to its technical advantages concerning the computation of statistical significance, but ω and π can also be reported for the effect size.

3.3   Determining Statistical Significance and the Effect Size

We say that two solvers perform the same on one instance if $\rho_{pb} = 0$, or if r_{pb} is not significantly different from 0 in sense of statistical testing. Also, for the measure of difference d_i between samples of random variables τ_i^1 and τ_i^2 we can take r_i — the estimate of ρ_{pb} for F_i. Statistical significance testing based on r_{pb} values is usually done after the Fisher transformation (see Sect. 2). To check the statistical significance of the overall test, for each r_i, value $z(r_i)$ is computed, and those values are averaged. Since all the $z(r_i)$ are asymptotically normally distributed, it is easy to see (using the properties of the normal distribution and asymptotics) that the average $\overline{z}$ is also asymptotically normally distributed:

$$\overline{z} \sim \mathcal{N}\left(\frac{1}{M} \sum_{i=0}^{M} z(\rho_i), \frac{1}{M^2} \sum_{i=1}^{M} \frac{var(r_i)}{(1 - r_i^2)^2} \right)$$

where ρ_i is the population parameter estimated by r_i. To see if the null hypothesis $\bar{\delta} = 0$ holds, one should check if the difference of obtained average $\bar{z}$ from $z(\bar{\delta}) = 0$ is statistically significant with respect to distribution of $\bar{z}$. The p value (two tailed) is $2(1 - \Phi(\bar{z}/\sqrt{var(\bar{z})}))$, where Φ is the distribution function of standard normal distribution. Note that we don't directly use the distribution Θ of $\bar{d}$ because the use of transformed values is more reliable.

The estimate of the effect size $\bar{d}$ is the average of values r_i, and its magnitude is interpreted in the way described in Sect. 2.

3.4   Ranking Several Solvers

If one is comparing several solvers, even if all pairwise comparison results are known one still needs a ranking method.

Important issue with application of statistical tests in general is their potential nontransitivity. Namely, there are examples of random variables A, B, and C such that $P(A < B) > \frac{1}{2}$ and $P(B < C) > \frac{1}{2}$ hold, but $P(A < C) > \frac{1}{2}$ does not. Note that this counterintuitive behavior is not a flaw of any test, but rather a natural probabilistic phenomenon. A popular example are Efron's dice [BH02].

There is still no proof that the proposed comparison procedure is transitive. As with Efron's dice it might be even meaningless to demand transitivity, but this should be a subject of a further study. To overcome this difficulty, one can use Kendal-Wei method for ranking based on pairwise comparisons [Ken55]. This method is designed for situations characterized by nontransitivity property.

4   Experimental Results

In this section we present two experiments. The first one is concerned with the number of shuffled variants appropriate for the application of the methodology, and the second one shows results of the application of the methodology. In both experiments we use the same 4 solvers and 2 benchmark sets as in Sect. 1. For the level of statistical significance α we take the usual value of 0.05. We sample from the runtime distributions by solving 50 shuffled variants of each formula with cutoff time of 1200 seconds. Though the shuffling is quite acceptable for the solvers used, one could also change the random seed. If all the shuffled variants of the benchmark were solved in less than 0.1 seconds[4] by both solvers or no shuffled variant was solved by any solver, the benchmark was discarded as uninformative. For function d we choose r_{pb}. The variance of r_{pb} is estimated by bootstrapping [Efr79] with 100000 bootstraps.[5]

First important question concerning the application of the proposed methodology is its computational cost reflected by the number of shuffled variants one has to use in order to obtain reliable estimates of the effect size and statistical

---

[4] At most 1 industrial and 30 graph coloring instances were discarded in any comparison on the basis of this criterion.

[5] Source code of software used for all the statistical calculations is available from http://www.matf.bg.ac.rs/~nikolic/solvercomparison/sc.zip

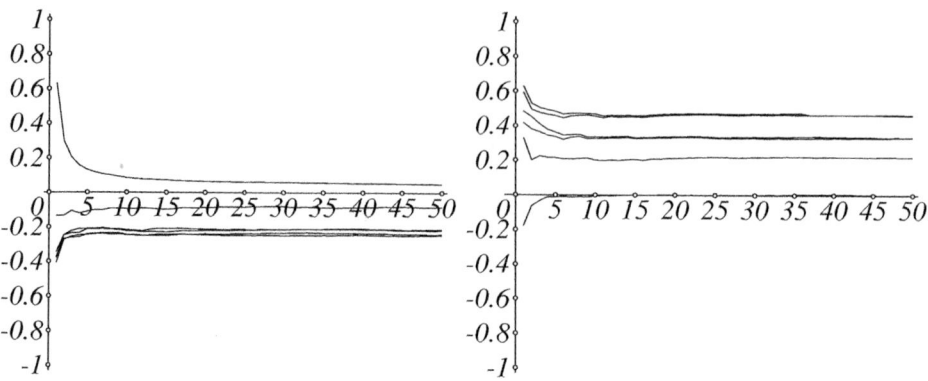

Fig. 2. Plots of r_{pb} for industrial (left) and graph coloring (right) benchmark sets as a function of the number of shuffled variants used

Table 2. Estimates of ρ_{pb} when comparing various solvers. Following labels are used A = MiniSAT 09z, B = minisat cumr r, C = minisat2, D = MiniSat2hack.

	Industrial				Graph coloring			
	A	B	C	D	A	B	C	D
A	-	-0.097	-0.249	-0.229	-	0.206	0.453	0.461
B	0.097	-	-0.241	-0.208	-0.206	-	0.327	0.333
C	0.249	0.241	-	0.072	-0.453	-0.327	-	-0.001
D	0.229	0.208	-0.072	-	-0.461	-0.333	0.001	-

significance. Also, increasing the number of shuffled variants leads to smaller p values due to larger sample size without the increase of the effect size. It is advised that the sample size is not increased beyond the point at which the effect size estimate stabilizes [Coh95]. To check the number of needed shuffled variants, for each two solvers, we plotted the value of r_{pb} as the number of used shuffled variants ranges from 1 to 50. The plot for each benchmark set is given in Fig. 2. The plots indicate that the number of shuffled variants that should be used is around 10 to 15. As expected, the results of the experiments based on the estimates of ω and π instead of ρ_{pb} are the same.

In Table 2 we present estimates of ρ_{pb} for comparisons of each pair of solvers using 15 shuffled variants. The obtained results are not surprising with respect to those shown in Table 1. In all the comparisons the p values (two tailed) are less than 0.001 except when comparing original MiniSAT version and MiniSat2hack on graph coloring instances when it is 0.945. Nevertheless, note that some statistically significant differences can be considered negligible with respect to guidelines provided in Sect. 2. Note that no problems with transitivity appeared. The ranking is easy to establish. It is ABDC on industrial and CDBA on graph coloring instances, where the same labels are used as in Table 2.

5   Related Work

There are already several papers concerning the comparison of SAT solvers. Le Berre and Simon recognize the importance of this question [LS04]. Also, the possibility that shuffling can change the order of solvers was noticed. It is suggested that the corpora could include shuffled variants of formulae. On the other hand, this paper is concerned with the usual way of solver comparison. Audemard and Simon further analyze the impact of the shuffling on the number of solved formulae, and conclude that it can be large [AS08].

Etzoni and Etzoni propose the use of statistical tests for censored data for evaluation of speedup learning systems, but the comparison of runtime distributions of instances is not discussed in their context [EE94]. Brglez et al. stress the importance of statistical approach for SAT solver comparison [BLS05, BO07]. Also the importance of runtime distributions for SAT solver comparison is recognized. Statistical tests are used to compare performances of two solvers, but only on one instance. Full methodology that could use a corpus of instances and combine results of testing on individual instances is not devised. Moreover, we exploit the notion of the effect size which is important for such methodology and propose the extension to ranking several solvers using method which takes the nontransitivity issue into account.

Pulina gives an excellent empirical analysis of ranking methods for systems used in automated reasoning and more importantly establishes reasonable properties that those ranking methods should possess [Pul06].

6   Conclusions and Future Work

We demonstrated that comparison methods that are widely used can be unreliable, and depend on variable naming, ordering of clauses and literals, and random seeds used (see Sect. 1). A new, statistically founded, methodology is proposed for comparison of SAT solvers. It is based on the comparison of runtime distributions instead of single solving times and uses standard effect size measures to quantify the difference between those distributions.

We showed that the needed number of shuffled variants to estimate the effect size between solvers is around 10 to 15. The testing corpora could be somewhat reduced to compensate for this increase of solving time, thus trading some benchmarks for thorough analysis. We regard this approach better, since the results presented in Sect. 1 do not suggest that the use of large corpora eliminates the significant chance effects on number of solved formulae. The new methodology is able to practically eliminate the chance effects from the comparison (up to p value) and provide information on statistical significance and effect size in the way usual for statistical testing which standard approach does not.

As for the future work, important issue is finding the assumptions that guarantee the transitivity of proposed comparison procedure, and checking if nontransitive effects can appear in SAT solving. Also, proposed ranking method is yet to be analyzed in the light of the criteria established by Pulina [Pul06].

Acknowledgements

The author is grateful to Predrag Janičić, Filip Marić, and Oliver Kullmann for their valuable suggestions and discussion about this work.

References

[AS08]   Audemard, G., Simon, L.: Experiments with Small Changes in Conflict-Driven Clause Learning Arghorithms. In: Proc. of the 14th International Conf. on Principles and Practice of Constraint Programming (2008)

[BLS05]   Brglez, F., Li, X.Y., Stallmann, M.: On SAT Instance Classes and a Method for Reliable Performance Experiments with SAT Solvers. In: Annals of Mathematics and Artificial Intelligence (2005)

[BO07]   Brglez, F., Osborne, J.: Performance Testing of Combinatorial Solvers With Isomorph Class Instances. In: ECS 2007: Experimental Computer Science on Experimental Computer Science (2007)

[BH02]   Brown, B., Hettmansperger, T.: Kruskal-Wallis, Multiple Comparisons and Efron Dice. Australian & New Zealand Journal of Statistics (2002)

[Coh88]   Cohen, J.: Statistical Power Analysis for the Behavioral Sciences. Lawrence Erlbaum Associates, Mahwah (1988)

[Coh95]   Cohen, P.: Empirical Methods for Artificial Intelligence. MIT Press, Cambridge (1995)

[Cra46]   Cramér, H.: Mathematical Methods of Statistics. Princeton Univeristy Press, Princeton (1946)

[DM61]   David, F., Mallows, C.: The Variance of Spearman's rho in normal samples. In: Biometrika (1961)

[DKS51]   David, S., Kendall, M., Stuart, A.: Some Questions of Distribution in the Theory of Rank Correlation. Biometrika (1951)

[Efr79]   Efron, B.: Bootstrap Methods: Another Look at Jackknife. The Annals of Statistics (1979)

[ES81]   Efron, B., Stein, C.: The Jackknife Estimate of Variance. The Annals of Statistics (1981)

[EE94]   Etzoni, O., Etzoni, R.: Statistical Methods for Analyzing Speedup Learning Experiments. Machine Learning (1994)

[FRV97]   Frost, D., Rish, I., Vila, L.: Summarizing CSP hardness with continuous probability distributions. In: Proc. of the 14th National Conf. on Artificial Intelligence (1997)

[Geh65]   Gehan, E.: A Generalized Wilcoxon Test for Comparing Arbitrarily Singly-Censored Samples. Biometrika (1965)

[GSCK00]   Gomes, C., Selman, B., Crato, N., Kautz, H.: Heavy-Tailed Phenomena in Satisfiability and Constraint Satisfaction Problems. Journal of Automated Reasoning (2000)

[GK05]   Grissom, R., Kimm, J.: Effect Sizes for Research: A Broad Practical Approach. Lawrence Erlbaum Associates, Mahwah (2005)

[Hoe48]   Hoeffding, W.: A Class of Statistics with Asymptotically Normal Distribution. The Annals of Mathematical Statistics (1948)

[Hot53]   Hotelling, H.: New Light on the Correlation Coefficient and its Transforms. Journal of the Royal Statistical Society (1953)

[Ken55]   Kendall, M.: Further Contributions to the Theory of Paired Comparisons. Biometrics (1955)

[LS04]   Le Berre, D., Simon, L.: The Essentials of the SAT 2003 Competition.
            In: Giunchiglia, E., Tacchella, A. (eds.) SAT 2003. LNCS, vol. 2919, pp.
            452–467. Springer, Heidelberg (2004)
[Leh51]   Lehmann, E.: Consistency and Unbiasedness of Certain Nonparametric
            Tests. In: The Annals of Mathematical Statistics (1951)
[Man67]   Mantel, N.: Ranking Procedures for Arbitrarily Restricted Observations.
            Biometrics (1967)
[Pul06]   Pulina, L.: Empirical evaluation of Scoring Methods. In: Proc. of the 3rd
            European Starting AI Researcher Symposium (2006)
[Ros91]   Rosenthal, R.: Meta-Analytic Procedures for Social Research. Sage, Thou-
            sand Oaks (1991)
[Zar05]   Zarpas, E.: Benchmarking SAT Solvers for Bounded Model Checking. The-
            ory and Applications of Satisfiability Testing (2005)

Appendix

Proof of Theorem 1.
Let $n_1 = |T^1|$, $n_2 = |T^2|$, and $N = n_1 + n_2$. The numbers of censored observations in each sample are denoted by c_1 and c_2, and $C = c_1 + c_2$. Let I^1 and I^2 be the sets of indices in the pooled sample of uncensored observations from samples T^1 and T^2 respectively. Let $I = I^1 \cup I^2$. The set of indices in the pooled sample of all the observations of the first sample is denoted by A_1.

First we show that the relation (1) holds. We will consider expressions $n_1 n_2 W_G$ and $S_R S_Y r_{pb}$ and will conclude that they are equal. We use Mantel's version of W_G [Man67] noting that it can be decomposed in terms of ranks of uncensored observations plus the term for censored observations.

$$n_1 n_2 W_G = \sum_{i \in A_1} U_i = \sum_{i \in I^1} [(R_i - 1) - (N - R_i)] + c_1(N - C)$$

$$= 2 \sum_{\in I^1} R_i - (n_1 - c_1)(N + 1) + c_1(N - C)$$

$$= 2 \sum_{\in I^1} R_i - (n_1 - 2c_1)(N + 1) - c_1(C + 1)$$

Let us consider $S_R S_Y r_{pb}$:

$$S_R S_Y r_{pb} = \sum_{i=1}^{N} (R_i - \overline{R})(Y_i - \overline{Y}) = \sum_{i=1}^{N} R_i Y_i - \sum_{i=1}^{N} R_i \overline{Y} - \sum_{i=1}^{N} \overline{R} Y_i + \sum_{i=1}^{N} \overline{R} \overline{Y}$$

where $\overline{R}$ and $\overline{Y}$ are the means of R_i and Y_i. Note that the last three sums are equal, and hence

$$S_R S_Y r_{pb} = \sum_{i=1}^{N} R_i Y_i - \sum_{i=1}^{N} R_i \overline{Y} = \sum_{i=1}^{N} R_i Y_i - E_1$$

where $E_1 = (N+1)(n_1 - n_2)/2$ and is obtained using the fact that the sum of ranks is constant and equals $N(N+1)/2$ and that $\overline{Y} = (n_1 - n_2)/N$. Separating censored and uncensored observations yields

$$S_R S_Y r_{pb} = \sum_{i \in I} R_i Y_i + E_2 - E_1 = \sum_{i \in I^1} R_i - \sum_{i \in I^2} R_i + E_2 - E_1$$

where $E_2 = (2N - C + 1)(c_1 - c_2)/2$ since $(2N - C + 1)/2$ is the average rank of the censored observations. Since all the uncensored observations are less than censored ones, and since the sum of their ranks is constant, the second sum can be expressed in terms of the first sum:

$$S_R S_Y r_{pb} = 2 \sum_{i \in I^1} R_i - (N - C)(N - C + 1)/2 + E_2 - E_1$$

After elementary calculations we obtain:

$$S_R S_Y r_{pb} = 2 \sum_{\in I^1} R_i - (n_1 - 2c_1)(N + 1) - c_1(C + 1)$$

thus proving the relation (1).

To prove the relation (2), we note that S_Y is constant, and that S_R is constant for fixed c_1 and c_2. For convenience, we will talk in terms of ratios $a_1 = c_1/n_1$ and $a_2 = c_1/n_2$. Using (1), the conditional variance of W_G is $var(W_G | a_1, a_2) = \frac{S_R^2 S_Y^2}{n_1^2 n_2^2} var(r_{pb})$. We need to prove $var(W_G)/var(W_G | a_1, a_2) \to 1$ when $n_1 \to \infty$. We will follow the reasoning of Gehan [Geh65]. By the law of total variance we have

$$var(W_G) = E_{l_1, l_2} var(W_G | l_1, l_2) + var_{l_1, l_2} E(W_G | l_1, l_2)$$

By the law of large numbers, a_1 and a_2 converge in probability to their expectations α_1 and α_2 when $n_1 \to \infty$. Since the probabilities of l_i such that $|l_i - \alpha_i| \geq \varepsilon$ vanish for all $\varepsilon > 0$ when $n_1 \to \infty$, it holds

$$\frac{n_1^{-3} E_{l_1, l_2} var(W_G | l_1, l_2)}{n_1^{-3} var(W_G | a_1, a_2)} \to 1$$

when $n_1 \to 1$. The last relation is obtained using the convergence theorems by Cramér and Slutsky [Cra46] which can be used since it is known that $n_1^{-3} var(W_G | a_1, a_2) = O(1)$ when $n_1 \to \infty$ [Geh65].

Regarding the second term in the expansion of unconditional variance, by definition

$$var_{l_1, l_2} E(W_G | l_1, l_2) = E_{l_1, l_2} E^2(W_G | l_1, l_2) - (E_{l_1, l_2} E(W_G | l_1, l_2))^2$$

which converges to 0 by similar reasoning as for the first term. This proves the convergence (2). $\qquad \square$

On the Relative Merits of Simple Local Search Methods for the MAX-SAT Problem

Denis Pankratov[1] and Allan Borodin[2]

[1] Department of Computer Science, University of Chicago
pankratov@cs.uchicago.edu
[2] Department of Computer Science, University of Toronto
bor@cs.toronto.edu

Abstract. Algorithms based on local search are popular for solving many optimization problems including the maximum satisfiability problem (MAX-SAT). With regard to MAX-SAT, the state of the art in performance for universal (i.e. non specialized solvers) seems to be variants of Simulated Annealing (SA) and MaxWalkSat (MWS), stochastic local search methods. Local search methods are conceptually simple, and they often provide near optimal solutions. In contrast, it is relatively rare that local search algorithms are analyzed with respect to the worst-case approximation ratios. In the first part of the paper, we build on Mastrolilli and Gambardella's work [14] and present a worst-case analysis of tabu search for the MAX-k-SAT problem. In the second part of the paper, we examine the experimental performance of determinstic local search algorithms (oblivious and non-oblivious local search, tabu search) in comparison to stochastic methods (SA and MWS) on random 3-CNF and random k-CNF formulas and on benchmarks from MAX-SAT competitions. For random MAX-3-SAT, tabu search consistently outperforms both oblivious and non-oblivious local search, but does not match the performance of SA and MWS. Initializing with non-oblivious local search improves both the performance and the running time of tabu search. The better performance of the various methods that escape local optima in comparison to the more basic oblivious and non-oblivious local search algorithms (that stop at the first local optimum encountered) comes at a cost, namely a significant increase in complexity (which we measure in terms of variable flips). The performance results observed for the unweighted MAX-3-SAT problem carry over to the weighted version of the problem, but now the better performance of MWS is more pronounced. In contrast, as we consider MAX-k-SAT as k is increased, MWS loses its advantage. Finally, on benchmark instances, it appears that simulated annealing and tabu search initialized with non-oblivious local search outperform the other methods on most instances.

1 Introduction

The maximum satisfiability problem (MAX-SAT) is of great interest in both theoretical[8] and applied computer science[18]. The MAX-SAT problem is NP-hard. Current state-of-the-art algorithms can solve the problem optimally within

O. Strichman and S. Szeider (Eds.): SAT 2010, LNCS 6175, pp. 223–236, 2010.
© Springer-Verlag Berlin Heidelberg 2010

a reasonable amount of time only for input instances of moderate size[5]. These methods are often based on branch and bound techniques with rather sophisticated rules that try to exploit the structure of the problem. MAX-SAT applications often involve instances of far larger scale than what exact solvers can handle. As a result, a number of approximation algorithms have been developed [6]. Local search based algorithms have gained popularity for their conceptual simplicity and approximation performance. Furthermore, local search methods can be used as a preprocessing step in a branch and bound method so as to eliminate non-productive branches.

In the exact MAX-k-SAT problem each clause is restricted to have exactly k literals, and the same variable cannot repeat within the same clause whether negated or not. In the weighted version of the problem, each clause has a real-valued positive weight, and the objective is to find a truth assignment that maximizes the total weight of satisfied clauses.

A "basic" local search algorithm for the MAX-SAT problem starts with an arbitrary truth assignment. The neighborhood of a solution consists of all the truth assignments obtained by flipping the truth value of one or a small number of the variables. At each step, the basic local search algorithm looks through the neighborhood for a truth assignment that increases the number of satisfied clauses. If it finds such an assignment, the algorithm flips the value of the corresponding variable(s) and continues. Otherwise, the algorithm terminates.

Many heuristic methods are aimed at improving the performance of the basic local search, such as tabu search, random restarts, plateau moves, boosting, and others [1]. In spite of the popularity and success of local search, these algorithms are rarely analyzed with respect to either worst-case or "average-case" performance. Mastrolilli and Gambardella seem to be the first to analyze the worst-case performance of tabu search for the unweighted exact MAX-2-SAT problem[14]. In the first part of this paper, we extend their work to the exact MAX-k-SAT problem. Tabu search guarantees a better approximation ratio than "oblivious local search" but loses significantly to the Khanna et al [13]) "non-oblivious local search" that uses a related potential function in the neighborhood search. In the second part of the paper, we study the experimental performance of local search methods for the MAX-SAT problem. Our experiments indicate that tabu search and the stochastic local search methods consistently outperform both the oblivious and non-oblivious local searches. However, initializing tabu search with the truth assignment obtained by non-oblivious local search leads to results more competitive with the stochastic methods. But perhaps of equal interest is the fact that tabu search and the stochastic methods require substantially more time, and if the goal is simply to obtain a reasonable approximation then the basic methods have an advantage in terms of significantly reduced time complexity.

2   Local Search Algorithms for MAX-k-SAT

The input for each of the following algorithms is a boolean formula in CNF with m clauses over n variables.

2.1   Oblivious Local Search

For a given truth assignment τ, its one-flip neighborhood is the set of all truth assignments at Hamming distance one from τ. Oblivious local search (OLS) starts with an arbitrary fixed truth assignment. At each step, it searches the one-flip neighborhood of the current assignment for neighbors that achieve a better value of the given objective function. If such a neighbor exists, the algorithm replaces the current truth assignment with a truth assignment from the one-flip neighborhood that satisfies the most number of clauses. Ties are broken arbitrarily. If such a neighbor does not exist, OLS terminates.

For unweighted MAX-SAT the objective function is simply the number of satisfied clauses. The running time of OLS is polynomial in this case, as each step improves the value of the objective function by at least one, and the optimal value is bounded above by m. For weighted MAX-SAT the objective function is the total weight of the satisfied clauses. Without any restrictions on weights, the running time of OLS is no longer necessarily polynomial. This can be remedied by insisting that improvements at each step are sufficiently large.

2.2   Non-oblivious Local Search

The idea behind non-oblivious local search (NOLS) [13] is to introduce a related potential function and use it in the neighborhood search. This potential function gives preference to the clauses satisfied by many literals, as they are likely to stay satisfied even if the algorithm flips many variables in the future. Let C_j denote the set of clauses satisfied by j literals. Then the potential function for MAX-2-SAT is $3/2|C_1| + 2|C_2|$. For the case of MAX-k-SAT with $k > 2$, the reader is referred to Section 6 titled "The Power of Non-Oblivious Local Search" of Khanna et al paper [13]. Replacing $|C_j|$ by the total weight of C_j in the potential function extends NOLS to the weighted case of MAX-SAT. The running time analysis of this algorithm is similar to that of OLS.

2.3   Tabu Search

Oblivious and non-oblivious local searches terminate as soon as they achieve a local optimum for the given objective (respectively, the related potential) function. Tabu search (TS) offers a determinstic method for attempting to improve upon the current local optimum. Each iteration of TS consists of two stages. In the first stage, given a current truth assignment, OLS is used to compute a local optimum. In the second stage, tabu search maintains an additional data structure - a list of size t. This list is referred to as a *taboo list*. and t is called *taboo tenure*. When TS reaches a local optimum, it tries to escape it as follows. The algorithm records (x_i, t_i) pairs for the last t steps in the list, where x_i is a variable flipped at step t_i. During some of these steps, the current truth assignment can worsen, during other steps it can improve. If at some point, the current truth assignment improves over the local optimum found in stage 1, then stage 1 is repeated starting from the improved truth assignment. If during stage 2, tabu search does not find a solution that improves over the current local optimum, the algorithm terminates.

Each step of TS in the escape phase consists of flipping a variable. The algorithm follows the following rules (in given order) to decide which variable to flip: 1) **aspiration condition** - if flipping a variable improves the best value of the objective function found so far, then the best such variable is chosen; 2) **taboo** - if there are variables that appear in unsatisfied clauses and that were not flipped in the last t steps, i.e. they are not in the taboo list, the algorithm chooses the best such variable; 3) **LRU** - if all variables that appear in unsatisfied clauses also appear in the taboo list, then the *least recently used* such variable is selected. In the above rules, the best variable means that flipping it results in the largest increase of the objective function, or smallest decrease in the objective function, if no variable can improve it. All ties are broken arbitrarily. Taboo tenure controls the number of allowed steps during the escape phase. Mastrolilli and Gambardella argued that n is a reasonable choice for taboo tenure. The proof of a worst-case approximation ratio of TS relies on this requirement.

The same algorithm works for the weighted MAX-k-SAT problem, except that the weights of clauses are used in consideration of which variable to flip. In the unweighted case, TS with taboo tenure of n runs in polynomial time, because it improves the objective by at least one for every n steps, and m is an upper bound on the value of the objective function. As before, in the weighted case we can guarantee strongly polynomial running time by considering only large enough improvement for the stopping conditions in phase 1.

2.4   Simulated Annealing

We present a version of simulated annealing (SA) that appears in [17] and was found to work well for the satisfiability problem. SA was motivated by an analogous physical process, and the parameters of this algorithm have a corresponding semantic meaning. SA keeps track of the current "temperature". Initially, the temperature is high and the algorithm explores the solution space uniformly at random. As the temperature starts to cool down, SA gradually starts to prefer solutions that achieve better values of the objective function, concentrating on more promising parts of the solution space. The rule that specifies how temperature changes with time is called *temperature schedule*. In our implementation it is specified by three parameters: 1) MT - initial maximum temperature, 2) DR - decay rate per step, and 3) mT - minimum temperature for stopping. At step s SA computes the current temperature $T = MT \exp(-s \cdot DR)$. If $T < mT$, SA terminates. Otherwise, values $p_i = 1/(1 + \exp(-\Delta(i)/T))$ are computed, where $\Delta(i)$ is the change in the objective function when variable i is flipped. SA then flips variable i with probability p_i. After all variables have been processed, SA moves to the next step $s + 1$. We use the parameters $MT = 0.3, mT = 0.01$, and $DR = 1/n$ as suggested in [17].

2.5   MaxWalkSat

There are many variants of MaxWalkSat (MWS) algorithms. In general, given any current truth assignment for the unweighted MAX-SAT problem, an unsatisfied clause is chosen uniformly at random among all unsatisfied clauses. Various heuristics are then used to select a literal from this clause, and the truth value of that literal is flipped. Our experiments indicate that overall, the "productsum" heuristic performs best, and hence we restrict our attention to it. Suppose MWS decides to choose a literal in a clause $C = z_1 \vee z_2 \ldots \vee z_k$. Let $b(i) =$ the number of clauses that become unsatisfied if the literal z_i is flipped. Then "productsum" assigns a value $v_i = \left(\prod_{j \neq i} b(j) \right) \left(\sum_{j \neq i} b(j) \right)$ for each literal z_i in clause C and flips literal z_i with probability $\frac{v_i}{\sum_{1 \leq j \leq k} v_j}$. In the weighted case, MWS considers clauses of highest weight to be "hard" clauses. Given any truth assignment, it chooses a random unsatisfied hard clause and applies "pickproductsum" heuristic to it with $b(i) =$ the weight of clauses that become unsatisfied if z_i is flipped. If all hard clauses are satisfied, MWS chooses a random unsatisfied clause and applies "pickproductsum" heuristic. After this step, some hard clauses might become unsatisfied, and MWS will try to fix them in the very next step.

3   Background

Unless otherwise stated, MAX-k-SAT will mean exact MAX-k-SAT. Oblivious local search with 1-flip neighborhood achieves the approximation ratio of $\frac{k}{k+1}$ for the unweighted (and weighted) MAX-k-SAT problem, and this ratio is tight [8]. Non-oblivious local search provides a better worst-case guarantee of $\frac{2^k-1}{2^k}$ for the same problem [13]. For the MAX-2-SAT problem, tabu search was shown to have the tight approximation ratio of $\frac{3}{4}$ [14]. This ratio matches that of NOLS for MAX-2-SAT, which raises a question as to whether the two algorithms have the same approximation ratio for MAX-k-SAT for all $k \geq 2$. This paper answers this question in negative, showing that tabu search has a weaker approximation guarantee than non-oblivious local search for MAX-k-SAT with $k > 2$.

A general inapproximability result says that if $P \neq NP$, then $\frac{2^k-1}{2^k}$ is (essentially) the best possible approximation ratio achievable by any polynomial time algorithm for MAX-k-SAT with $k > 2$ [9]. A simple randomized algorithm, that picks a truth assignment uniformly at random, satisfies $\frac{2^k-1}{2^k}$ of all clauses in the exact MAX-k-SAT formula in expectation. Derandomization of this algorithm leads to a simple greedy algorithm achieving the approximation ratio of $\frac{2^k-1}{2^k}$ [11]. The case $k = 2$ is special in the MAX-SAT world. Currently, the best approximation ratio for MAX-2-SAT is .931 (Feige and Goemans[7]) using an algorithm based on semidefinite programming relaxation and rounding. An inapproximability result for MAX-2-SAT states that for any $\epsilon > 0$ it is NP-hard to approximate MAX-2-SAT within a factor of $\frac{21}{22} + \epsilon \approx 0.955 + \epsilon$[9].

A natural extension of the 1-flip neighborhood is a larger p-flip neighborhood for $p > 1$. The size of this neighborhood is $\sum_{j=1}^{p} \binom{n}{j}$ for a formula over n variables. Even for "small" constant values of p, it still requires a substantial amount of time to search through the entire neighborhood, and experimentally the quality of solutions seems to be not much better than those obtained through a 1-flip neighborhood. From the worst-case point of view, [13] shows that oblivious local search with an $o(n)$-flip neighborhood has the tight approximation ratio of $\frac{2}{3}$ for MAX-2-SAT - the same as with a 1-flip neighborhood. In general, these larger neighborhoods are not practical, and so this paper focuses on algorithms with the 1-flip neighborhood.

The second part of this paper deals with an empirical evaluation of different algorithms based on local search. We consider both benchmark examples from the "Second Evaluation" of MAX-SAT solvers (see [10] for a detailed description of these benchmark instances) and random k-SAT instances. Random exact k-SAT instances were generated by choosing formulas uniformly at random with the clause density around the estimated phase transition. There is a discrepancy between what has been proven rigorously about the threshold values for k-SAT in contrast to what has been experimentally shown and justified by well motivated analysis. See [4], [15] and [3] for current results concerning threshold values. The situation for 3-SAT represents a glaring gap in our current knowledge, namely, the best lower bound on $\hat{c}_3$ (for which clause density $c < \hat{c}_3$ implies satisfiability with high probability) is a constructive bound $\hat{c}_3 > 3.52$ obtained by a myopic (i.e. greedy) algorithm [12]. The provable upper bound is $\hat{c}_3 < 4.51$. Experimentally, the conjectured threshold is approximately 4.24 (see [16]).

4   Worst-Case Analysis of Tabu Search

Our version of tabu search, as described in Section 2, uses the length of taboo list equal to the number of variables. Tabu search contains oblivious local search as a subroutine, so an analysis of OLS occurs as a part of worst-case analysis of TS. In both weighted and unweighted cases of the exact MAX-k-SAT problem, oblivious local search achieves a worst-case approximation ratio of $\frac{k}{k+1}$. In fact, Khanna et al [8] prove the following stronger result on the "totality ratio".

Lemma 1. *At a local optimum, oblivious local search satisfies at least $\frac{k}{k+1}$ of the total number of clauses in the formula.*

Khanna et al show that the $\frac{k}{k+1}$ bound is tight. Adapting the proof of Lemma 1 to tabu search, Mastrolilli and Gambardella showed a 3/4 approximation guarantee of TS for the MAX-2-SAT problem. We extend their result to the MAX-k-SAT problem for all $k \geq 2$.

Theorem 1. *Tabu search outputs a truth assignment that satisfies at least $\frac{k+1}{k+2}$ of the total number of clauses.*

Proof. Suppose, oblivious local search is given a formula ϕ in k-CNF form with m clauses over n variables. The initial truth assignment is X^0. Let X^s be the truth assignment output by tabu search and let C_j^t denote the set of clauses that have exactly j literals satisfied by X^s at step t. By the halting condition, the algorithm terminated at step $s + n$. There exist t, such that $0 < t \le n$, and each variable from an unsatisfied clause at step $s + t$ was flipped exactly once during the last t steps. To prove this claim, consider two possibilities at step $s + n$. If n variables were flipped during last n steps, then the claim follows trivially with $t = n$. If less than n variables were flipped between steps s and $s + n$, then at least one variable was flipped at least twice during last n steps. In this case, choose a variable that is flipped the second time at the earliest step and let that step be $s + d + 1$.

Then $t = d$ satisfies the claim. To see that, consider step $s + d$. This is the step immediately before the chosen variable was flipped the second time. The algorithm had to repeat a variable, because all variables from unsatisfied clauses were in the taboo list at step $s + d$. In particular, each of these variables was flipped at least once during the d last steps, and each of them could not be flipped more than once, since $s + d + 1$ is the earliest step, when a variable is flipped for the second time.

Taking t as in the above claim, a clause with an unsatisfied literal at step s is satisfied at step $s + t$. Then a clause at step $s + t$ is unsatisfied by all literals only if it is satisfied by all literals at step s, i.e. $C_0^{s+t} \subseteq C_k^s$. This provides a lower bound on the number of C_k-clauses at the solution: $|C_0^s| \le |C_0^{s+t}| \le |C_k^s|$, where the first inequality follows because the solution does not improve in between steps s and $s + n$. Together with X^s being a local optimum and the Lemma 1 , we get $m = \Sigma_{i=0}^k |C_i^s| \ge |C_0^s| + |C_1^s| + |C_k^s| \ge |C_0^s| + k |C_0^s| + |C_0^s| = (2 + k) |C_0^s|$. Thus $|C_0^s| \le \frac{m}{k+2}$ and the theorem follows.

The approximation ratio guaranteed by TS for MAX-2-SAT matches that of NOLS, suggesting that tabu search might have the same worst-case performance as NOLS. However, we show that for $k > 2$, although TS improves over OLS, it has a significantly weaker approximation guarantee than NOLS.

Theorem 2. *The worst-case approximation (and totality) ratio of tabu search with tabu tenure n is at most $\frac{3k-3}{3k-2}$ of the total number of clauses.*

Proof. Fix k. The goal is to construct a satisfiable formula, such that the truth assignment that tabu search finds satisfies $\frac{3k-3}{3k-2}$ of the total number of clauses in the formula. The formula is over $2k - 1$ variables, which we denote $x_1, x_2, \ldots, x_{2k-1}$. The formula consists of 5 sets of clauses:

$$S_1 = \{\bar{x}_1 \vee \bar{x}_2 \vee \ldots \vee \bar{x}_k\},$$
$$S_2 = \bigcup_{i=1}^k \{x_i \vee \bigvee_{j=i+1}^{k+i-1} \bar{x}_j\},$$
$$S_3 = \bigcup_{i=k+1}^{2k-2} \{x_i \vee \bigvee_{j=i+1}^{2k-1} \bar{x}_j \vee \bigvee_{j=1}^{i-k} x_j\},$$
$$S_4 = \bigcup_{i=1}^{k-2} \{\bar{x}_i \vee \bigvee_{j=i+1}^{i+k-1} x_j\},$$
$$S_5 = \{x_1 \vee x_2 \vee \ldots \vee x_{k-1} \vee x_{2k-1}\}.$$

The adversary chooses the initial truth assignment to be all-variables-true. In case of a tie, the adversary chooses a variable to flip. Initially, one clause from S_1

is unsatisfied, and all the other clauses in the formula are satisfied. During the next $n = 2k - 1$ steps, TS does not improve upon the initial truth assignment. To show this, we trace the execution of tabu search step by step.

In Table 1, allowed variables to flip are the variables that occur in unsatisfied clauses, but not in the taboo list. The chosen variable is the one chosen by the adversary. It is straightforward to verify that at each step exactly one clause is unsatisfied. Moreover, flipping any of the allowed variables does not change this condition, so the aspiration conditions never hold. During the execution of tabu search, the truth values of variables will be flipped in order $x_1, x_2, \ldots, x_{2k-1}$. In general, S_1 contains an initially unsatisfied clause. Clauses from S_2 are required for flipping truth assignments of variables $x_1, \ldots, x_k$. Clauses from S_3 are required for flipping truth values of variables $x_{k+1}, \ldots, x_{2k-1}$. Clauses from S_4 guarantee that aspiration conditions are never met. Finally, the clause from S_5 is unsatisfied after the execution of tabu search.

Table 1. Execution of tabu search step by step

Step No	Allowed variables to flip	Chosen variable	Taboo list
0	$x_1, x_2, \ldots, x_k$	x_1	$\emptyset$
$1 \leq i \leq k$	$x_{i+1}, x_{i+2}, \ldots, x_{i+k-1}$	x_{i+1}	$\{x_1, \ldots, x_i\}$
$k + 1 \leq i \leq 2k - 2$	$x_{i+1}, x_{i+2}, \ldots, x_{2k-1}$	x_{i+1}	$\{x_1, \ldots, x_k, \ldots, x_i\}$

This proves the theorem, since TS stops after n steps, the formula contains $3k - 2$ clauses, and an optimal truth assignment satisfies all the clauses. For example, the truth assignment that assigns value true to x_1 and false to all the other variables is an optimal one, as can be readily checked.

5   Experimental Results

In addition to the three deterministic and two randomized algorithms from section 2, we consider tabu search first initialized with a truth assignment found by non-oblivious local search[1]. We consider a system-independent definition of the running time of a local search algorithm, namely simply counting the number of variable flips. The complexity of our determinstic algorithms is identified when a local optimum is reached. In contrast, for simulated annealing (SA), the complexity is bounded by the setting of parameters, and for MaxWalkSat (MWS), the stopping time is determined by an ad-hoc limit on the number of flips. All algorithms will immediately terminate if a satisfying assignment is found.

---

[1] We excluded experimental results of a simple greedy algorithm based on de randomizing the naive randomized method since the greedy algorithm did not compare favorably to any of the other methods. Furthermore initializing other methods using this greedy algorithm did not substantially improve performance.

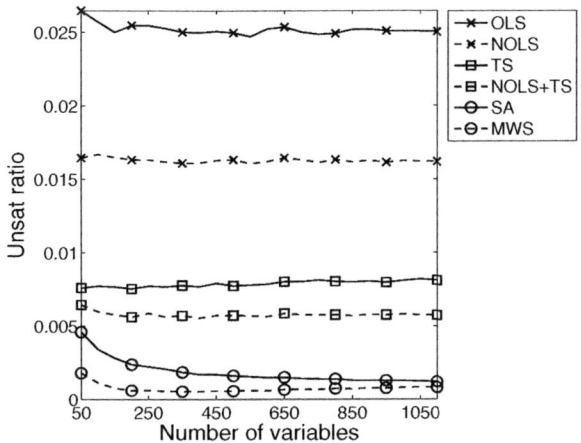

Fig. 1. Average performance on random instances of exact MAX-3-SAT

The relative performance of all algorithms is evaluated with respect to both benchmarks and random exact MAX-k-SAT instances[2]. We first compare the performance of algorithms for random instances of the unweighted MAX-3-SAT problem. The number of variables in a formula varies from 50 to 1100 in increments of 50. The number of clauses is always chosen to be slightly above the conjectured phase transition for MAX-3-SAT, i.e. $m = 4.25n$. For a given n, the performance of each algorithm is averaged over 500 trials. Each trial is an execution of the algorithm on a random formula starting at the all-variables-false truth assignment. For large n, it is not feasible to compute the exact solution and the true approximation ratio. Instead, we calculate the "totality ratio" of the satisfied clauses by a given algorithm to the total number of clauses. This is only a lower bound on the approximation ratio. Given our choice of clause density around the phase transition, the formulas are "almost" satisfiable, so the true maximum is around m, and the computed bound is a good estimate of the true approximation ratio. All algorithms are executed on the same formulas with the same initial truth assignment which allows for a relative comparison of performance. Given that the totality ratios are close to 1, we compare performance in terms *unsat ratio*, the "unsatisfiability ratio", defined as the ratio of the number of unsatisfied clauses to the total number of clauses.

Figure 1 presents the performance results for random MAX-3-SAT instances. All the techniques are clearly separated from each other in terms of their performance. The behavior of non-oblivious local search (NOLS) and its oblivious

---

[2] We generate a random formula with m clauses over n variables as follows. For each of the clauses, we select k out of n variables uniformly at random without replacement, and negate each of them with probability $1/2$. A more recent set of benchmarks called "maxsat_crafted" and additional experimental results can be found at http://people.cs.uchicago.edu/\char126\relaxpankratov/maxsat.html.

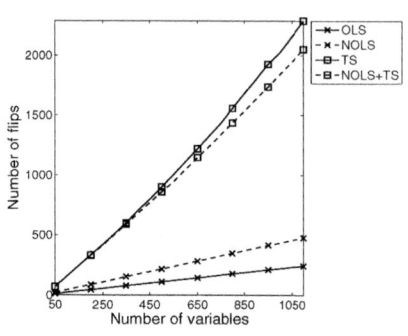
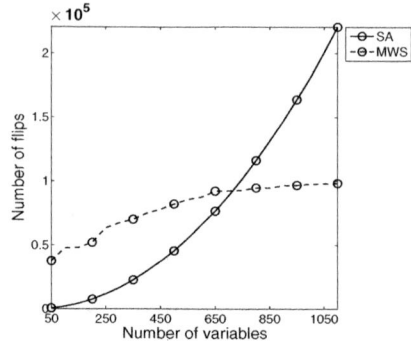

(a) Average completion time for the four determinstic algorithms.

(b) Average completion time for the two stochastic algorithms.

Fig. 2. Average completion time for executing on random instances of MAX-3-SAT

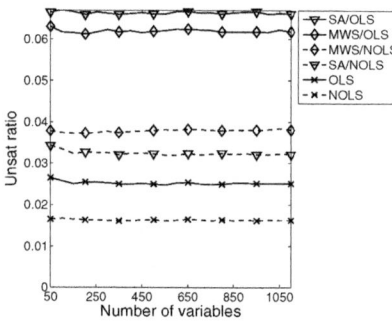
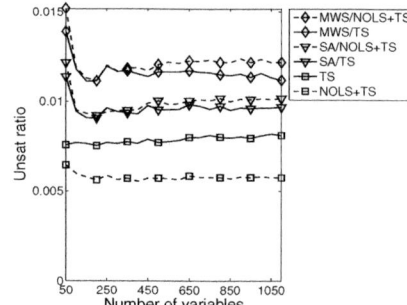

(a) Average performance when MWS and SA are normalized with respect to OLS and NOLS

(b) Average performance when MWS and SA are normalized with respect to TS and NOLS+TS

Fig. 3. In this experiment, algorithms run for exactly the same number of flips as the specified deterministic algorithm

counterpart (OLS) matches their relative standings in the worst-case scenario. However, in spite of a weaker worst-case guarantee, tabu search (TS) beats NOLS very comfortably. In addition, if TS is initialized with a truth assignment found by NOLS, the resulting hybrid method (NOLS+TS) outperforms plain tabu search. Simulated annealing and MaxWalkSat are the overall leaders and they get very close (on average) to the optimal 0 unsat ratio. The fact that for SA and MSW the unsat ratio is high for small n is due to the relatively small number of total clauses. For $n \geq 150$, the unsat ratio for MWS is at most .00082. As we will see in Figures 2 and 3 the better performance of the SA and MSW algorithms comes at a greater computational cost.

Techniques providing better results tend to require more time. An exception is the hybrid algorithm, which finds better truth assignments than regular TS

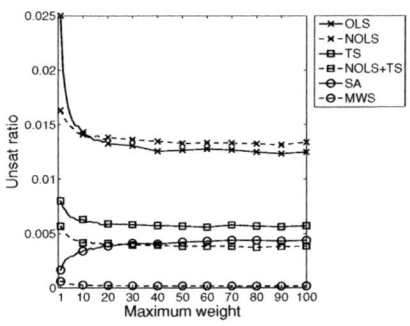

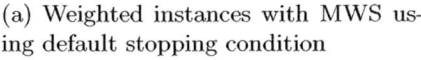

(a) Weighted instances with MWS using default stopping condition

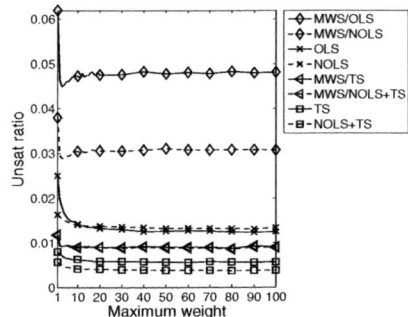

(b) Weighted instances with normalized MWS matching deterministic complexity

Fig. 4. Average performance when executing on random weighted instances of exact MAX-3-SAT

and for large n uses slightly fewer computations. The running time for all our determinstic techniques scales well with the size of the formula. The running time of simulated annealing for the given temperature schedule blows up dramatically, and MaxWalkSat was given a fixed stopping time of 100,000 flips. The average running time of MWS is less than 100,000 flips for small n values, because the method obtains a satisfying assignment for many instances. Figure 3 depicts the normalized performance of algorithms relative to the four deterministic methods. The notation "A/B" means that algorithm A is terminated when it uses the number of flips made by B. The normalized results indicate that NOLS and the hybrid method are efficient choices when only a "good" approximation is needed.

For weighted MAX-3-SAT experiment, we fix $n = 500$ and $m = 4.25n$. A random formula is generated, and for each clause an integer weight value is chosen uniformly at random between one and a prescribed maximum weight value. The unsat ratio in figure 4 now refers to the ratio of the weight of unsatisfied clauses to the total weight of all clauses. As before, the performance of each algorithm is averaged over 500 trials. The performance of MWS now becomes dramatically better than the performance of the other algorithms. As explained in section 2.5, MWS is designed to focus on weighted clauses and is successful in this regard. The unsat ratio for MWS is for the most part decreasing as a function of the max weight W having (for example) ratio .000245 at $W = 10$ and ratio .000153 at $W = 100$. The hybrid method becomes the best of the deterministic algorithms and also has the best performance with respect to normalized performance.

The performance of all deterministic methods and MWS improves as the maximum weight attainable in a formula increases from 1 to 30. In contrast, over the same range the performance of SA declines, and NOLS+TS starts to slightly outperform SA, but remains significantly worse than MWS. Another phenomenon concerns relative performances of oblivious and non-oblivious local searches. OLS outperforms NOLS in formulas with large enough weights. As the

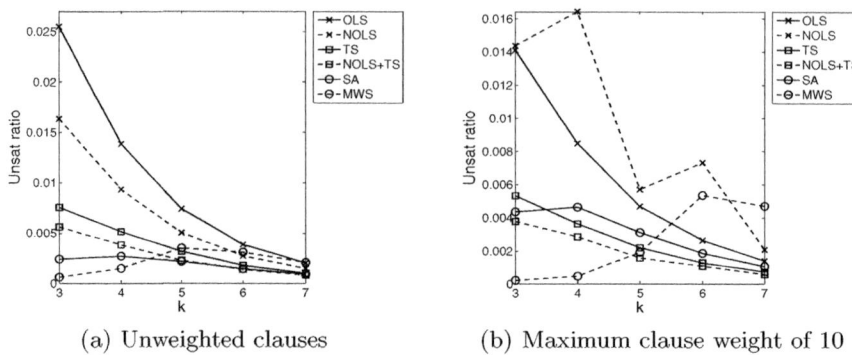

(a) Unweighted clauses   (b) Maximum clause weight of 10

Fig. 5. Average performance when executing on random instances around the threshold of exact k-SAT $(3 \leq k \leq 7)$

weights of clauses grow, the scaling weights used in the potential function have less and less effect, to the point that they hurt the performance of the algorithm.

Next we consider random MAX-k-SAT instances for $k > 3$. We choose the number of clauses to be $m = c_k n$, where c_k is slightly above the estimated threshold [15] for random k-SAT and $n = 200$ is fixed. Achlioptas et al [2] analyze how random 3-SAT instances behave very differently from k-SAT instances around the threshold value as k increases. Figure 5 also demonstrates a dramatic change in the relative performance of algorithms as k increases. For $k \geq 5$, SA and the hybrid method outperform all other methods. As further evidence that the performance of MWS suffers as k increases, in figure 5 we consider weighted instances of MAX-k-SAT with weights chosen uniformly in the range $[1, 10]$. Again in marked contrast to the weighted MAX-3-SAT case, MWS is outperformed at $k \geq 5$, and TS and the hybrid method yield the best performance. The alternating performance (between even and odd k) of NOLS is an interesting phenomena that is overcome when followed by tabu search. In general as k increases, we expect the unsat ratios of the algorithms under consideration to decrease, because even a random assignment in expectation satisfies all but a fraction $1/2^k$ of the clauses in an exact k-SAT formula.

Finally, we consider the relative performance of algorithms on benchmark instances. In contrast to many of the results concerning random instances, MWS does not fare as well as SA or our hybrid algorithm. We ran the six algorithms on the benchmarks from the Second Evaluation of MAX-SAT Solvers (MAX-SAT 2007) and recorded how many times one technique improved over another one. The benchmarks contain instances generated in many different ways. Some are random just like the ones considered in the previous experiments, others were obtained by encoding different problems (for example, MAX-CUT) as an instance of the MAX-SAT problem. In table 2 we see two off-diagonal zeros where one technique is subsumed by the other, namely, OLS is a part of TS, and NOLS is a part of the hybrid algorithm. All the other off-diagonal entries are non-zero. For some instances even OLS, arguably the weakest of the considered

Table 2. MAX-SAT 2007 benchmark results. Total number of instances is 815. The tallies in the table show for how many instances a technique from the column improves over the corresponding technique from the row.

	OLS	NOLS	TS	NOLS+TS	SA	MWS
OLS	0	457	741	744	730	567
NOLS	160	0	720	750	705	504
TS	0	21	0	246	316	205
NOLS+TS	8	0	152	0	259	179
SA	30	50	189	219	0	185
MWS	205	261	453	478	455	0

algorithms, improves over SA and MWS, arguably the strongest of the algorithms for random 3-SAT instances. The hybrid algorithm improves over the NOLS in most instances, which shows the usefulness of the tabu phase. As for the two major rivals, SA and the hybrid algorithm, their performances are similar. The hybrid method improves over OLS, NOLS and MWS slightly more often than does SA, while SA improves over tabu search more often than the hybrid method.

6   Summary and Future Work

Beyond improved performance, the NOBL+TS hybrid method is somewhat faster than TS by itself. Our experiments indicate that no one method dominates in all MAX-SAT instances. For small k, e.g. $k = 3$, MaxWalkSat has the best performance, and especially so in the weighted case. However, this performance comes at a cost of higher computational complexity. As k increases, MWS loses its advantage and is overtaken in performance by all the other methods with the hybrid method and simulated annealing performing best. The performance of the hybrid method and SA is further witnessed in the benchmark experiments.

We conclude with several open questions. A tight bound on the approxima-tion or totality ratio of tabu search is still open. For all local search methods, rather than worst-case approximation (unsat) ratios, it would be insightful to be able to prove expected ratios where the expectation is taken over random initial assignments. A challenging direction is to provide theoretical results correspond-ing to the experiments. For example, what is the expected approximation ratio achieved by any of the deterministic local search based methods under a uniform random model of k-SAT formulas with clause densities near the hypothesized threshold? In particular, for densities above the known algorithmic lower bound [12] can anything be said about the expected MAX-SAT approximation? If the length of the taboo list is infinite, tabu search enters a cycle. What is the ex-pected number of steps that tabu search makes before entering a cycle? Is there a theoretical explanation why NOLS seems to provide such a substantial im-provement when used to initialize tabu search but does not seem to help (for example) MaxWalkSat.

References

1. Aarts, E., Lenstra, J. (eds.): Local Search in Combinatorial Optimization, 2nd edn. Princeton University Press, Princeton (2003)
2. Achlioptas, D., Naor, A., Peres, Y.: Rigorous location of phase transitions in hard optimization problems. Nature 435, 759–764 (2005)
3. Achlioptas, D., Naor, A., Peres, Y.: On the maximum satisfiability of random formulas. JACM 54(2) (2007)
4. Achlioptas, D., Peres, Y.: The threshold for random k-sat is $2^k log2 - o(k)$. JAMS 17(2), 947–973 (2004)
5. Argerlich, J., Li, C.M., Manya, F., Planes, J.: The first and second max-sat evaluations. Journal on Satisfiability, Boolean Modeling and Computation 4, 251–278 (2008)
6. Asano, T., Williamson, D.P.: Improved approximation algorithms for max sat. J. Algorithms 42(1), 173–202 (2002)
7. Feige, U., Goemans, M.: Approximating the value of two power proof systems, with applications to max 2sat and max dicut. In: ISTCS 1995: Proceedings of the 3rd Israel Symposium on the Theory of Computing Systems (ISTCS 1995), Washington, DC, USA, p. 182. IEEE Computer Society, Los Alamitos (1995)
8. Hansen, P., Jaumard, B.: Algorithms for the maximum satisfiability problem. Computing 44(4), 279–303 (1990)
9. Håstad, J.: Some optimal inapproximability results. J. ACM 48(4), 798–859 (2001)
10. Heras, F., Larrosa, J., de Givry, S., Schiex, T.: 2006 and 2007 max-sat evaluations: Contributed instances. JSAT 4(2-4), 239–250 (2008)
11. Johnson, D.S.: Approximation algorithms for combinatorial problems. In: STOC 1973: Proceedings of the fifth annual ACM symposium on Theory of computing, pp. 38–49. ACM, New York (1973)
12. Kaporis, A., Kirousis, L.M., Lalas, E.G.: The probabilistic analysis of a greedy satisfiability algorithm. Random Structures & Algorithms 28(4), 444–480 (2006)
13. Khanna, S., Motwani, R., Sudan, M., Vazirani, U.: On syntactic versus computational views of approximability. SIAM J. Comput. 28(1), 164–191 (1999)
14. Mastrolilli, M., Gambardella, L.M.: Max-2-sat: How good is tabu search in the worst-case? In: AAAI, pp. 173–178 (2004)
15. Mertens, S., Mezard, M., Zecchina, R.: Threshold values for random k-sat from the cavity method. Random Structures & Algorithms 28(3), 340–373 (2006)
16. Pennock, D.M., Stout, Q.F.: Exploiting a theory of phase transitions in three-satisfiability problems. In: Proc. AAAI 1996, pp. 253–258. AAAI Press/MIT Press (1996)
17. Spears, W.M.: Simulated annealing for hard satisfiability problems. In: Workshop, pp. 533–558. American Mathematical Society, Providence (1993)
18. Xu, H., Rutenbar, R.A., Sakallah, K.: sub-sat: a formulation for relaxed boolean satisfiability with applications in routing. In: ISPD 2002: Proceedings of the 2002 international symposium on Physical design, pp. 182–187. ACM, New York (2002)

The Seventh QBF Solvers Evaluation (QBFEVAL'10)[*]

Claudia Peschiera[1], Luca Pulina[1], Armando Tacchella[1],
Uwe Bubeck[2], Oliver Kullmann[3], and Inês Lynce[4]

[1] DIST, University of Genoa – 16145 Genova, Italy
{*name.surname*}@unige.it
[2] Computer Science Inst., University of Paderborn – D33095 Paderborn, Germany
bubeck@upb.de
[3] Computer Science Dept., Swansea University – SA2 8PP Swansea, United Kingdom
O.Kullmann@swansea.ac.uk
[4] INESC-ID/IST, Technical University of Lisbon – 1000-029 Lisbon, Portugal
ines@sat.inesc-id.pt

Abstract. In this paper we report about QBFEVAL'10, the seventh in a series of events established with the aim of assessing the advancements in reasoning about quantified Boolean formulas (QBFs). The paper discusses the results obtained and the experimental setup, from the criteria used to select QBF instances to the evaluation infrastructure. We also discuss the current state-of-the-art in light of past challenges and we envision future research directions that are motivated by the results of QBFEVAL'10.

1 Introduction

For almost a decade now, competitive events in the field of Boolean reasoning have influenced related research agendas and shaped the course of tool developments. Nowadays, organized evaluations are popular for several subfields of Boolean reasoning, including propositional satisfiability (SAT) [1,2], quantified Boolean formula (QBF), and satisfiability modulo theory (SMT) solving [3]. While in all these events the organizers are trying to answer the question "Which solver should I use?" [4] using transparent and fair evaluation methods, researchers and practitioners look at the outcomes as the preferred way to understand the current state of the art in each subfield. Often, it is not just the best solver for the given application which is sought, but also some feeling about what makes a solver suitable for one kind of problem or another. It is therefore important, in any such event, to motivate and explain thoroughly the testing methods which led to the latest results. Whenever possible, the organizers should try to answer the question "Which solver is best, for which task and why?", which is far more relevant than just assessing the winner(s) of the event.

In this spirit, we report about the 2010 evaluation of QBF solvers (QBFEVAL'10), the seventh in a series of events established with the aim of assessing the advancements

---

[*] The authors would like to thank all the participants to the QBF competition and evaluations for submitting their work, and in particular Martina Seidl for her contribution related to the input format of non-prenex non-CNF formulas.

O. Strichman and S. Szeider (Eds.): SAT 2010, LNCS 6175, pp. 237–250, 2010.

Table 1. QBFEVAL'10 at a glance. "# Solvers" and "# Formulas" denote the total amount of solvers and formulas involved in a given track, respectively.

	Track				
	MAIN	2QBF	SH	RND	NPNCNF
# Solvers	11	9	10	9	2
# Formulas	568	200	50	550	478

in QBF reasoning. An important expansion with respect to the last evaluation (QBFE-VAL'08) is the introduction of five different tracks, each with its own rules, solvers and instances. Overall, QBFEVAL'10 received 13 solver submissions from 8 different developing teams. Also, a pool of 136 formulas was submitted about module extraction in description logics [5]. These formulas have been made available both with a prenex CNF encoding and a non-prenex non-CNF encoding.

Table 1 shows QBFEVAL'10 at a glance. The main competition (MAIN in the following) is comprised only of prenex CNF formulas obtained by encoding various automated reasoning tasks into QBF. This track is competitive, i.e., we declare a winner in the end, whereas all the remaining tracks are non-competitive. This is because evaluation of QBFs in prenex CNF is a fairly mature area, faring more than ten years of active research into algorithms, heuristics and optimizations. On the other hand, we preferred to have non-competitive tracks for formulas with a single $\forall\exists$ alternation (2QBF), "small but hard" instances (SH), and random generated formulas (RND), because these tracks are meant to experiment with specific features for which solvers may have (not) been specifically optimized. All of the above tracks feature formulas in prenex CNF, and the intersection between the test sets of these tracks is empty, i.e., there is no formula tested in two different tracks. In the case of non-prenex non-CNF encodings (NPNCNF), since this is only the third time in which non-prenex non-CNF solvers are evaluated, and the research contributions in this area are currently still limited in number, we felt that the topic is not mature enough for a proper competition. We devote part of Section 2 to describe in some detail the five tracks, including solvers and formulas.

As in every event of this kind, it is important to choose carefully the test set(s) in order to get meaningful results while completing the evaluation in a reasonable time. In order to fulfil these two requirements, in the MAIN track we have extracted the final test set by sampling the pool of instances available to us – currently, more than 15000 instances and 551 families – to extract a much smaller, yet representative, test set. In all the remaining tracks we chose the test set according to the topic of the track, and we tried to cover as much as possible the variety of the instances available to us, either by sampling (2QBF, SH) or by controlled generation (RND). In the NPNCNF track, we considered all the instances available to us. Finally, we ran the solvers on a farm of identical PCs, imposing different resource bounds according to the track. Section 2 is chiefly devoted to the description of the above selection methods and the computing infrastructure.

We present an analysis of QBFEVAL'10 data in two stages. In the first stage – Section 3 – we consider all the solvers and the instances in each track to give a rough, but complete, picture of the state of the art in QBF. By analyzing the results for problems and discrepancies among the solvers, we were able to isolate some solvers which turned out to be problematic, and we have removed them from subsequent analysis. In the second stage – Section 4 – we analyze the results with the aim of understanding the relative strengths and weaknesses of the various solvers, as well as the reasons why some formulas turn to be harder than others. Here we try to extract a narrow, but crisp, picture of the current state of the art. We also try to put things in perspective, by comparing the current results with past challenges [6] and envisioning possible future developments. For the MAIN track, we also provide the final ranking of the solvers according to the YASMv2 scoring method [7]. Finally, in Section 5 we conclude with an analysis of the evaluation and suggestions for future events of this kind.

2 Setup: Solvers, Instances and Infrastructure

Table 2 summarizes the solvers submitted to QBFEVAL'10. The salient features of the participants are briefly described in the following. The input format description is assumed to be prenex CNF format, unless otherwise specified.

AIGSOLVE [8] uses And-Inverter Graphs (AIGs) as the main data structure, and AIG-based operations to reason about the input formula. The solver includes preliminary phases devoted to simplification, structure extraction and early quantification of the input formula.

AQME'10 [9] is a multi-engine solver, i.e., a tool using Machine Learning techniques to select among its reasoning engines the one which is more likely to yield optimal results. The reasoning engines of AQME'10 are a subset of those submitted to QBFEVAL'06, namely 2CLSQ, QUANTOR2.11, QUBE3.1, SKIZZO, and SSOLVE. Engine selection is performed according to the adaptive strategy described in [9].

CIRQIT2.1 [10] is a solver for non-prenex non-CNF formulas using a circuit-based data structure to represent and reason about the input formula. Search is performed on the internal representation using unit and Don't Care propagation. It performs term- and clause-learning, and it leverages a variable-state independent heuristic.

DEPQBF is a search-based solver leveraging compact dependency graphs to represent the prefix, instead of the standard linear representation – see [11]. The difference of DEPQBF-PRE from the basic version is the use of QUANTOR3.1 as a preprocessor.

Table 2. The QBFEVAL'10 systems. The table is structured as follows. The first column ("Solver") reports the name of the solver, the second column ("Track") indicates the track in which the given solver is involved, while the last column ("Author(s)") reports solvers' authors.

Solver	Track	Author(s)
AIGSOLVE	MAIN, SH	F. Pigorsch, C. Scholl
AQME'10	MAIN, 2QBF, SH, RND	L. Pulina, A. Tacchella
CIRQIT2.1	NPNCNF	A. Goultiaeva, V. Iverson, F. Bacchus
DEPQBF	MAIN, 2QBF, SH, RND	F. Lonsing
DEPQBF-PRE	MAIN, 2QBF, SH, RND	F. Lonsing
NENOFEX	MAIN, 2QBF, SH, RND	F. Lonsing, A. Biere
QMAIGA	MAIN	S. Reimer, F. Pigorsch, M. Lewis, B. Becker, C. Scholl
QPRO	NPNCNF	U. Egly, M. Seidl, S. Woltran
QUANTOR3.1	MAIN, 2QBF, SH, RND	A. Biere
QUBE7	MAIN, 2QBF, SH, RND	E. Giunchiglia, P. Marin, M. Narizzano
QUBE7-C	MAIN, 2QBF, SH, RND	E. Giunchiglia, P. Marin, M. Narizzano
QUBE7-M	MAIN, 2QBF, SH, RND	E. Giunchiglia, P. Marin, M. Narizzano
STRUQS'10	MAIN, 2QBF, SH, RND	L. Pulina, A. Tacchella

NENOFEX [12] is an expansion-based solver which operates on negation normal form (NNF) formulas. NNF formulas are represented as structurally restricted trees, and expansions are scheduled based on expansion cost estimates.

QMAIGA merges two "orthogonal" approaches. Its core is AIGSOLVE, but when AIGSOLVE is stuck in a sub-problem, the search-based solver QMiraXT [13] takes over the entire solution process.

QPRO [14] is a search-based solver for non-prenex non-CNF formulas implementing dependency-directed backtracking.

QUANTOR3.1 [15] is based on variable elimination and expansion, plus a number of features, such as equivalence reasoning, subsumption checking, pure literal detection, unit propagation, and also a scheduler for the elimination step.

QUBE7 is based on the composition of two reasoning tools: the preprocessor sQueezeBF [16], combining various techniques for reducing the size of the input QBF, and the search-based solver QUBE3.1 [17].

STRUQS'10 [18] main feature is a dynamic combination of search – with solution- and conflict-backjumping – and variable-elimination. The key point in this approach is to implicitly leverage graph abstractions of QBFs to yield structural features which support an effective decision between search and variable elimination.

Further details about the solvers can be found in the short papers submitted by their authors and made available through the QBFEVAL'10 website [19]. Concerning the formulas, there was a single submission by Roman Kontchakov. The "Kontchakov" suite, as it is named in the following, consists of 136 formulas obtained by encoding minimal query inseparability module extraction in DL-Lite [5]. The formulas have been made available both with a prenex CNF encoding and their corresponding non-prenex non-CNF encoding. All the remaining formulas used in the evaluation have been extracted from QBFLIB [20]. These include the "Wintersteiger" suite – 372 encodings concerning ranking function synthesis problems [21] – submitted after QBFEVAL'08, and thus not extensively tested so far. The complete test set used in QBFEVAL'10 can be downloaded from the QBFEVAL'10 website [19]. In the following, we describe in some detail the choice of the test sets for each of the five tracks. Tables 6 and 7 in Section 4 show further details about the selected formulas.

In order to construct a meaningful comparison between solvers, we constructed the test set for the MAIN track according to the following guidelines:

1. Balance the mix of different kinds of formulas, considering both syntactic features, e.g., number of variables and quantifier alternations, and structural features, e.g., density of the associated variable dependency graph.
2. Balance the empirical hardness of formulas so that the whole spectrum between easy and hard formulas is covered, where we consider QBFEVAL'08 results as a yardstick to assess formula hardness.
3. Balance between true and false formulas, again considering QBFEVAL'08 results.
4. Balanced the mix of formulas coming from different application domains, so that no application domain is neglected or overrepresented.
5. Finally, ensure that no more than 10% of the total test set comes from a single submitter whenever the formula submitter is also authoring a competing solver.

To satisfy requirement (1) above, we start with a pool of 2734 formulas extracted from QBFLIB, i.e., all the publicly available prenex CNF QBFs which have not been generated using some probabilistic model, excluding the ones eligible for 2QBF and SH tracks (see below). For each formula, we compute the same set of syntactic features used in AQME'10 [9] to automate engine selection, plus an approximation of quantified treewidth as described in [22]. Then, we extract a representative subset of the original pool by (i) clustering the formulas according to their features, and (ii) random sampling without repetitions the members of each cluster. The result of the above procedure is a pool of 432 formulas to which we add formulas from the "Kontchakov" suite to yield a final test set of 568 formulas. It turns out that such test set satisfies also requirements (2-5) above. In particular, the hardness mix turns out to be balanced – at least considering the formulas which were tested in QBFEVAL'08. If we consider requirement (3), we can see that the final selection consists of 166 and 170 formulas which are known to be true and false, respectively. Among the 232 formulas whose truth value has not been established, we have newly submitted formulas, hard formulas from QBFEVAL'08 (48), and formulas that were not used in the previous evaluation (48). As for requirement (4), we notice that the final test set is composed by 197 instances of formal verification problems, 101 instances of planning problems, 136 instances of module extraction problems and 134 instances of miscellaneous problems, so there is already some balance among application domains. In the following, for the sake of simplicity, we will consider module extraction problems as part of the miscellaneous category. Finally, also requirement (5) is implicitly satisfied by our selection of formulas.

2QBFs, i.e., formulas with a single $\forall\exists$ alternation in the prefix, arise frequently in applications such as conformant planning, symbolic diameter calculation of finite state machines and, more in general, problems having Σ_2^p complexity. If we consider the pool of prenex CNF formulas that we used to extract the test set for the MAIN track, we see that about 40% of them are 2QBFs. Clearly, any general-purpose QBF solver can deal with 2QBFs, but they are also appealing for special-purpose approaches, e.g., cooperation of two state-of-the-art SAT solvers or encoding to other logics, like, e.g., Disjunctive Logic Programming. Moreover, it could be the case that even general-purpose solvers behave differently on 2QBFs with respect to the unbound-alternation case. Indeed, no formulas were submitted specifically for this track. Therefore, we selected 200 formulas from QBFLIB, trying to meet the same requirements mentioned previously for the MAIN track. Even if we followed the same procedure outline above, we did not manage to get an overall balance between the formulas. This is because two suites – "Basler" and "Wintersteiger", both made up by encoding of formal verification problems – are numerically dominating the category. However, we know from QBFEVAL'08 that most of the formulas in the suite "Basler" were solved by either 1 or 2 solvers, so they represent a reasonably difficult test set. The suite "Wintersteiger" is interesting in its own right, since it was submitted after QBFEVAL'08 and it has never been evaluated extensively.

Because solving QBFs is a hard combinatorial problem, we expect that while heuristic-based solvers can deal with QBFs of increasing "size", there will always be "small" instances that turn out to be extremely hard to solve in practice. The focus of this track is precisely on "small but hard" instances, i.e., relatively small QBFs that resisted solution attempts in previous QBF evaluations. Clearly, a key factor in selecting these formulas is deciding on a "size" parameter. The formulas composing the test set for the SH track have been picked up from QBFLIB according to the following procedure. We focus on formulas that no participant was able to solve in QBFEVAL'08. We rank such formulas in ascending order according to (i) number of variables, (ii) number of clauses and (iii) total number of literals. We consider only the first 100 entries in each ranking, and then each formula is scored using the Borda method. The position of the formula in each of the rankings above is considered as a preference expressed by a voter, and the individual preferences are added up to yield the final score. The 50 highest ranking formulas are selected in the end.

As for the RND track, since no new formulas/generators were submitted specifically for this track, we composed the test set considering QBFEVAL'05 and '06 experience – the last two events in which were ran random formulas. In particular, we selected 55 instances with ten samples each, obtaining a total amount of 550 formulas. Our selection considers both generators – the "Chen-Interian" probabilistic model [23] – and families of formulas available in QBFLIB. In the latter case, we made a selection from NestedCounterFactual category, Miscellanea category (family "ASP_Program_Inclusion"), and Planning category (suite "Narizzano" and family "Strategic_Companies"). Finally, the test set of the NPNCNF is composed by 342 formulas already available in QBFLIB at the time of the submission, together with 132 newly submitted ones – the suite "Kontchakov" mentioned above.

In the MAIN track, the only competitive one, formulas have been preprocessed using a satisfiability-preserving shuffling of the variables in the prefix, literals in the clauses, and clauses in the matrix, respectively. For each track, the CPU time limit is set to 1200 seconds for all the tracks, except the SH track where we allowed 43200 seconds (12 hours) of runtime. To prevent memory swapping, we also set a memory limit at 2GB. If a solver exceeds the resource bounds while attempting to solve a formula, it is killed and the corresponding result is left undefined. In the MAIN track, we rank the solvers using the YASMv2 [7] in order to declare the winner. As shown in [7], YASMv2 is more robust than other common scoring methods, including simple criteria based on the number of problems solved and the CPU time spent solving them. The evaluation runs on a farm of 9 identical PCs locally available at the University of Genoa. The PCs are equipped with a processor Intel Core2Duo running at 2.13 GHz, with 4 GB of RAM, and running GNU Linux Debian 2.6.18.5.

3   Evaluation: First Stage Results

Table 3 presents the raw results of the evaluation, considering all the prenex CNF solvers and the four tracks where prenex CNF QBFs were tested. In the following, when we say that "solver A *dominates* solver B" we mean that the set of problems solved by B is a subset of those solved by A. Looking at the results, we can see that in the MAIN track all the competitors were able to solve at least 33% of the test set. On the other hand, NENOFEX, QUANTOR3.1 and STRUQS'10 were not able to solve more than 50% of the instances. In terms of number of problems solved, the best solver is AQME'10, which can solve about 76% of the test set. This is interesting if we consider that AQME'10 combines solvers that used to be state-of-the-art four years ago. However, we can also see that some solvers that do not exploit a multi-engine paradigm like, e.g., QUBE7, DEPQBF and QMAIGA are getting close to the performances of AQME'10.

Table 3. First stage results for all prenex CNF solvers. For each track, we report the number of formulas solved within the time limit ("#") and the total CPU time ("Time") spent on the solved instances. Results in boldface are those of the best three solvers in each track – according to number of problems solved and total time only. NA means that the solver did not participate in a track. Solvers marked with an asterisk are those whose output gave rise to discrepancies.

Solver	MAIN		2QBF		SH		RND	
	#	Time	#	Time	#	Time	#	Time
AIGSOLVE	329	22786.60	NA	NA	**37**	1140.01	NA	NA
AQME'10	**434**	33346.60	128	2323.11	11	30132.40	**407**	20078.90
DEPQBF	370	21515.30	24	690.42	4	41448.00	342	12895.10
DEPQBF-PRE	356	18995.90	51	877.02	4	33371.90	**343**	9438.62
NENOFEX	225	13786.90	50	3545.65	3	30194.20	149	34502.80
QMAIGA	361	43058.10	NA	NA	NA	NA	NA	NA
QUANTOR3.1	205	6711.37	48	3689.30	5	57960.90	134	2830.97
QUBE7*	**410**	52142.10	**173**	1981.88	9	80570.70	**359**	27092.90
QUBE7-C*	389	34926.80	**172**	3340.89	**16**	27207.10	339	29495.20
QUBE7-M*	**393**	40786.30	**171**	4109.16	**16**	21458.50	340	29393.30
STRUQS'10	240	32839.70	132	1399.30	5	26257.30	117	15480.40

This indicates a clear progress in the field over the past four years. The performance of the solvers is also quite diverse: there are 229 instances – 40% of the test set – separating the strongest solver, from the weakest one.

Still looking at Table 3, in the track 2QBF we can see that QUBE7, QUBE7-M and QUBE7-C are able to solve more than 85% of the test set, while STRUQS'10 and AQME'10 both solve slightly more than 60% of the test set. On the other hand, all the remaining solvers cannot do better than a mere 25%. This indicates that given the current state of the art in QBF reasoning, the performance demand of these encodings is still exceeding the capabilities of most solvers. As for the SH track, we see that AIG-SOLVE was able to solve a considerable number of previously open problems, whereas all the remaining solvers did not perform particularly well. On one hand, this indicates that AIGSOLVE is particularly suited for this kind of problems – probably due to its internals which are fairly different from all the other solvers that participated in the track. On the other, it also shows a clear progress over previous QBFEVAL events. Looking at the results on random instances, we can see that all the solvers, with the only exception of STRUQS'10 and QUANTOR3.1, were able to conquer at least 25% of the instances in this category, and six solvers were able to conquer more than 50% of the instances. Only AQME'10 is able to solve 74% of the instances. Overall it seems that the choice of parameters for the generation of random instances yielded a performance demand within the capabilities of most solvers. Considering the contestants which solved more than 50% of the test set, but excluding the multi-engine solver, we see that the performance of the solvers is similar: only 19 instances separate the worst solver (QUBE7-M) from the best (QUBE7). Finally, if we consider the NPNCNF results – not shown in Table 3 – we have that CIRQIT2.1 solved 291 instances and QPRO solved 153 instances, i.e., 61% and 32% of the test set, respectively. Even if CIRQIT2.1 solves much more instances than QPRO, it does not dominate it, because QPRO is able to solve 30 instances uniquely.

As we have anticipated in Section 1, a few discrepancies in the results of the solvers were detected during the analysis of the first stage results. A total of 3 discrepancies were detected in the MAIN track, and 4 discrepancies in the SH track. For each of the discrepancies we reran the solvers reporting a result different from the majority of the other solvers and/or the expected result of the instance. We also inspected the instances, looking for weird syntax and other pitfalls that may lead a correct solver to report an incorrect result. In particular, the instances which gave rise to discrepancies belong to three different suites: "Letombe", "Katz" and "Mneimneh-Sakallah". We ran on such instances state-of-the-art QBF certifiers in order to get a definite answer on them. The only instance for which we obtained a push-button result is par8-4-90 (suite "Letombe"), certified by QBD [24] as FALSE. By performing some manually-assisted reasoning, we found out that test3_quant2 is FALSE as well. At the end of this analysis we excluded from the second stage the following solvers: QuBE7, QuBE7-M and QuBE7-C, responsible for all discrepancies detected, and answering TRUE on the formulas above. Clearly, for instances that were conquered by just one solver, and for which we do not know the satisfiability status in advance, the possibility that the solver is wrong still exists, but we consider this as unavoidable given the current state of the art.

4 Evaluation: Second Stage Results

Table 4 shows the results obtained computing YASMv2 on the competitive track. Look-ing at the table, we declare DEPQBF as the winner of QBFEVAL'10, followed by DEPQBF-PRE, and AQME'10. Notice that, even if AQME'10 is able to solve more prob-lems than both DEPQBF and DEPQBF-PRE, the setup time of the engine-selection strat-egy may considerably increase the runtime in most short-to-solve instances, and YASMv2 is designed to penalize this behavior. In the remainder of the section, we look in detail at each track, considering both solver- and instance-centric views.

Solver-centric view. Table 5 shows the results of the MAIN track dividing the formulas into three categories. As we can see, in terms of number of problems solved, AQME'10 is the strongest solver: it leads the count in Formal Verification and Planning, and it is second best in Miscellanea category. Overall, the strongest solver in Formal Verifica-tion category is able to solve 68% of the total number of instances, while DEPQBF-PRE is able to solve about 82% of the Miscellanea category, and AQME'10 solves 80% of instances in Planning category. Focusing on Formal Verification category, 68 instances separate the strongest solver from the weakest one. If we consider the problems that are

Table 4. Final ranking of QBFEVAL'10. The first column ("Solver") reports the solvers partic-ipating in MAIN, while the second one ("Points") is filled with the points computed by using YASMv2.

Solver	Points		Solver	Points
DEPQBF	2896.68		AIGSOLVE	2037.22
DEPQBF-PRE	2508.96		QUANTOR3.1	1235.14
AQME'10	2467.96		STRUQS'10	947.83
QMAIGA	2117.55		NENOFEX	829.11

Table 5. MAIN track second stage results. "Category" reports the application domain, and for each solver, the table shows the number of instances solved ("#") and the total CPU time spent to solve them ("Time"). Total number of formulas solved ("Total") is also split into true, false, and uniquely solved formulas ("True", "False" and "Unique", respectively). A dash means that a solver did not solve any instance in the related group. Solvers are sorted according to the number of instances solved, and, in case of a tie, according to CPU time.

Category	Solver	Total		True		False		Unique	
		#	Time	#	Time	#	Time	#	Time
Formal Verification (197)	AQME'10	134	9056.00	49	3938.94	85	5117.06	3	639.97
	QMAIGA	130	10611.28	55	5374.21	75	5237.07	–	–
	AIGSOLVE	123	7129.58	56	4189.91	67	2939.67	2	1006.30
	STRUQS'10	92	16501.98	39	8722.64	53	7779.34	6	2205.45
	DEPQBF	89	4109.93	17	959.61	72	3150.31	5	505.70
	QUANTOR3.1	73	4236.95	31	3096.88	42	1140.07	–	–
	NENOFEX	70	7928.03	27	4744.43	43	3183.60	3	350.89
	DEPQBF-PRE	65	4195.32	20	1032.27	45	3163.05	–	–
Miscellanea (270)	DEPQBF-PRE	222	12179.63	117	10735.33	105	1444.31	–	–
	AQME'10	219	20210.17	95	11698.52	124	8511.64	1	5.22
	DEPQBF	208	14063.50	110	12234.42	98	1829.08	–	–
	QMAIGA	162	27280.91	89	13087.86	73	14193.05	1	859.00
	AIGSOLVE	145	13569.84	81	6386.08	64	7183.76	–	–
	STRUQS'10	90	11597.60	39	3751.63	51	7845.97	–	–
	NENOFEX	88	4786.58	45	2985.23	43	1801.35	–	–
	QUANTOR3.1	62	1087.27	31	678.05	31	409.22	–	–
Planning (101)	AQME'10	81	2824.92	40	188.10	41	2636.81	2	892.65
	DEPQBF	73	3341.86	37	577.74	36	2764.12	1	434.81
	QUANTOR3.1	70	1387.15	38	355.70	32	1031.46	1	585.96
	DEPQBF-PRE	69	2620.92	35	686.16	34	1934.76	–	–
	QMAIGA	69	5165.87	36	2234.56	33	2931.31	–	–
	NENOFEX	67	1072.33	37	512.21	30	560.12	–	–
	AIGSOLVE	61	2087.13	34	1515.49	27	571.64	–	–
	STRUQS'10	58	4740.17	31	1331.23	27	3408.94	–	–

uniquely solved, and the five best solvers, then we see that no solver is dominated by the others, with the noticeable exception of QMAIGA (dominated by AQME'10). Finally, we report that DEPQBF yields the smallest average CPU time (about 46s). In the Miscellanea category, the first thing to be observed is that the strongest solver is DEPQBF-PRE, which ranks last in the Formal Verification category. The first three solvers are pretty much in the same capability ballpark: only 14 instances separate the first one from the third one. Finally, considering the Planning category, we can see that only 23 instances separate the strongest solver from the weakest one. We also report that QUANTOR3.1 seems to perform better in the Planning category with respect to the other ones.

Considering now the 2QBF[1] track, we have that the test set is mostly composed by Formal Verification formulas (177 out of 200). STRUQS'10 turns out to be the strongest solver in this category in the 2QBF track, with 126 instances solved (71% of the test set). AQME'10 trails with 111 instances solved, and NENOFEX is third best with 44 instances solved. STRUQS'10 is very close to dominate all remaining solvers: it uniquely solves 20 formulas, but both AQME'10 and NENOFEX are also able to uniquely solve 1 formula. If we compare this result with the main track, where AQME'10, QMAIGA and AIGSOLVE are stronger than STRUQS'10, then we may conjecture that this is due to

---

[1] For the lack of space, we do not show tables about the non-competitive tracks, and we made them available on-line at [19].

STRUQS'10 becoming less effective when the number of alternations increases. As for the remaining formulas in the 2QBF track, we have that STRUQS'10, QUANTOR3.1, AQME'10, and NENOFEX were able to solve all the 6 instances in the "Miscellanea" category, whereas DEPQBF-PRE is able to solve all 17 formulas in the "Planning" category, and it also dominates all the other solvers. In particular, QUANTOR3.1, NENOFEX, and STRUQS'10 were not able to solve any formula in this category. These formulas are all sortnetsortXX.AE (Family "Sorting Network", suite "Rintanen"), which, according to previous QBFEVAL results, are best solved by search-based solvers. Also in the SH track Formal Verification formulas are prevalent (42 out of 50). The main result of this track is that AIGSOLVE is able to globally solve 37 open problems of QBFE-VAL'08. In particular, 35 out of 37 fall in the Formal Verification category, mainly belonging in "Biere" and "Katz" suite.

Looking at the RND track, and focusing on the Chen-Interian family, we report that AQME'10 is the strongest solver, and it is able to solve about 60% of the test set. It also dominates all remaining solvers. It is followed by a group of three solvers, namely DEPQBF-PRE, DEPQBF, and STRUQS'10, which are able to solve the same set of 100 formulas. The remaining solvers are not able to deal with more than 25% of the test set. There are also 61 formulas uniquely solved by AQME'10 (by using SSOLVE), which are the ones supposedly close to the phase transition in the random generation model. No other solver is able to return a solution in this range of formulas. Concerning other random formulas, we report that in the family "ASP_Program_Inclusion", DEPQBF-PRE, AQME'10 (mainly using QUBE3.1), and DEPQBFare able to solve 100% of the test set, while the remaining three are not able to solve any formula. We also report that in the family "NestedCounterFactual" the strongest solver is DEPQBF-PRE, able to cope with all the test set. The other solvers able to solve at least 50% of the test set are DEPQBF, which solves 79 out of 80 instances, and AQME'10, with 58 out of 80 solved instances. Finally, regarding the Planning category, the picture is very close to the one described for the Chen-Interian category. The strongest solver is AQME'10, coping with 92% of the dataset, followed by DEPQBF-PRE and DEPQBF, having solved both 77% of the dataset. In this case, AQME'10 also dominates all the remaining solvers. The noticeable difference with "Chen-Interian" family is that now STRUQS'10 ranks last, solving only 15 formulas.

Concluding, in the NPNCNF track there are two categories of formulas, namely "Formal Verification", composed by 342 out of 478 instances, and "Miscellanea", composed by the suite Kontchakov. For both categories, the strongest solver is CIRQIT2.1. Concerning "Formal Verification", CIRQIT2.1 solves 205 formulas, against 72 solved by QPRO, which is also dominated by CIRQIT2.1. The picture slightly changes in the case of "Miscellanea" because, even if CIRQIT2.1 is still the strongest solver (86 solved instances), being the gap with QPRO of only 5 formulas. Finally, for the Miscellanea category we also report that CIRQIT2.1 and QPRO uniquely solve 35 and 30 formulas, respectively. Therefore, no solver dominates the other in this category.

Instance-centric view. In Table 6 we show the classification of formulas included in MAIN according to the solvers admitted to the second phase. In the table, the number of instances solved and the cumulative time taken for each family is computed considering the "SOTA solver", i.e., the ideal solver that always fares the best time among

Table 6. Classification of MAIN formulas considering second stage data. The table consists of seven columns where for each family of instances we report the name of the family in alphabetical order (column "Family"), the number of instances included in the family, and the number of instances solved (group "Overall", columns "N", "#", respectively), the CPU time taken to solve the instances (column "Time"), the number of easy, medium and medium-hard instances (group "Hardness", columns "EA", "ME", "MH").

Family	Overall		Time	Hardness			Family	Overall		Time	Hardness		
	N	#		EA	ME	MH		N	#		EA	ME	MH
Abduction	52	50	25.19	14	35	1	k_grz_p	3	3	3.16	1	2	–
Adder	15	15	568.79	–	12	3	k_lin_n	5	5	283.95	–	5	–
blackbox-01X-QBF	59	53	807.66	8	39	6	k_lin_p	4	4	0.40	2	2	–
blackbox_design	2	2	82.35	–	1	1	k_path_n	3	3	0.12	–	3	–
Blocks	5	5	16.05	1	4	–	k_path_p	4	4	0.14	–	4	–
BMC	18	17	64.16	7	10	–	k_ph_n	6	6	7.86	6	–	–
C432	4	4	0.52	1	3	–	k_ph_p	4	2	6.11	2	–	–
C499	2	2	0.90	–	2	–	k_poly_n	4	4	0.09	–	4	–
C5315	7	3	3.80	1	2	–	k_poly_p	2	2	0.06	–	2	–
C6288	4	2	21.09	–	1	1	k_t4p_n	4	4	5.62	–	4	–
C880	1	1	0.20	–	1	–	k_t4p_p	5	5	3.01	–	5	–
Chain	1	1	0.02	–	1	–	Logn	1	1	1.12	–	1	–
circuits	3	3	18.00	1	2	–	mqm	136	136	7953.95	–	136	–
comp	2	2	0.03	2	–	–	s1196	1	–	–	–	–	–
conformant_planning	15	10	1042.32	2	7	1	s1269	1	–	–	–	–	–
Connect4	11	9	125.77	6	3	–	s27	1	1	0.02	–	1	–
Counter	4	4	937.36	–	3	1	s298	4	4	158.09	–	4	–
Debug	5	4	462.59	–	4	–	s3330	2	–	–	–	–	–
evader-pursuer-4x4-logarithmic	3	2	2.86	–	2	–	s386	1	1	715.71	–	1	–
evader-pursuer-4x4-standard	7	1	5.75	–	1	–	s499	2	2	160.23	–	2	–
evader-pursuer-6x6-logarithmic	4	3	721.25	–	2	1	s510	3	2	403.96	–	2	–
evader-pursuer-6x6-standard	2	2	892.65	–	–	2	s713	2	1	60.28	–	1	–
evader-pursuer-8x8-logarithmic	6	4	40.12	–	4	–	s820	2	–	–	–	–	–
FPGA_PLB_FIT_FAST	2	2	0.05	1	1	–	Sorting_networks	6	6	51.65	2	4	–
FPGA_PLB_FIT_SLOW	1	1	1.61	–	1	–	SzymanskiP	2	2	527.78	–	1	1
Impl	1	1	0.00	1	–	–	term1	3	3	0.34	1	2	–
jmc_quant	2	2	1.15	–	2	–	tipdiam	14	14	102.15	1	13	–
jmc_quant_squaring	1	1	8.38	–	1	–	tipfixpoint	24	22	2282.68	4	12	6
k_branch_n	4	4	870.06	–	3	1	Toilet	4	4	0.49	3	1	–
k_branch_p	7	7	380.20	–	7	–	ToiletA	10	10	0.72	7	3	–
k_d4_n	10	10	15.63	–	10	–	ToiletC	23	23	1.32	22	1	–
k_d4_p	5	5	1.63	1	4	–	ToiletG	4	4	0.01	4	–	–
k_dum_n	2	2	0.04	–	2	–	VonNeumann	2	2	5.92	1	1	–
k_dum_p	4	4	0.74	–	4	–	z4ml	1	1	0.00	1	–	–
k_grz_n	4	4	4.86	2	2	–							

all the participants. An instance is thus solved if at least one of the solvers solves it, and the time taken is the best among the times of the solvers that solved the instances. The instances are classified according to their hardness with the following criteria: easy instances are those solved by all the solvers, medium instances are those non-easy instances that could still be solved by at least two solvers, medium-hard instances are those solved by one reasoner only, and hard instances are those that remained unsolved.

According to the data summarized in Table 6, MAIN consisted of 568 instances, 523 of which have been solved, resulting in 105 easy, 393 medium, 25 medium-hard, and 45 hard instances. These results indicates that the selected instances are not trivial for current state-of-the-art QBF – only 18% of the test set are easy instances. At the same time, the test set is not overwhelming, since most of the non-easy instances (about

Table 7. Classification of formulas related to 2QBF, SH, RND, and NPNCNF tracks, considering second stage data. The table is organized as Table 6, with the only difference that in the leftmost column ("Track") is reported the related track.

Track	Family	Overall N	#	Time	Hardness EA	ME	MH	Family	Overall N	#	Time	Hardness EA	ME	MH
2QBF	irqlkeapclte	7	1	0.46	–	1	–	Sorting_networks	17	17	107.19	–	11	6
	MutexP	7	7	5.08	3	4	–	terminator	57	51	427.64	–	38	13
	Qshifter	6	6	2.51	2	4	–	wmiforward	46	27	159.21	11	15	1
	RankingFunctions	60	55	72.50	–	47	8							
SH	Abduction	2	2	142.85	–	2	–	k_branch_n	2	2	9.50	–	2	–
	C499	3	3	17.92	–	–	3	Sorting_networks	4	3	5133.28	–	2	1
	C880	6	4	54.19	–	–	4	tipdiam	1	1	0.44	–	–	1
	circuits	1	1	187.02	–	1	–	tipfixpoint	13	9	102.25	–	4	5
	Counter	4	4	429.15	–	1	3	uclid	2	1	12.09	–	–	1
	jmc_quant	5	5	6.61	–	–	5	wmiforward	2	2	0.67	–	–	2
	jmc_quant_squaring	5	5	32.44	–	–	5							
RND	ASP_Program_Inclusion	40	40	47.99	–	40	–	RobotsD3	30	29	259.31	–	29	–
	CounterFactual	80	80	205.06	2	77	1	RobotsD4	30	30	202.22	5	22	3
	q2k3k3	270	161	12317.15	30	70	61	RobotsD5	30	30	290	7	23	–
	RobotsD2	30	29	1292.02	–	26	3	Strategic_Companies	40	30	1300.85	3	12	15
NPNCNF	mqm	136	116	14943.18	51	–	65	QLTL_safety	250	116	38217.74	6	–	110
	NuSMV_diam	92	89	385.37	66	–	23							

69%) are solved by at least two solvers. Some "old" instances are still pretty hard for current state-of-the-art solvers, like the ones coming from the Mneimneh-Sakallah suite (families s641, s1196, s1269, and s3330). As a final consideration, we report that the main contributors to the SOTA solver, in percentage, were DEPQBF, DEPQBF-PRE, and QUANTOR3.1, with 28%, 18%, and 15%, respectively. This is the main explanation of the differences in the ranking positions between Table 3 and Table 4.

Concerning the 2QBF track, looking at the Table 7, we can see that, out of 200 instances, 164 have been solved, resulting in 16 easy, 120 medium, 28 medium-hard, and 36 hard instances. In particular, we report that the families in the suite "Basler" are the ones with the highest number of open formulas, while 92% of the formulas in the new family "RankingFunctions" (suite "Wintersteiger") are solved. Notice that 100% of the formulas in "legacy" families, i.e., "MutexP", "Qshifter", and "Sorting_networks" are solved by the SOTA solver. Finally, the three main contributors to the SOTA solver are STRUQS'10, AQME'10 and DEPQBF-PRE with 54%, 21%, and 20%, respectively. As regards the SH track, we can see in Table 7 that, out of 50 instances, 42 have been solved, resulting in 12 medium, 30 medium-hard, and 8 hard instances. These 8 open instances are mostly located in the "tipfixpoint" family (4). Notice that 1 hard instance is in the "uclid" family, that was submitted to the first QBFEVAL in 2003. The SOTA solver is mainly composed by AIGSOLVE, while other contributions do not exceed 5%. Looking now at the classification of formulas included in RND, in the related section of Table 7 we can see that, out of 550 instances, 429 have been solved, resulting in 47 easy, 299 medium, 83 medium-hard, and 121 hard instances. We report that the main contributors to the SOTA solver are AQME'10 and DEPQBF-PRE, with 42% for each one. Finally, concerning NPNCNF track, in Table 7 we can see that, out of 478 instances, 321 have been solved, resulting in 123 easy, 198 solved by only one solver, and 157 hard instances. The contribution of CIRQIT2.1 and QPRO to the SOTA solver is 64% and 36%, respectively. Noticeably, there has been a leap forward in solving the "QLTL_safety". In QBFEVAL'08, only 6 instances were solved[2], while we now report 116 solved formulas with "only" a factor $2\times$ in the CPU time limit. Another interesting observation is about "mqm" formulas. If we compare the total amount of solved formulas with the one reported in Table 6, we can see that, even if the results achieved using prenex CNF solvers are better than those achieved with non-prenex non-CNF solvers, the two values are comparable.

5 Conclusions

The final balance of QBFEVAL'10 can be summarized as follows:

– 13 solvers participated, 11 requiring QBFs in prenex CNF format.
– 136 formulas obtained by encoding minimal query inseparability module extraction in DL-Lite were submitted, both with an encoding in prenex CNF and their corresponding non-prenex non-CNF.
– State-of-the-art solvers for each track have been identified; also, a total of 367 challenging hard instances have been identified to set the reference point for future developments in the field.

All the information contained in this paper can be retrieved at the QBFEVAL'10 web portal [19]. Despite some long-standing limitations it is our opinion that the development of QBF solvers has reached its maturity, in the sense that QBF solvers are now dependable and effective tools which have a concrete potential for applications. The main question to be addressed by future research is whether the current state-of-the-art solvers, alone or in combinations among them, can solve industrial-sized and practically relevant problems. While QBF-based automation techniques can be regarded as a promising research direction, the lack of applications may discourage further developments, so it is important for researchers in the field to come out with "killer applications" wherein QBF solvers have an edge over competing technologies, e.g., SAT solvers or BDDs.

References

1. Le Berre, D., Roussel, O., Simon, L.: The SAT 2009 Competition (2009), www.satcompetition.org/2009
2. Sinz, C.: Sat-race (2008), baldur.iti.uka.de/sat-race-2008
3. Barrett, C., Deters, M., Oliveras, A., Stump, A.: Design and results of the 4th annual satisfiability modulo theories competition (SMT-COMP 2008) (to appear, 2008)
4. Long, D., Fox, M.: The 3rd International Planning Competition: Results and Analysis. Artificial Intelligence Research 20, 1–59 (2003)
5. Kontchakov, R., Pulina, L., Sattler, U., Schneider, T., Selmer, P., Wolter, F., Zakharyaschev, M.: Minimal Module Extraction from DL-Lite Ontologies using QBF Solvers. In: Proc. of IJCAI 2009, pp. 836–841 (2009)

---

[2] We are not tacking into account PQBF, reported as problematic solver in QBFEVAL'08.

6. Le Berre, D., Simon, L., Tacchella, A.: Challenges in the QBF arena: the SAT'03 evaluation of QBF solvers. In: Giunchiglia, E., Tacchella, A. (eds.) SAT 2003. LNCS, vol. 2919, pp. 468–485. Springer, Heidelberg (2004)
7. Narizzano, M., Pulina, L., Tacchella, A.: Ranking and Reputation Sytems in the QBF competition. In: Proc. of AI*IA 2007, pp. 97–108 (2007)
8. Pigorsch, F., Scholl, C.: Exploiting structure in an aig based qbf solver. In: Proc. of DATE 2009, pp. 1596–1601 (2009)
9. Pulina, L., Tacchella, A.: A self-adaptive multi-engine solver for quantified Boolean formulas. Constraints 14(1), 80–116 (2009)
10. Goultiaeva, A., Iverson, V., Bacchus, F.: Beyond CNF: A Circuit-Based QBF Solver. In: Kullmann, O. (ed.) SAT 2009. LNCS, vol. 5584, pp. 412–426. Springer, Heidelberg (2009)
11. Lonsing, F., Biere, A.: Efficiently Representing Existential Dependency Sets for Expansion-based QBF Solvers. Electronic Notes in Theoretical Computer Science 251, 83–95 (2009)
12. Lonsing, F., Biere, A.: Nenofex: Expanding nnf for qbf solving. In: Kleine Büning, H., Zhao, X. (eds.) SAT 2008. LNCS, vol. 4996, pp. 196–210. Springer, Heidelberg (2008)
13. Lewis, M., Schubert, T., Becker, B.: QMiraXT–A Multithreaded QBF Solver. In: Methoden und Beschreibungssprachen zur Modellierung und Verifikation von Schaltungen und Systemen (2009)
14. Egly, U., Seidl, M., Woltran, S.: A solver for QBFs in negation normal form. Constraints 14(1), 38–79 (2009)
15. Biere, A.: Resolve and Expand. In: Hoos, H., Mitchell, D.G. (eds.) SAT 2004. LNCS, vol. 3542, pp. 59–70. Springer, Heidelberg (2005)
16. Giunchiglia, E., Marin, P., Narizzano, M.: Preprocessing Techniques for QBFs. In: Proc. of 15th RCRA workshop (2008)
17. Giunchiglia, E., Narizzano, M., Tacchella, A.: Clause-Term Resolution and Learning in Quantified Boolean Logic Satisfiability. Artificial Intelligence Research 26, 371–416 (2006)
18. Pulina, L., Tacchella, A.: A structural approach to reasoning with quantified Boolean formulas. In: Proc. of IJCAI 2009, pp. 596–602 (2009)
19. Peschiera, C., Pulina, L., Tacchella, A.: Seventh QBF solvers evaluation, QBFEVAL (2010), http://www.qbfeval.org/2010
20. Giunchiglia, E., Narizzano, M., Pulina, L., Tacchella, A.: Quantified Boolean Formulas satisfiability library, QBFLIB (2001), http://www.qbflib.org
21. Cook, B., Kroening, D., Rümmer, P., Wintersteiger, C.M.: Ranking function synthesis for bit-vector relations. In: Proc. of TACAS 2010 (to appear, 2010)
22. Pulina, L., Tacchella, A.: Treewidth: a useful marker of empirical hardness in quantified Boolean logic encodings. In: Cervesato, I., Veith, H., Voronkov, A. (eds.) LPAR 2008. LNCS (LNAI), vol. 5330, pp. 528–542. Springer, Heidelberg (2008)
23. Chen, H., Interian, Y.: A Model for Generating Random Quantified Boolean Formulas. In: Proc. of IJCAI 2005, pp. 66–71 (2005)
24. Yu, Y., Malik, S.: Verifying the Correctness of Quantified Boolean Formula(QBF) Solvers: Theory and Practice. In: Proc. of ASP-DAC 2005, pp. 1047–1051 (2005)

Complexity Results for Linear XSAT-Problems

Stefan Porschen*, Tatjana Schmidt, and Ewald Speckenmeyer

Institut für Informatik, Universität zu Köln, Pohligstr. 1, D-50969 Köln, Germany
{porschen,schmidt,esp}@informatik.uni-koeln.de

Abstract. We investigate the computational complexity of the exact satisfiability problem (XSAT) restricted to certain subclasses of *linear* CNF formulas. These classes are defined through restricting the number of occurrences of variables and are therefore interesting because the complexity status does not follow from Schaefer's theorem [14,7]. Specifically we prove that XSAT remains NP-complete for linear formulas which are monotone and all variables occur exactly l times. We also present some complexity results for *exact linear* formulas left open in [9]. Concretely, we show that XSAT for this class is NP-complete, in contrast to SAT or NAE-SAT. This can also be established when clauses have length at least k, for fixed integer $k \geq 3$. However, the XSAT-complexity for exact linear formulas with clause length exactly k remains open, but we provide its polynomial-time behaviour at least for every positive integer $k \leq 6$.

Keywords: exact satisfiability, linear formula, NP-completeness, finite projective plane, regularity.

1 Introduction

Recently in [12] the propositional satisfiability problem (SAT) was studied when restricted to the class of *linear* formulas in conjunctive normal form (CNF). By definition, each pair of distinct clauses of a linear formula has at most one variable in common. Thus linear formulas overlap only sparsely and there is some evidence that linear formulas form the algorithmically hard kernel for CNF-SAT making this class specifically interesting also regarding other variants of SAT. Here we essentially focus on a well-known variant of SAT, namely, the exact satisfiability problem (XSAT). A truth assignment solving XSAT has to set exactly one literal in each clause of the input formula to 1, and the other literals to 0. XSAT is NP-complete for unrestricted CNF formulas [14]. Moreover it remains NP-complete also for formulas without negated variables (monotone formulas). Monotone XSAT essentially is closely related to the well-known NP-complete problems Set Partitioning and Exact Hitting Set [4] having many applications in combinatorial optimization. As shown in [9], XSAT is NP-complete for monotone and k-uniform linear formulas. In the present paper we investigate the computational complexity of XSAT restricted to some subclasses of linear formulas

---

* The first author was partially supported by the DFG project under grant No. SP 317/7-1.

O. Strichman and S. Szeider (Eds.): SAT 2010, LNCS 6175, pp. 251–263, 2010.

defined through bounding the number of occurrences of variables. Moreover we fill-in some gaps of [9] regarding *exact linear* formulas. Exact linear means that every two distinct clauses share exactly one variable.

Recall that Schaefer's theorem in [14] classifies generalized satisfiability problems (including XSAT) w.r.t. their complexity. However, this dichotomy theorem does not automatically apply if restrictions on the number of occurrences of variables in CNF formulas are made. E.g. in [7] it is shown that whereas unrestricted k-SAT is NP-complete, for $k \geq 3$, it can be solved easily (i.e. all formulas are satisfiable) if each clause has length exactly k and no variable occurs in more than $f(k)$ clauses; it becomes NP-complete if variables are allowed to occur at most $f(k) + 1$ times. Here $f(k)$ asymptotically grows as $\lfloor 2^k/(e \cdot k) \rfloor$; this bound has meanwhile been improved by other authors.

Here we show the NP-completeness of XSAT for CNF formulas which are l-regular meaning that every variable occurs exactly l times, where $l \geq 3$ is a fixed integer. On that basis we can provide also the NP-completeness of XSAT for the subclass of linear and l-regular formulas. This result carries over to the monotone case.

Using some connections to finite projective planes we can also show that XSAT remains NP-complete for linear and l-regular formulas that in addition are l-uniform (all clauses have the same length l) whenever $l = q + 1$, where q is a prime power. Thus XSAT most likely is NP-complete also for other values of $l \geq 3$.

The XSAT-complexity for monotone and exact linear formulas remained open in [9]. Here we show its NP-completeness which we can also establish for the subclass where clauses have length *at least* k, $k \geq 3$. This result is surprising, since both SAT and not-all-equal SAT are polynomial-time solvable for exact linear formulas [12,9]. However, a difficulty arises when trying to transfer the NP-completeness proof to the case where, in addition, all clauses are required to have length *exactly* k, for arbitrary $k \geq 3$. It might be possible that XSAT for these classes is polynomial-time solvable, which we can only show for $k \in \{3, 4, 5, 6\}$ so far. Finally, we draw some nice conclusions regarding the complexity of Exact Hitting Set and Set Partitioning for linear and regular, respectively exact linear, hypergraphs. Recall that in a linear, respectively exact linear, hypergraph each pair of distinct hyperedges shares at most one, respectively exactly one, vertex [1].

2   Basic Definitions and Notations

Let CNF denote the set of duplicate-free conjunctive normal form formulas over propositional variables $x \in \{0, 1\}$. A *positive (negative)* literal is a (negated) variable. For a formula C, clause c, by $V(C), V(c)$ we denote the set of its variables (neglecting negations), respectively. In this paper we assume that clauses neither contain duplicate literals nor a pair of complementary literals. Clauses, respectively formulas are regarded as sets of literals, respectively clauses. Thus by $|c|$ we denote the number of literals in a clause c, and by $|C|$ the number of clauses in formula C. XSAT is the variant of SAT asking for a truth assignment setting exactly one literal in each clause of a CNF formula to 1 and all other literal(s) to 0; such a truth assignment is called an x-model. C is called *x-(un)satisfiable* if it has an (has no) x-model. The counterpart not-all-equal SAT (NAE-SAT) is defined similarly, but for a solution it is required that in each clause at least one literal is set to 1, and at least one literal is set to 0.

A CNF formula C is called *linear* if for all $c_1, c_2 \in C : c_1 \neq c_2$ we have $|V(c_1) \cap V(c_2)| \leq 1$. C is called *exact linear* if for all $c_1, c_2 \in C : c_1 \neq c_2$ we have $|V(c_1) \cap V(c_2)| = 1$. A *monotone* formula has no negated variables. A formula C is *k-uniform* if all its clauses have length exactly k; it is *l-regular* if each of its variables occurs exactly l times in C. By $w_C(x)$ we denote the number of occurrences of variable x in C (disregarding negations). Let LCNF denote the class of linear formulas and XLCNF the class of exact linear formulas. Let $\mathcal{C} \in \{\text{CNF}, \text{LCNF}, \text{XLCNF}\}$ be fixed. Then $k\text{-}\mathcal{C}$, $(\geq k)\text{-}\mathcal{C}$, $(\leq k)\text{-}\mathcal{C}$ denotes the subclass of formulas in $\mathcal{C}$ with the additional property that all clauses have length exactly k, at least k, at most k, respectively. Similarly $\mathcal{C}^l$, $\mathcal{C}^{\geq l}$, $\mathcal{C}^{\leq l}$ denotes the subclass of formulas in $\mathcal{C}$ with the additional property that all variables occur exactly l times, at least l times, at most l times, respectively. $\mathcal{C}_+$ denotes the subclass of monotone formulas in $\mathcal{C}$. Let C be an x-satisfiable formula. A variable $y \in V(C)$ is called an *x-backbone variable* of C, if y has the same value in each x-model of C. We call a monotone, k-uniform and exact linear formula that in addition is k-regular a *k-block formula*; such formulas are closely related to finite projective planes [12].

3   XSAT on Linear Formulas with Regularity Conditions

In this section we focus on the following classes:

$$\text{CNF}_+ \supset k\text{-CNF}_+ \supset k\text{-CNF}_+^{\leq l} \supset k\text{-CNF}_+^l \supset (\leq l)\text{-LCNF}_+^{\geq l} \supset (\leq l)\text{-LCNF}_+^l$$
$$\text{CNF}_+ \supset \text{CNF}_+^l \supset \text{LCNF}_+^l \supset (\leq l)\text{-LCNF}_+^l$$

Clearly, the same inclusion relations are also valid for the non-monotone counterparts. XSAT and NAESAT can be solved in polynomial time for all classes mentioned above with parameters k and l at most 2. This even holds for the variable-weighted optimisation versions of these problems [11]. The remaining cases are believed to be NP-complete. Whereas for several of them concrete reductions will be provided in the sequel, for others the NP-completeness proofs are left open. Specifically, the classes $k - \text{LCNF}^l$ can be verified to be NP-complete only when regularity parameter l and uniformity parameter k are equal and in addition $l - 1$ is a prime power. This clearly provides evidence that NP-completeness holds for all values of $l = k$, but we have no rigorous proof. The same lack is present if l and k have distinct values.

We start with non-linear CNF classes serving as reduction base for what follows. To that end, recall that for CNF_+ and $k\text{-CNF}_+$ the NP-completeness of XSAT is well known. For LCNF_+, $k\text{-LCNF}_+$ and $(\geq k)\text{-LCNF}_+$ the NP-completeness was shown in [9]. For the remaining classes we prove the

NP-completeness of XSAT in this section. The next two results treat the non-linear case, which will be referred to later.

Theorem 1. XSAT *remains* NP-*complete for* k-CNF$_{+}^{\leq l}$ *and* k-CNF$^{\leq l}$, $k, l \geq 3$.

Proof. We provide a polynomial-time reduction from k-CNF$_{+}$-XSAT (which is NP-complete [4]) to k-CNF$_{+}^{\leq l}$ establishing NP-completeness of the latter and thus of k-CNF$^{\leq l}$. To that end, let C be an arbitrary formula in k-CNF$_{+}$. For each $x \in V(C)$ with $w_C(x) > l$, we introduce $p := w_C(x) - (l - 1)$ new variables $x_1, x_2, \ldots, x_p$. Let the first $l - 1$ occurrences of x remain unchanged and replace the p remaining occurrences of x by the variables $x_1, x_2, \ldots, x_p$. Let C' be the resulting formula. Next we introduce new, pairwise different variables a_{ij}, for $i = 1, \ldots, p$, $j = 1, \ldots, k - 1$, and add the following clauses to C' which ensure XSAT-equivalence of the newly introduced variables $x_1, x_2, \ldots, x_p$ with x.

$$(x \vee a_{11} \vee a_{12} \vee \ldots \vee a_{1,k-1}) \wedge (x_1 \vee a_{11} \vee a_{12} \vee \ldots \vee a_{1,k-1})$$
$$\wedge (x_1 \vee a_{21} \vee a_{22} \vee \ldots \vee a_{2,k-1}) \wedge (x_2 \vee a_{21} \vee a_{22} \vee \ldots \vee a_{2,k-1})$$
$$\wedge \ldots$$
$$\wedge (x_{p-1} \vee a_{p1} \vee a_{p2} \vee \ldots \vee a_{p,k-1}) \wedge (x_p \vee a_{p1} \vee a_{p2} \vee \ldots \vee a_{p,k-1})$$

Hence C and C' are XSAT-equivalent, and obviously no variable occurs more than l times in C'. $\qquad\square$

Theorem 2. XSAT *remains* NP-*complete for* k-CNF$_{+}^{l}$ *and* k-CNF^{l}, $k, l \geq 3$.

Proof. We provide a polynomial-time reduction from k-CNF$_{+}^{\leq l}$-XSAT (which is NP-complete) to k-CNF$_{+}^{l}$-XSAT similar to the technique in [7]. Let C be an arbitrary formula in k-CNF$_{+}^{\leq l}$ with variable set $V(C) = \{x_1, \ldots, x_n\}$. We introduce l pairwise variable-disjoint copies $C_1, \ldots, C_l$ of C, such that the variables in C_i are $\{x_1^i, \ldots, x_n^i\}$, for $i = 1, \ldots, l$. For each $x_j \in V(C)$ with $w_C(x_j) < l$ we construct the formulas $D_{x_j,1}, \ldots, D_{x_j,l-w_C(x_j)}$ with

$$D_{x_j,i} = \bigwedge_{r=1}^{l} \left(x_j^r \vee a_{i,1} \vee \ldots \vee a_{i,k-1} \right)$$

for $1 \leq i \leq l - w_C(x_j)$. Note that $a_{i,j}$ occurs exactly l times in $D_{x_j,i}$ and nowhere else, for $i = 1, \ldots, l - w_C(x_j), j = 1, \ldots, k - 1$. Defining

$$C' = \bigwedge_{i=1}^{l} C_i \wedge \bigwedge_{x_j \in V(C)} \bigwedge_{i=1}^{l-w_C(x_j)} D_{x_j,i}$$

we observe that x_j^i occurs $w_C(x_j)$ times in C_i and once in each $D_{x_j,i}$, for $i = 1, \ldots, l - w_C(x_j)$. Thus each x_j^i occurs l times in C'. So C' belongs to k-CNF$_{+}^{l}$.

$C \in$ XSAT if and only if $C' \in$ XSAT: Let C be x-satisfiable, then we can use a fixed x-model t of C to x-satisfy the copies $C_1, \ldots, C_l$ of C. If $t(x_j) = 1$, we set $x_j^r = 1$, for $r = 1, \ldots, l$, and $a_{i,1} = \ldots = a_{i,k-1} = 0$ yielding an x-model for $D_{x_j,i}$, for all $x_j \in V(C)$ and $1 \leq i \leq w_C(x_j)$. If $t(x_j) = 0$, we set $x_j^i = 0$, for $i = 1, \ldots, l$ and we assign $a_{i,1} = 1$ as well as $a_{i,2} = \ldots = a_{i,k-1} = 0$ yielding a x-model for $D_{x_j,i}$, for all $x_j \in V(C)$ and $1 \leq i \leq w_C(x_j)$. The reverse direction is obvious. □

The situation is different from the SAT case where k and l are, for $k, l \geq 3$, such that k-CNF^l-SAT is polynomial-time solvable as mentioned in the introduction [7]. Now we are able to attack the linear and l-regular classes.

Theorem 3. XSAT *remains* NP-*complete for* LCNF$_+^l$, *and* LCNF^l, $l \geq 3$.

Proof. We provide a polynomial-time reduction from CNF$_+^l$-XSAT (which is NP-complete) to LCNF$_+^l$-XSAT. Let C be an arbitrary formula in CNF$_+^l$. Is C not linear, we proceed as follows for each variable $x_i \in V(C)$: Since each fixed variable $x_i \in V(C)$ has exactly l occurrences in C, namely in the clauses $c_{j_1}, \ldots, c_{j_l}$, we introduce a new variable $y_{x_i}^{j_s} \notin V(C)$, for each such occurrence $2 \leq s \leq l$, except for the first occurrence of x_i in c_{j_1}. Then we replace each occurrence of x_i in c_{j_s} (except in c_{j_1}) with $y_{x_i}^{j_s}$, for $2 \leq s \leq l$. Let C' be the resulting formula. Then C' is obviously linear, monotone and each variable occurs exactly once in C'. For each $x_i \in V(C)$, we introduce new, pairwise different variables $z_1^{x_i}, \ldots, z_{l-1}^{x_i} \notin V(C')$. Next we add the following 2-clauses to C' providing XSAT-equivalence of the variables $x_i, y_{x_i}^{j_2}, \ldots, y_{x_i}^{j_l}$:

$$
\begin{aligned}
P_{x_i} = &(x_i \vee z_1^{x_i}) \wedge (x_i \vee z_2^{x_i}) \wedge \ldots \wedge (x_i \vee z_{l-1}^{x_i}) \\
&\wedge (y_{x_i}^{j_2} \vee z_1^{x_i}) \wedge (y_{x_i}^{j_2} \vee z_2^{x_i}) \wedge \ldots \wedge (y_{x_i}^{j_2} \vee z_{l-1}^{x_i}) \\
&\wedge (y_{x_i}^{j_3} \vee z_1^{x_i}) \wedge (y_{x_i}^{j_3} \vee z_2^{x_i}) \wedge \ldots \wedge (y_{x_i}^{j_3} \vee z_{l-1}^{x_i}) \\
&\wedge \ldots \\
&\wedge (y_{x_i}^{j_l} \vee z_1^{x_i}) \wedge (y_{x_i}^{j_l} \vee z_2^{x_i}) \wedge \ldots \wedge (y_{x_i}^{j_l} \vee z_{l-1}^{x_i})
\end{aligned}
$$

Observe that $C'' := \bigwedge_{x_i \in V(C)} P_{x_i} \wedge C'$ is l-regular: $w_{C''}(z_r^{x_i}) = l$, for $r = 1, \ldots, l-1$, $w_{C''}(y_{x_i}^{j_s}) = l$, for $s = 2, \ldots, l$, and $w_{C''}(x_i) = l$, for each $x_i \in V(C)$. Since every variable appears only once in C', C' is linear. Obviously P is linear; and the variables $z_1^{x_i}, \ldots, z_{l-1}^{x_i}$ do not occur in C', so $P \wedge C' = C''$ is linear, too. C is XSAT-equivalent with C'', because of the XSAT-equivalence of the variables $x_i, y_{x_i}^{j_2}, \ldots, y_{x_i}^{j_l}$ and P_{x_i} are always x-satisfiable, for all i. □

Finding a concrete reduction for the NP-completeness proof of XSAT for k-LCNF$_+^l$ is a tricky problem. We are only able to show NP-completeness of XSAT for k-LCNF^l where $l = q + 1$ and q is a prime power. This results from the fact that we can exploit block formula patterns providing backbone-variables. A k-block formula directly corresponds to a finite projective plane of order $k - 1$. Unfortunately it is a hard open question to decide whether a projective plane exists for a given $k \in \mathbb{N}$ [13]. However, it is a well known fact in combinatorics that for prime power orders the corresponding projective planes can easily be computed.

Theorem 4. XSAT *remains* NP-*complete for* l-LCNF^l, *for* $l = q + 1$, *where* q *is a prime power.*

Proof. We provide a polynomial-time reduction from l-CNF$^l_+$ to l-LCNF^l, for $l = q + 1$, where q is a prime power. Let $C \in l$-CNF$^l_+$ be an arbitrary formula and $V(C) = \{x_1, x_2, \ldots, x_n\}$ the set of its variables. For each variable $x_i \in V(C)$ we proceed like in the beginning of the proof of Theorem 3 obtaining the corresponding formulas C' and $P = \bigwedge_{x_i \in V(C)} P_{x_i}$. Next we enlarge each 2-clause of P by exactly $l - 2$ many x-backbone variables all of which must be assigned to 0 and obtain l-clauses this way. The x-backbone variables are provided via l-block formulas as follows: Each such l-block formula B_l is l-regular, l-uniform and exists whenever $l = q + 1$, for q prime power [12]. B_l is not x-satisfiable but removing an arbitrary clause of B_l yields a x-satisfiable formula, where the variables of the removed clauses are x-backbone variables which have to be set to 1 [9]. Therefore we provide $n(l - 1)(l - 2)$ many l-block formulas which are pairwise variable-disjoint. Removing a clause of each of them in total yields $n(l - 1)(l - 2)l$ distinct x-backbone variables. Since in P we have $n(l - 1)l$ many 2-clauses, each of which needs $l - 2$ variables to become an l-clause, this fits perfectly. Let P' be the formula obtained from P this way. Then $C'' := C' \wedge P'$ is l-regular and l-uniform by construction. Moreover, C'' is x-satisfiable if, and only if, C is x-satisfiable because P provides the XSAT-equivalence of the original variables with the replaced ones which is preserved by P' through the x-backbone 0 variables. Note that C'' is non-monotone as we have to negate the x-backbone variables when adding them to the clauses of P.   □

This result provides evidence that NP-completeness also holds for all values of $l \geq 3$. Unfortunately the proof does not easily transfer to the monotone case due to the fact that we have not been able to find suitable formulas providing backbone variables which must be set to 0. However considering the monotone case we are able to treat the following larger classes.

Theorem 5. XSAT *is* NP-*complete for* $(\leq l)$-LCNF$^l_+$, $(\leq l)$-LCNF^l, $l \geq 3$.

Proof. We provide a polynomial time reduction from l-CNF$^l_+$-XSAT (which is NP-complete according to Theorem 2) to $(\leq l)$-LCNF$^l_+$-XSAT. Let C be an arbitrary formula in l-CNF$^l_+$. If C is not linear, we proceed as in the beginning of the proof of Theorem 3 obtaining the corresponding formulas C' and $P = \bigwedge_{x_i \in V(C)} P_{x_i}$. Again $C'' := \bigwedge_{x_i \in V(C)} P_{x_i} \wedge C'$ is l-regular and linear by construction. Moreover each clause of C'' has a clause length of at most l, because in C' each clause has a length of exactly l and P consists of 2-clauses only. C'' is XSAT-equivalent with C ensured by P.   □

4   XSAT for Exact Linear Formulas

Exact linear formulas are quite small instances since the number of clauses never exceeds the number of variables [8,12]. Moreover it is known that SAT is linear-time solvable for XLCNF [12]. According to [9] also NAE-SAT restricted to

XLCNF is polynomial-time solvable. Treating the XSAT-complexity of XLCNF remained open in [9]. This section is mainly devoted to consider XSAT for exact linear formulas. Besides proving the NP-completeness of XLCNF-XSAT we also provide polynomial-time subclasses. The next result is remarkable in the sense that XSAT is the problem with the smallest search space among XSAT, NAESAT and SAT, but has the highest complexity (under the assumption $NP \neq P$) on the rather small class of exact linear formulas.

Theorem 6. XSAT *remains* NP-*complete for* XLCNF$_+$ *and* XLCNF.

Proof. We give a polynomial-time reduction from LCNF$_+$-XSAT to XLCNF$_+$-XSAT. Let $C = c_1 \wedge c_2 \wedge \ldots \wedge c_m \in$ LCNF$_+$ be an arbitrary formula that is not exact linear, otherwise we are done. As long as there is a pair of clauses $c_i, c_j \in C$, $i, j \in \{1, \ldots, m\}$ which do not share a variable, introduce a new variable z that does not occur in the current formula and augment both c_i and c_j by the variable z. The resulting formula C' obviously is exact linear. Let Z denote the collection of newly introduced variables this way. Next, we add at least $m+1$ further clauses collected in D whereas C' is modified to $\tilde{C}'$ so that the resulting formula $C'' := \tilde{C}' \wedge D$ stays exact linear and becomes XSAT-equivalent to C.

The construction of D and the modification of C' proceeds hand in hand: Initially, D is empty. As long as there is a variable $z \in Z$ not occurring in any clause of D, add a new clause d to D containing z and a new distinguished variable u (which is required to be contained in each clause of D). For each clause c_i of the current formula C' such that $V(d) \cap V(c_i) = \varnothing$ introduce a new variable $w_{d,i}$ and add it to d and c_i. Let W denote the collection of all these newly introduced variables.

When all variables in Z occur in D, but D still contains less than $m+1$ clauses then add sufficiently many new clauses to D each containing u. Each such new clause e is filled-up by m new variables $y_{e,1}, \ldots, y_{e,m}$ such that $y_{e,r}$ is added to W and to the rth clause of the first m clauses, $1 \leq r \leq m$. Finally, all newly introduced variables in $Z \cup W$ occur in D and in the final version $\tilde{C}'$ of C' and the formula is exact linear.

Let C be x-satisfiable with x-model t. Obviously, t can be extended to all variables of C'' by setting all newly introduced variables of $W \cup Z$ to 0 and $u = 1$. This yields a x-model for C''.

Let C be x-unsatisfiable, and assume that C'' is x-satisfiable. Then $\tilde{C}'$ can only be x-satisfied by setting at least one variable $x \in Z \cup W$ to 1. As each variable of $Z \cup W$ also occurs in D, there must be a clause $d_i \in D$ with $x \in d_i$. Hence $u = 0$ in d_i, and thus, to x-satisfy D, exactly one variable from $Z \cup W$ must be set to 1 in each of its clauses. As D has at least $m+1$ clauses, there must be at least $m + 1$ distinct variables from $Z \cup W$ set to 1 in D. Since all these variables occur in $\tilde{C}'$, but $\tilde{C}'$ has exactly m clauses, the pigeonhole principle implies that there is a clause in $\tilde{C}'$ containing at least two variables set to 1. This yields a contradiction, hence C'' is x-unsatisfiable, too.

To illustrate this reduction, consider the input formula

$$C = (x_1 \vee x_2 \vee x_3) \wedge (x_4 \vee x_5 \vee x_6) \wedge (x_1 \vee x_7 \vee x_8) \in \text{LCNF}_+$$

First we obtain C' by making the clauses of C exact linear introducing the new variables $Z = \{z_1, z_2\}$:

$$C' = (x_1 \vee x_2 \vee x_3 \vee z_1)$$
$$\wedge (x_4 \vee x_5 \vee x_6 \vee z_1 \vee z_2)$$
$$\wedge (x_1 \vee x_7 \vee x_8 \vee z_2)$$

Next we add clauses $D = \{d_1, d_2\}$ each containing a fixed variable u such that all variables in Z occur in the new clauses. To preserve exact linearity we need to introduce new variables $W = \{w_1, w_2\}$:

$$(x_1 \vee x_2 \vee x_3 \vee z_1 \vee w_2)$$
$$\wedge (x_4 \vee x_5 \vee x_6 \vee z_1 \vee z_2)$$
$$\wedge (x_1 \vee x_7 \vee x_8 \vee z_2 \vee w_1)$$
$$\wedge \underbrace{(u \vee z_1 \vee w_1)}_{=:d_1}$$
$$\wedge \underbrace{(u \vee z_2 \vee w_2)}_{=:d_2}$$

In this example D has only two clauses, so we have to add two more clauses e_1, e_2 to ensure XSAT-equivalence and preserve exact linearity, finally yielding $W = \{w_1, w_2, y_1, \ldots, y_6\}$, and:

$$C'' = (x_1 \vee x_2 \vee x_3 \vee z_1 \vee w_2 \vee y_1 \vee y_4)$$
$$\wedge (x_4 \vee x_5 \vee x_6 \vee z_1 \vee z_2 \vee y_2 \vee y_5)$$
$$\wedge (x_1 \vee x_7 \vee x_8 \vee z_2 \vee w_1 \vee y_3 \vee y_6)$$
$$\wedge (u \vee z_1 \vee w_1)$$
$$\wedge (u \vee z_2 \vee w_2)$$
$$\wedge \underbrace{(u \vee y_1 \vee y_2 \vee y_3)}_{=:e_1}$$
$$\wedge \underbrace{(u \vee y_4 \vee y_5 \vee y_6)}_{=:e_2} \quad \in \text{XLCNF}_+ \qquad \square$$

It is not hard to see that the result above sharpens the long-standing NP-hardness result for clique packing of a graph maximizing the number of covered edges of Hell and Kirkpatrick [5]. Recently Chataigner et al. have provided nice approximation (hardness) results regarding the clique packing problem [3]. It could be interesting to investigate whether similar approximation results can be gained for XSAT on (X)LCNF in the future.

Next we are interested in XSAT for $(\geq k)$-XLCNF$_+$, with $k \geq 3$. To prove its NP-completeness, we need first to consider the class $(\geq |C|)$-LCNF$_+$ consisting of all monotone and linear formulas C such that each clause has at least length $|C|$.

Lemma 1. *Every formula in $(\geq |C|)$-LCNF$_+$ is x-satisfiable.*

Proof. Let C be a formula in $(\geq |C|)$-LCNF$_+$ with $m := |C|$ clauses and assume there is a clause $c_0 \in C$ containing at least the variables $x_1, \ldots, x_m$ such that $w_C(x_i) \geq 2$, for $1 \leq i \leq m$. Due to linearity this implies that there are clauses c_i, $1 \leq i \leq m$, such that $x_i \in V(c_i)$, thus $|C| \geq |\{c_0, c_1, \ldots, c_m\}| \geq m + 1$ yielding a contradiction. It follows that each clause c of C contains at least one literal which occurs only once in C. Hence, setting exactly these variables to 1 x-satisfies C. □

Let $(k, |C| - 1)$-LCNF$_+$, $k \geq 3$, denote the class of all monotone and linear formulas C such that each clause has length k at least and $|C| - 1$ at most. According to Theorem 4 in [9] XSAT is NP-complete for $(\geq k)$-LCNF$_+$, for each fixed $k \geq 3$. According to Lemma 1 $(\geq |C|)$-LCNF$_+$ behaves trivially for XSAT. Since XSAT is NP-complete for k-LCNF$_+$ and k-LCNF$_+ \subseteq (k, |C| - 1)$-LCNF$_+$, XSAT is NP-complete for $(k, |C| - 1)$-LCNF$_+$, too.

We were not able to establish the NP-completeness for k-XLCNF$_+$, i.e. uniform formulas, regarding XSAT. However, on behalf of the NP-completeness of $(k, |C| - 1)$-LCNF$_+$-XSAT just shown, we can provide the next result by using the same technique as in the proof of Theorem 6 starting with a formula in $(k, |C| - 1)$-LCNF$_+$.

Theorem 7. *XSAT remains NP-complete for $(\geq k)$-XLCNF$_+$, for each $k \geq 3$.*

There are some difficulties in establishing the complexity of XSAT for arbitrary k-uniform, exact linear formulas. We instead present the polynomial-time solvability of XSAT for the k-uniform subclasses k-XLCNF$_+$, where $k \leq 6$. For that purpose we introduce several lemmas.

Lemma 2. *Let $C \in k-\text{XLCNF}_+$. If there is a variable $x \in V(C)$ with $w(x) > k$, then $x \in V(c)$, for all $c \in C$.*

Proof. Suppose there is a clause $c_i \in C$ that is not an x-clause. Then c_i must share exactly one variable with each x-clause. There are more than k many x-clauses, but c_i is only k-uniform. Hence C contains only x-clauses. □

Lemma 3. *[9] The class k-XLCNF$_+^k$ is not x-satisfiable.*

Lemma 4. *Let $C \in k$-XLCNF$_+$ containing a clause $c = (x_1 \vee x_2 \vee \ldots \vee x_k) \in C$ such that $w(x_1) = w(x_2) = \ldots = w(x_k) = k - 2$. Then we can decide XSAT for C in polynomial time.*

Proof. Let $C \in k$-XLCNF$_+$ and $c_1 = (x_1 \vee x_2 \vee \ldots \vee x_k) \in C$ with $w(x_1) = w(x_2) = \ldots = w(x_k) = k-2$. XSAT-evaluating C according to the setting $x_i = 1$, for any fixed $i \in \{1, \ldots, k\}$, yields a formula $C[x_i]$ in 2-LCNF$_+$ because of exact linearity and k-uniformity. Therefore XSAT for $C[x_i]$ can be decided in linear-time [11]. Hence, in the worst case we have to check every such formula $C[x_i]$, $1 \leq i \leq k$, yielding a polynomial-time worst-case running time of $O(k \cdot ||C||)$. □

Lemma 5. *Let $C \in k$-XLCNF$_+$ and let $x \in V(C)$ be a variable with $w(x) = k - 1$ in C. Then C is x-satisfiable.*

Proof. Let $C \in k\text{-XLCNF}_+$ and $x \in V(C)$ with $w(x) = k - 1$. Let $c_1, \ldots, c_{k-1}$ be the clauses containing x. We set $x = 1$ in $c_1, \ldots, c_{k-1}$ and assign 0 to the other variables in these clauses. This way we x-satisfy the clauses $c_1, \ldots, c_{k-1}$ and remove these from the formula C. Now we consider the remaining clauses $c_j \in C - \{c_1, \ldots, c_{k-1}\}$, which satisfy $V(c_j) \cap (V(c_i) - \{x\}) \neq \emptyset$, for all $i = 1, \ldots, k-1$. Hence each of the remaining clauses contains $k-1$ distinct variables already assigned to 0. When we remove these from all of the remaining clauses $c_j \in C - \{c_1, \ldots, c_{k-1}\}$ the remaining formula consists of unit clauses only, and thus C is x-satisfiable. □

Lemma 6. *[11] Let $C \in \text{CNF}$ with $w(x) \leq 2$ for all $x \in V(C)$. Then we can decide XSAT for C in polynomial time.*

No we are ready to establish a Theorem which may be useful to design better exact exponential algorithms for k-SAT: The case resolved by the polynomial-time algorithm may occur in some branching rule of the exponential algorithm being designed.

Theorem 8. *The classes $k\text{-XLCNF}_+$, for $k \in \{1, 2, 3, 4, 5, 6\}$, can be x-solved in polynomial time.*

Proof. We only treat case $k = 6$. Let $C \in 6\text{-XLCNF}_+$ be an arbitrary formula with variable set $V(C)$. We set $w_C(x) := w(x)$ since C is fixed, and provide a case analysis guided by the number of occurrences of variables in $V(C)$.

- If there is a variable $x \in V(C)$ with $w(x) \geq 7$, then $x \in c$ for all $c \in C$ according to Lemma 2. In this case we set $x = 1$ and assign 0 to all the other variables in $V(C)$. This way we get an x-model for C.
- If $w(x) = 6$, for all $x \in V(C)$, then C is x-unsatisfiable according to Lemma 3.
- If there is a variable $x \in V(C)$ with $w(x) = 5$ then C is x-satisfiable according to Lemma 5.
- If $w(x) = 4$, for all $x \in V(C)$, then we can decide XSAT for C in polynomial time according to Lemma 4.
- If there a variable $x \in V(C)$ with $w(x) = 6$ as well as a variable $y \in V(C)$ with $w(y) \neq 6$: If C only consists of clauses containing x, then we set $x = 1$ and all other variables in $V(C)$ to 0 obtaining an x-model for C. Otherwise, there is a variable $x \in V(C)$ with $w(x) \geq 4$. Setting x to 1, and all other variables in the clauses containing x to 0 yields a formula only containing clauses of length ≤ 2. Such a formula can be checked for XSAT in polynomial time. This way we treat each variable $z \in V(C)$ with $w(z) \geq 4$ until an x-model is found. In the negative case we proceed as follows:
  (a) If there is no variable $x \in V(C)$ with $w(x) = 3$, we set all the variables x with $w(x) \geq 4$ to 0. Hence the resulting formula C' contains only variables occurring ≤ 2 times in C' and by using Lemma 6 we can solve C' in polynomial time.
  (b) If there is a variable $x \in V(C)$ with $w(x) = 3$, then after having set x to 1 and all other variables to 0 in the clauses containing x, we obtain a

formula C' which is in 3-LCNF$_+$. If there is no further variable occurring three times in C', we set all variables occurring ≥ 4 times to 0 and can decide x-satisfiability of C' in polynomial time according to Lemma 6. Otherwise, there is a variable y with $w(y) = 3$ in C'. Then we set $y = 1$ and to 0 all other variables in the clauses containing y. Now all y-clauses are x-satisfied and there are at most three clauses in the remaining formula that do not share any variable with any of the clauses containing y. Hence all clauses, except for at most three, do share at least one variable with one clauses containing y. Since these variables are all set to 0, the remaining formula only contains clauses of length two at most (except for at most three clauses) which can be decided for XSAT in polynomial time.

- If $w(x) \leq 3$, for all $x \in V(C)$, and there $x \in V(C)$ with $w(x) = 3$: After having set x to 1 and all other variables in the clauses containing x to 0, the remaining formula C' is in 3-LCNF$_+$. If all variables occur at most twice in C', we can decide x-satisfiability of C' in polynomial time according to Lemma 6. Otherwise there is still a variable y with $w(y) = 3$ in C'. In that case we set $y = 1$ and all variables in the clauses containing y to 0. Now all of y-clauses are x-satisfied and there are at most three clauses which do not share any variable with at least one of the y-clauses. Hence we can decide XSAT in polynomial time as above.
- If $w(x) \leq 2$, for all $x \in V(C)$, we can decide XSAT for C in polynomial time according to Lemma 6. □

5   Concluding Remarks and Open Problems

XSAT has been shown to be NP-complete when restricted to certain linear k-uniform and l-regular CNF classes, for $k, l \geq 3$.

Our results imply the NP-completeness of some subversions of the well-known combinatorial optimization problems Set Partitioning and Exact Hitting Set on regular and linear hypergraphs. Recall that Set Partitioning takes as input a finite hypergraph with vertex set M and a set of hyperedges $\mathcal{M}$ (i.e. subsets of M). It asks for a subfamily $\mathcal{T}$ of $\mathcal{M}$ such that each element of M occurs in exactly one member of $\mathcal{T}$. It is easy to see that monotone XSAT coincides with Set Partitioning when the clauses overtake the roles of vertices in M and the variables are regarded as the hyperedges in $\mathcal{M}$ in such a way that a variable contains all clauses in which it occurs.

Exact Hitting Set, however, is just the same as monotone XSAT only translated to the hypergraph (or set system) terminology; implying that Exact Hitting Set remains NP-complete for linear, l-regular hypergraphs. And a simple dualization argument implies that the same is true for Set Partitioning on that specific class of hypergraphs.

Regarding the exact linear case, we have shown that XSAT for unrestricted exact linear formulas is NP-complete in contrast to SAT, respectively NAE-SAT, which both are polynomial-time solvable on this class. Therefore it easily follows that Exact Hitting Set is NP-complete for exact linear hypergraphs. In summary we have:

Theorem 9. *(1) Exact Hitting Set and Set Partitioning both are NP-complete for linear, l-regular hypergraphs.*
(2) Exact Hitting Set is NP-complete for exact linear hypergraphs.

Observe that Set Partitioning for exact linear hypergraphs is trivial in the sense that it has no solution either the input hypergraph consists of one hyperedge only.

There are several problems left open for future work, so the complexity status for XSAT restricted to the classes k-LCNF$_+^l$, for arbitrary values of $k, l \geq 3$. Moreover, we do not know the complexity of XSAT restricted to k-uniform exact linear formulas, for $k \geq 7$. The same lack is present for the case where the number of occurrences is bounded, meaning that the XSAT-complexity of XLCNF$_+^l$ and k-XLCNF$_+^l$ are open, for $k, l \geq 3$.

Finally from the point of view of exact algorithmics, it would be desirable to gain progress for XSAT restricted to linear formulas beyond the so far best bound of $O(2^{0.2325 \cdot n})$, for unrestricted CNF-XSAT over n variables, provided by Byskov et al. [2]. Note that for NAE-SAT, such progress seems to be hard to achieve since NAE-SAT can be shown to be as hard as SAT itself for unrestricted CNF formulas [6,10]. It is open whether NAE-SAT or even SAT can be solved in less than 2^n steps for linear formulas.

Acknowledgement. We want to thank the anonymous referees for their valuable comments.

References

1. Berge, C.: Hypergraphs. North-Holland, Amsterdam (1989)
2. Byskov, J.M., Ammitzboll Madsen, B., Skjernaa, B.: New Algorithms for Exact Satisfiability. Theoretical Comp. Sci. 332, 515–541 (2005)
3. Chataigner, F., Manic, G., Wakabayashi, Y., Yuster, R.: Approximation algorithms and hardness results for the clique packing problem. Discrete Appl. Math. 157, 1396–1406 (2009)
4. Garey, M.R., Johnson, D.S.: Computers and Intractability: A Guide to the Theory of NP-Completeness. W.H. Freeman and Company, San Francisco (1979)
5. Hell, P., Kirkpatrick, D.G.: On the complexity of general k-factor problems. SIAM J. Comput. 12, 601–609 (1983)
6. Knuth, D.E.: Axioms and Hulls. LNCS, vol. 606. Springer, Heidelberg (1992)
7. Kratochvil, J., Savicky, P., Tusa, Z.: One more occurrence of variables makes satisfiability jump from trivial to NP-complete. SIAM J. Comput. 22, 203–210 (1993)
8. Palisse, R.: A short proof of Fisher's inequality. Discrete Math. 111, 421–422 (1993)
9. Porschen, S., Schmidt, T.: On Some SAT-Variants over Linear Formulas. In: Nielsen, M., Kucera, A., Miltersen, P.B., Palamidessi, C., Tuma, P., Valencia, F.D. (eds.) SOFSEM 2009. LNCS, vol. 5404, pp. 449–460. Springer, Heidelberg (2009)

10. Porschen, S., Randerath, H., Speckenmeyer, E.: Linear-Time Algorithms for some Not-All-Equal Satisfiability Problems. In: Giunchiglia, E., Tacchella, A. (eds.) SAT 2003. LNCS, vol. 2919, pp. 172–187. Springer, Heidelberg (2004)
11. Porschen, S., Speckenmeyer, E.: Algorithms for Variable-Weighted 2-SAT and Dual Problems. In: Marques-Silva, J., Sakallah, K.A. (eds.) SAT 2007. LNCS, vol. 4501, pp. 173–186. Springer, Heidelberg (2007)
12. Porschen, S., Speckenmeyer, E., Zhao, X.: Linear CNF formulas and satisfiability. Discrete Appl. Math. 157, 1046–1068 (2009)
13. Ryser, H.J.: Combinatorial Mathematics. Carus Mathematical Monographs 14, Mathematical Association of America (1963)
14. Schaefer, T.J.: The complexity of satisfiability problems. In: Proc. STOC 1978, pp. 216–226. ACM, New York (1978)

Bounds on Threshold of Regular Random k-SAT

Vishwambhar Rathi[1,2], Erik Aurell[1,3], Lars Rasmussen[1,2], and Mikael Skoglund[1,2]

[1] KTH Linnaeus Centre ACCESS, KTH-Royal Institute of Technology,
Stockholm, Sweden
[2] School of Electrical Engineering, KTH-Royal Institute of Technology,
Stockholm, Sweden
[3] Dept. Information and Computer Science, TKK-Helsinki University of Technology,
Espoo, Finland
{vish,eaurell}@kth.se,
{lars.rasmussen,skoglund}@ee.kth.se

Abstract. We consider the regular model of formula generation in conjunctive normal form (CNF) introduced by Boufkhad et. al. in [6]. In [6], it was shown that the threshold for regular random 2-SAT is equal to unity. Also, upper and lower bound on the threshold for regular random 3-SAT were derived. Using the first moment method, we derive an upper bound on the threshold for regular random k-SAT for any $k \geq 3$ and show that for large k the threshold is upper bounded by $2^k \ln(2)$. We also derive upper bounds on the threshold for Not-All-Equal (NAE) satisfiability for $k \geq 3$ and show that for large k, the NAE-satisfiability threshold is upper bounded by $2^{k-1} \ln(2)$. For both satisfiability and NAE-satisfiability, the obtained upper bound matches with the corresponding bound for the uniform model of formula generation [9, 1].

For the uniform model, in a series of break through papers Achlioptas, Moore, and Peres showed that a careful application of the second moment method yields a significantly better lower bound on threshold as compared to any rigorously proven algorithmic bound [3, 1]. The second moment method shows the existence of a satisfying assignment with uniform positive probability (w.u.p.p.). Thanks to the result of Friedgut for uniform model [10], existence of a satisfying assignment w.u.p.p. translates to existence of a satisfying assignment with high probability (w.h.p.). Thus, the second moment method gives a lower bound on the threshold. As there is no known Friedgut type result for regular random model, we assume that for regular random model existence of a satisfying assignments w.u.p.p. translates to existence of a satisfying assignments w.h.p. We derive the second moment of the number of satisfying assignments for regular random k-SAT for $k \geq 3$. There are two aspects in deriving the lower bound using the second moment method. The first aspect is given any k, numerically evaluate the lower bound on the threshold. The second aspect is to derive the lower bound as a function of k for large enough k. We address the first aspect and evaluate the lower bound on threshold. The numerical evaluation suggests that as k increases the obtained lower bound on the satisfiability threshold of a regular random formula converges to the lower bound obtained for the uniform model. Similarly, we obtain lower bounds on the NAE-satisfiability threshold of the regular random formulas and observe that the obtained lower bound seems to converge to the corresponding lower bound for the uniform model as k increases.

O. Strichman and S. Szeider (Eds.): SAT 2010, LNCS 6175, pp. 264–277, 2010.

1   Regular Formulas and Motivation

A clause is a disjunction (OR) of k variables. A formula is a conjunction (AND) of a finite set of clauses. A k-SAT formula is a formula where each clause is a disjunction of k literals. A *legal* clause is one in which there are no repeated or complementary literals. Using the terminology of [6], we say that a formula is *simple* if it consists of only legal clauses. A *configuration* formula is not necessarily legal. A satisfying (SAT) assignment of a formula is a truth assignment of variables for which the formula evaluates to true. A Not-All-Equal (NAE) satisfying assignment is a truth assignment such that every clause is connected to at least one true literal and at least one false literal. We denote the number of variables by n, the number of clauses by m, and the clause density, i.e. the ratio of clauses to variables, by $\alpha = \frac{m}{n}$. We denote the binary entropy function by $h(\cdot)$, $h(x) \triangleq -x\ln(x) - (1-x)\ln(1-x)$, where the logarithm is the natural logarithm.

The popular, uniform k-SAT model generates a formula by selecting uniformly and independently m-clauses from the set of all $2^k \binom{n}{k}$ k-clauses. In this model, the literal degree can vary. We are interested in the model where the literal degree is almost constant, which was introduced in [6]. Suppose each literal has degree r. Then $2nr = km$, which gives $\alpha = 2r/k$. Hence α can only take values from a discrete set of possible values. To circumvent this, we allow each literal to take two possible values for a degree. For a given α, let $r = \frac{k\alpha}{2}$ and $\mathtt{r} = \lfloor r \rfloor$. Each literal has degree either $\mathtt{r}$ or $\mathtt{r}+1$. Also a literal and its negation have the same degree. Thus, we can speak of the degree of a variable which is the same as the degree of its literals. Let the number of variables with degree d be n_d, $d \in \{\mathtt{r}, \mathtt{r}+1\}$. Let $X_1, \ldots, X_{n_\mathtt{r}}$ be the variables which have degree $\mathtt{r}$ and $X_{n_\mathtt{r}+1}, \ldots, X_n$ be the variables with degree $\mathtt{r}+1$. Then,

$$n_\mathtt{r} = n + \mathtt{r}n - \left\lfloor \frac{k\alpha n}{2} \right\rfloor, \quad n_{\mathtt{r}+1} = \left\lfloor \frac{k\alpha n}{2} \right\rfloor - \mathtt{r}n$$

and $n_\mathtt{r} + n_{\mathtt{r}+1} = n$. As we are interested in the asymptotic setting, we will ignore the floor in the sequel. We denote the fraction of variables with degree $\mathtt{r}$ (resp. $\mathtt{r}+1$) by $\Lambda_\mathtt{r}$ (resp. $\Lambda_{\mathtt{r}+1}$) which is given by

$$\Lambda_\mathtt{r} = 1 + \mathtt{r} - \frac{k\alpha}{2}, \quad \Lambda_{\mathtt{r}+1} = \frac{k\alpha}{2} - \mathtt{r}. \tag{1}$$

When $\Lambda_\mathtt{r}$ or $\Lambda_{\mathtt{r}+1}$ is zero, we refer to such formulas as *strictly* regular random formulas. This implies that there is no variation in literal degree. If $\Lambda_\mathtt{r}, \Lambda_{\mathtt{r}+1} > 0$, then we say that the formulas are 2-regular random formulas.

A formula is represented by a bipartite graph. The left vertices represent the literals and right vertices represent the clauses. A literal is connected to a clause if it appears in the clause. There are $k\alpha n$ edges coming out from all the literals and $k\alpha n$ edges coming out from the clauses. We assign the labels from the set $\mathcal{E} = \{1, \ldots, k\alpha n\}$ to edges on both sides of the bipartite graph. In order to generate a formula, we generate a random permutation Π on $\mathcal{E}$. Now we connect an edge i on the literal node side to an edge $\Pi(i)$ on the clause node side. This gives rise to a regular random k-SAT formula. Note that not all the formulas generated by this procedure are simple. However, it was shown in [6] that the threshold is the same for this collection of formulas and the collection of simple formulas. Thus, we can work with the collection of configuration formulas generated by this procedure.

The regular random k-SAT formulas are of interest because such instances are computationally harder than the uniform k-SAT instances. This was experimentally observed in [6], where the authors also derived upper and lower bounds for regular random 3-SAT. The upper bound was derived using the first moment method. The lower bound was derived by analyzing a greedy algorithm proposed in [13]. To the best of our knowledge, there are no known upper and lower bounds on the thresholds for regular random formulas for $k > 3$.

Using the first moment method, we compute an upper bound α_u^* on the satisfiability threshold α^* for regular random formulas for $k \geq 3$. We show that $\alpha^* \leq 2^k \ln(2)$, which coincides with the upper bound for the uniform model. We also apply the first moment method to obtain an upper bound $\alpha_{u,\text{NAE}}^*$ on the NAE-satisfiability threshold α_{NAE}^* of regular random formulas. We show that $\alpha_{\text{NAE}}^* \leq 2^{k-1} \ln(2)$ which coincides with the corresponding bound for the uniform model.

In order to derive a lower bound α_l^* on the threshold, we apply the second moment method to the number of satisfying assignments. The second moment method shows the existence of a satisfying assignment with uniform positive probability (w.u.p.p.). Due to the result of Friedgut for uniform model [10], existence of a satisfying assignment w.u.p.p. translates to existence of a satisfying assignment with high probability (w.h.p.). Thus, the second moment method gives lower bound on the threshold for uniform model. As there is no known Friedgut type result for regular random model, we assume that for regular random model existence of a satisfying assignments w.u.p.p. translates to existence of a satisfying assignments w.h.p. This permits us to say that second moment method gives valid lower bound on the threshold. We compute the second moment of the number of satisfying assignments for regular random model. Similar to the case of the uniform model, we show that for the second moment method to succeed the term corresponding to overlap $n/2$ should dominate other overlap terms. We observe that the obtained lower bound α_l^* converges to the corresponding lower bound of the uniform model, which is $2^k \ln(2) - (k+1)\frac{\ln(2)}{2} - 1$ as k increases. Similarly, by computing the second moment of the number of NAE-satisfying assignments we obtain that $\alpha_{l,\text{NAE}}$ converges to the corresponding bound $2^{k-1} \ln(2) - O(1)$ for the uniform model. The lower bounds are not obtained explicitly as computing the second moment requires finding all the positive solutions of a system of polynomial equations. For small values of k, this can be done exactly. However, for large values of k we resort to a numerical approach. Our main contribution is that we obtain almost matching lower and upper bounds on the satisfiability (resp. NAE-satisfiability) threshold for the regular random formulas. Thus, we answer in affirmative the following question posed in [1]: *Does the second moment method perform well for problems that are symmetric "on average"? For example, does it perform well for regular random k-SAT where every literal appears an equal number of times?*.

In the next section, we obtain an upper bound on the satisfiability threshold and NAE satisfiability threshold.

2   Upper Bound on Threshold via First Moment

Let X be a non-negative integer-valued random variable and $E(X)$ be its expectation. Then the first moment method gives: $P(X > 0) \leq E(X)$. Note that by choosing X to be the number of solutions of a random formula, we can obtain an upper bound on the threshold α^* beyond which no solution exists with probability one. This upper bound corresponds to the largest value of α at which the average number of solutions goes to zero as n tends to infinity. In the following lemma, we derive the first moment of the number of SAT solutions of the regular random k-SAT for $k \geq 3$.

Lemma 1. *Let $N(n,\alpha)$ (resp. $N_{NAE}(n,\alpha)$) be the number of satisfying (resp. NAE satisfying) assignments for a randomly generated regular k-SAT formula. Then[1],*

$$E(N(n,\alpha)) = 2^n \frac{\left(\left(\frac{k\alpha n}{2}\right)!\right)^2}{(k\alpha n)!} \mathrm{coef}\left(\left(\frac{p(x)}{x}\right)^{\alpha n}, x^{\frac{k\alpha n}{2} - \alpha n}\right), \tag{2}$$

$$E(N_{NAE}(n,\alpha)) = 2^n \frac{\left(\left(\frac{k\alpha n}{2}\right)!\right)^2}{(k\alpha n)!} \mathrm{coef}\left(\left(\frac{p_{NAE}(x)}{x}\right)^{\alpha n}, x^{\frac{k\alpha n}{2} - \alpha n}\right), \tag{3}$$

where

$$p(x) = (1+x)^k - 1, \quad p_{NAE}(x) = (1+x)^k - 1 - x^k, \tag{4}$$

and $\mathrm{coef}\left(p(x)^{\alpha n}, x^{\frac{k\alpha n}{2}}\right)$ denotes the coefficient of $x^{\frac{k\alpha n}{2}}$ in the expansion of $p(x)^{\alpha n}$.

Proof. Due to symmetry of the formula generation, any assignment of variables has the same probability of being a solution. This implies

$$E(N(n,\alpha)) = 2^n P(X = \{0,\dots,0\} \text{ is a solution}).$$

The probability of the all-zero vector being a solution is given by

$$P(X = \{0,\dots,0\} \text{ is a solution}) =$$
$$\frac{\text{Number of formulas for which } X = \{0,\dots,0\} \text{ is a solution}}{\text{Total number of formulas}}.$$

The total number of formulas is given by $(k\alpha n)!$. The total number of formulas for which the all-zero assignment is a solution is given by

$$\left(\left(\frac{k\alpha n}{2}\right)!\right)^2 \mathrm{coef}\left(p(x)^m, x^{\frac{k\alpha n}{2}}\right).$$

The factorial terms correspond to permuting the edges among true and false literals. Note that there are equal numbers of true and false literals. The generating function $p(x)$ corresponds to placing at least one positive literal in a clause. With these results and observing that

$$\mathrm{coef}\left(p(x)^{\alpha n}, x^{\frac{k\alpha n}{2}}\right) = \mathrm{coef}\left(\left(\frac{p(x)}{x}\right)^{\alpha n}, x^{\frac{k\alpha n}{2} - \alpha n}\right),$$

---

[1] We assume that $k\alpha n$ is an even integer.

we obtain (2). The derivation for $E(N_{NAE}(n,\alpha))$ is identical except that the generating function for clauses is given by $p_{NAE}(x)$.   □

We now state the Hayman method to approximate the coef-term which is asymptotically correct [11].

Lemma 2 (Hayman Method). *Let* $q(y) = \sum_i q_i y^i$ *be a polynomial with non-negative coefficients such that* $q_0 \neq 0$ *and* $q_1 \neq 0$*. Define*

$$a_q(y) = y \frac{dq(y)}{dy} \frac{1}{q(y)}, \qquad b_q(y) = y \frac{da_q(y)}{dy}. \tag{5}$$

Then,

$$\mathrm{coef}(q(y)^n, y^{\omega n}) = \frac{q(y_\omega)^n}{(y_\omega)^{\omega n} \sqrt{2\pi n b_q(y_\omega)}} (1 + o(1)), \tag{6}$$

where y_ω *is the unique positive solution of the saddle point equation* $a_q(y) = \omega$.

We now use Lemma 2 to compute the expectation of the total number of solutions.

Lemma 3. *Let* $N(n,\alpha)$ *(resp.* $N_{NAE}(n,\alpha)$*) denote the total number of satisfying (resp. NAE satisfying) assignments of a regular random k-SAT formula. Let* $q(x) = \frac{p(x)}{x}$*,* $q_{NAE}(x) = \frac{p_{NAE}}{x}$*, where* $p(x)$ *and* $p_{NAE}(x)$ *is defined in (4). Then,*

$$E(N(n,\alpha)) = \sqrt{\frac{k}{4b_q(x_k)}} e^{n\left(\ln(2) - k\alpha \ln(2) + \alpha \ln(q(x_k)) - \left(\frac{k\alpha}{2} - \alpha\right) \ln(x_k)\right)} (1 + o(1)), \tag{7}$$

$$E(N_{NAE}(n,\alpha)) = \frac{\sqrt{k} e^{n\left(\ln(2)(1 - k\alpha) + \alpha \ln\left(q_{NAE}(x_{k,NAE})\right) - \left(\frac{k\alpha}{2} - \alpha\right) \ln\left(x_{k,NAE}\right)\right)}}{\sqrt{4b_{q_{NAE}}(x_{k,NAE})}} (1 + o(1)), \tag{8}$$

where x_k *(resp.* $x_{k,NAE}$*) is the positive solution of* $a_q(x) = \frac{k}{2} - 1$ *(resp.* $a_{q_{NAE}}(x) = \frac{k}{2} - 1$*). The quantity* $a_q(x)$*,* $a_{q_{NAE}}(x)$*,* $b_q(x)$*, and* $b_{q_{NAE}}(x)$ *are defined according to (5).*

In the following lemma we derive explicit upper bounds on the satisfiability and NAE satisfiability thresholds for $k \geq 3$.

Lemma 4 (Upper bound). *Let* α^* *(resp.* α^*_{NAE}*) be the satisfiability (resp. NAE satisfiability) threshold for the regular random k-SAT formulas. Define* α^*_u *(resp.* $\alpha^*_{u,NAE}$*) to be the upper bound on* α^* *(resp.* α^*_{NAE}*) obtained by the first moment method. Then,*

$$\alpha^* \leq \alpha^*_u \leq 2^k \ln(2)(1 + o_k(1)), \qquad \alpha^*_{NAE} \leq \alpha^*_{u,NAE} = 2^{k-1} \ln(2) - \frac{\ln(2)}{2} - o_k(1). \tag{9}$$

Proof. We observe that the solution x_k of the saddle point equation $a_q(x) = \frac{k}{2} - 1$ satisfies: $x_k = \mathrm{argmin}_{x>0} \frac{q(x)}{x^{\frac{k}{2}-1}}$, where $a_q(x)$ is defined according to (5). This implies that we obtain the following upper bound on the growth rate of $E(N(n,\alpha))$ for any $x > 0$,

$$\lim_{n \to \infty} \frac{\ln(E(N(n,\alpha)))}{n} \leq \ln(2) - k\alpha \ln(2) + \alpha \ln(q(x)) - \left(\frac{k\alpha}{2} - \alpha\right) \ln(x). \tag{10}$$

We substitute $x = 1 - \frac{1}{2^k}$ in (10). Then we use the series expansion of $\ln(1-x)$, $1/i \geq 1/2^i$, and $-1/i \geq -1$ to obtain the following upper bound on the threshold,

$$\alpha^* \leq \frac{2^k \ln(2)}{\frac{1}{\left(1-\frac{1}{2^{k+1}}\right)^k} + \frac{k}{2^{k+4}} + \frac{1}{2^{k+2}\left(1-\frac{1}{2^{k+1}}\right)^{2k}} - \frac{1}{2^{k+1}}}. \tag{11}$$

The summation of the last three terms in the denominator of (11) is positive. This can be easily seen for $k \geq 8$. For $3 \leq k < 8$, it can be verified by explicit calculation. Dropping this summation in (11), we obtain the desired upper bound on the threshold. To derive the bound for NAE satisfiability, we note that $x_{k,\text{NAE}} = 1$ for $k \geq 3$. By substituting this in the exponent of $E(N_{\text{NAE}})$ and equating it to zero, we obtain the desired expression for $\alpha^*_{u,\text{NAE}}$. $\qquad\Box$

In the next section we use the second moment method to obtain lower bounds on the satisfiability and NAE satisfiability thresholds of regular random k-SAT.

3 Second Moment

A lower bound on the threshold can be obtained by the second moment method. The second moment method is governed by the following equation

$$P(X > 0) \geq \frac{E(X)^2}{E(X^2)}. \tag{12}$$

In this section we compute the second moment of $N(n, \alpha)$ and $N_{\text{NAE}}(n, \alpha)$. Our computation of the second moment is inspired by the computation of the second moment for the weight and stopping set distributions of regular LDPC codes in [14, 15] (see also [4]). We compute the second moment in the next lemma.

Lemma 5. *Let $N(n, \alpha)$ be the number of satisfying solutions to a regular random k-SAT formula. Define the function $f(x_1, x_2, x_3)$ by*

$$f(x_1, x_2, x_3) = (1 + x_1 + x_2 + x_3)^k - (1 + x_1)^k - (1 + x_3)^k + 1. \tag{13}$$

If the regular random formulas are strictly regular, then

$$E\left(N(n,\alpha)^2\right) =$$
$$\sum_{i=0}^{n} 2^n \binom{n}{i} \frac{((\mathbf{r}(n-i))!)^2 ((\mathbf{r}i)!)^2 \operatorname{coef}\left(f(x_1,x_2,x_3)^{\alpha n}, x_1^{\mathbf{r}(n-i)} x_2^{\mathbf{r}i} x_3^{\mathbf{r}(n-i)}\right)}{(k\alpha n)!}. \tag{14}$$

If the regular random formulas are 2-regular, then

$$E\left(N(n,\alpha)^2\right) = \sum_{i_{\mathbf{r}}=0}^{n_{\mathbf{r}}} \sum_{i_{\mathbf{r}+1}=0}^{n_{\mathbf{r}+1}} 2^n \binom{n_{\mathbf{r}}}{i_{\mathbf{r}}} \binom{n_{\mathbf{r}+1}}{i_{\mathbf{r}+1}} \left(\left(\frac{k\alpha n}{2} - \mathbf{r}i_{\mathbf{r}} - (\mathbf{r}+1)i_{\mathbf{r}+1}\right)!\right)^2$$
$$\frac{((\mathbf{r}i_{\mathbf{r}} + (\mathbf{r}+1)i_{\mathbf{r}+1})!)^2}{(k\alpha n)!} \operatorname{coef}\left(f(x_1,x_2,x_3)^{\alpha n}, (x_1 x_3)^{\frac{k\alpha n}{2} - \mathbf{r}i_{\mathbf{r}} - (\mathbf{r}+1)i_{\mathbf{r}+1}} x_2^{\mathbf{r}i_{\mathbf{r}} + (\mathbf{r}+1)i_{\mathbf{r}+1}}\right). \tag{15}$$

For both the strictly regular and the 2-regular case, the expression for $E\left(N_{NAE}(n,\alpha)^2\right)$ is the same as that for $E\left(N(n,\alpha)^2\right)$ except replacing the generating function $f(x_1,x_2,x_3)$ by $f_{NAE}(x_1,x_2,x_3)$, which is given by

$$f_{NAE}(x_1,x_2,x_3) = (1+x_1+x_2+x_3)^k -$$
$$\left((1+x_1)^k + (1+x_3)^k - 1 + (x_1+x_2)^k - x_1^k + (x_2+x_3)^k - x_2^k - x_3^k\right). \quad (16)$$

Proof. Let $\mathbb{1}_{XY}$ be the indicator variable which evaluates to 1 if the truth assignments X and Y satisfy a randomly regular k-SAT formula. Then,

$$E(N(n,\alpha)^2) = \sum_{X,Y\in\{0,1\}^n} E\left(\mathbb{1}_{XY}\right) = 2^n \sum_{Y\in\{0,1\}^n} P\left(\mathbf{0} \text{ and } Y \text{ are solutions}\right).$$

The last simplification uses the fact that the number of formulas which are satisfied by both X and Y depends only on the number of variables on which X and Y agree. Thus, we fix X to be the all-zero vector.

We now consider the strictly regular case. The probability that the all-zero truth assignment and the truth assignment Y both are solutions of a randomly chosen regular formula depends only on the *overlap*, i.e., the number of variables where the two truth assignments agree. Thus for a given overlap i, we can fix Y to be equal to zero in the first i variables and equal to 1 in the remaining variables. This gives,

$$E(N(n,\alpha)^2) = \sum_{i=0}^{n} 2^n \binom{n}{i} P\left(\mathbf{0} \text{ and } Y \text{ are solutions}\right). \quad (17)$$

In order to evaluate the probability that both $\mathbf{0}$ and Y are solutions for a given overlap i, we observe that there are four different types of edges connecting the literals and the clauses. There are $r(n-i)$ **type 1** edges which are connected to true literals w.r.t. the $\mathbf{0}$ truth assignment and false w.r.t. to the Y truth assignment. The ri **type 2** edges are connected to true literals w.r.t. both the truth assignments. There are $r(n-i)$ **type 3** edges which are connected to false literals w.r.t. the $\mathbf{0}$ truth assignment and true literals w.r.t. to the Y truth assignment. The ri **type 4** edges are connected to false literals w.r.t. both the truth assignments. Let $f(x_1,x_2,x_3)$ be the generating function counting the number of possible edge connections to a clause, where the power of x_i gives the number of edges of type i, $i \in \{1,2,3\}$. A clause is satisfied if it is connected to at least one type 2 edge. Otherwise, it is satisfied if it is connected to at least one type 1 and at least one type 3 edge. Then the generating function $f(x_1,x_2,x_3)$ is given as in (13). Using this, we obtain

$$P\left(\mathbf{0} \text{ and } Y \text{ are solutions}\right) =$$
$$\frac{((r(n-i))!)^2\,((ri)!)^2\,\mathrm{coef}\left(f(x_1,x_2,x_3)^{\alpha n}, x_1^{r(n-i)} x_2^{ri} x_3^{r(n-i)}\right)}{(k\alpha n)!}, \quad (18)$$

where $(k\alpha n)!$ is the total number of formulas. Consider a given formula which is satisfied by both truth assignments $\mathbf{0}$ and Y. If we permute the positions of type 1 edges on the clause side, we obtain another formula having $\mathbf{0}$ and Y as solutions. The argument holds true for the type i edges, $i \in \{2,3,4\}$. This explains the term $(\mathbf{r}(n-i))!$ in (18) which corresponds to permuting the type 1 edges (it is squared because of the same contribution from type 3 edges). Similarly, $(\mathbf{r}i)!^2$ corresponds to permuting type 2 and type 4 edges. Combining (17) and (18), we obtain the desired expression for the second moment of the number of solutions as given in (14).

We now consider the two regular case. Note that in this case the equivalent equation corresponding to (17) is

$$E(N(n,\alpha)^2) = \sum_{i_{\mathbf{r}}=0}^{n_{\mathbf{r}}} \sum_{i_{\mathbf{r}+1}=0}^{n_{\mathbf{r}+1}} 2^n \binom{n_{\mathbf{r}}}{i_{\mathbf{r}}} \binom{n_{\mathbf{r}+1}}{i_{\mathbf{r}+1}} P((\mathbf{0},Y) \text{ is a solution}), \qquad (19)$$

where $i_{\mathbf{r}}$ (resp. $i_{\mathbf{r}+1}$) is the variable corresponding to the overlap between truth assignments $\mathbf{0}$ and Y among variables with degree $\mathbf{r}$ (resp. $\mathbf{r}+1$). Similarly, the equivalent of (18) is given by

$$P(\mathbf{0} \text{ and } Y \text{ are solutions}) = ((\mathbf{r}(n_{\mathbf{r}}-i_{\mathbf{r}}) + (\mathbf{r}+1)(n_{\mathbf{r}+1}-i_{\mathbf{r}+1}))!)^2$$
$$\times \frac{((\mathbf{r}i_{\mathbf{r}} + (\mathbf{r}+1)i_{\mathbf{r}+1})!)^2}{(k\alpha n)!}$$
$$\times \text{coef}\left(f(x_1,x_2,x_3)^{\alpha n}, (x_1x_3)^{\mathbf{r}(n_{\mathbf{r}}-i_{\mathbf{r}})+(\mathbf{r}+1)(n_{\mathbf{r}+1}-i_{\mathbf{r}+1})} x_2^{\mathbf{r}i_{\mathbf{r}}+(\mathbf{r}+1)i_{\mathbf{r}+1}}\right). \qquad (20)$$

Combining (19) and (20), and observing that $\mathbf{r}n_{\mathbf{r}} + (\mathbf{r}+1)n_{\mathbf{r}+1} = \frac{k\alpha n}{2}$, we obtain (15). The derivation of $E\left(N_{\text{NAE}}(n,\alpha)^2\right)$ is identical except the generating function for NAE-satisfiability of a clause is different. This can be easily derived by observing that a clause is not NAE-satisfied for the following edge connections. Consider the case when a clause is connected to only one type of edge, then it is not NAE-satisfied. Next consider the case when a clause is connected to two types of edges. Then the combinations of type 1 and type 4, type 3 and type 4, type 1 and type 2, or type 2 and type 3 do not NAE-satisfies a clause. This gives the generating function $f_{\text{NAE}}(x_1,x_2,x_3)$ defined in (16).   □

In order to evaluate the second moment, we now present the multidimensional saddle point method in the next lemma [5]. A detailed technical exposition of the multidimensional saddle point method can be found in Appendix D of [18].

Theorem 1. *Let $\underline{i} := (i_1,i_2,i_3)$, $\underline{j} := (j_1,j_2,j_3)$, and $\underline{x} = (x_1,x_2,x_3)$*

$$0 < \lim_{n \to \infty} i_1/n, \quad 0 < \lim_{n \to \infty} i_2/n, \quad 0 < \lim_{n \to \infty} i_3/n.$$

Let further $f(\underline{x})$ be as defined in (13) and $\underline{t} = (t_1,t_2,t_3)$ be a positive solution of the saddle point equations $a_f(\underline{x}) \triangleq \left\{ x_i \frac{\partial \ln(f(x_1,x_2,x_3))}{\partial x_i} \right\}_{i=1}^3 = \frac{i}{\alpha n}$. Then $\text{coef}\left(f(\underline{x})^{\alpha n}, \underline{x}^{\underline{i}}\right)$ can be approximated as ,

$$\text{coef}\left(f(\underline{x})^{\alpha n}, \underline{x}^{\underline{i}}\right) = \frac{f(\underline{t})^{\alpha n}}{(\underline{t})^{\underline{i}} \sqrt{(2\pi\alpha n)^3 |B(\underline{t})|}} (1+o(1)),$$

using the saddle point method for multivariate polynomials, where $B(\underline{x})$ is a 3×3 matrix whose elements are given by $B_{i,j} = x_j \frac{\partial a_{fi}(x_1,x_2,x_3)}{\partial x_j} = B_{j,i}$ and $a_{fi}(\underline{x})$ is the ith coordinate of $a_f(\underline{x})$. Also, $\mathrm{coef}\left(f(\underline{x})^{\alpha n}, \underline{x}^{\underline{j}}\right)$ can be approximated in terms of $\mathrm{coef}\left(f(\underline{x})^{\alpha n}, \underline{x}^{\underline{i}}\right)$. This approximation is called the local limit theorem of $\underline{j}$ around $\underline{i}$. Explicitly, if $\underline{u} := \frac{1}{\sqrt{\alpha n}}(\underline{j} - \underline{i})$ and $\|\underline{u}\| = O((\ln n)^{\frac{1}{3}})$, then

$$\mathrm{coef}\left(f(\underline{x})^{\alpha n}, \underline{x}^{\underline{j}}\right) = \underline{t}^{\underline{i}-\underline{j}} \exp\left(-\frac{1}{2}\underline{u} \cdot B(\underline{t})^{-1} \cdot \underline{u}^T\right) \mathrm{coef}\left(f(\underline{x})^{\alpha n}, \underline{x}^{\underline{i}}\right)(1 + o(1)).$$

Because of the relative simplicity of the expression for the second moment, we explain its computation in detail for the strictly regular case. Then we will show how the arguments can be easily extended to the 2-regular case. The derivation for the NAE-satisfiability is identical for both cases.

Theorem 2. *Consider the strictly regular random k-SAT model with literal degree* $\mathbf{r}$. *Let* $S(i)$ *denote the* i^{th} *summation term in (14), and* $\gamma = i/n$. *If* $S(n/2)$ *is the dominant term i.e.,*

$$\lim_{n \to \infty} \frac{\ln\left(S\left(\frac{n}{2}\right)\right)}{n} > \lim_{n \to \infty} \frac{\ln\left(S(\gamma n)\right)}{n}, \quad \gamma \in [0, 1], \gamma \neq \frac{1}{2}, \tag{21}$$

then with positive probability a randomly chosen formula has a satisfying assignment, i.e.

$$\lim_{n \to \infty} P\left(N(\alpha, n) > 0\right) \geq \frac{2\sqrt{|B_f(x_k, x_k^2, x_k)|}}{\sigma_s b_q(x_k)\sqrt{k}}, \tag{22}$$

where x_k *is the solution of the saddle point equation* $a_q(x) = \frac{k}{2} - 1$ *defined in Lemma 3,* $a_q(x)$ *and* $b_q(x)$ *are defined according to (5),* $B_f(x_k, x_k^2, x_k)$ *is defined as in Theorem 1, and the "normalized variance"* σ_s^2 *of the summation term around* $S\left(\frac{n}{2}\right)$ *is given by*

$$\sigma_s^2 = \frac{1}{4 + \frac{kr}{2}([-1, 1, -1] \cdot B_f(x_k, x_k^2, x_K)^{-1} \cdot [-1, 1, -1]^T) - 8r}. \tag{23}$$

Let $\mathbf{r}^*$ *be the largest literal degree for which* $S(n/2)$ *is the dominant term, i.e. (21) holds, then the threshold* α^* *is lower bounded by* $\alpha^* \geq \alpha_l^* \triangleq \frac{2r^*}{k}$.

Proof. From (14) and Theorem 1, the growth rate of $S(\gamma n)$ is given by,

$$s(\gamma) \triangleq \lim_{n \to \infty} \frac{\ln\left(S(\gamma n)\right)}{n} =$$
$$(1 - k\alpha)(\ln(2) + h(\gamma)) + \alpha \ln(f(t_1, t_2, t_3)) - r(1 - \gamma)(\ln(t_1) + \ln(t_3)) - r\gamma \ln(t_2), \tag{24}$$

where t_1, t_2, t_3 is a positive solution of the saddle point equations as defined in Theorem 1,

$$a_f(\underline{t}) \triangleq \left\{ t_1 \frac{\partial \ln\left(f(t_1, t_2, t_3)\right)}{\partial t_1}, \quad t_2 \frac{\partial \ln\left(f(t_1, t_2, t_3)\right)}{\partial t_2}, \quad t_3 \frac{\partial \ln\left(f(t_1, t_2, t_3)\right)}{\partial t_3} \right\} =$$
$$\left\{ \frac{k}{2}(1 - \gamma), \frac{k}{2}\gamma, \frac{k}{2}(1 - \gamma) \right\}. \tag{25}$$

In order to compute the maximum exponent of the summation terms, we compute its derivative and equate it to zero,

$$\frac{ds(\gamma)}{d\gamma} = (1 - k\alpha)\ln\left(\frac{1-\gamma}{\gamma}\right) + r\ln(t_1) - r\ln(t_2) + r\ln(t_3) = 0. \qquad (26)$$

Note that the derivatives of t_1, t_2 and t_3 w.r.t. γ vanish as they satisfy the saddle point equation. Every positive solution (t_1, t_2, t_3) of (25) satisfies $t_1 = t_3$ as (25) and $f(t_1, t_2, t_3)$ are symmetric in t_1 and t_3. If $\gamma = 1/2$ is a maximum, then the vanishing derivative in (26) and equality of t_1 and t_3 imply $t_2 = t_1^2$. We substitute $\gamma = 1/2$, $t_1 = t_3$, and $t_2 = t_1^2$ in (25). This reduces (25) to the saddle point equation corresponding to the polynomial $q(x)$ defined in Lemma 3 whose solution is denote by x_k. Then by observing $f(x_k, x_k^2, x_k) = p(x_k)^2$, we have

$$S(n/2) = \frac{k^{3/2}}{2^{7/2}\sqrt{\pi n}\sqrt{|B_f(x_k, x_k^2, x_k)|}} e^{n(2\ln(2)(1-k\alpha)+2\alpha\ln(p(x_k))-k\alpha\ln(x_k))}(1+o(1)). \qquad (27)$$

Using the relation that $q(x) = \frac{p(x)}{x}$, we note that the exponent of $S(n/2)$ is twice the exponent of the first moment of the total number of solutions as given in (7). In order to compute the sum over $S(\gamma n)$, we now use Laplace's method, a detailed discussion of which can be found in [12,7,8]. We want to approximate the term $S(n/2 + \Delta i)$ in terms of $S(n/2)$. For the coef terms, we make use of the local limit theorem given in Theorem 1 and for the factorial terms we make use of Stirling's approximation. This gives,

$$S(n/2 + \Delta i) = S(n/2)e^{-\frac{\Delta i^2}{2n\sigma_s^2}}(1+o(1)), \quad \text{where } \Delta i = O(n^{1/2}\ln(n)^{1/3}). \qquad (28)$$

Note that in the exponent on the R.H.S. of (28), the linear terms in Δi are absent as the derivative of the exponent vanishes at $\gamma = 1/2$. As the deviation around the term $S(n/2)$ is $\Theta(\sqrt{n})$ and the approximation is valid for $\Delta i = O(\sqrt{n}\ln(n)^{1/3})$, the dominant contribution comes from $-\Theta(\sqrt{n}) \leq \Delta i \leq \Theta(\sqrt{n})$. We are now ready to obtain the estimate for the second moment.

$$E(N^2(\alpha, n)) \stackrel{(28)}{=} \sum_{\Delta i = -c\sqrt{n}}^{c\sqrt{n}} S(n/2)e^{-\frac{\Delta i^2}{2n\sigma_s^2}}(1+o(1)), \qquad (29)$$

$$= S(n/2)\int_{\delta=-\infty}^{\infty} e^{-\frac{\delta^2}{2n\sigma_s^2}}d\delta(1+o(1)). \qquad (30)$$

$$= S(n/2)\sqrt{2\pi n\sigma_s^2}(1+o(1)). \qquad (31)$$

We can replace the sum by an integral by choosing sufficiently large c. Using the second moment method given in (12) and combining Lemma 3, (27), and (31), we obtain

$$P(N(\alpha, n) > 0) \geq \frac{E(N(\alpha, n)^2)}{E(N(\alpha, n)^2)} = \frac{2\sqrt{|B_f(x_k, x_k^2, x_k)|}}{\sigma_s b_q(x_k)\sqrt{k}}(1+o(1)). \qquad (32)$$

Letting n go to infinity, we obtain (22). Clearly, if the supremum of the growth rate of $S(\gamma n)$ is not achieved at $\gamma = 1/2$, then the lower bound given by the second moment method converges to zero. This gives the desired lower bound on the threshold.   □

We can easily extend this result to the 2-regular case. In the following theorem we accomplish this task. Due to space limitation, we omit explanation of some steps which can be found in [17].

Theorem 3. *Consider the 2-regular random k-SAT model where the number of variables with degree r (resp. $r + 1$) is $n_r = \Lambda_r n$ (resp. $n_{r+1} = \Lambda_{r+1} n$). Let $S(i_r, i_{r+1}) \triangleq S(\gamma_r n_r, \gamma_{r+1} n_{r+1})$ be the summation term on the R.H.S. of (15) corresponding to overlap i_r(resp. i_{r+1}) on the degree r(resp. $r + 1$) literals. Let $g(\gamma_r, \gamma_{r+1})$ be the growth rate of $S(\gamma_r n_r, \gamma_{r+1} n_{r+1})$ i.e. $g(\gamma_r, \gamma_{r+1}) \triangleq \lim_{n\to\infty} \frac{\ln(S(n_r \gamma_r, n_{r+1}))}{n}$. If*

$$g\left(\frac{1}{2},\frac{1}{2}\right) > g(\gamma_r, \gamma_{r+1}), \gamma_r \in [0,1], \gamma_{r+1} \in [0,1], \gamma_r \neq \frac{1}{2}, \gamma_{r+1} \neq \frac{1}{2},$$

then with positive probability a randomly chosen formula has a solution. More precisely,

$$\lim_{n\to\infty} P(N(\alpha,n) > 0) \geq \frac{\sqrt{|B_f(x_k, x_k^2, x_k)|\Lambda_r \Lambda_{r+1}}}{b_q(x_k)\sqrt{k|\Sigma|}}. \tag{33}$$

The definition of x_k, $b_q(x_k)$, and $B_f(x_k, x_k^2, x_k)$ is same as in the Theorem 2. The 2×2 matrix Σ is defined via,

$$C_f = [-1,1,-1].(B_f(x_k, x_k^2, x_k))^{-1}.[-1,1,-1]^T, \quad A = \frac{4}{\Lambda_r} + 2r^2\left(\frac{C_f}{2\alpha} - \frac{4}{k\alpha}\right),$$

$$B = \frac{4}{\Lambda_{r+1}} + \frac{2(r+1)^2}{\alpha}\left(\frac{C_f}{2} - \frac{4}{k}\right), C = \frac{2r(r+1)}{\alpha}\left(\frac{C_f}{2} - \frac{4}{k}\right), \text{ then } \Sigma = \begin{bmatrix} A & B \\ B & C \end{bmatrix}^{-1}. \tag{34}$$

The threshold α^* is lower bounded by α_l^*, where α_l^* is defined by

$$\alpha_l^* = \sup\left\{\alpha : g\left(\frac{1}{2},\frac{1}{2}\right) > g(\gamma_r, \gamma_{r+1}), \gamma_r \in [0,1], \gamma_{r+1} \in [0,1], \gamma_r \neq \frac{1}{2}, \gamma_{r+1} \neq \frac{1}{2}\right\}.$$

Proof. Define $\Gamma(\gamma_r, \gamma_{r+1}) = r\Lambda_r\gamma_r + (r+1)\Lambda_{r+1}\gamma_{r+1}$. Then by using Theorem 1 and Stirling's approximation, we obtain

$$g(\gamma_r, \gamma_{r+1}) = \ln(2) + \Lambda_r h(\gamma_r) + \Lambda_{r+1} h(\gamma_{r+1})$$

$$+ (k\alpha - 2\Gamma(\gamma_r, \gamma_{r+1}))\ln\left(\frac{k\alpha}{2} - \Gamma(\gamma_r, \gamma_{r+1})\right)$$

$$+ 2(\Gamma(\gamma_r, \gamma_{r+1}))\ln(\Gamma(\gamma_r, \gamma_{r+1})) - k\alpha\ln(k\alpha)$$

$$+ \alpha\ln\left(f((t))\right) - \left(\frac{k\alpha}{2} - \Gamma(\gamma_r, \gamma_{r+1})\right)\ln(t_1 t_3) - (\Gamma(\gamma_r, \gamma_{r+1}))\ln(t_2), \tag{35}$$

where $\underline{t} = \{t_1, t_2, t_3\}$ is a positive solution of the saddle point point equations as given in Theorem 1,

$$a_f(\underline{t}) = \left\{ \frac{k}{2} - \frac{\Gamma(\gamma_r, \gamma_{r+1})}{\alpha}, \frac{\Gamma(\gamma_r, \gamma_{r+1})}{\alpha}, \frac{k}{2} - \frac{\Gamma(\gamma_r, \gamma_{r+1})}{\alpha} \right\}, \qquad (36)$$

corresponding to the coefficient term of power of $f(x_1, x_2, x_3)$. In order to obtain the maximum exponent, we take the partial derivatives of $g(\gamma_r, \gamma_{r+1})$ with respect to γ_r and γ_{r+1} and equate them to zero. This gives the following equations.

$$\ln\left(\frac{1-\gamma_r}{\gamma_r}\right) - 2r\ln\left(\frac{k\alpha}{2} - \Gamma(\gamma_r, \gamma_{r+1})\right) + 2r\ln\left(\Gamma(\gamma_r, \gamma_{r+1})\right) + r\ln\left(\frac{t_1 t_3}{t_2}\right) = 0,$$

$$\ln\left(\frac{1-\gamma_{r+1}}{\gamma_{r+1}}\right) - 2(r+1)\ln\left(\frac{k\alpha}{2} - \Gamma(\gamma_r, \gamma_{r+1})\right)$$

$$+ 2(r+1)\ln\left(\Gamma(\gamma_r, \gamma_{r+1})\right) + (r+1)\ln\left(\frac{t_1 t_3}{t_2}\right) = 0. \quad (37)$$

Note that $t_1 = t_3 = x_k$, $t_2 = x_k^2$, $\gamma_r = \frac{1}{2}$, and $\gamma_{r+1} = \frac{1}{2}$ is a solution of (36), (37), which corresponds to $S\left(\frac{n_r}{2}, \frac{n_{r+1}}{2}\right)$, where x_k is the solution of the saddle point equation corresponding to $q(x)$ defined in Lemma 3. We recall that for the second moment method to work, the maximum exponent should be equal to twice the exponent of the average number of solutions. Indeed by the proposed solution, the term $S\left(\frac{n_r}{2}, \frac{n_{r+1}}{2}\right)$ has an exponent which is twice that of the average number of solutions. If this is also the maximum, then we have the desired result. Assuming that $S\left(\frac{n_r}{2}, \frac{n_{r+1}}{2}\right)$ has the maximum exponent, we now compute the second moment of the total number of solutions. By using Stirling's approximation and the local limit result of Theorem 1, we obtain

$$\frac{S\left(i_r + \Delta i_r, i_{r+1} + \Delta i_{r+1}\right)}{S(i_r, i_{r+1})} = e^{-\frac{1}{2n}[\Delta i_r, \Delta i_{r+1}] \cdot \Sigma^{-1} \cdot [\Delta i_r, \Delta i_{r+1}]^T}(1 + o(1)), \qquad (38)$$

where the matrix Σ is defined in (34). Using the same series of arguments as in Theorem 2, we obtain

$$E\left(N(\alpha, n)^2\right) = \sum_{\Delta i_r, \Delta i_{r+1}} S\left(\frac{n_r}{2}, \frac{n_{r+1}}{2}\right) e^{-\frac{1}{2n}[\Delta i_r, \Delta i_{r+1}] \cdot \Sigma^{-1} \cdot [\Delta i_r, \Delta i_{r+1}]^T}(1 + o(1)), \quad (39)$$

$$= S\left(\frac{n_r}{2}, \frac{n_{r+1}}{2}\right) \int_{-\infty}^{\infty}\int_{-\infty}^{\infty} e^{-\frac{1}{2n}[x_r, x_{r+1}] \cdot \Sigma^{-1} \cdot [x_r, x_{r+1}]^T} dx_r dx_{r+1}(1 + o(1)), \quad (40)$$

$$= \frac{\sqrt{|\Sigma|}k^{\frac{3}{2}}}{4\sqrt{|B_f(x_k, x_k^2, x_k)|\Lambda_r\Lambda_{r+1}}} e^{2n\left(\ln(2) - k\alpha\ln(2) + \alpha\ln(p(x_k)) - \frac{k\alpha}{2}\ln(x_k)\right)}(1 + o(1)). \quad (41)$$

By using the second moment method, we obtain the bound given in (22). Note that the second moment method fails if the term $S\left(\frac{n_r}{2}, \frac{n_{r+1}}{2}\right)$ is not the dominant term. This gives the lower bound α_l^* on α^*. $\qquad \square$

In the next section we discuss the obtained lower and upper bounds on the satisfiability threshold and NAE-satisfiability threshold.

4   Bounds on Threshold

In Table 1, lower bounds and upper bounds for the satisfiability threshold are given. The upper bound is computed by the first moment method. As expected, we obtain the same upper bound for regular random 3-SAT as given in [6]. The lower bound is derived by the second moment method for strictly regular random k-SAT. In order to apply the second moment method, we have to verify that $s(\gamma)$, defined in (24), attains its maximum at $\gamma = \frac{1}{2}$ over the unit interval. This requires that $\gamma = \frac{1}{2}$ is a positive solution of the system of equations consisting of (25) and (26) and it corresponds to a global maximum over $\gamma \in [0,1]$. Also, σ_s^2 defined in (23) should be positive. The system of equations (25), (26) is equivalent to a system of polynomial equations. For small value of k, we can solve this system of polynomial equations and verify the desired conditions. In Table 1 this has been done for $k = 3, 4$. The obtained lower bound for 3-SAT is 2.667 which is an improvement over the algorithmic lower bound 2.46 given in [6]. For larger values of k, the degree of monomials in (26) grows exponentially in k. Thus, solving (25) and (26) becomes computationally difficult. However, $s(\gamma)$ can be easily computed as its computation requires solving only (25), where the maximum monomial degree is only k. Thus, the desired condition for maximum of $s(\gamma)$ at $\gamma = \frac{1}{2}$ can be verified numerically in an efficient manner.

Note that the difference between the lower bound obtained by applying the second moment method to the strictly regular case can differ by at most $2/k$ from the corresponding lower bound for the 2-regular case. We observe that as k increases the lower bound seems to converge to $2^k \ln(2) - (k+1)\frac{\ln(2)}{2} - 1$, which is the lower bound for the uniform model.

We observe similar behavior for the NAE-satisfiability bounds. As expected, we observe that the upper bound on the NAE-satisfiability threshold for the regular random model converges to $2^{k-1} \ln(2) - \frac{\ln(2)}{2}$. The lower bound obtained by applying the second moment method to the regular random model seems to converge to the value obtained for the uniform model. Thus, the NAE-threshold for the regular random model is $2^{k-1} \ln(2) - O(1)$. This suggests that as k increases, the threshold of the regular model does not differ much from the uniform model.

Table 1. Bounds on the satisfiability threshold for strictly regular random k-SAT. α_l^* and r^* are defined in Theorem 2. The upper bound α_u^* is obtained by the first moment method. $\alpha_{l,\text{uni}}^* = 2^k \ln(2) - (k+1)\frac{\ln(2)}{2} - 1$ is lower bound for uniform model obtained in [3]. The quantities $r_{\text{NAE}}^*, \alpha_{l,\text{NAE}}, \alpha_{u,\text{NAE}}$ are analogously defined for the NAE-satisfiability.

k	r^*	α_l^*	α_u^*	$\alpha_{l,\text{uni}}^* - \alpha_l^*$	r_{NAE}^*	$\alpha_{l,\text{NAE}}^*$	$\alpha_{u,\text{NAE}}^*$
3	4	2.667	3.78222	0.492216	3	2	2.40942
4	16	8	9.10776	0.357487	8	4	5.19089
7	296	84.571	85.8791	0.378822	152	43.4286	44.0139
10	3524	704.8	705.9533	0.170403	1770	354	354.545
15	170298	22706.4	22707.5	0.101635	85167	11355.6	11356.2
17	772182	90844.94	90845.9	0.007749	386114	45425.2	45425.7

Our immediate future work is to derive explicit lower bounds for the regular random k-SAT model for large values of k as was done for the uniform model in [3, 1]. The challenge is that the function $s(\gamma)$ depends on the solution of the system of polynomial equations given in (25). Thus determining the maximum requires determining the behavior of the positive solution of this system of polynomial equations. Another interesting direction is the maximum satisfiability of regular random formulas. For the uniform model, the maximum satisfiability problem was addressed in [2] using the second moment method. In [16], authors have derived lower and upper bounds on the maximum satisfiability threshold of regular random formulas.

References

1. AChlioptas, D., Moore, C.: Random k-SAT: Two moments suffice to cross a sharp threshold. SIAM J. COMPUT. 36, 740–762 (2006)
2. Achlioptas, D., Naor, A., Peres, Y.: On the maximum satisfiability of random formulas. Journal of the Association of Computing Machinary (JACM) 54 (2007)
3. AChlioptas, D., Peres, Y.: The threshold for random k-SAT is $2^k \ln(2) - O(k)$. Journal of the American Mathematical Society 17, 947–973 (2004)
4. Barak, O., Burshtein, D.: Lower bounds on the spectrum and error rate of LDPC code ensembles. In: International Symposium on Information Theory, Adelaide, Australia (2002)
5. Bender, E.A., Richmond, L.B.: Central and local limit theorems applied to asymptotic enumeration II: Multivariate generating functions. J. Combin. Theory, Ser. A 34, 255–265 (1983)
6. Boufkhad, Y., Dubois, O., Interian, Y., Selman, B.: Regular random k-SAT: Properties of balanced formulas. Journal of Automated Reasoning (2005)
7. Bruijn, N.G.D.: Asymptotic Methods in Analysis. North Holland, Amsterdam (1981)
8. Flajolet, P., Sedgewick, R.: Analytic Combinatorics. Cambridge University Press, Cambridge (2009)
9. Franco, J., Paull, M.: Probabilistic analysis of the Davis Putnam procedure for solving the satisfiability problem. Discrete Appl. Math. 5, 77–87 (1983)
10. Friedgut, E.: Sharp threshold for graph properties, and the k-SAT problem. Journal of the American Mathematical Society 17, 947–973 (2004)
11. Gardy, D.: Some results on the asymptotic behavior of coefficients of large powers of functions. Discrete Mathematics 139, 189–217 (1995)
12. Henrici, P.: Applied and Computation Complex Analysis, vol. 2. John Wiley, Chichester (1974)
13. Kaporis, A.C., Kirousis, L.M., Lalas, E.G.: The probabilistic analysis of a greedy satisfying algorithm. In: 10th Annual European Symposium on Algorithms, vol. Ser. A 34 (2002)
14. Rathi, V.: On the asymptotic weight and stopping set distributions of regular LDPC ensembles. IEEE Trans. Inform. Theory 52, 4212–4218 (2006)
15. Rathi, V.: Non-binary LDPC codes and EXIT like functions, PhD thesis, Swiss Federal Institute of Technology (EPFL), Lausanne (2008)
16. Rathi, V., Aurell, E., Rasmussen, L., Skoglund, M.: Bounds on maximum satisfiability threshold of regular random k-SAT. arXiv:1004.2425 (submitted)
17. Rathi, V., Aurell, E., Rasmussen, L., Skoglund, M.: Satisfiability and maximum satisfiability of regular random k-sat: Bounds on thresholds (in preparation for journal submission)
18. Richardson, T., Urbanke, R.: Modern Coding Theory. Cambridge University Press, Cambridge (2008)

Dynamic Scoring Functions with Variable Expressions: New SLS Methods for Solving SAT

Dave A.D. Tompkins and Holger H. Hoos

Department of Computer Science
University of British Columbia
{davet,hoos}@cs.ubc.ca

Abstract. We introduce a new conceptual model for representing and designing Stochastic Local Search (SLS) algorithms for the propositional satisfiability
problem (SAT). Our model can be seen as a generalization of existing variable
weighting, scoring and selection schemes; it is based upon the concept of Variable Expressions (VEs), which use properties of variables in dynamic scoring
functions. Algorithms in our model are constructed from conceptually separated
components: variable filters, scoring functions (VEs), variable selection mechanisms and algorithm controllers. To explore the potential of our model we introduce the Design Architecture for Variable Expressions (DAVE), a software
framework that allows users to specify arbitrarily complex algorithms at run-
time. Using DAVE, we can easily specify rich design spaces of SLS algorithms
and subsequently explore these using an automated algorithm configuration tool.
We demonstrate that by following this approach, we can achieve significant improvements over previous state-of-the-art SLS-based SAT solvers on software
verification benchmark instances from the literature.

1 Introduction

The propositional satisfiability problem (SAT) is an important subject of study in many
areas of computer science and is a prototypical $\mathcal{NP}$-complete problem. Among the
best known methods currently available for solving certain types of SAT instances are
Stochastic Local Search (SLS) procedures; these are typically incomplete, *i.e.*, they cannot determine with certainty that a given propositional formula is unsatisfiable, but they
often find models of satisfiable formulae surprisingly effectively [9]. SLS algorithms
for SAT typically start by randomly assigning to every variable appearing in a given
formula a value of either true or false; then, in each subsequent *search step* a variable is
selected to have its truth assignment *flipped* from true to false or vice versa. The method
of selecting the variable to be flipped in each step is usually guided by a *scoring function*
that minimizes the number of currently unsatisfied clauses.

In this work, we propose a new conceptual model for specifying SLS algorithms for
SAT, and provide a software framework to aid in the development of new algorithms.
Our model was developed to provide a clean conceptual separation between the scoring
function(s) and the *Variable-Selection Mechanism (VSM)* of an algorithm. We introduce
the concept of *Variable Expressions (VEs)* to generalize scoring functions; while VEs
are ultimately used for variable selection, they can transcend the traditional notion of
score. VEs are mathematical expressions that compute numerical values from one or

O. Strichman and S. Szeider (Eds.): SAT 2010, LNCS 6175, pp. 278–292, 2010.

more *properties* of a variable in combination with constants, operators and functions. The variable properties that can appear in VEs include well-known concepts from the literature, such as GSAT's score property [17] and the age property used by NOV-ELTY and WALKSAT/TABU [14]. A VE can be a simple property (*e.g.*, $\langle$age$\rangle$) or any mathematical expression with one or more properties, such as $\langle$score$ + 3 \cdot \log(age)\rangle$. Most existing SLS algorithms for SAT select variables based on scoring functions that correspond to a single, rather simplistic VE; in this paper we present evidence that potentially complex VEs can be very effective.

To explore the potential of our model, we introduce the *Design Architecture for Variable Expressions (DAVE)*, a software extension of our versatile UBCSAT architecture [18]. No programming is required to develop new algorithms in DAVE; the complete algorithm specification (including arbitrarily complex VEs) can be provided at run-time. We provided this flexibility in DAVE from the outset, with the goal of leveraging existing automated algorithm configuration tools (henceforth, *configurators*) such as PARAMILS [11]. With the combination of DAVE and a configurator, designers have an unprecedented amount of flexibility and power to help automate the design of new high-performance SLS algorithms and algorithm hybrids.

The remainder of the paper is structured as follows. In Section 2, we describe our experimental methodologies. In Section 3, we introduce more advanced VEs and demonstrate their efficacy. In Section 4, we present our general conceptual model and briefly discuss its implementation (DAVE). In Section 5, we introduce a new, highly parametric algorithm named VE-SAMPLER to demonstrate how DAVE facilitates the automated design of SLS algorithms. In Section 6, we discuss related work from the literature, and in Section 7, we summarize the contributions made in this work and outline directions for future research.

2 Experimental Details and Methodology

In the experiments presented throughout this study, we used the PARAMILS automated algorithm configurator by Hutter *et al.* [11] to optimize the parameter settings of various SLS-based SAT algorithms for performance on a particular instance set. To ensure that our results generalize to instances other than those used during the optimization process, we randomly split each set into two halves, a *training* set and a *test* set, where an optimal configuration is found by conducting experiments on the training set. Instances in the test set were only used for the final performance measurements presented in this paper.

In our experiments, we mostly focused on the CBMC software verification instance set generated, and used as a benchmark, by KhudaBukhsh *et al.* [12]. The instances were generated by a Bounded Model Checking (BMC) tool [4] and were pre-processed with SATELITE [5]. This set is interesting to us primarily because it has some of the structural properties of larger and more complicated software verification problems (that are still somewhat intractable for SLS solvers). For example, many of the complete solvers from the 2009 SAT Competition (such as PICOSAT [3]) can solve the hardest CBMC instance in less than one second, whereas well-known state-of-the-art SLS solvers from the competition such as ADAPTG^{2}WSAT and GNOVELTY$^+$ require over an hour to solve the same instance. At the same time, a significant number of the instances can be solved by SLS algorithms within a low enough time to allow for extensive experiments. In Section 5 we also provide for the first time experimental data for SLS algorithms on the software verification benchmark set SWV generated by the CALYSTO static checker [2] and used as a benchmark for complete solvers by Hutter *et al.* [10].

A more detailed description of our experimental methodologies, PARAMILS settings, specifications of our run-time environment, further details of our instance sets and algorithm configurations in DAVE can be found in a supplementary online appendix, available at the UBCSAT website [19].

3 Advanced Variable Expressions

Various variable properties and VEs play a prominent role in SLS-based solvers known from the literature. Perhaps the most popular VE currently used by SLS algorithms is ⟨score⟩, which is equivalent to the VE ⟨make − break⟩ where the properties make and break measure the number of clauses that would become satisfied and unsatisfied, respectively, if the variable were to be flipped. The WALKSAT/SKC algorithm [16] was the first algorithm to use the even simpler VE ⟨break⟩ for scoring variables and also introduced a Boolean freebie property that is true if, and only if, break equals zero. Algorithms with dynamic clause penalties, such as SAPS, use a *(penalized)* property penScore that reflects the dynamic clause penalty values (weights). The G^2WSAT algorithm uses a Boolean promising property that indicates a positive score property value, but only under certain circumstances (see [13] for details).

Another variable property that is prominently used in existing SLS algorithms for SAT is age; it is defined as the number of search steps that have occurred since the given variable was last flipped. The age property is closely related to the flips property (*a.k.a. flipcount*) used by the HSAT algorithm [7] as a *tie-breaking* mechanism; the flips property measures how many times a variable has been flipped. An interesting and effective combination of the freebie, break, age and flips properties is used in the VW2 algorithm [15].

3.1 Deconstructing VW2

In many ways, Prestwich's VW2 algorithm [15] provided the starting point for our work on VEs, and we describe VW2 in the following.[1] Each variable is assigned a *weight* (which we call the vw2w property) initialized to zero. At each search step the flip candidates are those variables that appear in a randomly selected unsatisfied clause. If there are any candidates with a freebie value of one, one of those is selected; otherwise, with probability p, a candidate is selected uniformly at random, and in the remaining cases (*i.e.*, with probability $(1 - p)$), the candidate is selected with the smallest value of the VE:

$$\text{break} + c \cdot (\text{vw2w} - \overline{\text{vw2w}}) , \tag{1}$$

---

[1] For consistency with other parts of our study, we chose to use our notations instead of Prestwich's when describing VW2.

where the constant c is a parameter and $\overline{\mathsf{vw2w}}$ denotes the average of the $\mathsf{vw2w}$ property across all variables. When a variable is flipped, its $\mathsf{vw2w}$ property is updated according to:

$$\mathsf{vw2w} := (1 - s) \cdot (\mathsf{vw2w} + 1) + s \cdot \mathbf{step} \,, \qquad (2)$$

where s is another constant parameter, and $\mathbf{step}$ is the current step iteration value.

A variant of VW2 that we call VW2-SAT05 received the bronze medal in the satisfiable random category of the 2005 SAT competition. This variant eliminates the three VW2 parameters (s, c, p) by setting p to zero and introducing a randomized mechanism to change the behaviour of c and s during the search; it has been included recently in the SATENSTEIN-LS [12] and HYBRID [20] algorithms. However, in our experiments, we found that the original VW2 procedure with parameter settings optimized for a given set of benchmark instances will often outperform VW2-SAT05. In particular, we observed this performance difference on the CBMC software verification instances described in Section 2. In experiments not presented here (see [19] for details), we found that VW2 with parameters $(s, c, p) = (0, 0.01, 0.2)$ is the best-performing SLS-based SAT algorithm currently known for CBMC, which motivated us to study it in more depth.

Upon closer examination of the VW2 VE shown in Equation 1 above, we noticed that the $\overline{\mathsf{vw2w}}$ term can be removed without changing the behaviour of VW2, since this term is constant over all variables and therefore does not affect the variable selection. In the $\mathsf{vw2w}$ property update procedure, the s parameter is a *smoothing* parameter. if s is set to one, the VE becomes equivalent to $\langle \mathsf{break} - c \cdot \mathsf{age} \rangle$. If s is set to zero, as in the optimal setting for CBMC, the variable property $\mathsf{vw2w}$ becomes equivalent to $\langle \mathsf{break} + c \cdot \mathsf{flips} \rangle$.

For very small values of c, it may appear as though the $\mathsf{vw2w}$ property acted as a tie-breaking mechanism, and Prestwich observed that when s is zero, VW2 behaves like HSAT [7]. While it may be easy to dismiss the mechanics of VW2 as a simple tie-breaking scheme, this simplification does not seem justified when considering the parameter settings obtained for VW2 and the length of typical runs required for solving CBMC instances. In our analysis of VW2 on the hardest CBMC instance, we observed that for over half of the search steps the break and flips properties were interacting in a complex way, and VW2 was making trade-offs between satisfying additional clauses (intensification) and changing the values of rarely flipped variables (diversification).

3.2 VW2+VE: Modifying the VE in VW2

Considering this type of complementarity in the role of the break and flips properties and the strong performance of VW2, it seemed promising to explore different ways of constructing a VE based on those two properties. Because the difference in scale between the two properties becomes increasingly larger as the search progresses, we decided to *normalize* the values of these properties to the interval $[0, 1]$. We achieved this using the formula $\frac{\mathsf{p}}{\max(\mathsf{p})}$, where $\max(\mathsf{p})$ refers to the maximum value of the property p for all flip candidates, which for VW2 would be those variables in the currently selected clause.

In addition to normalizing the property values, we also allowed for *non-linear* interaction between the two properties. Our motivation was that the *relative* difference in magnitude between two different property values could have an important impact on

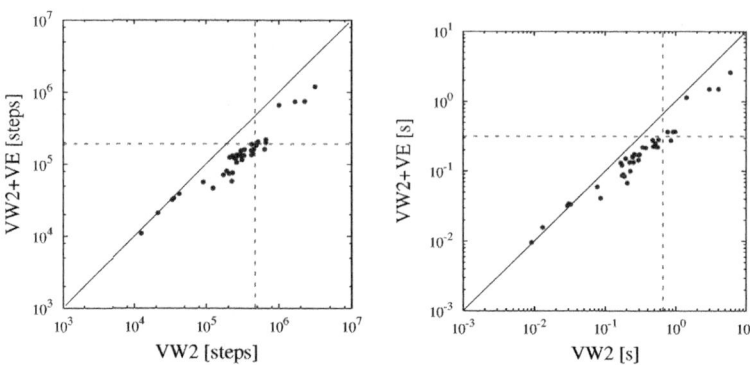

Fig. 1. VW2+VE *vs* VW2 on CBMC. Each point corresponds to the median run-length (left) and run-time (right) from 25 runs on an instance in the CBMC test set. The mean values of those medians are indicated by the dashed lines. The ratio of the means (which we denote as the speedup factor *s.f.*) is 2.47 (left) and 2.10 (right).

the behaviour of the algorithm. Since the values have already been normalized, we used a simple polynomial transformation on the normalized values of the flips property, to obtain the generalized VE:

$$\frac{\text{break}}{\max(\text{break})} + c \cdot \left(\frac{\text{flips}}{\max(\text{flips})}\right)^a , \tag{3}$$

which we used to replace the scoring function of VW2. We refer to the resulting variant of VW2, in which we also disabled smoothing, as VW2+VE. Automated configuration of this algorithm for our CBMC training set using PARAMILS yielded the parameter configuration $(c, a, p) = (0.95, 8, 0.05)$ (see [19] for details).

As can be seen from Figure 1, the use of this generalized VE leads to improved performance in terms of local search steps required for solving the CBMC instances (as always, we show results for the test set, which is disjoint from the training set used for parameter optimization). However, the VE is more complex, and evaluating it requires an additional initial iteration to determine the maximum values. This leads to a less pronounced improvement in terms of time performance, which is illustrated in Figure 1 (right). Still, VW2+VE performs better than VW2 on the CBMC benchmark, which – based on our earlier findings – makes it the best SLS-based SAT algorithm for that benchmark currently available.

3.3 Normalization in VEs

In VW2+VE, we normalized the break and flips properties so they would fall within the interval $[0, 1]$. We will now generalize this further, using from here on the notation $\|x\|$ in VEs to indicate that the value x has been normalized using one of several different methods. The method used in VW2+VE, $\|x\| = \frac{x}{\max(x)}$, preserves ratios between the values being normalized. Alternatively, a flat normalization $\|x\| = \frac{x - \min(x)}{\max(x) - \min(x)}$

forces the maximum and minimum to be one and zero, respectively, and a summation normalization $\|x\| = \frac{x}{\text{sum}(x)}$ forces the sum of the values to be one. Of course, numerous other normalizations are possible, including non-linear normalizations and normalizations more suitable for both positive and negative values.

In the literature, some scoring functions are designed to select variables with the *minimum* value (such as VW2's), whereas others select the variable with the *maximum* value (such as the traditional ⟨make − break⟩). Both cases are common, and which one should be used is usually obvious from the context; however, this may not always be the case as we consider more complicated VEs. To address this issue, we first note that the question of favouring minimum or maximum values already arises for variable properties: for example, a small value of flips is considered favourable, while the opposite is true for age. To facilitate the construction of more complex VEs, we will require that all properties be transformed to favour maximum values. To this end, we revise our notation for normalization so that $\|p\|$ will indicate that p has been normalized and transformed (if necessary). A simple transformation and normalization would be $(1 - \|p\|)$, and we found that $\|\max(p) + \min(p) - p\|$ worked quite effectively in practice.

When normalizing the make and break properties, we observed that they can also be normalized *w.r.t.* the number of clauses in which the variable appears. We will introduce the variable properties relMake and relBreak to correspond to the *relative* number (fraction) of clauses that become satisfied or unsatisfied, respectively, as a result of flipping a given variable. For example, if the positive literal x occurs in numPosOcc clauses and the negative literal $\neg x$ occurs in numNegOcc clauses, then the value of relMake is equivalent to ⟨make/numPosOcc⟩ when x is false and ⟨make/numNegOcc⟩ when x is true. While for randomly generated instances with uniform structure, normalizing the score in this manner would have no material effect, for structured formulae, such as the CBMC instances, there is often large variability in the number of clauses each variable appears in, and consequently, this normalization can make a substantial difference. Ansótegui *et al.* explored the *scale-free* structure of industrial instances and the impact of this structure on complete solvers [1], and we believe that there is potential for SLS algorithms to exploit this structure as well.

Another observation we made is that existing algorithms combine make and break symmetrically, but there may be an advantage to constructing VEs in which they are weighted differently. We therefore consider the generalized VE ⟨$c_1 \cdot$ make − $c_2 \cdot$ break⟩, which uses simple scaling to weight the two variable properties differently. We note that WALKSAT/SKC [16] can be seen as using a special case of this VE where $c_1 = 0$. While it is possible that in many cases choosing $c_1 = 1$ may lead to the best performance, there is no reason to assume that this would always be the case.

Finally, we observed that the summation normalization $(\frac{x}{\text{sum}(x)})$ behaved quite differently than the one we used in VW2+VE $(\frac{x}{\max(x)})$, even though at first glance it would appear that they should only differ by a constant factor. However, that constant factor is the *clause length*, which is constant for any particular search step, but can differ between search steps. In other words, we discovered that normalization *w.r.t.* the clause length can be beneficial, and we believe that such normalizations merit further study.

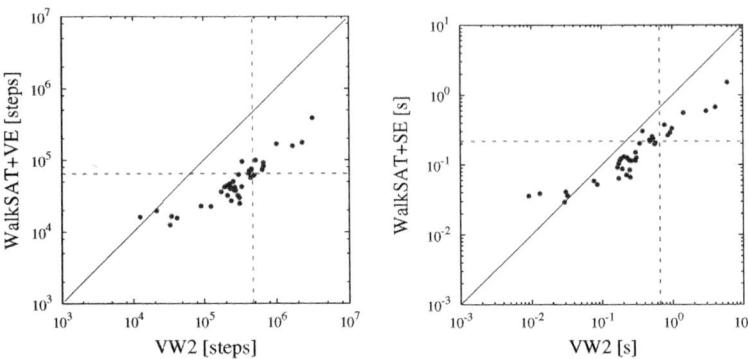

Fig. 2. WALKSAT+VE *vs* VW2 on CBMC. The *s.f.* is 7.25 (left) and 3.07 (right). (The data is presented analogously to that in Figure 1).

3.4   WALKSAT+VE: Modifying the VE in WALKSAT

To investigate the potential latent in the generalizations introduced up to this point, we constructed a new SLS algorithm we call WALKSAT+VE. This algorithm is obtained from the original WALKSAT/SKC algorithm by replacing the VE ⟨break⟩ with the following VE that makes use of scaling, normalizations and non-linear transformations:

$$c_1 \cdot \|\mathsf{make}\|^{a_1} + c_2 \cdot \|\mathsf{relMake}\|^{a_2} + c_3 \cdot \|\mathsf{break}\|^{a_3} + c_4 \cdot \|\mathsf{relBreak}\|^{a_4} \ . \qquad (4)$$

Whereas VW2+VE benefited from the flips property providing diversification, this VE uses only greedy components (make and break) and a standard random walk mechanism. To test the effectiveness of our new algorithm, we ran PARAMILS to optimize the values of the constants and the normalization parameters (hidden in the $\|\mathsf{p}\|$ notation) on the CBMC training set (see [19] for details).

The performance of the configuration thus obtained on the CBMC test set is illustrated in Figure 2. Our new WALKSAT variant significantly outperforms the previously best known SLS algorithm for this benchmark (VW2) and solves it more than 1 000 times faster than WALKSAT/SKC. These results are especially impressive when examining step performance, but because of the complexity involved with this advanced VE, the results *w.r.t.* time performance are somewhat less impressive, but still significant. We were genuinely surprised that with this relatively modest modification to the venerable, but rather dated WALKSAT/SKC algorithm, we were able to outperform all known SLS algorithms. This experiment clearly demonstrate the potential of complex VEs as a basis for the development of new, high-performance SLS algorithms.

4   Modeling and Designing SLS Algorithms with VEs

Now that we have motivated our interest in VEs, we will present our VE-based model. Our model, as illustrated in Figure 3, includes an *algorithm controller* and three core

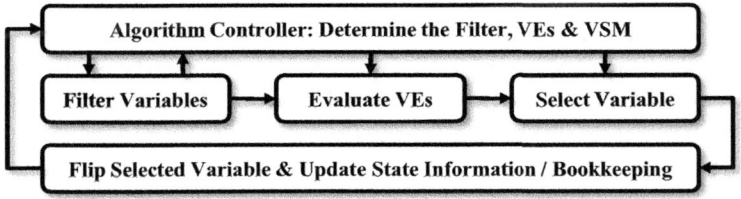

Fig. 3. Our conceptual SLS algorithm model

stages: a variable filter stage, a VE evaluation stage and a variable selection stage. There is a final stage that simply flips the selected variable and updates the state information resulting from the flip (*e.g.*, property values) and any algorithm state information (such as the noise value in algorithms with adaptive noise). We will first describe the three core stages and then describe the algorithm controller.

The **Variable Filter Stage** outputs a list of variables that are candidates to be flipped in this search step. For example, the clause-based filter used in WALKSAT/SKC [16] and VW2 [15] selects an unsatisfied clause uniformly at random, and then only the variables that appear in that clause are flip candidates. Other examples include the GSAT algorithm [17], which considers all variables, the SAPS algorithm, which includes all variables that appear in unsatisfied clauses, and the G^2WSAT algorithm [13], which includes a filter that only allows variables with a promising property value of one.

The **VE Evaluation Stage** is very straightforward. The input is the list of n flip candidates from the filter stage and k VEs from the controller, and the output is an array of $n \times k$ values where each of the VEs are evaluated for each candidate.

The **Variable Selection Stage** makes the final decision as to which of the candidates will be flipped, based on the array of values from the VE evaluation stage. For simplicity, we will assume that a single candidate is selected and flipped in each step, but in practice, the VSM could select zero or many candidates. For most existing SLS algorithms, the variable selection mechanism (VSM) is a simple max (or min) operation, where the candidate with the maximum value of the first VE is selected; additional VEs can be used for tie-breaking, and any remaining ties will be broken randomly. The NOVELTY algorithm [14] is an example of an algorithm with a VSM that incorporates multiple VEs (score and age).

The **Algorithm Controller** controls the behaviour at each step by determining the components of each of the three stages: the filter, the set of VEs and the VSM. The controller may use the same components for every step, make independent random decisions for each step or it may use a more sophisticated decision mechanism. The GSAT algorithm [17] represented in our model uses a simple controller, where the components are the same at every step: no filter (consider all variables), use a simple VE of ⟨score⟩ and a max VSM. The GWSAT algorithm added a random walk to GSAT, and is represented in our model by a randomized controller that with some probability selects an alternate filter (only variables that appear in unsatisfied clauses) and a VSM that selects candidates randomly. In Figure 3, we indicate control flow from the filter back to the controller to allow for controllers that may wish to re-filter the variables or defer the determination of the VEs or VSM until after the filter results are known. For example, as a form of clause normalization (see Section 3.3), a controller could use a random-clause-based filter and choose VEs based on the length of the selected clause.

In our model, complex controllers can be constructed that do not directly decide the components for the three stages, but instead utilize a number sub-controllers. Since each sub-controller can correspond to a unique algorithm (or the same algorithm with different parameter settings), this allows the construction of *hybrid* algorithms. A hybrid algorithm can switch between different algorithms randomly, periodically, when some criteria is met (*e.g.*, search stagnation is detected) or according to some other customized mechanism. G^2WSAT is one such hybrid algorithm, where if any variables have a promising property of one, a GSAT-based step occurs, otherwise, a WALK-SAT-based step occurs [13].

Now that we have presented our highly flexible model, we will briefly outline our Design Architecture for Variable Expressions (DAVE), based on our versatile UBC-SAT architecture [18]. (For a complete and up-to-date description of DAVE, consult the UBCSAT website [19].) One of the design goals of DAVE was to reduce (and potentially eliminate) the programming component of algorithm design by allowing the entire algorithm behaviour to be specified at run-time. The user can specify the algorithm controller (and sub-controllers), the filter(s), the VE(s) and the VSM(s). The only programming required is to introduce *new* variable properties, controllers, filters or VSMs. Because the *configuration space* of DAVE is actually an *algorithm specification space*, when we use DAVE in combination with an automated configurator, we can find optimized algorithm specifications automatically. To further facilitate the use of a configurator, DAVE supports a sophisticated macro-based syntax that allows controllers, filters, VEs, and VSMs to be highly paramaterized.

In DAVE, most variable properties depend on the current value of the variable. We use the notation p' to correspond to the property value for the negation of a given variable. For example, the flips property in DAVE is actually half of the total flip count (flips + flips$'$); similarly, age$'$ ignores the most recent flip and measures the number of search steps that have occurred since the flip prior to the most recent flip.

The only other implementation detail of DAVE that we will address here, as it is specifically relevant to the presentation and understanding of the performance results we report later, is the interpreted nature of the algorithms specified in DAVE. Since DAVE receives the algorithm specification and VEs at run-time, the code is not natively compiled, but instead, each operation is individually interpreted and executed. This means that an algorithm in DAVE will not achieve the same performance as the equivalent algorithm in compiled source code *w.r.t.* CPU time, which is why we encourage measuring DAVE algorithms by their step performance where there is no such penalty. In preliminary experiments, we have seen algorithms in DAVE run 1.5-3 times slower than their native implementations, where the speed of DAVE is often more a function of the number of operators used in the VE, as opposed to the true complexity of the algorithm. This is one reason why we present DAVE as a design architecture that facilitates the exploration of new algorithmic ideas; it is our intent that new and robust algorithms that are developed in DAVE will subsequently be incorporated directly in UBCSAT as stand-alone optimized algorithms. We are currently in the preliminary stages of developing a software tool that can automatically generate fast, native source code that will implement an algorithm specified in DAVE.

5   VE-SAMPLER: Exploring New SLS Methods Using DAVE

In this section we introduce a new algorithm framework we call VE-SAMPLER. VE-SAMPLER uses a randomized controller that selects between six sub-controllers, where each sub-controller is selected with a probability proportional to a configurable weight. Each of the six sub-controllers uses a simple max VSM, and has a configurable clause-based filter, where the unsatisfied clause selected is either random, the clause unsatisfied the longest, or the clause most frequently unsatisfied. The VE of the first sub-controller is ⟨freebie⟩, similar to the random walk in WALKSAT/SKC [16]; the max VSM will select all freebie candidates, or all candidates if no freebies exist, and then break ties randomly. The VEs for the other five sub-controllers are all of the form:

$$\|\mathsf{p1}\|^{a_1} + \mathrm{clw}(s, m, l) \cdot \|\mathsf{p2}\|^{a_2} \ , \tag{5}$$

where $\mathsf{p1}$ and $\mathsf{p2}$ are configurable, and correspond to variable properties (or a ratio of properties) selected from lists we describe below. The clw function represents a simple mechanism we created to addresses clause normalization (briefly discussed in Section 3.3) in a practical, yet interesting way; the three configurable parameters of $\mathrm{clw}(s, m, l)$ correspond to scaling coefficients that depend on whether the clause length is small (< 3), medium $(= 3)$, or large (> 3); *i.e.*, if the clause length is two then $\mathrm{clw}(s, m, l) = s$.

The VE described in Equation 5 is similar to the VEs in VW2+VE and WALKSAT+VE *w.r.t.* the normalization and non-linear transformation used. We chose to use only two properties to avoid the reduction in CPU time performance we saw with four properties in WALKSAT+VE; however, we believe that our approach of using multiple VEs via a controller can provide a similar level of algorithm robustness without significantly degrading per-step time complexity. Of the five sub-controllers, one was configured to have only *greedy* properties similar to WALKSAT+VE, while the remaining four were configured to have one greedy property ($\mathsf{p1}$) and one *diversification* property ($\mathsf{p2}$) similar to VW2+VE. The five greedy properties available were score, make, relMake, break and relBreak.

We wanted diversification properties that were independent of the greedy variable properties and required little or no computational overhead to maintain. For VE-SAMPLER we created the following new properties: flitCount is incremented every search step where the variable (with its current value) has appeared in the list of flip candidates, relFiltCount is similar, but increases by $1/\mathbf{clauselen}$, and goodFlips and badFlips are incremented every time the variable (with its current value) is flipped and the number of satisfied clauses goes up or down, respectively. In total, there were thirteen diversification properties (or ratios of properties) available in VE-SAMPLER:

flips,	age/flips,	relFiltCount,	goodFlips/flips,
age,	age′/age,	relFiltCount/flips,	goodFlips/goodFlips′,
age′,	filtCount,	relFiltCount/relFiltCount′,	goodFlips/badFlips

and rand, which draws a number uniformly at random from the interval $[0, 1]$. While some of these properties are based on prior evidence and intuition, others are simply interesting ideas that we thought might be effective.

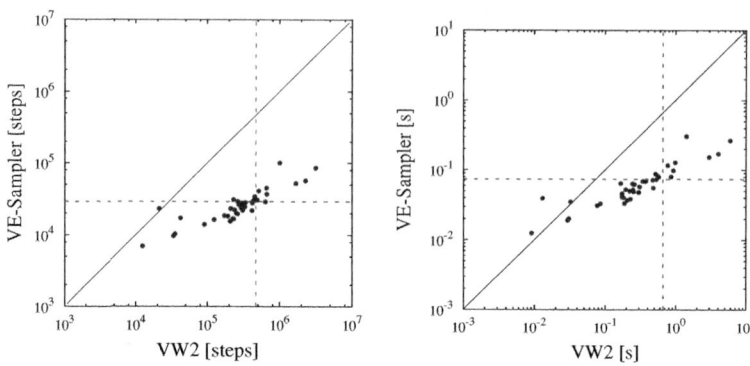

Fig. 4. VE-SAMPLER *vs* VW2 on CBMC. The *s.f.* is 16.2 (left) and 9.0 (right). (The data is presented analogously to that in Figure 1).

Algorithm	CBMC			SWV (partial)				SWV (full)	
	Steps $\times 10^3$	Time		Steps $\times 10^3$	Time		% Compl.	PAR	% Compl.
		sec.	*s.f.*		sec.	*s.f.*			
VW2-SAT05	3 577	6.22	0.11	10 089	19.20	0.16	100	3 008	50.1
VW2	467	0.66	*ref.*	1 555	3.10	*ref.*	100	3 042	49.3
SATENSTEIN-LS	228	0.80	0.82	1 465	12.50	0.25	100	3 040	49.5
VE-SAMPLER	**29**	**0.07**	**9.00**	**245**	**0.90**	**3.61**	100	**2 664**	**50.7**

Fig. 5. Experimental Results for VE-SAMPLER. Values shown are the means of the median run-length and run-time from (left) 25 runs on instances from the CBMC test set and (right) 10 runs on instances from SWV. The *s.f.* is the ratio of the time *w.r.t.* VW2. All algorithms completed 100% of the CBMC instances. The PAR (Penalized Average Run-time) is the average from all runs on all instances, where incomplete runs after 600 seconds are penalized by a factor of 10 (6 000 seconds) (see [12] for details). All algorithms (except the parameterless VW2-SAT05) were optimized by PARAMILS.

Our goal with VE-SAMPLER was to make very few decisions at design time and to configure the resulting, highly paramaterized algorithm automatically for optimized performance [8]. In total, VE-SAMPLER has over 10^{50} possible configurations, which, to the best of our knowledge, is the largest design space searched using PARAMILS [11] so far. We present the results of our PARAMILS-configured VE-SAMPLER in Figures 4–5. We compared VE-SAMPLER against the SLS-based solvers VW2 [15] and SATENSTEIN-LS (see Section 6), both also configured with PARAMILS (see [19] for details). The results we present were obtained using a compiled version of VE-SAMPLER, where the original version, implemented in DAVE, was approximately 1.5 times slower.

VE-SAMPLER performs substantially better than VW2 and SATENSTEIN-LS on our CBMC test set, especially in terms of search steps. On the much more challenging real-world software verification instances from the SVW set, VE-SAMPLER also performs significantly better than VW2 and SATENSTEIN-LS. We note that none of the

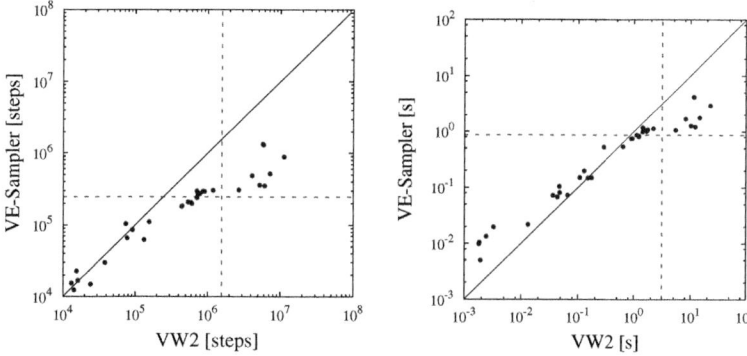

Fig. 6. VE-SAMPLER vs VW2 on SWV (partial). Each point corresponds to the median run-length (left) and run-time (right) from 10 runs on an instance in the SWV (partial) test set. The *s.f.* is 6.36 (left) and 3.61 (right).

SLS algorithms we are aware of can solve more than about half of the complete set of SWV instances within our 600 second cutoff, but VE-SAMPLER does solve the other half of the instances more efficiently than any other SLS algorithm. While the results in Figure 5 are impressive and represent the current state-of-the-art in SLS-based SAT solvers on these types of instances, the complete solver PICOSAT [3] is twice as fast as VE-SAMPLER on CBMC, seven times as fast on SWV (partial) and can solve any instance from the full SWV set in just a few CPU seconds. Thus, while we have considerably reduced the performance gap between SLS-based and DPLL-based SAT solvers on these software verification instances, there is still much room for improvement.

When studying the VE-SAMPLER configurations found by PARAMILS, we observed that configurations with similarly good performance often had substantially different configurations. This might suggest that VE-SAMPLER is somewhat robust *w.r.t.* its configuration, and that PARAMILS was far from finding the true optimal configuration of VE-SAMPLER (with over 10^{50} possible configurations, this is not surprising). We also observed configurations where two sub-controllers would be configured to use the same variable properties, but to be quite different otherwise. This was the case in the configurations featured in the results above, where the final CBMC configuration heavily weighted two sub-controllers with the properties relMake and age′, and the final SWV configuration heavily weighted two sub-controllers with the properties break and flips (see [19] for details). We believe this suggests that a hybrid algorithm including multiple configurations of the same underlying mechanism can achieve very robust performance.

6   Related Work

The manner in which SLS algorithm hybrids can be implemented in DAVE can be seen as a generalization of the HYBRID algorithm by Wei *et al.* [20]. HYBRID implements a clever heuristic to select between the algorithms VW2-SAT05 and ADAPTG^{2}WSAT at each search step. Their heuristic corresponds to a specific algorithm controller in our model, and once implemented in DAVE, it becomes a universal controller that can be used to select between *any* two algorithms. Furthermore, the selection of the algorithms to be hybridized can be achieved by using an automated configurator.

DAVE is conceptually related to the SATENSTEIN-LS solver by KhudaBukhsh *et al.* [12], which also extends UBCSAT, albeit in a different direction. SATENSTEIN-LS incorporates proven components from over two dozen existing SLS algorithms, including GNOVELTY$^+$, ADAPTG^{2}WSAT$_+$, SAPS and PAWS (see [12] for details) and can be configured to instantiate any of those algorithms, as well as many complex hybrids. SATENSTEIN-LS is very efficient when properly configured and is the best known SLS algorithm on several benchmark sets [12]. Whereas the SATENSTEIN-LS authors liken their generated algorithms to Frankenstein's monster, stitched together from existing algorithm parts, we believe that our model is more akin to a mad scientist experimenting with algorithmic DNA. The significant difference is that SATENSTEIN-LS has a bounded configuration space, whereas DAVE is a design environment that supports arbitrarily complex algorithms in a potentially unbounded space.

In that latter respect, DAVE is similar in nature to the Composite heuristic Learning Algorithm for SAT Search (CLASS) by Fukunaga [6]. CLASS is a genetic programming system that constructs new variable selection heuristics. Our work with VEs is somewhat orthogonal to the research direction underlying CLASS; our goal has been to decouple the scoring functions (VEs) from the VSMs and focus on the VEs, whereas in CLASS they are tightly coupled. There is potential for combining the strategies of DAVE and CLASS, and we are considering incorporating a CLASS-like syntax for VSMs into a future version of DAVE. Conversely, CLASS could be extended by incorporating our concept of VEs.

7   Conclusions and Future Work

In this work, we have proposed a new conceptual model for SLS algorithms based on variable expressions (VEs), and we demonstrated that algorithms with complex VEs can be very effective in practice. We created a new software framework for designing new SLS algorithms and algorithm hybrids in our model, and we demonstrated that by combining our software with an automated algorithm configuration tool, it was quite easy to construct a new algorithms that is nine times faster than the existing state-of-the-art SLS-based SAT solvers on a set of software verifications known from the literature.

Apart from the previously mentioned work on CLASS-based VSMs (Section 6) and the automated generation of source code from DAVE configurations (Section 4), we see several other promising directions for future work. We expect that there are more variable properties that can be effectively incorporated into VEs, as well as more sophisticated ways of combining variable properties beyond the simple normalization, scaling and non-linear transformations we presented in this work; we especially believe that there are more effective ways to handle clause normalization. Now that we have conceptually separated the components of algorithm controllers, filters, VEs and VSMs, we believe that algorithm designers will be able to focus on those individual components;

with the ability to quickly and automatically test their ideas in DAVE, we anticipate rapid development in each of these areas. Overall, we believe that the utilization of rich and flexible design environments such as DAVE in combination with powerful automated configuration tools will make it possible to achieve further, substantial progress in the state of the art in SLS-based SAT solving.

Acknowledgements

We would like to thank the anonymous reviewers for their valuable feedback. This research has been enabled by the use of computing resources provided by WestGrid and Compute/Calcul Canada. Furthermore, HH acknowledges funding received through the NSERC Discovery Grants Program.

References

1. Ansótegui, C., Bonet, M.L., Levy, J.: On the structure of industrial SAT instances. In: Gent, I.P. (ed.) CP 2009. LNCS, vol. 5732, pp. 127–141. Springer, Heidelberg (2009)
2. Babić, D., Hu, A.J.: Calysto: Scalable and precise extended static checking. In: Proceedings of ICSE 2008, pp. 211–220 (2008)
3. Biere, A.: PicoSAT essentials. Journal on Satisfiability, Boolean Modeling and Computation 4, 75–97 (2008)
4. Clarke, E., Kroening, D., Lerda, F.: A tool for checking ANSI-C programs. In: Jensen, K., Podelski, A. (eds.) TACAS 2004. LNCS, vol. 2988, pp. 168–176. Springer, Heidelberg (2004)
5. Eén, N., Biere, A.: Effective preprocessing in SAT through variable and clause elimination. In: Bacchus, F., Walsh, T. (eds.) SAT 2005. LNCS, vol. 3569, pp. 61–75. Springer, Heidelberg (2005)
6. Fukunaga, A.S.: Automated discovery of local search heuristics for satisfiability testing. Evolutionary Computation 16(1), 31–61 (2008)
7. Gent, I.P., Walsh, T.: Towards an understanding of hill-climbing procedures for SAT. In: Proceedings of AAAI 1993, pp. 28–33 (1993)
8. Hoos, H.H.: Computer-aided design of high-performance algorithms. Tech. Rep. TR-2008-16, Department of Computer Science, University of British Columbia (2008)
9. Hoos, H.H., Stützle, T.: Stochastic Local Search: Foundations and Applications. Morgan Kaufmann, San Francisco (2005)
10. Hutter, F., Babić, D., Hoos, H.H., Hu, A.J.: Boosting verification by automatic tuning of decision procedures. In: Proceedings of FMCAD 2007, pp. 27–34 (2007)
11. Hutter, F., Hoos, H.H., Leyton-Brown, K., Stützle, T.: ParamILS: An automatic algorithm configuration framework. Journal of Artificial Intelligence Research 36, 267–306 (2009)
12. KhudaBukhsh, A.R., Xu, L., Hoos, H.H., Leyton-Brown, K.: SATenstein: Automatically building local search SAT solvers from components. In: IJCAI 2009, pp. 517–524 (2009)
13. Li, C.M., Huang, W.Q.: Diversification and determinism in local search for satisfiability. In: Bacchus, F., Walsh, T. (eds.) SAT 2005. LNCS, vol. 3569, pp. 158–172. Springer, Heidelberg (2005)
14. McAllester, D., Selman, B., Kautz, H.: Evidence for invariants in local search. In: Proceedings of AAAI 1997, pp. 321–326 (1997)

15. Prestwich, S.: Random walk with continuously smoothed variable weights. In: Bacchus, F., Walsh, T. (eds.) SAT 2005. LNCS, vol. 3569, pp. 203–215. Springer, Heidelberg (2005)
16. Selman, B., Kautz, H.A., Cohen, B.: Noise strategies for improving local search. In: Proceedings of AAAI 1994, pp. 337–343 (1994)
17. Selman, B., Levesque, H., Mitchell, D.: A new method for solving hard satisfiability problems. In: Proceedings of AAAI 1992, pp. 459–465 (1992)
18. Tompkins, D.A.D., Hoos, H.H.: UBCSAT: An implementation and experimentation environment for SLS algorithms for SAT and MAX-SAT. In: Hoos, H., Mitchell, D.G. (eds.) SAT 2004. LNCS, vol. 3542, pp. 306–320. Springer, Heidelberg (2005)
19. UBCSAT Website, http://www.satlib.org/ubcsat
20. Wei, W., Li, C.M., Zhang, H.: A switching criterion for intensification and diversification in local search for SAT. Journal on Satisfiability, Boolean Modeling and Computation 4, 219–237 (2008)

Improved Local Search for Circuit Satisfiability

Anton Belov and Zbigniew Stachniak*

Department of Computer Science and Engineering,
York University, Toronto, Canada
{antonb,zbigniew}@cse.yorku.ca

Abstract. In this paper we describe a significant improvement to the justification-based local search algorithm for circuit satisfiability proposed by Järvisalo et al [7]. By carefully combining local search with Boolean Constraint Propagation we achieve multiple orders of magnitude reduction in run-time on industrial and structured benchmark circuits, and, in some cases, performance comparable with complete SAT solvers.

1 Introduction

It is known that Boolean Constraint Propagation (BCP) can enhance the performance of non-clausal SLS SAT solvers significantly, particularly on industrial and structured SAT instances (cf. [1]). In this paper we report the results of the study on augmenting stochastic local search algorithm BC SLS for circuit satisfiability, proposed by Järvisalo, Junttila and Niemelä in [7], with a range of strategies for employing BCP. Informally speaking, these strategies take into consideration some BCP features such as the direction of propagation, the way conflicts are handled during the propagation, or how far and how often the changes of truth-values determined by BCP are to be propagated through the circuit. We show that by carefully combining BC SLS with BCP we can achieve considerable improvements in performance of local search on industrial and structured circuits, and even outperform a modern complete CNF SAT solver on some industrial instances.

2 Preliminaries

In this paper we use a slightly modified terminology of [7], which we briefly describe below. The reader is referred to [7] for further details and examples.

A *Boolean circuit* C is a directed acyclic graph in which each vertex with in-degree $k > 0$ is associated with a Boolean function $\{0,1\}^k \mapsto \{0,1\}$. Vertices of C are called *gates*. The *fanin* of a gate $g \in C$, in symbols $FI(g)$, is the set of its direct predecessors in C. Similarly, the *fanout* of g, in symbols $FO(g)$, is the set of its direct successors in C. A gate with the empty fanin is an *input gate*, a gate with an empty fanout is an *output gate*. By $g = f(g_1, \ldots, g_k)$ we denote the fact that $FI(g) = \{g_1, \ldots, g_k\}$, and g is associated with the Boolean function f. Note that we do not restrict our attention to any

---

* Research of both authors supported by grants from the Natural Sciences and Engineering Research Council of Canada and Ministry of Training, Colleges and Universities of Ontario.

O. Strichman and S. Szeider (Eds.): SAT 2010, LNCS 6175, pp. 293–299, 2010.

particular family of circuits, though in our experimental study we focused on circuits composed only of inverters and 2-input AND gates (AIGs).

A *(partial) truth-value assignment* for a circuit C is a (partial) function τ from the set of gates in C to $\{0, 1\}$. By $dom(\tau)$ we denote the set of gates assigned by τ (the domain of τ). An assignment is *consistent* with C if for each gate $g = f(g_1, \ldots, g_k)$ we have $\tau(g) = f(\tau(g_1), \ldots, \tau(g_k))$. A *constrained Boolean circuit* is a circuit C together with a partial assignment α for C, in symbols C^α. Then, a *satisfying assignment* for C^α is an assignment τ consistent with C that extends α. The constrained circuit C^α is *satisfiable* if it has a satisfying assignment.

Given a gate $g = f(g_1, \ldots, g_k)$ and a truth-value $v \in \{0, 1\}$, a *justification* for the pair $\langle g, v \rangle$ is a (possibly partial) assignment σ to $FI(g)$, such that for every assignment σ' that extends σ to $FI(g)$, we have $f(\sigma'(g_1), \ldots, \sigma'(g_k)) = v$. In other words, a justification is a partial assignment to $FI(g)$ which guarantees that g evaluates to v. A justification σ for $\langle g, v \rangle$ is *minimal* if no proper subset of σ is a justification for $\langle g, v \rangle$.

Let C^α be a constrained Boolean circuit, and τ be an assignment for C that extends α. A gate $g \in C$ is *justified* under τ if either (i) g is an input gate, or (ii) $\tau(g) = f(\tau(g_1), \ldots, \tau(g_k))$. By $unjust(C^\alpha, \tau)$ we will denote the set of all gates in C not justified under τ. Thus, τ satisfies C^α if and only if $unjust(C^\alpha, \tau) = \emptyset$.

A *justification cone* of C^α under τ, in symbols $jcone(C^\alpha, \tau)$, is a transitive fanin of the constrained gates up to and including either an input or an unjustified gate. Formally, $jcone(C^\alpha, \tau)$ is the smallest set S of gates in C such that (i) $dom(\alpha) \subseteq S$, and (ii) if $g \in S$ and g is justified under τ, then for all minimal justifications $\sigma \subseteq \tau$ for $\langle g, \tau(g) \rangle$, $dom(\sigma) \subseteq S$. The *justification frontier* of C^α under τ, in symbols $jfront(C^\alpha, \tau)$, is the set of all unjustified gates in $jcone(C^\alpha, \tau)$. A gate $g \in C$ is said to be *interesting* if its either in $jfront(C^\alpha, \tau)$ or in the transitive fanin of some gate in $jfront(C^\alpha, \tau)$. A gate g is said to be an *(observability) don't care* if g is neither in justification cone, nor interesting. When $jfront(C^\alpha, \tau)$ is empty all gates that are not observability don't care are justified, and so τ can be easily modified to satisfy C^α by assigning the don't care gates the truth-values implied by the inputs in their transitive fanin.

3   Justification-Based Local Search

In [7] Järvisalo, Junttila and Niemelä proposed a stochastic local search algorithm BC SLS for circuit satisfiability. The algorithm starts with a random assignment τ to C^α that extends α, and proceeds in the following manner: as long as $jfront(C^\alpha, \tau)$ is not empty, a random gate g is selected from $jfront(C^\alpha, \tau)$, and one of the minimal justifications for g is either chosen randomly, or greedily in order to minimize an objective function (the number of interesting gates). Occasionally, an "upward" move is taken, that is the truth-value of the gate g itself is flipped (thereby g becomes justified, but some gates in its fanout become unjustified). The algorithm's termination condition is $jfront(C^\alpha, \tau) = \emptyset$, as it is sufficient to establish the satisfiability of C^α.

4   Incomplete Circuit BCP

In the complete, backtracking-based, circuit-SAT solvers (such as [9]), Boolean Constraint Propagation (BCP) is initiated when a truth-value is assigned to an unassigned

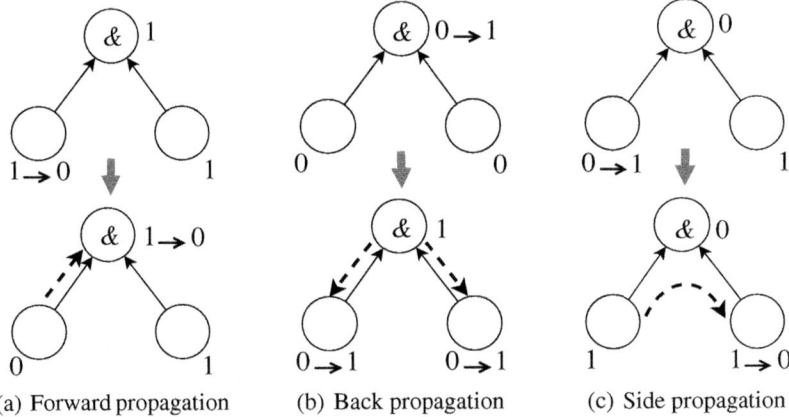

(a) Forward propagation   (b) Back propagation   (c) Side propagation

Fig. 1. Examples of different types of propagation in the incomplete BCP. The short horizontal arrows represent flips, the dashed arrows represent the direction of propagation.

gate g. As a result, some of the unassigned gates in g's neighbourhood (i.e. fanin and fanout) may acquire truth-values implied by g's new truth-value and the truth-values of the assigned gates in g's neighbourhood. The process continues recursively until either no new assignments can be made, or some previously assigned gate gets a conflicting truth-value.

SLS-based incomplete SAT solvers always maintain a complete assignment to the variables or, in the case of circuits, to gates. In this setting, the BCP is initiated by a *change* of the truth-value currently assigned to some gate (a flip). Furthermore, in such SLS solvers, the BCP algorithm does not need to be complete, i.e., not all the consequences of a flip must be propagated throughout the circuit. For example, we may decide not to propagate the effect of a flip to fanins of a gate. We will refer to this limited form of propagation as *incomplete* BCP. In the rest of this section, we will discuss some of the possible "degrees of freedom" available with incomplete circuit BCP such as the direction of propagation, the way the conflicts are handled during the propagation, and how far and how often the changes are propagated.

Directing BCP. In the context of incomplete BCP we may distinguish between the following types of propagation based on its direction: in *forward propagation* the flip of one or more gates in fanin of a gate is propagated to the gate itself; in *back propagation* the flip of a gate is propagated to one or more gates in its fanin; in *side propagation* the flip of one or more gates in fanin of a gate is propagated to other gates in its fanin and the truth-value of the gate itself doesn't change. An example of the different types of propagation applied to a 2-input AND gate is shown in Fig. 1.

Treatment of conflicts. In the complete setting the BCP is aborted when a gate gets assigned a conflicting truth-value. In the incomplete BCP, unless the gate is constrained, we have a choice of either keeping the truth-value assigned to the gate initially and stopping the propagation through this gate (*"stop at conflict"*), or overwriting the old truth-value with the new one and continuing the propagation (*"propagate through conflict"*).

Extent and frequency of BCP. The power of incomplete BCP can be further controlled by allowing the changes to propagate only to the gates that are close (in some sense, for example number of logic levels) to the gate whose truth-value is flipped. Another possibility is to limit the frequency of BCP.

5 CRSat: Justification-Based Local Search with Incomplete BCP

The empirical studies reported in [1] provide some evidence that BCP can significantly enhance the performance of SLS SAT solvers, particularly on industrial SAT instances. In this section we propose a justification-based local search algorithm CRSat in an attempt to demonstrate possible benefits of employing incomplete BCP in SLS procedures for circuit satisfiability. The resulting algorithm is presented in Alg. 1.

Alg. 1. CRSAT([*in*] C^α, [*out*] τ)

1: $\tau \leftarrow$ random truth-value assignment to inputs of C^α
2: $\tau \leftarrow$ FORWARD-BCP($C^\alpha, \tau, dom(\tau)$)
3: $steps \leftarrow 0$
4: **while** $steps{+}{+} <$ MAX_STEPS **do**
5:     **if** $unjust(C^\alpha, \tau) == \emptyset$ **then**
6:         **return** SAT
7:     **end if**
8:     $g \leftarrow$ SELECT-UNJUST-GATE(C^α, τ)   ▷ pick unjustified gate
9:     $Justs \leftarrow$ all minimal justifications for $\langle g, \tau(g) \rangle$
10:     **with-probability** wp **do**
11:         $\sigma \leftarrow$ random justification from $Justs$ ▷ random walk
12:     **otherwise**
13:         $\sigma \leftarrow$ SELECT-BEST-JUSTIFICATION($C^\alpha, \tau, g, Justs$)   ▷ greedy move
14:     **end with-probability**
15:     $\tau \leftarrow \sigma \cup (\tau \setminus \tau|_{dom(\sigma)})$ ▷ (multi)flip
16:     $\tau \leftarrow$ INCOMPLETE-BCP($C^\alpha, \tau, dom(\sigma)$)   ▷ propagate
17: **end while**
18: **return** UNDECIDED

The algorithm starts with a random partial assignment τ to the inputs of C^α, which is immediately extended to all gates using forward propagation (line 2). This step is not strictly necessary, however in our preliminary experiments we noticed a slight performance improvement when it is performed. To narrow down the effects of incomplete BCP, we keep the local search component of our algorithm extremely simple: we do not maintain the justification frontier and the set of interesting gates, and do not make upward moves. CRSat selects an unjustified gate g from $unjust(C^\alpha, \tau)$ (line 8) uniformly at random, and attempts to justify g by choosing a justification σ from the set of all minimal justifications $Justs$ (lines 9–14). The objective function used to evaluate justifications during the greedy moves is the number of unjustified gates. The search step of CRSat consists of modifying the current truth-value assignment τ so that it agrees with σ on gates in $dom(\sigma)$ (line 15), and propagating the consequences of this modification using some form of the incomplete BCP (line 16) – in our initial implementation, we use forward propagation with propagation through conflicts. We also experimented with back and side propagations.

6 Experimental Evaluation and Final Remarks

To evaluate the effectiveness of CRSat, we have implemented the algorithm in a prototype circuit-SAT solver `crsat`. The base solver uses forward propagation only. The two extended versions, `crsat+back` and `crsat+side`, add the back propagation and the side propagation, respectively. The side propagation is performed with the probability 0.01, and the solvers do not back- and side- propagate through conflicts. `crsat` maintains the circuits as and-inverter graphs (AIGs) – this allows to implement efficient table-driven circuit BCP, as in [9].

The evaluation of the solvers is based on the analysis of run-length and run-time distributions (cf. [6]) on variety of circuits from different benchmark sets. The distributions were obtained over 100 tries with near optimal noise and the infinite cutoff. The first set of benchmarks are the circuits that encode various bounded model checking problems – these benchmarks were used in [7] to evaluate BC SLS. The second set of benchmarks are the circuits that encode the MDP problem [3]. The third and the fourth sets are M.Velev's microprocessor verification suites SSS-SAT-1.0 and VLIW-SAT-1.1 [11]. All benchmark circuits were translated to n-ary AND/OR basis using a modified Boolean circuit simplifier `bc2cnf` [8], converted to AIG by representing n-ary gates as balanced binary trees, and simplified using 2-input subformula hashing. All experiments were performed on Intel Core 2 Duo, 3.00 GHz.

During the preliminary evaluation, we noticed that while `crsat` solves the problems used in [7] with ease, it times out on all but the simplest of these problems when the BCP is disabled. This result indicates that BCP plays a key role in our algorithm, and, also, emphasizes the importance of using the circuit structure in local search: as opposed to BC SLS, CRSat does not maintain justification frontier and so it can not take advantage of the observability don't care and interesting gates to speed-up the search. To compare the performance of CRSat with that of BC SLS we implemented BC SLS on AIGs, as the original implementation in [7] uses a different datastructure – the resulting solver is called `bcsls-aig`. The results of our comparative study, presented in Table 1, show that incomplete BCP can significantly improve the performance of circuit SLS. Interestingly, the application of back propagation consistently reduces the number of search steps, while the effect of side propagation in the current form is not yet clear.

Motivated by the performance CRSat on the VLIW-SAT-1.1 suite, we obtained the run-times of the complete SAT solver `minisat` [4] (version 2.0) on CNF instances obtained from the AIG versions of the benchmarks (produced from ISCAS, as described above), as well as the original CNF versions of the benchmarks. As `minisat` uses randomness, we have obtained the run-time distributions for the solver over 100 runs with different random seeds. The results are presented in Table 2. While keeping in

Table 1. Median number of steps and median run-time in seconds for the SLS algorithms. "–" indicates no solutions were found in 20 tries with 300 seconds timeout each try.

Instance	bcsls-aig		crsat		crsat+back		crsat+side	
	steps	time	flips	time	steps	time	steps	time
dp_12.fsa-b9-s	136166	2.40	1103	0.01	755	**0.01**	1316	0.01
elevator_2-b8-s	53566	2.24	5281	0.11	1192	**0.07**	5017	0.10
mmgt_3.fsa-b9-p	132047	1.77	11762	0.15	3748	**0.12**	9063	0.14
speed_1.fsa-b15-s	1199186	19.42	1907	0.02	792	**0.01**	2393	0.02
parity8-27	17661	0.35	2780	0.01	858	0.02	1657	**0.01**
parity12-35	1188736	28.27	98285	**0.62**	87700	1.77	167683	1.35
parity12-47	456861	8.74	35285	0.28	13478	0.27	34716	**0.25**
parity12-98	1417260	26.68	124996	1.03	87283	1.61	114981	**0.97**
2dlx_*_005	–		7517	**1.06**	5388	2.57	26981	2.58
2dlx_*_017	–		245	**0.04**	198	0.04	1309	0.14
2dlx_*_025	–		1198	**0.16**	327	0.18	1882	0.18
2dlx_*_049	–		8415	0.49	4911	1.17	6427	**0.42**
9vliw_*_009	–		1410	**0.34**	1328	0.39	2575	1.08
9vliw_*_090	–		2956	**1.67**	2038	2.96	6280	3.27
9vliw_*_092	–		10808	**7.07**	2484	7.62	26429	12.96
9vliw_*_097	–		1211	**0.26**	1196	0.27	2637	1.28

Table 2. Median run-time for crsat and minisat in seconds on VLIW-SAT-1.1 benchmarks, 100 tries

Solver / original format	9vliw_*_007	9vliw_*_009	9vliw_*_017	9vliw_*_090	9vliw_*_092	9vliw_*_097
crsat / AIG	0.57	**0.34**	11.79	**1.67**	7.07	**0.26**
minisat / AIG	0.78	0.85	8.56	5.94	0.99	0.58
minisat / CNF	**0.27**	1.75	**0.47**	1.77	**0.60**	10.54

mind that the instances were "cherry-picked" for the evaluation of crsat, some of the results seem to be quite encouraging.

The experimental results reported in this paper indicate that the use of the incomplete BCP provides a general direction for enhancing the performance of SLS solvers for circuit-SAT. Further investigation is required to understand and take full advantage of the influence of the different variants of the incomplete BCP on the dynamic behaviour of circuit SLS. The techniques proposed in this paper are related to the previous research on SAT procedures that combine SLS with BCP (e.g. [5]) and on SLS procedures for circuit-SAT (e.g. [10]). This relationship is explored in full in [2].

Acknowledgments. We thank the anonymous referees for very helpful comments.

References

1. Belov, A., Stachniak, Z.: Improving Variable Selection Process in Stochastic Local Search for Propositional Satisfiability. In: Kullmann, O. (ed.) SAT 2009. LNCS, vol. 5584, pp. 258–264. Springer, Heidelberg (2009)
2. Belov, A.: Stochastic Local Search for Non-Clausal and Circuit Satisfiability. In: preparation PhD Thesis, York University, Toronto, Canada (2010)
3. Crawford, J.M., Kearns, M.J., Shapire, R.E.: The Minimal Disagreement Parity Problem as Hard Satisfiability Problem. Tech. report CIRL and ATT Bell Labs (1994)

4. Eén, N., Sörensson, N.: An Extensible SAT-solver. In: Giunchiglia, E., Tacchella, A. (eds.) SAT 2003. LNCS, vol. 2919, pp. 502–518. Springer, Heidelberg (2004)
5. Hirsch, E.A., Kojevnikov, A.: UnitWalk: A New SAT Solver that Uses Local Search Guided by Unit Clause Elimination. Annals of Mathematics and Artificial Intelligence (2005)
6. Hoos, H.H., Stutzle, T.: Local Search Algorithms for SAT: An Empirical Evaluation. J. of Automated Reasoning (2000)
7. Järvisalo, M., Junttila, T., Niemelä, I.: Justification-Based Non-Clausal Local Search for SAT. In: Proc. of ECAI 2008 (2008)
8. Junttila, T.: The BC Package and a File Format for Constrained Boolean Circuits, http://www.tcs.hut.fi/~tjunttil/bcsat/
9. Kuehlmann, A., Ganai, M., Paruthi, V.: Circuit-based Boolean Reasoning. In: Proc. of DAC 2001 (2001)
10. Pham, D.N., Thornton, J., Sattar, A.: Building Structure into Local Search for SAT. In: Proc. of IJCAI 2007 (2007)
11. Velev, M.N., http://www.ece.cmu.edu/~mvelev

A System for Solving Constraint Satisfaction Problems with SMT

Miquel Bofill, Josep Suy, and Mateu Villaret[*]

Departament d'Informàtica i Matemàtica Aplicada
Universitat de Girona
E-17003 Girona, Spain
{mbofill,suy,villaret}@ima.udg.edu

Abstract. SAT Modulo Theories (SMT) consists of deciding the sat-isfiability of a formula with respect to a decidable background theory, such as linear integer arithmetic, bit-vectors, etc, in first-order logic with equality. SMT has its roots in the field of verification. It is known that the SAT technology offers an interesting, efficient and scalable method for constraint solving, as many experimentations have shown. Although there already exist some results pointing out the adequacy of SMT tech-niques for constraint solving, there are no available tools to extensively explore such adequacy. In this paper we introduce a tool for translating FlatZinc (MiniZinc intermediate code) instances of constraint satisfac-tion problems to the standard SMT-LIB language. It can be used for deciding satisfiability as well as for optimization. The tool determines the required logic for solving each instance. The obtained results suggest that SMT can be effectively used to solve CSPs.

1 Introduction

Over the last decade there have been important advances in the Boolean satisfia-bility (SAT) solving techniques, to the point that nowadays modern SAT solvers can tackle real-world problem instances with millions of variables. Hence, SAT solvers have become a viable engine for solving combinatorial discrete prob-lems. For instance, in [2], an application that compiles specifications written in a declarative modeling language into SAT is shown to give promising results. In-teresting comparisons between SAT and Constraint Satisfaction Problems (CSP) encodings and techniques can be found in [13].

SAT techniques have been adapted for more expressive logics. For instance, in the case of *Satisfiability Modulo Theories (SMT)*, the problem is to decide the satisfiability of a formula with respect to a decidable background theory (or combinations of them) in first order logic with equality [9,11]. Some of these theories are (quantifier free) *Linear Integer Arithmetic* (QF_LIA), *Inte-ger Difference Logic* (QF_IDL), *Linear Real Arithmetic* (QF_LRA), *Uninter-preted Functions* (QF_UF), *Non-linear Integer Arithmetic* (QF_NIA), etc [10].

---

[*] All authors partially supported by the Spanish Ministry of Science and Innovation through the project SuRoS (ref. TIN2008-04547/TIN).

O. Strichman and S. Szeider (Eds.): SAT 2010, LNCS 6175, pp. 300–305, 2010.

Usually, SMT solvers deal with problems with thousands of clauses like, e.g., $x + 3 < y \lor y = f(f(x+2)) \lor g(y) \leq 1$, containing atoms over combined theories, and involving functions with no predefined interpretation, i.e., uninterpreted functions. Adaptations of SAT techniques to the SMT framework have been described in [12]. Although most SMT solvers are restricted to decidable quantifier free fragments of their logics, this suffices for many applications. The main application area of SMT is hardware and software verification. Nevertheless, there are already promising results in the direction of adapting SMT techniques for solving CSPs (see e.g. [1]) even in the case of combinatorial optimization (see e.g. [7]). Fundamental challenges on SMT for constraint programming and optimization are detailed in [8].

Since the beginning of Constraint Programming (CP), its *holy grail* has been to obtain a declarative language allowing users to easily specify their problem and forget about the techniques required to solve it. Among many others [2,4,6], *MiniZinc* [6] is proposed to be a standard CP modeling language. CSP models and data are written in the MiniZinc language which, after compilation, result into CSP instances codified in a sort of intermediate code called *FlatZinc*. Several solvers, such as Gecode, ECL^{i}PS^e and SICStus Prolog, provide specialized backends for this intermediate language.

In this paper we introduce a tool called fzn2smt[1] for solving FlatZinc CSP instances through SMT. Our work is similar to that of [1], where a compiler from a declarative language to the standard SMT-LIB language [10] was developed, and to that of FzNTini [5], that solves FlatZinc CSP instances through SAT. As FzNTini, our system fzn2smt does not only solve decision problems but also optimization problems, and uses FlatZinc as input language, supporting all its standard data types and constraints. The logic required for solving each instance is determined automatically by fzn2smt during the translation.

2   Architecture of the Tool

The architecture of fzn2smt is depicted in Fig. 1 throughout the process of compiling and solving. The input of the compiler is a FlatZinc instance, which is translated into an SMT instance (in the standard SMT-LIB format) and fed into an SMT solver. Due to the large number of existing SMT solvers, each one supporting different combinations of theories, the user can choose which solver to use (default being Yices [3]).

FlatZinc has three solving options: solve satisfy, solve minimize x and solve maximize x, where x is an integer variable. Since currently optimization is not supported in the SMT-LIB language, we have naively implemented it by means of iterative calls performing a binary search on the domain of the variable to optimize. Moreover, since there is no standard output model in the SMT-LIB language[2], we need a specialized *recover module* for each solver in order to obtain the answers in the standard FlatZinc output format. In this work we have only

---

[1] fzn2smt can be downloaded from http://ima.udg.edu/Recerca/GrupESLIP.html

[2] There are even solvers that only return sat, unsat or unknown.

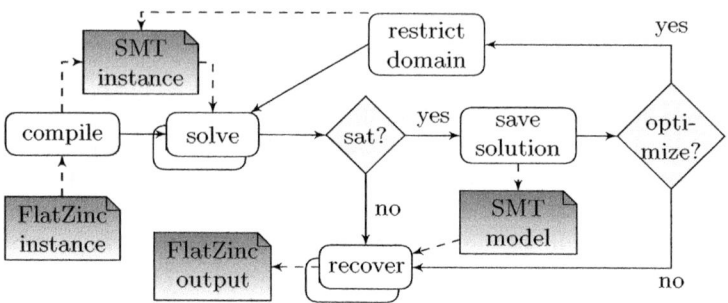

Fig. 1. The compiling and solving process of `fzn2smt`

built the one for Yices. As a byproduct, `fzn2smt` also generates the corresponding SMT instance and model.

3   Some Hints on the Translation

FlatZinc has two categories of data: constants and variables. It provides three scalar types (booleans, integers and floats) and two compound types (sets of integers and one-dimensional integer-indexed arrays). Scalar type domains can be bounded by an interval or a list. Our translation goes as follows:

- Scalar types are translated into their equivalent in SMT (`boolean` to `Bool`, `integer` to `Int` and `float` to `Real`). Constraining a variable to its domain results in a disjunction of equalities when the domain is an explicit enumeration of values, or into a conjunction of two inequalities when the domain is described as a range. Instantiated data (i.e., constants and instantiated variables) is always replaced by its value.
- Arrays have two parametric possible translations:
    - Using QF_UF. Each array is translated into an uninterpreted function with the same name. E.g., `array[1..8] of var int:p` is translated into the uninterpreted function `p()` of type `Int` $\rightarrow$ `Int`. Accesses to the array `p` such as `p[i]`, where `i` is a constant, are translated into `p(i)`. For undetermined references to arrays, FlatZinc provides the constraint `array_int_element(i,t,p)`, that we transform into `p(i) = t` $\land$ $1 \leq i \leq 8$. Constraints referring to all positions of an array are expanded: e.g., `array_bool_and(p,res)` becomes `res = and(p(1)...p(8))`.
    - Decomposing the vector into as many base type variables as elements are in the vector. E.g., `array[1..8] of var int:p` is translated into `p_1,...,p_8` integer variables.

We want to remark that the SMT theory of arrays involves read and write operations and, hence, is intended to be used for modelling state change of programs with arrays. However, since there are only read operations in CP models (i.e., there is no notion of state) it suffices to use an uninterpreted function for every array, translating every read operation of the form read(a,i) into a(i). This is enough since the theory of equality and un-interpreted functions guarantees that a(i)=a(j) whenever i=j. Moreover, deciding satisfiability of sets of equalities involving uninterpreted functions is far more cheaper than using the arrays theory. Nevertheless, preliminary ex-perimentation on FlatZinc instances has shown us that decomposing arrays into their elements provides a better performance than the UF approach.

– Sets can only be defined on a range or a list of integers. For this reason, we simply use a Boolean variable for every possible element, indicating whether it belongs to the set or not. In order to make it easier for some set restrictions, each Boolean variable has a partner 0-1 integer variable. Arrays of sets are always expanded.

We have implemented the translation into SMT of all FlatZinc constraints. During the compilation process it is crucial to detect the logic needed. For example, the int_times(a,b,c) constraint states a*b=c. This falls into linear arithmetic if a or b are constants or instantiated variables. Otherwise, non-linear arithmetic is necessary. Since few SMT solvers support non-linear arithmetic, when a and b are variables, but bounded by an interval or a list, we linearize this constraint by enumerating the possible values of the variable with smaller domain as follows: a=1→1*b=c, a=2→2*b=c, ...

4   Benchmarking and Conclusions

We have run the most of the FlatZinc instances of the problems provided with the MiniZinc 1.0.3 distribution[3]. In Table 1 we report the number of solved instances (within parenthesis) and the total time spent in them for each solver. The times are the sum of the translation, when needed (e.g., fzn2smt translates from FlatZinc to the SMT-LIB format), plus the solving time. We indicate in boldface the cases with more solved instances, breaking ties by total time. The experiments have been performed on an Intel Core i5 CPU at 2.66 GHz, with 4GB of RAM, running openSUSE 11.2 (kernel 2.6.31).

From these experiments we can observe that fzn2smt is the system that solves more instances. Although SMT solvers are black-box and our experimentation is not exhaustive, looking at the results more carefully we observe that fzn2smt outperforms other systems in problems with many constraints and non-trivial arithmetic (cars, carseq, cutstock, jobshop, nsp, rcpsp). In [5] it was already shown that FznTini, based on (plain) SAT encoding, was competitive with specialized systems. By adding theories we go one step further. It is worthy to notice that Gecode, G12 and ECL^{i}PS^e, since are search based systems, can take profit of annotations in order to use particular strategies orienting the search, whilst SAT and SMT do not. Surprisingly, our naive binary search approach to optimization is also giving good performance.

---

[3] We have omitted the 2DBinPacking and QCP benchmarks due to errors in the translation from MiniZinc to FlatZinc, some instances of nsp due to domain errors in specification, and the instances of debruijn which required more than 5 minutes for translation from MiniZinc to FlatZinc.

Table 1. Performance of different tools on FlatZinc instances. 's' stands for satisfaction and 'o' for optimization. # stands for the number of instances run of each benchmark. Time in seconds(solved). Timeout is 300 seconds for each instance.

Problem		#	G12	ECL^{i}PS^e	Gecode	FzNTini	fzn2smt
alpha	s	1	**0.01(1)**	0.47(1)	0.07(1)	0.60(1)	0.66(1)
areas	s	4	0.13(4)	2.04(4)	**0.05(4)**	0.38(4)	4.75(4)
bibd	s	14	118.16(12)	4.18(7)	35.58(7)	**353.78(13)**	79.48(12)
cars	s	79	0.02(1)	0.81(1)	295.64(3)	1.21(1)	**2271.17(21)**
carseq	s	81	0.47(2)	3.47(2)	0.12(2)	0.06(2)	**2502.95(45)**
curriculum	s	3	0.27(2)	95.09(2)	261.82(1)	**8.70(3)**	9.97(3)
cutstock	o	121	0.01(1)	0.49(1)	0.01(1)	1.53(1)	**1066.29(20)**
debruijn_binary	s	7	37.93(7)	12.40(6)	**10.33(7)**	0.09(1)	0.65(1)
eq	s	1	**0.01(1)**	0.49(1)	**0.01(1)**	20.14(1)	0.47(1)
golfers	s	9	113.7(5)	0.59(1)	**22.64(5)**	9.51(2)	26.30(2)
golomb	o	10	**251.18(9)**	85.84(8)	23.83(8)	23.87(6)	41.88(6)
jobshop	o	74	0.11(1)	1.92(2)	22.14(2)	514.70(4)	**1113.04(22)**
kakuro	s	6	**0.06(6)**	2.97(6)	**0.06(6)**	-(0)	4.67(6)
knights	s	4	**0.05(4)**	1.99(4)	0.27(4)	0.32(4)	3.08(4)
langford	s	25	52.19(20)	121.85(20)	**34.54(20)**	310.24(18)	50.52(20)
latin-squares	s	7	**5.98(6)**	12.54(6)	15.35(6)	129.40(3)	7.54(4)
magicseq	s	9	25.17(7)	21.56(7)	**7.39(7)**	0.30(3)	11.01(4)
nmseq	s	10	281.73(6)	-(0)	**171.03(7)**	-(0)	1.42(1)
nsp	s	20	-(0)	-(0)	0.09(1)	402.22(14)	**70.28(15)**
pentominoes	s	7	113.68(4)	27.88(2)	**208.52(5)**	12.81(1)	4.85(1)
photo	o	2	0.10(2)	1.07(2)	**0.04(2)**	0.08(2)	0.78(2)
quasigroup7	s	10	**1.37(5)**	293.67(3)	3.65(5)	380.28(3)	31.51(5)
queens	s	7	88.88(7)	**36.24(7)**	0.52(3)	94.76(4)	54.61(5)
radiation	o	9	**207.90(7)**	54(6)	231.41(7)	-(0)	584.25(7)
rcpsp	o	10	11.96(2)	49.85(4)	22.54(5)	-(0)	**581.13(8)**
schur_numbers	s	3	1.30(3)	1.03(2)	**0.30(3)**	0.02(2)	1.45(3)
search_stress	s	1	**0.01(1)**	0.54(1)	**0.01(1)**	0.16(1)	0.44(1)
shortest_path	o	10	4.36(4)	292.15(6)	**141.34(7)**	-(0)	45.79(6)
slow_convergence	s	10	68.74(10)	12.84(7)	**11.03(10)**	36.98(4)	222.37(10)
steiner-triples	s	6	0.13(2)	0.50(1)	0.01(1)	65.71(2)	**96.33(5)**
still_life	o	10	17.35(8)	80.03(8)	134.12(9)	60.29(8)	**25.20(9)**
talent_scheduling	o	11	10.43(3)	-(0)	**4.03(3)**	128.10(2)	45.71(3)
template_design	o	7	126.20(2)	1.40(1)	56.99(2)	14.91(1)	**37.37(2)**
tents	s	3	0.10(3)	1.52(3)	**0.04(3)**	0.30(3)	2.50(3)
trucking_hl	o	5	**3.82(5)**	-(0)	41.56(5)	-(0)	3.90(5)
Total		596	1543(163)	1221(132)	1757(164)	2571(114)	**9004(267)**

The obtained results suggest that SMT can be effectively used for CSP solving in a broad sense, i.e., not just for specialized problems. Hence the tool could serve for getting a big enough picture of the suitability of SMT solvers w.r.t. CSP solving in general, and to compare the performance of state-of-the-art SMT solvers outside the SMT competition.

We have not used any MiniZinc global constraint (such as, e.g., `cumulative`, `alldifferent`, ...) since they are not supported by current SMT solvers. We think that developing specialized solvers for such theories in SMT is a promising research line. Finally, we think that better results could be obtained if translating directly from the MiniZinc language to SMT. In doing so, most clever translations could be possible and probably less variables could be generated.

References

1. Bofill, M., Palahi, M., Suy, J., Villaret, M.: SIMPLY: a Compiler from a CSP Modeling Language to the SMT-LIB Format. In: Proceedings of the 8th Intl. Workshop on Constraint Modelling and Reformulation, pp. 30–44 (2009)
2. Cadoli, M., Mancini, T., Patrizi, F.: SAT as an Effective Solving Technology for Constraint Problems. In: Esposito, F., Raś, Z.W., Malerba, D., Semeraro, G. (eds.) ISMIS 2006. LNCS (LNAI), vol. 4203, pp. 540–549. Springer, Heidelberg (2006)
3. Dutertre, B., de Moura, L.: The Yices SMT solver (August 2006), Tool paper at http://yices.csl.sri.com/tool-paper.pdf
4. Frisch, A., Harvey, W., Jefferson, C., Martínez-Hernández, B., Miguel, I.: Essence: A Constraint Language for Specifying Combinatorial Problems. Constraints 13(3), 268–306 (2008)
5. Huang, J.: Universal Booleanization of Constraint Models. In: Stuckey, P.J. (ed.) CP 2008. LNCS, vol. 5202, pp. 144–158. Springer, Heidelberg (2008)
6. Nethercote, N., Stuckey, P., Becket, R., Brand, S., Duck, G., Tack, G.: MiniZinc: Towards a Standard CP Modelling Language. In: Bessière, C. (ed.) CP 2007. LNCS, vol. 4741, pp. 529–543. Springer, Heidelberg (2007)
7. Nieuwenhuis, R., Oliveras, A.: On SAT Modulo Theories and Optimization Problems. In: Biere, A., Gomes, C.P. (eds.) SAT 2006. LNCS, vol. 4121, pp. 156–169. Springer, Heidelberg (2006)
8. Nieuwenhuis, R., Oliveras, A., Rodríguez-Carbonell, E., Rubio, A.: Challenges in Satisfiability Modulo Theories. In: Baader, F. (ed.) RTA 2007. LNCS, vol. 4533, pp. 2–18. Springer, Heidelberg (2007)
9. Nieuwenhuis, R., Oliveras, A., Tinelli, C.: Solving SAT and SAT Modulo Theories: From an Abstract Davis–Putnam–Logemann–Loveland Procedure to DPLL(T). Journal of the ACM 53(6), 937–977 (2006)
10. Ranise, S., Tinelli, C.: The SMT-LIB Standard: Version 1.2. Tech. rep., Dept. of Comp. Science, University of Iowa (2006), http://www.SMT-LIB.org
11. Sebastiani, R.: Lazy Satisfiability Modulo Theories. Journal on Satisfiability, Boolean Modeling and Computation 3(3-4), 141–224 (2007)
12. Sheini, H., Sakallah, K.: From Propositional Satisfiability to Satisfiability Modulo Theories. In: Biere, A., Gomes, C.P. (eds.) SAT 2006. LNCS, vol. 4121, pp. 1–9. Springer, Heidelberg (2006)
13. Walsh, T.: SAT vs CSP. In: Dechter, R. (ed.) CP 2000. LNCS, vol. 1894, pp. 441–456. Springer, Heidelberg (2000)

Two Techniques for Minimizing Resolution Proofs

Scott Cotton

CNRS/Verimag

Abstract. Some SAT-solvers are equipped with the ability to produce resolution proofs for problems which are unsatisfiable. Such proofs are used in a variety of contexts, including finding minimal unsatisfiable sets of clauses, interpolant generation, configuration management, and proof replay in interactive theorem provers. In all of these settings, the size of the proof may be prohibitively large for subsequent processing. We suggest some new methods for resolution proof minimization. First, we identify a simple and effective method of extracting shared structure from a proof using structural hashing. Second, we suggest a heuristically-guided proof rewriting technique based on variable valuations. Our findings indicate structural sharing reduces proof length significantly and efficiently, and that our valuation-based rewriting method can give substantial further reductions but is currently limited to smaller proofs.

1 Introduction

Applications ranging from proof replay in automatic theorem provers [2] to automotive configuration management [21] make use of *resolution* proofs generated by SAT solvers [16,17,18]. At the same time, such generated proofs can be quite large and their size can become problematic for the application at hand. As a result, there has been some interest in the problem of minimizing or compressing a given resolution proof, especially those generated by SAT solvers [23,1,20,3,2]. It is also interesting to ask whether, or in what circumstances, effective proof minimization may lead to faster SAT solving processes. However, such *potential* applications will require more effective proof minimization methods and tools. Aside from computing unsat cores, the best existing methods can either efficiently trim proofs by a handful of percentage points [3], or they are complex [2], or they do not scale to larger problems [20,1]. The techniques introduced in this paper bring us a small step closer to effective proof minimization by providing one simple, fairly scalable method, and one method which often reduces proofs substantially and sometimes *can* find proofs several times smaller than the original, but is limited to smaller proofs.

1.1 Background

The problem of instrumenting a conflict driven SAT solver to generate resolution proofs is well studied [24,14,12,7,13]. In [12], a solution was given to an

O. Strichman and S. Szeider (Eds.): SAT 2010, LNCS 6175, pp. 306–312, 2010.

exponential blow-up in the size of proofs generated by zchaff [19]. Later, [22] demonstrated that conflict clause minimization in the style of [10] reduces solution time, and hence likely also reduces proof size. Additionally, [13] gave an algorithm for conflict clause minimization and showed that it reduces solution time on unsatisfiable problems slightly and can also be used to generate proof traces which are easier to check. The algorithm of [13] was also discovered by the author and reported in [9]. The computation of an unsatisfiable core from a generated proof, originally presented in [23], constitutes a simple and effective kind of minimal proof extraction.

Subsequently, several proof post-processors [20,1,2,3] were developed, and this paper presents another. The work in [20] focuses on finding common proof structure across derivations of learned clauses while that of [1] is based on subsumption. Reported experiments for both methods were limited to very small proofs. The work in [2] improves on [20] in part by memoizing shared proof sequences, and scales to proofs with millions of inferences. One of our minimization techniques also memoizes shared proof sequences; it uses a simpler method and appears to scale similarly. The post-processor in [3] presents a reduction technique based on double resolutions (where one resolution step dominates another and both pivot on the same variable) and another based on identifying how derived unit clauses can be applied to derivations occurring earlier in the SAT solving time line. These techniques reduced proofs generated by zchaff [19] on a set of bounded model checking problems by 13%.

Resolution Proofs. In the following, a *literal* is a propositional variable or its negation. A *clause* is a finite disjunction $(l_1 \vee l_2 \vee \ldots \vee l_k)$ of literals. Since disjunction is commutative and associative, we do not distinguish between clauses which contain the same literals but in different orders. For convenience, we also do not distinguish between clauses which contain some literals more than once and their reduced counterparts which contain exactly one representative of each literal. The empty clause is false, here denoted $\perp$.

Propositional resolution [8] is a proof rule which allows the derivation of a clause from a pair of clauses. In particular, suppose that x is a variable and C, D are clauses. Then the resolution rule is

$$\frac{(x \vee C), (\neg x \vee D)}{(C \vee D)}$$

We refer to x as the *pivot* variable, to the clause $(C \vee D)$ as the *resolvent* and to the clauses $(x \vee C)$ and $(\neg x \vee D)$ as *antecedents*.

Given a conjunction of clauses $\phi \stackrel{\circ}{=} C_1 \wedge C_2 \wedge \ldots \wedge C_m$, A *resolution proof* of unsatisfiability for ϕ is a *sequence* of clauses $(D_1, D_2, \ldots, D_n)$ such that

1. Each D_i with $1 \leq i \leq n$ is either a conjunct of ϕ (an *input*) or is the result applying the resolution rule to two antecedents D_j, D_k with $1 \leq j, k < i$ (a *derived clause*).
2. There is some $i \in [1..n]$ such that $D_i = \perp$.

A proof may be viewed as an acyclic graph in which nodes are clauses and edges connect a derived clause to its antecedents. Thus clauses which are derived have in-degree 2 while inputs have in-degree 0. We refer to such a graph as a *resolution graph*.

In the rest of this paper, we only consider proofs in which every clause is non-tautological, meaning that no clause contains both the disjunct x and the disjunct $\neg x$ for any variable x.

2   Two Minimization Techniques

2.1   Structural Minimization

Because clauses are non-tautological, all derived clauses in a resolution proof are determined by their antecedents. Hence, we can unambiguously consider proofs in terms of the input clauses together with the *structure* of the resolution graph, *i.e.* by identifying a derived clause not by its literals but rather by its antecedents. A simple way of minimizing a proof based on the structure of the resolution graph is to identify distinct nodes in the resolution sequence which are derived from the same antecedents and then merge them together. However, proofs generated by conflict-driven SAT solvers such as [24,14,7] offer more opportunities for identifying shared structure. In particular, such proofs come in the form of a sequence of (derivations of) learned clauses; but there are many possible derivations of each learned clause. As a result, we are free to find derivations of learned clauses which are small and which share structure.

To illustrate our method we assume that with each learned clause w in a proof, we can associate a graph $G_w \stackrel{\circ}{=} (A_w, \rightarrow_w)$. The graph G_w mirrors the implication graph used by the solver to derive w as follows. A_w consists of all the antecedent clauses of w and there is an edge $c \rightarrow_w d$ when, in the corresponding implication graph, c is a unit clause with unit literal l and the complement of l appears in the clause d. As illustrated in [4], such a graph defines a set of regular input resolution derivations [1] of w. We are interested in identifying sub-derivations which can be re-used in the derivation of many learned clauses without increasing the number of resolution steps required to derive any learned clause. Such a sub-derivation is characterized in [20] as an *isolateable subgraph* of G_w.

The method we propose maintains a resolution proof as structural hash in a manner similar to that used for circuits [11]. Each derived clause is given a hash value based on the identifiers of its antecedents, so it takes constant time to check whether a given resolution step is present in the proof, and constant time to add a resolution step to the proof.

To generate a binary resolution proof from a sequence of learned clause derivations, we read the learned clauses in topological order, and generate G_w for each

---

[1] Regular resolution denotes a resolution graph in which the sequence of pivot variables associated with every path does not allow repeating variables. Input resolution is resolution in which each derived clause has at most one derived clause as an antecedent.

learned clause w. Then for each graph G_w, we identify a sequence of resolution steps which will derive w. Each resolution step is added to the proof only if necessary, *i.e* if there is not already a node in the resolution graph with the same antecedents. To maximize the potential for sharing without increasing the number of resolutions steps required to derive w, we make use of the following observations.

1. G_w is guaranteed to have a clause c with out-degree 1.
2. For any $c \to_w d$ in G_w, if c has degree 1, then we may replace c and d in G_w with their resolvent r. Upon replacement, all edges coming into c or d are redirected to r, with duplicates removed, and all edges emanating from d are replaced with a corresponding edge from r.
3. Replacing c, d with r as above results in a new graph which either preserves property 1 or which has one node and that node is w. Moreover, the new graph again defines a set of regular input resolution sequences deriving w in the same manner as G_w, but each such derivation contains one less resolution step.

To generate a derivation of w, our method maintains a queue of nodes in G_w which have degree 1, initially sorted topologically in $\to_w$. Resolution steps are performed for each member of the queue in FIFO order and G_w is updated accordingly after each step until w is derived.

2.2 Proof Rewriting with Variable Valuations

The second minimization technique presented here is based on using variable valuations to rewrite proofs. Given a proof of unsatisfiability π and a variable x, it is easy to re-arrange the proof in such a way as to produce a proof of x and a proof of $\neg x$. Taken together with one additional resolution step, we arrive at a new proof of unsatisfiability, which might be smaller than the original. We call this operation *splitting*.

The method we apply generates a sequence of proofs $\pi_1 \pi_2 \ldots$, where each proof π_{i+1} is rewritten by splitting π_i. If a proof π_j is too large, we set π_{j+1} to the smallest proof in $\pi_1 \ldots \pi_{j-1}$. The process continues for as long as desired, and outputs a smallest proof from the sequence upon termination.

We propose a heuristic for variable selection based on the number of times a variable appears as a pivot in a resolution step together with the *additivity* of the corresponding resolution steps. Given a resolvent r with antecedents p, n the expression

$$add(r) \doteq \max(|r| - \max\{|p|, |n|\}, 0)$$

describes, in the worst case, how many literals are added to the proof as a result of the resolution step. Variables are scored by summing the additivity of all the resolution steps in a proof π which pivot along v together with the number of resolutions steps pivoting on v. We denote this value $add(v, \pi)$. Given such an estimation, we select variable v at step i with probability

$$\frac{add(v, \pi_i)}{\Sigma_x add(x, \pi_i)}$$

We define splitting in a manner similar to the *restriction* operator of [5]. Let π be a resolution proof of unsatisfiability and l a literal. Let $p \oplus_x n$ denote the resolvent of clauses p and n where x occurs in p and $\neg x$ occurs in n. Then define the map π_l map on nodes in the resolution graph of π:

$$\pi_l(c) \doteq \begin{cases} c & \text{if } c \text{ is an input} \\ \pi_l(p) & \text{if } c = p \oplus_x n \text{ and } (l = x \text{ or } x \notin \pi_l(p)) \\ \pi_l(n) & \text{if } c = p \oplus_x n \text{ and } (l = \neg x \text{ or } \neg x \notin \pi_l(n)) \\ \pi_l(p) \oplus_x \pi_l(n) & \text{if } x \in \pi_l(p) \text{ and } \neg x \in \pi_l(n) \end{cases}$$

Let o be the empty clause in π. To split on x from proof π, we compute $\pi_x(o)$ then $\pi_{\neg x}(o)$ and resolve the resulting two clauses.

3 Experimental Summary

We evaluated the two techniques on proofs generated by PicoSAT [7] version 913. Since both techniques require an in-memory representation of the entire proof, we selected unsatisfiable problems which picosat could solve within 10 seconds from a variety of well known benchmarks, as well as some small benchmarks from [15]. The proof checker tracecheck [6] verified all but one of the outputs of the resulting proofs, crashing on one large problem.

Structural minimization was applied to the unsat cores of 141 problems with a total of $469, 778$ core original clauses and $697, 286$ core learned clauses for a total of $77, 907, 715$ resolution steps. The resulting structurally reduced binary resolution graphs contained $30, 199, 342$ resolution steps, giving proofs 39% the size of the core proofs in a total of 357 seconds.

Variable splitting was applied to the structurally minimized binary resolution graphs of the 49 largest problems which contained less than $300, 000$ resolution steps, for a total of $1, 875, 020$ resolution steps. We used a timeout of 300 seconds per problem. The reduced proofs contained $1, 743, 311$ resolution steps or 93% of the structurally minimized size. With each problem weighted equally, the reduced proofs were on average 68% the size of the structurally reduced proofs. Smaller problems gave proofs as small as as 15% the size of the structurally minimized proof. The method compared favorably to selecting variables at random.

All results were run on a Sun JVM version 1.6.0_17 on a dual-core 3GHz Debian Linux machine with $4G$ RAM. The source code, proofs, and detailed results are available from the author's web page.

4 Conclusion and Future Work

We have presented and evaluated two minimization techniques for resolution proofs generated by conflict driven SAT solvers. In the future, we plan to investigate the use of extended resolution and the possibility of applying variable splitting to sub-proofs of larger proofs.

Acknowledgements. The author would like to thank the anonymous reviewers for many helpful comments.

References

1. Amjad, H.: Compressing propositional refutations. Electron. Notes Theor. Comput. Sci. 185, 3–15 (2007)
2. Amjad, H.: Data compression for proof replay. J. of Automated Reasoning (December 2008)
3. Bar-Ilan, O., Fuhrmann, O., Hoory, S., Shacham, O., Strichman, O.: Linear-time reductions of resolution proofs. In: HVC 2008, pp. 114–128 (2009)
4. Beame, P., Kautz, H., Sabharwal, A.: Understanding the Power of Clause Learning. J. of Artificial Intelligence Research (2004)
5. Ben-sasson, E., Wigderson, A.: Short proofs are narrow - resolution made simple. J. of the ACM, 517–526 (2001)
6. Biere, A.: tracecheck, http://fmv.jku.at/booleforce/index.html
7. Biere, A.: Picosat essentials. J. on Satisfiability, Boolean Modeling, and Computation, 75–97 (2008)
8. Blake, A.: Canonical Expressions in Boolean Algebra. PhD thesis, University of Chicago (1937)
9. Cotton, S.: On Some Problems in Satisfiability Solving. PhD thesis, University Joseph Fourier, Grenoble I (2009)
10. Eén, N., Sörensson, N.: MiniSat: A SAT Solver with Conflict-clause Minimization. In: Theory and Applications of Satisfiability Testing, SAT (2005)
11. Ganai, M.K., Kuehlmann, A.: On-the-fly compression of logical circuits. In: International Workshop on Logic Synthesis (2000)
12. Van Gelder, A.: Verifying propositional unsatisfiability: Pitfalls to avoid. In: Marques-Silva, J., Sakallah, K.A. (eds.) SAT 2007. LNCS, vol. 4501, pp. 328–333. Springer, Heidelberg (2007)
13. Van Gelder, A.: Improved conflict-clause minimization leads to improved propositional proof traces. In: Kullmann, O. (ed.) SAT 2009. LNCS, vol. 5584, pp. 141–146. Springer, Heidelberg (2009)
14. Goldberg, E., Novikov, Y.: Verification of proofs of unsatisfiability for cnf formulas. In: Design, Automation and Test in Europe, DATE (2003)
15. Küchlin, W., Sinz, C.: Proving consistency assertions for automotive product data management. J. Automated Reasoning 24(1-2) (February 2000)
16. Lynce, I., Marques-Silva, J.: On computing minimum unsatisfiable cores. In: Theory and Applications of Satisfiability Testing, SAT (2004)
17. McMillan, K., Amla, N.: Automatic abstraction without counterexamples. Tools and Algorithms for the Construction and Analysis of Systems, 2–17 (2003)
18. McMillan, K.L.: Interpolation and SAT-Based Model Checking. In: Hunt Jr., W.A., Somenzi, F. (eds.) CAV 2003. LNCS, vol. 2725, pp. 1–13. Springer, Heidelberg (2003)

19. Moskewicz, M.W., Madigan, C.F., Zhao, Y., Zhang, L., Malik, S.: Chaff: Engineering an Efficient SAT Solver. In: DAC 2001 (2001)
20. Sinz, C.: Compressing propositional proofs by common subproof extraction. In: Moreno Díaz, R., Pichler, F., Quesada Arencibia, A. (eds.) EUROCAST 2007. LNCS, vol. 4739, pp. 547–555. Springer, Heidelberg (2007)
21. Sinz, C., Kaiser, A., Küchlin, W.: Formal methods for the validation of automotive product configuration data. Artif. Intell. Eng. Des. Anal. Manuf. 17 (2003)
22. Sörensson, N., Biere, A.: Minimizing learned clauses. In: Kullmann, O. (ed.) SAT 2009. LNCS, vol. 5584, pp. 237–243. Springer, Heidelberg (2009)
23. Zhang, L., Malik, S.: Extracting small unsatisfiable cores from unsatisfiable boolean formulas. In: Theory and Applications of Satisfiability Testing, SAT (2003)
24. Zhang, L., Malik, S.: Validating sat solvers using an independent resolution-based checker: Practical implementations and other applications. In: Design, Automation and Test in Europe, DATE (2003)

On Moderately Exponential Time for SAT

Evgeny Dantsin and Alexander Wolpert

Department of Computer Science, Roosevelt University
430 S. Michigan Av., Chicago, IL 60605, USA
{edantsin,awolpert}@roosevelt.edu

Abstract. Can SAT be solved in "moderately exponential" time, i.e., in time $p(|F|) \, 2^{cn}$ for some polynomial p and some constant $c < 1$, where F is a CNF formula of size $|F|$ over n variables? This challenging question is far from being resolved. In this paper, we relate the question of moderately exponential complexity of SAT to the question of moderately exponential complexity of problems defined by existential second-order sentences. Namely, we extend the class SNP (Strict NP) that consists of Boolean queries defined by existential second-order sentences where the first-order part has a universal prefix. The extension is obtained by allowing a $\forall \ldots \forall \exists \ldots \exists$ prefix in the first-order part. We prove that if SAT can be solved in moderately exponential time then all problems in the extended class can also be solved in moderately exponential time.

1 Introduction

All known algorithms for SAT take exponential time in the worst case. However, there may be a significant difference in the efficiency between exponential-time algorithms, compare an algorithm running in time $\mathcal{O}(2^n)$ with algorithms running in time $\mathcal{O}(2^{n/100})$ or in time $\mathcal{O}(2^{\sqrt{n}})$. How large is the exponent for SAT?

1.1 Moderately Exponential Complexity

Can SAT be solved in moderately exponential time? To state this question formally, we need to specify a parameter with respect to which we measure the complexity of SAT. It is standard in complexity theory to use the instance size as a complexity parameter. However, for problems in NP, there is another approach that suggests the choice of a *certificate-size parameter*, i.e., a parameter that characterizes the size of a certificate rather than the instance size, see for example [PP09] for the motivation and see Section 3 for precise definitions.

When considering SAT, a certificate for a formula with n variables is a satisfying assignment, and it is straightforward to describe it using n bits. Therefore, a natural certificate-size parameter for SAT is the number of variables in the input formula. Throughout the paper we regard SAT as a parameterized problem (SAT, n) where n is the number of variables in the input CNF formula.

Let $(\mathcal{A}, p)$ be a parameterized problem, where p is a specified certificate-size parameter. Since a certificate can be found by enumerating 2^p bit strings, a

O. Strichman and S. Szeider (Eds.): SAT 2010, LNCS 6175, pp. 313–325, 2010.

trivial upper bound for $\mathcal{A}$ is 2^p up to a factor polynomial in the instance size. We say that $(\mathcal{A}, p)$ can be solved in *moderately exponential time* if there is a polynomial q and a constant $c < 1$ such that

$$\mathcal{A} \in \mathsf{RTIME}(q(|x|)\, 2^{cp})$$

where $|x|$ is the instance size. We define ME to be the class of all parameterized problems that can be solved in moderately exponential time.

Why is ME defined in terms of RTIME rather than DTIME? All results on ME proved and mentioned in this paper hold for both possible definitions, i.e., for probabilistic time as well as for deterministic time, so we could choose any of them. We have chosen probabilistic time just to conform with the definitions in [CIP06], where moderately exponential complexity bounds are defined in terms of RTIME.

1.2   Relevant Results

A challenging open question is whether there exists a constant $c > 0$ such that no algorithm solves 3-SAT in time $\mathcal{O}(2^{cn})$ where n is the number of variables in the input formula. Impagliazzo, Paturi, and Zane in their seminal papers [IP01, IPZ01] obtained deep results that shed more light on this and similar questions.

The question whether $(\text{SAT}, n) \in \mathsf{ME}$ is also open, and the state of the art of research in this field is far from answering this question even under plausible complexity assumptions. However, we know that some restrictions of (SAT, n) are in ME. In particular,

- $(k\text{-SAT}, n) \in \mathsf{ME}$ for any constant k [PPSZ98, Sch99, DGH$^+$02];
- $(\text{SAT}_\Delta, n) \in \mathsf{ME}$ for any constant Δ [AS03, DH09], where SAT_Δ is the restriction of SAT to formulas of clause density at most Δ (the clause density is the ratio of the number of clauses to the number of variables).

Impagliazzo, Paturi, and Zane [IP01, IPZ01] showed the relationship between ME and SNP (Strict NP), the class defined in [KV87] and used in [PY91] for proving inapproximability of certain optimization problems. The class SNP consists of Boolean queries defined by existential second-order sentences of the form

$$\exists T_1 \ldots \exists T_r \forall x_1 \ldots \forall x_s\, \phi \tag{1}$$

where $T_1, \ldots, T_r$ are relation symbols, $x_1, \ldots, x_s$ are first-order variables, and ϕ is a quantifier-free formula. For any problem in SNP, there is a natural certificate-size parameter expressed through the cardinality of the universe and arities of second-order variables, see Section 3 for details. Thus, SNP can be regarded as the class of parameterized problems.

It is shown in [IPZ01] that, loosely speaking, $(k\text{-SAT}, n)$ is complete for SNP via reductions that allow a linear increase of the parameter (the polynomial-time many-one *strong* reducibility). However, as it is seen from the proof, the reduction does not change the parameter at all. Thus, it follows from the existence of moderately exponential time algorithms for k-SAT that any parameterized problem in SNP can be solved in moderately exponential time, i.e., SNP $\subseteq$ ME.

A number of results relevant to the $(\text{SAT}, n) \in$ ME question were obtained assuming the Exponential Time Hypothesis (ETH) defined in [IP01]. Informally, ETH says that k-SAT cannot be solved in subexponential time. Formally, the hypothesis is stated in terms of the following sequence $\{s_k\}$: for any integer $k \geq 3$, let s_k be the infimum of all $c > 0$ such that

$$k\text{-SAT} \in \mathsf{RTIME}(2^{cn}).$$

ETH says that $s_k > 0$ for all $k \geq 3$. Using sparsification, it was proved in [IPZ01] that ETH is equivalent to $s_3 > 0$. Also, ETH can be re-stated in terms of SNP: the class SNP contains a problem that cannot be solved in subexponential time.

Assuming ETH, it was proved in [IP01] that the sequence $\{s_k\}$ increases infinitely often when k grows. Let $s_\infty = \lim_{k \to \infty} s_k$. The value of s_∞ is still unknown even under plausible complexity assumptions. The Strong Exponential Time Hypothesis (SETH) states that $s_\infty = 1$ [IP01, CIP09]. Note that SETH implies $(\text{SAT}, n) \notin$ ME.

It was shown in [CIP06] that $\{s_k\}$ is interwoven with a similar sequence defined for SAT_Δ. Let t_Δ be the infimum of all $c > 0$ such that $\text{SAT}_\Delta \in \mathsf{RTIME}(2^{cn})$. Then

- for any $\Delta > 0$, there is $k \geq 3$ such that $t_\Delta \leq s_k$;
- for any $k \geq 3$ there is $\Delta > 0$ such that $s_k \leq t_\Delta$.

Therefore, $s_\infty = t_\infty$ where t_∞ denotes the limit of $\{t_\Delta\}_{\Delta \in \mathbb{N}}$ as $\Delta \to \infty$.

For a number of important computational problems, new insights into their complexity were obtained using ETH as a complexity assumption. In particular, Calabro, Impagliazzo, and Paturi [CIP09] compared the complexity of k-SAT and the complexity of the Decision-Unique-k-SAT problem in which we are asked whether a given k-CNF formula has exactly one satisfying assignment. It turned out that, assuming ETH, these problems have the same complexity. Marx [Mar07] analyzed the complexity of binary CSP with respect to the number-of-variables parameter. He proved that if the complexity of this problem is $f(G) n^{o(k/\lg k)}$ then ETH fails, where G is the underlying primal graph, f is an arbitrary function, and k is the treewidth of G. Another result connecting CSP and ETH was proved by Traxler [Tra08] who showed that, assuming ETH, $(k, 2)$-CSP has complexity k^{cn}, where c is a constant not depending on k.

The currently best upper bound for SAT is

$$|F|^{\mathcal{O}(1)} 2^{n(1 - 1/\mathcal{O}(\lg \frac{m}{n}))}$$

where m is the number of clauses [CIP06, DH09]. This result neither supports nor opposes the $(\text{SAT}, n) \in$ ME conjecture. Williams [Wil07] notes that while many researchers are skeptical that $(\text{SAT}, n) \in$ ME, he has not found much evidence for this skepticism. He argues that $(\text{SAT}, n) \in$ ME is consistent with ETH. He also presents three hypotheses (about the complexities of three other computational problems) and proves that if any of them are true then $(\text{SAT}, n) \in$ ME.

1.3   Our Results

Implication of $(\text{SAT}, n) \in \mathsf{ME}$. When trying to prove or disprove $(\text{SAT}, n) \in \mathsf{ME}$, it can be helpful to find implications of both alternatives. In this paper, we describe a class of parameterized problems such that if $(\text{SAT}, n) \in \mathsf{ME}$ then all problems in the class are also in ME. This class is obtained from SNP in two steps. First, we extend SNP by allowing existential first-order quantifiers after universal ones. The extension is denoted by SNP_1. The second step is to close SNP_1 under polynomial-time reductions that do not increase the parameter. The closure is denoted by $\mathcal{R}(\mathsf{SNP}_1)$. Thus, we prove

$$(\text{SAT}, n) \in \mathsf{ME} \quad \Rightarrow \quad \mathcal{R}(\mathsf{SNP}_1) \subseteq \mathsf{ME}. \tag{2}$$

More exactly, the underlying reducibility is a special case of strong many-one reducibility in [IPZ01]. While a strong reduction ρ from $(\mathcal{A}, p)$ to $(\mathcal{B}, q)$ allows a linear increase of the parameter, i.e., $q(\rho(x)) = \mathcal{O}(p(x))$, our reductions (called *admissible* reductions) allow only an additive increase by a constant, i.e., $q(\rho(x)) = p(x) + \mathcal{O}(1)$. Thus, if there is a polynomial-time admissible reduction from $(\mathcal{A}, p)$ to $(\mathcal{B}, q)$ and $(\mathcal{B}, q)$ is in ME then $(\mathcal{A}, p)$ is also in ME.

To prove implication (2), we restrict SAT to formulas of polynomial size. Namely, for any $d > 0$, we restrict SAT to formulas that have at most n^d clauses where n is the number of variables. This restriction is denoted by $\text{SAT}^{(d)}$. It is natural to think of $\text{SAT}^{(d)}$ as a generalization of SAT_Δ.

Obviously, if $(\text{SAT}, n) \in \mathsf{ME}$ then $(\text{SAT}^{(d)}, n) \in \mathsf{ME}$ for all d. In Section 4, we prove that for every problem in $\mathcal{R}(\mathsf{SNP}_1)$, there exists a constant d such that $(\mathcal{A}, p)$ is reducible to $(\text{SAT}^{(d)}, n)$ via a polynomial-time admissible reduction. Thus, if $(\text{SAT}, n) \in \mathsf{ME}$ then $(\text{SAT}^{(d)}, n) \in \mathsf{ME}$ for all d and, hence, every problem in $\mathcal{R}(\mathsf{SNP}_1)$ is also in ME.

In addition to this "hardness" result, we prove "inclusion": $(\text{SAT}^{(d)}, n)$ in $\mathcal{R}(\mathsf{SNP}_1)$ for all d. Thus, we do not give a single complete problem for $\mathcal{R}(\mathsf{SNP}_1)$, but all $\text{SAT}^{(d)}$ collectively play the role of a complete problem.

Restrictions of SAT *to formulas of polynomial size.* A natural question about $\text{SAT}^{(d)}$ is whether $(\text{SAT}^{(d)}, n) \in \mathsf{ME}$. We do not answer this question but we relate the complexity of $(\text{SAT}^{(d)}, n)$ to the complexities of $(k\text{-SAT}, n)$ and (SAT_Δ, n) in Section 2. Similar to bounds s_k and t_Δ, we define bounds σ_d that characterize the complexity of $\text{SAT}^{(d)}$. Namely, σ_d is the infimum of all $c > 0$ such that $\text{SAT}^{(d)} \in \mathsf{RTIME}(2^{cn})$. If $d > 1$, we have

$$s_k \leq \sigma_d \quad \text{and} \quad t_\Delta \leq \sigma_d$$

for all k and Δ. Moreover, if ETH is true, these inequalities are strict: k-SAT and SAT_Δ can be solved strictly faster than $\text{SAT}^{(d)}$ for any $d > 1$. Assuming SETH, we have

$$s_\infty = t_\infty = \sigma_\infty = 1$$

where σ_∞ is the limit of $\{\sigma_d\}_{d \in \mathbb{N}}$ as $d \to \infty$. If SETH fails, it is still unknown which of the following options hold:

- $\sigma_\infty < 1$, which means $(\text{SAT}^{(d)}, n) \in \text{ME}$;
- $\sigma_\infty = 1$, which means $(\text{SAT}, n) \notin \text{ME}$.

Expressibility of (SAT, n) *as an existential second-order sentence.* While implication (2) can be useful for proving $(\text{SAT}, n) \notin \text{ME}$, Theorem 2 in Section 3 may give guidance if we try to prove the opposite, $(\text{SAT}, n) \in \text{ME}$. For instance, suppose we have a Boolean query defined by an existential second-order sentence and proved to be in ME. Then we could prove $(\text{SAT}, n) \in \text{ME}$ by reducing (SAT, n) to this query via a polynomial-time admissible reduction. What type of such a reduction we can hope for? Theorem 2 shows that the reduction cannot be one-one. Moreover, not only must it be many-one, but "many" must be "exponentially many". This is because instances of SAT can be CNF formulas whose size is super-polynomial in n.

2   Restrictions of SAT to Formulas of Polynomial Size

In this section, we consider three restrictions of SAT to formulas whose size is polynomial in the number of variables. Two of them are well known:

- k-SAT is the restriction to formulas where every clause has at most k literals;
- SAT_Δ is the restriction to formulas of clause density at most Δ (the clause density is the ratio of the number of clauses to the number of variables).

The third restriction is defined below.

Definition 1. *For any $d > 0$, we define $\text{SAT}^{(d)}$ to be the restriction of SAT to CNF formulas with at most n^d clauses, where n is the number of variables.*

To compare the complexities of these restrictions, we use families $\{s_k\}$, $\{t_\Delta\}$, $\{\sigma_d\}$ of constants where

- s_k is the infimum of all $c > 0$ such that k-SAT $\in \text{RTIME}(2^{cn})$;
- t_Δ is the infimum of all $c > 0$ such that $\text{SAT}_\Delta \in \text{RTIME}(2^{cn})$;
- σ_d is the infimum of all $c > 0$ such that $\text{SAT}^{(d)} \in \text{RTIME}(2^{cn})$;

The sequences $\{s_k\}_{k\in\mathbb{N}}$, $\{t_\Delta\}_{\Delta\in\mathbb{N}}$, $\{\sigma_d\}_{d\in\mathbb{N}}$ are nondecreasing and upper bounded by 1; their limits are denoted by $s_\infty, t_\infty, \sigma_\infty$ respectively. It was shown in [CIP06] that $s_\infty = t_\infty$. The following observation shows that $t_\infty \leq \sigma_d$ for any $d > 1$ (and therefore $s_\infty \leq \sigma_d$ for any $d > 1$).

Theorem 1. *For all $\Delta > 0$ and for all $d > 1$, we have $t_\Delta \leq \sigma_d$.*

Proof. Informally, the class of formulas of constant clause density is contained in some class of formulas with polynomially many clauses (for sufficiently large n). Namely, let $\Delta > 0$ and $d > 1$ be fixed. Then there exists n_0 such that for any formula with m clauses over $n \geq n_0$ variables, if $m \leq \Delta n$ then $m \leq n^d$. Therefore, any algorithm that solves $\text{SAT}^{(d)}$ can be used to solve SAT_Δ with (asymptotically) the same complexity. Hence, $t_\Delta \leq \sigma_d$.   □

Corollary 1. *For all k and for all $d > 1$, we have $s_k \leq \sigma_d$.*

Proof. It is shown in [CIP06] that the sequences $\{s_k\}_{k \in \mathbb{N}}$ and $\{t_\Delta\}_{\Delta \in \mathbb{N}}$ are interwoven: for any k there exists Δ such that $s_k \leq t_\Delta$ and vice versa. □

The Exponential Time Hypothesis (ETH) states that $s_3 > 0$. It turns out that, assuming ETH, k-SAT and SAT$_\Delta$ can be solved faster than SAT$^{(d)}$ for all k, Δ and for all $d > 1$:

Corollary 2. *If ETH is true then $s_k < \sigma_d$ and $t_\Delta < \sigma_d$ for all k, Δ and for all $d > 1$.*

Proof. If ETH is true then both sequences $\{s_k\}_{k \in \mathbb{N}}$ and $\{t_\Delta\}_{\Delta \in \mathbb{N}}$ increase infinitely often [IP01]. Combining this with Theorem 1 and Corollary 1, we obtain the claim. □

The Strong Exponential Time Hypothesis (SETH) states that the sequence $\{s_k\}_{k \in \mathbb{N}}$ converges to 1 [IP01, CIP09]. Hence, assuming SETH, SAT cannot be solved in moderately exponential time. It follows from Theorem 1 that this is true even if we restrict SAT to formulas with at most n^d clauses where $d > 1$:

Corollary 3. *If SETH is true then SAT$^{(d)}$ cannot be solved in moderately exponential time for any $d > 1$.*

Proof. By Theorem 1, $s_\infty \leq \sigma_d$ for any $d > 1$. Hence, if $\sigma_d < 1$ for some $d > 1$ then $s_\infty < 1$, i.e., SETH fails. □

3   Parameterized Problems Defined by Existential Second-Order Sentences

3.1   Parameterized Problems and Admissible Reductions

Let $\mathcal{A}$ be a decision problem in the class NP (we identify decision problems with languages over $\{0, 1\}$). Since $\mathcal{A} \in$ NP, there exists a polynomial-time verifier V for $\mathcal{A}$. An instance x is a *"yes" instance* of $\mathcal{A}$ if and only if V accepts $\langle x, y \rangle$ where y is a *certificate* for x. Let p be a polynomial-time computable function that bounds shortest certificates, i.e., for any "yes" instance x, there is a certificate y with $|y| \leq p(x)$. Any such function p is called a *certificate-size parameter* of $\mathcal{A}$ (we say simply *parameter* since we consider only certificate-size parameters in this paper). By a *parameterized problem* we mean a pair $(\mathcal{A}, p)$ where $\mathcal{A}$ is a problem with a specified parameter p.

Obvious examples of problems parameterized by certificate-size parameters are (SAT, n) where n is the number of variables in the instance, the set cover problem (in its decision version) with parameter m where m is the number of sets, the graph isomorphism problem with parameter $n \lg n$ where n is the number of vertices, etc.

Definition 2. *Let $(\mathcal{A}, p)$ and $(\mathcal{B}, q)$ be parameterized problems. Let ρ be a many-one reduction from $\mathcal{A}$ to $\mathcal{B}$. The reduction ρ is called an* admissible reduction *from $(\mathcal{A}, p)$ to $(\mathcal{B}, q)$ if there is a constant c such that for any instance x of $\mathcal{A}$, we have $q(\rho(x)) \leq p(x) + c$. If there is an admissible reduction from $(\mathcal{A}, p)$ to $(\mathcal{B}, q)$, we write $(\mathcal{A}, p) \leq_a (\mathcal{B}, q)$. If, in addition, this reduction is computable in polynomial time (in the size of instances of $\mathcal{A}$), we write $(\mathcal{A}, p) \leq_{pa} (\mathcal{B}, q)$*

Thus, admissible reducibility is a special case of *strong many-one reducibility* [IPZ01] where $q(\rho(x))$ is allowed to be $\mathcal{O}(p(x))$. Also, if an admissible reduction ρ is computable in polynomial time then ρ is a special case of *efficient reducibility* [CP09], i.e. Turing reducibility that allows oracle calls with parameters not exceeding $p(x) + \mathcal{O}(1)$.

Clearly, if $(\mathcal{A}, p) \leq_{pa} (\mathcal{B}, q)$ and $(\mathcal{B}, q)$ is in ME then $(\mathcal{A}, p)$ is also in ME.

3.2   Parameterized Boolean Queries

Loosely speaking, a Boolean query is a decision problem whose instances are structures with a given vocabulary [Imm99]. We consider *relational vocabularies*, i.e., tuples of the form $\tau = \langle R_1, \ldots, R_k, c_1, \ldots, c_l \rangle$ where $R_1, \ldots, R_k$ are relation symbols and $c_1, \ldots, c_l$ are constant symbols. A *structure* with vocabulary τ is a tuple $\langle U, R_1^U, \ldots, R_k^U, c_1^U, \ldots, c_l^U \rangle$ where U is a nonempty set called the *universe* of this structure, each R_i^U is a relation (of the arity assigned to R_i) over U, and each c_i^U is a specified element in U. Given a vocabulary τ, a *Boolean query* is a mapping from the set of all structures with τ to { "yes", "no" }. When an encoding of structures is fixed (relations are usually represented by tables), any Boolean query can be viewed as a decision problem, see [Imm99] on binary encodings of structures.

We consider Boolean queries defined by existential second-order sentences. Let τ be a vocabulary and ϕ be a quantifier-free first-order formula over τ. Consider an existential second-order sentence Φ of the form

$$\exists T_1 \ldots \exists T_r Q_1 x_1 \ldots Q_s x_s \, \phi \tag{3}$$

where $T_1, \ldots, T_r$ are relation symbols from τ, $Q_1, \ldots, Q_s$ are quantifiers, and $x_1, \ldots, x_s$ are first-order variables. The formula Φ defines the following Boolean query: for any structure S with the vocabulary τ, the query maps S to "yes" if and only if $S \models \Phi$. We denote this query by Φ^*.

Since any Boolean query defined by an existential second-order sentence is in NP [Fag74], it is natural to think of the class of such queries as a class of parameterized problems with certificate-size parameters. What could be a certificate-size parameter for Φ^*? For any structure S such that $S \models \Phi$, the tables corresponding to values of $T_1, \ldots, T_r$ can be viewed as a certificate for this "yes" instance. How many bits are needed to describe such a certificate? It is easy to see that the tables can be described using $\sum_{i=1}^r |U|^{a_i}$ bits, where U is the universe of S and each a_i is the arity of the relation symbol T_i. The function that maps every structure for Φ to the sum $\sum_{i=1}^r |U|^{a_i}$ is called the *standard parameter* of Φ. We denote this function by p_Φ. Note that if Φ has a single second-order quantifier $\exists T$ where T is a monadic relation symbol then p_Φ is just the cardinality of the universe U. We write (Φ^*, p_Φ) to denote the parameterized problem where Φ^* is the Boolean query and p_Φ is the specified parameter.

3.3   SAT and the Definability by Existential Second-Order Sentences

It is well known how to reduce SAT to a Boolean query defined by an existential second-order sentence. For example, consider a mapping ρ that maps every CNF formula F with clauses $C_1, \ldots, C_m$ over variables $v_1, \ldots, v_n$ to the following structure S. The universe of S consists of $\max\{m, n\}$ elements numbered by $1, \ldots, \max\{m, n\}$. The clauses are represented by two binary relations P and N such that $P(i, j)$ means that C_i contains a positive occurrence of v_j and, similarly, $N(i, j)$ means that C_i contains a negative occurrence of v_j. Let Φ be the following sentence:

$$\exists T \forall x \exists y \left[(P(x, y) \wedge T(y)) \vee (N(x, y) \wedge \neg T(y)) \right].$$

where T is a monadic relation symbol with the meaning: "$T(j)$ means that v_j is set to true". Since F is satisfiable if and only if $S \models \Phi$, the mapping ρ is a reduction from SAT to Φ^*. Moreover, according to Definition 2, this reduction is also an admissible reduction from $(\text{SAT}, \max\{m, n\})$ to (Φ, p_Φ).

An admissible reduction ρ is called *one-one* if ρ is a one-to-one mapping. Notice that the above reduction ρ from $(\text{SAT}, \max\{m, n\})$ to (Φ, p_Φ) is one-one. What other parameterized version of SAT can be reduced via one-one admissible reductions to Boolean queries defined by existential second-order sentences? In particular, can we represent (SAT, n) in a similar way? The following observation shows that it is not possible.

Theorem 2. *There is no existential second-order sentence Φ such that (SAT, n) can be reduced to (Φ^*, p_Φ) via a one-one admissible reduction.*

Proof. Suppose indirectly that there exists an existential second-order sentence Φ such that (SAT, n) is reducible to (Φ^*, p_Φ) via a one-one admissible reduction ρ. Let $\mathcal{F}_n$ be the set of all CNF formulas over n variables. We show that the cardinality of $\mathcal{F}_n$ is larger than the cardinality of the image $\rho(\mathcal{F}_n)$, which contradicts the assumption that ρ is one-to-one.

The number of all possible clauses over n variables is 3^n (for each variable v and for each clause C, there are three options: either C contains v, or C contains $\neg v$, or C contains neither v nor $\neg v$). Therefore, the set $\mathcal{F}_n$ consists of 2^{3^n} CNF formulas.

To count the cardinality of $\rho(\mathcal{F}_n)$, recall that ρ is admissible, which implies $p_\Phi \leq n + c$ for some constant c. This inequality upper-bounds the cardinality of a universe. Indeed, if Φ has r second-order variables for relations of arities $a_1, \ldots, a_r$, then we have

$$n + c \geq p_\Phi(S) = \sum_{i=1}^{r} |U|^{a_i} \geq |U|$$

for any structure S with universe U. Therefore, any structure for Φ such that $p_\Phi \leq n + c$ has a universe of cardinality at most $n + c$. How many such structures are there? Any table in such a structure contains at most $|U|^a$ rows where a is the maximum arity of relation symbols appearing in Φ. Therefore, there can be at most $2^{(n+c)^a}$ possible tables for each relation. If Φ has k relation symbols, the number of possible structures is at most $2^{k(n+c)^a}$.

Hence, ρ maps a set of cardinality 2^{3^n} to a set of cardinality $2^{k(n+c)^a}$ and it cannot be one-to-one. $\qquad\square$

Let R be a restriction of SAT and let Φ be an existential second-order sentence. The above proof shows that (R, n) is reducible to (Φ, p_Φ) via a one-one admissible reduction only if for any fixed n, the restriction R has at most $2^{n^{\mathcal{O}(1)}}$ instances. Notice that the restrictions k-SAT, SAT$_\Delta$, SAT$^{(d)}$ have this property.

4   Moderately Exponential Complexity and Strict NP

The class SNP (Strict NP) consists of Boolean queries defined by existential second-order sentences where the first-order part has a universal prefix [Pap94]. This class was defined in [KV87] and used in [PY91] for proving inapproximability of certain optimization problems. We "parameterize" this class and close it under the $\leq_{pa}$ reducibility. This closure is denoted by $\mathcal{R}(\mathsf{SNP})$.

Definition 3. *The class* SNP *is the set of Boolean queries defined by sentences of the form*

$$\exists T_1 \ldots \exists T_r \forall x_1 \ldots \forall x_s \, \phi \tag{4}$$

where $T_1, \ldots, T_r$ are relation symbols, $x_1, \ldots, x_s$ are first-order variables, and ϕ is a quantifier-free formula. The class $\mathcal{R}(\mathsf{SNP})$ is defined to be the set of parameterized problems $(\mathcal{A}, p)$ such that $(\mathcal{A}, p) \leq_{pa} (\Phi^*, p_\Phi)$ for some sentence Φ is of the form (4).

The relationship between SNP and parameterized versions of k-SAT is described in [IPZ01]. In our terms, this relationship can be re-stated as follows.

Theorem 3 ([IPZ01]). *For any $k \geq 3$, the parameterized problem $(k$-SAT$, n)$ is complete for $\mathcal{R}(\mathsf{SNP})$ via admissible reductions computable in polynomial time.*

Proof. It is shown in [IPZ01] that $(k$-SAT$, n)$ is complete for the parameterized version of SNP via strong many-one reductions. However, it is clear from the proof that this reducibility is in fact the admissible reducibility computable in polynomial time. $\qquad\square$

Corollary 4. $\mathcal{R}(\mathsf{SNP}) \subseteq \mathsf{ME}$.

Proof. This follows from the fact that $(k$-SAT$, n) \in \mathsf{ME}$ [PPSZ98, Sch99] and the definition of the $\leq_{pa}$ reducibility.

In this section, we extend SNP by allowing existential first-order quantifiers after universal ones. The extended class is denoted by $\mathsf{SNP_1}$. Similar to $\mathcal{R}(\mathsf{SNP})$, we define $\mathcal{R}(\mathsf{SNP_1})$ and we prove that $(\mathrm{SAT}^{(d)}, n)$ is complete for $\mathcal{R}(\mathsf{SNP_1})$ via the $\leq_{ps}$ reducibility. Therefore, if $(\mathrm{SAT}^{(d)}, n) \in \mathsf{ME}$ then the entire class $\mathcal{R}(\mathsf{SNP_1})$ is also in ME.

Definition 4. *The class* $\mathsf{SNP_1}$ *is the set of Boolean queries defined by sentences of the form*

$$\exists T_1 \ldots \exists T_r \forall x_1 \ldots \forall x_s \exists y_1 \ldots \exists y_t \, \phi \tag{5}$$

where $y_1, \ldots, y_t$ *are first-order variables and the other is the same as in (4). The class* $\mathcal{R}(\mathsf{SNP_1})$ *is defined to be the set of parameterized problems* $(\mathcal{A}, p)$ *such that* $(\mathcal{A}, p) \leq_{pa} (\Phi^*, p_\Phi)$ *for some sentence* Φ *is of the form (5).*

The following two theorems prove the "inclusion" and "hardness" respectively.

Theorem 4. *For all* $d > 0$, *we have* $(\mathrm{SAT}^{(d)}, n) \in \mathcal{R}(\mathsf{SNP_1})$.

Proof. We describe a reduction ρ from $(\mathrm{SAT}^{(d)}, n)$ to (Φ^*, p_Φ), where

$$\Phi = \exists T \forall x_1 \ldots \forall x_d \exists y \, \phi$$

with T being a monadic relation symbol, $x_1, \ldots, x_d, y$ being first-order variables, and ϕ being a quantifier-free formula. Let F be a CNF formula with clauses $C_1, \ldots, C_m$ over variables $v_1, \ldots, v_n$, where $m \leq n^d$. The reduction ρ maps F to the following structure S. The universe of S consists of elements $\{1, \ldots, n\}$ used to encode variables: variable v_i is encoded by element i. Since there are at most n^d clauses, they can be encoded by d-tuples $\langle k_1, \ldots, k_d \rangle$ where each k_j is an element from the universe. The vocabulary has two relation symbols P and N of arity $d + 1$ each and one monadic relation symbol T with the following interpretation:

 - $P(x_1, \ldots, x_d, y)$ means that the clause encoded by $\langle x_1, \ldots, x_d \rangle$ contains a positive occurrence of the variable encoded by y;
 - N is a similar relation symbol for negative occurrences;
 - $T(y)$ means that the variable encoded by y is set to true.

It is easy to see that F is satisfiable if and only if $S \models \Phi$ where Φ is

$$\exists T \forall x_1 \ldots \forall x_d \exists y \, [(P(x_1, \ldots, x_d, y) \wedge T(y)) \vee (N(x_1, \ldots, x_d, y) \wedge \neg T(y))].$$

Since the cardinality of the universe is n, we have $n = p_\Phi(S)$ and therefore ρ is admissible. Obviously, ρ is polynomial-time computable. □

Theorem 5. *For any parameterized problem* $(\mathcal{A}, p)$ *in* $\mathcal{R}(\mathsf{SNP_1})$, *there exists a constant* d *such that* $(\mathcal{A}, p) \leq_{pa} (\mathrm{SAT}^{(d)}, n)$.

Proof. Since $(\mathcal{A}, p) \in \mathcal{R}(\mathsf{SNP_1})$, there exists a sentence Φ of the form (5) such that $(\mathcal{A}, p) \leq_{pa} (\Phi^*, p_\Phi)$. We describe a reduction ρ from (Φ^*, p_Φ) to $(\mathrm{SAT}^{(d)}, n)$ for some d.

Let $a_1, \ldots, a_r$ be the arities of the second-order variables in Φ. Let S be a structure for Φ with a universe U. The reduction ρ constructs a CNF formula F that has the following properties:

1. F is satisfiable if and only if $S \models \Phi$;
2. F is a CNF formula with n^d clauses over n variables, where

$$n = p_\Phi(S) = \sum_{i=1}^{r} |U|^{a_i}.$$

The construction proceeds in four steps. The first step is to get rid of first-order quantifiers in Φ. Each subformula $\forall x \, \psi(x)$ is replaced by

$$\psi(c_1) \wedge \ldots \wedge \psi(c_{|U|})$$

where c_i's are constants for all elements of the universe. Similarly, each subformula $\exists x \, \psi(x)$ is replaced by

$$\psi(c_1) \vee \ldots \vee \psi(c_{|U|}).$$

The resulting formula (denoted by F_1) has the form

$$\exists T_1 \ldots \exists T_r \left[\bigwedge_i \left(\bigvee_j \phi_{ij} \right) \right]$$

where each ϕ_{ij} has no first-order variables.

The second step is to convert F_1 into an equivalent Boolean formula F_2 (not in CNF). All second-order quantifiers are eliminated. Each atom in F_1 with a relation symbol different from $T_1, \ldots, T_r$ is replaced by true or false depending on its value in S. Each atom $T_i(d_1, \ldots, d_{a_i})$, where d_j's are constants, is replaced by a Boolean variable v indexed by the tuple $\langle i, d_1, \ldots, d_{a_i} \rangle$. Notice that the number of all such tuples is exactly

$$\sum_{i=1}^{r} |U|^{a_i} = n.$$

We write $v_1, \ldots, v_n$ to denote these Boolean variables. The resulting formula (denoted by F_2) has the form $\bigwedge_i (\bigvee_j \psi_{ij})$, where each ψ_{ij} is a Boolean formula over variables $v_1, \ldots, v_n$. Observe that the size of each ψ_{ij} is a constant not exceeding the size of ϕ in (5).

The third step is to convert F_2 into a CNF formula F_3. To do this, we replace each ψ_{ij} by an equivalent DNF formula of constant size. Then the resulting formula F_3 has the form

$$\bigwedge_{i=1}^{n^{\mathcal{O}(1)}} \bigvee_{j=1}^{\mathcal{O}(n)} \bigwedge_{k=1}^{\mathcal{O}(1)} u_{ijk}$$

where each u_{ijk} is one of the Boolean variables $v_1, \ldots, v_n$.

The last step is to obtain a required CNF formula F in which the number of clauses is $\mathcal{O}(n^s)$ where s is the number of universal quantifiers in Φ. This can be done using distributivity.

To complete the proof, it remains to note that the reduction ρ takes polynomial time. □

Corollary 5. *If* $(\text{SAT}^{(d)}, n)$ *is in* ME *for all* d, *then all problems of the class* $\mathcal{R}(\text{SNP}_1)$ *are also in* ME.

Proof. Immediately follows from Theorem 5 and the definition of admissible reducibility.   □

Note that the statement "$(\text{SAT}^{(d)}, n)$ is in ME for all d" does not imply automatically the statement "$(\text{SAT}, n) \in$ ME", but the converse implication is obviously true. Therefore, for those who are interested in SAT (rather than $\text{SAT}^{(d)}$), the following weaker result could look more natural:

Corollary 6. *If* (SAT, n) *is in* ME *then all problems of the class* $\mathcal{R}(\text{SNP}_1)$ *are also in* ME.

Acknowledgment

We are grateful to the anonymous referees for their careful analysis and valuable suggestions. We implemented most of the suggestions in the final version.

References

[AS03]   Arvind, V., Schuler, R.: The quantum query complexity of 0-1 knapsack and associated claw problems. In: Ibaraki, T., Katoh, N., Ono, H. (eds.) ISAAC 2003. LNCS, vol. 2906, pp. 168–177. Springer, Heidelberg (2003)

[CIP06]   Calabro, C., Impagliazzo, R., Paturi, R.: A duality between clause width and clause density for SAT. In: Proceedings of the 21st Annual IEEE Conference on Computational Complexity, CCC 2006, pp. 252–260. IEEE Computer Society, Los Alamitos (2006)

[CIP09]   Calabro, C., Impagliazzo, R., Paturi, R.: The complexity of satisfiability of small depth circuits. In: IWPEC 2009. LNCS, vol. 5917, pp. 75–85. Springer, Heidelberg (2009)

[CP09]   Calabro, C., Paturi, R.: k-SAT is no harder than Decision-Unique-k-SAT. In: Frid, A., Morozov, A., Rybalchenko, A., Wagner, K.W. (eds.) Computer Science - Theory and Applications. LNCS, vol. 5675, pp. 59–70. Springer, Heidelberg (2009)

[DGH+02]   Dantsin, E., Goerdt, A., Hirsch, E.A., Kannan, R., Kleinberg, J., Papadimitriou, C.H., Raghavan, P., Schöning, U.: A deterministic $(2 - 2/(k+1))^n$ algorithm for k-SAT based on local search. Theoretical Computer Science 289(1), 69–83 (2002)

[DH09]   Dantsin, E., Hirsch, E.A.: Worst-case upper bounds. In: Handbook of Satisfiability, ch. 12, pp. 403–424. IOS Press, Amsterdam (2009)

[Fag74]   Fagin, R.: Generalized first-order spectra and polynomial-time recognizable sets. In: Karp, R. (ed.) Complexity of Computation, SIAM–AMS Proceedings, vol. 7, pp. 27–41. AMS, Providence (1974)

[Imm99]   Immerman, N.: Descriptive Complexity. Springer, Heidelberg (1999)

[IP01]   Impagliazzo, R., Paturi, R.: On the complexity of k-SAT. Journal of Computer and System Sciences 62(2), 367–375 (2001)

[IPZ01]   Impagliazzo, R., Paturi, R., Zane, F.: Which problems have strongly ex-
            ponential complexity. Journal of Computer and System Sciences 63(4),
            512–530 (2001)
[KV87]   Kolaitis, P.G., Vardi, M.Y.: The decision problem for the probabilities of
            higher-order properties. In: Proceedings of the 19th Annual ACM Sym-
            posium on Theory of Computing, STOC 1987, pp. 425–435. ACM, New
            York (1987)
[Mar07]   Marx, D.: Can you beat treewidth? In: Proceedings of the 48th Annual
            IEEE Symposium on Foundations of Computer Science, FOCS 2007, pp.
            169–179 (2007)
[Pap94]   Papadimitriou, C.H.: Computational Complexity. Addison-Wesley, Read-
            ing (1994)
[PP09]   Paturi, R., Pudlák, P.: On the complexity of circuit satisfiability (2009),
            http://www.math.cas.cz/~pudlak/csat.pdf
[PPSZ98]   Paturi, R., Pudlák, P., Saks, M.E., Zane, F.: An improved exponential-time
            algorithm for k-SAT. In: Proceedings of the 39th Annual IEEE Symposium
            on Foundations of Computer Science, FOCS 1998, pp. 628–637 (1998)
[PY91]   Papadimitriou, C.H., Yannakakis, M.: Optimization, approximation and
            complexity classes. Journal of Computer and System Sciences 43, 425–440
            (1991)
[Sch99]   Schöning, U.: A probabilistic algorithm for k-SAT and constraint satisfac-
            tion problems. In: Proceedings of the 40th Annual IEEE Symposium on
            Foundations of Computer Science, FOCS 1999, pp. 410–414 (1999)
[Tra08]   Traxler, P.: The time complexity of constraint satisfaction. In: Grohe,
            M., Niedermeier, R. (eds.) IWPEC 2008. LNCS, vol. 5018, pp. 190–201.
            Springer, Heidelberg (2008)
[Wil07]   Williams, R.: Algorithms and resource requirements for fundamental prob-
            lems. PhD thesis, Carnegie-Melon University (2007)

Minimising Deterministic Büchi Automata Precisely Using SAT Solving*

Rüdiger Ehlers

Reactive Systems Group
Saarland University, Germany
ehlers@react.cs.uni-saarland.de

Abstract. We show how deterministic Büchi automata can be fully minimised by reduction to the satisfiability (SAT) problem, yielding the first automated method for this task. Size reduction of such ω-automata is an important step in probabilistic model checking as well as synthesis of finite-state systems. Our experiments demonstrate that state-of-the-art SAT solvers are capable of solving the resulting satisfiability problem instances quickly, making the approach presented valuable in practice.

1 Introduction

The success of techniques for formal verification is closely connected to advances in the efficient solution of the satisfiability (SAT) problem. In the area of *bounded model checking* [4], it has been shown that checking whether a given system has a bounded-length witness for erroneous behaviour can efficiently and effectively be performed by reduction to the SAT problem. Likewise, in software verification, approaches involving *satisfiability modulo theory* solvers [2] have emerged, which rely on the underlying SAT techniques.

However, there are many areas in formal verification that do not benefit from advances in SAT solving yet. Taking for example probabilistic model checking, where many of its variants are PSPACE-hard or require computing some quantitative result [1], it is by no means obvious how the success of modern SAT techniques can be transferred to this area. In this paper, we present some progress towards closing this gap by reporting how SAT solving can be used for the full minimisation of deterministic automata over infinite words, which arise in intermediate steps of many formal methods.

One particular application of this automaton type is the verification of *Markov decision processes* against properties stated in *linear-time temporal logic (LTL)* [1]. Here, the classical model checking procedure requires converting the LTL formula to a deterministic automaton. A similar situation arises in common approaches to *synthesis of finite state systems* from LTL specifications [15].

---

* This work was supported by the German Research Foundation (DFG) within the program "Performance Guarantees for Computer Systems" and the Transregional Collaborative Research Center "Automatic Verification and Analysis of Complex Systems" (SFB/TR 14 AVACS).

O. Strichman and S. Szeider (Eds.): SAT 2010, LNCS 6175, pp. 326–332, 2010.

In both application areas, the minimisation of the number of states of the deterministic automata involved results in significant speed-ups of the actual model checking and synthesis tasks. Current tools for computing automata equivalent to LTL formulas use state space reduction techniques such as *bisimulation quotienting* [8] that have been developed for the minimisation of *non-deterministic* Büchi automata which are suitable for model checking non-deterministic systems. As the universality problem, and thus, the minimisation problem, of this class of automata is PSPACE-hard, the techniques used are typically incomplete and thus do not guarantee to find a minimal automaton. To the best of our knowledge, special techniques for the minimisation of deterministic automata have only been considered for *weak deterministic automata* [13], which however cannot even represent basic liveness properties.

In this paper, we lift the applicability of (complete) automaton minimisation to a more expressive class of deterministic automata. For conciseness, we restrict ourselves to deterministic *Büchi* automata (DBA) here. While DBA are not expressive enough to represent all LTL specifications (and are strictly less expressive than for example deterministic parity automata), they can represent most properties that appear in hardware specifications [6] and it is assured that in these cases, the smallest DBA is not larger than the smallest deterministic parity, Rabin or Streett automaton for the given property [12].

In both application areas mentioned above, the specification is usually a Boolean combination of a set of individual properties, which allows the precise minimisation of the specification parts representable as DBA with our technique before the automaton for the overall specification is composed. Additionally, some modern approaches for fast synthesis from LTL properties achieve a remarkable speedup by even requiring the specification parts to be representable and given as DBAs (in a certain encoding) [14].

In the following, we show that the problem of deciding whether there exists a smaller deterministic Büchi automaton for some given such automaton is contained in NP and can efficiently be reduced to the SAT problem. We evaluate whether currently available SAT solvers are capable of dealing with such problem instances. By succesively building and solving the resulting SAT instances, the minimal automaton is found. Our experiments suggest that for practical applications, the current state of the art in SAT solving is sufficient for the successful usage of the this minimisation technique.

2   Preliminaries

A deterministic Büchi automaton (DBA) is a 5-tuple $\mathcal{A} = (Q, \Sigma, \delta, q_0, F)$ with a finite set of states Q, a finite alphabet Σ, a transition function $\delta : Q \times \Sigma \to Q$, an initial state $q_0 \in Q$ and a set of accepting states $F \subseteq Q$. For the scope of this paper, we assume that δ is a *total* function.

We say that some infinite word $w = w_0 w_1 \ldots \in \Sigma^\omega$ is accepted by $\mathcal{A}$ if the run π induced by w on $\mathcal{A}$ is *accepting*. We define $\pi = \pi_0 \pi_1 \ldots$ such that $\pi_0 = q_0$ and for all $i \in \mathbb{N}_0$, $\delta(\pi_i, w_i) = \pi_{i+1}$. We say that π is accepting if and only if

$\{q \in Q \mid \exists^\infty j \in \mathbb{N} : \pi_j = q\} \cap F \neq \emptyset$, i.e., some accepting state occurs infinitely often on π. We define the language $\mathcal{L}(\mathcal{A})$ of $\mathcal{A}$ to consist of all words in Σ^ω that induce accepting runs. We denote the size of an automaton by $|Q|$. A DBA is minimal if there exists no other DBA with the same alphabet that has less states and accepts the same language. We define the language of a state $q \in Q$ to be $\mathcal{L}(q) = \mathcal{L}(\mathcal{A}_q)$ for $\mathcal{A}_q = (Q, \Sigma, \delta, q, F)$.

For space reasons, we do not describe the logic LTL here, but rather refer to [4,1]. A word can either satisfy an LTL formula or not. We define the language of an LTL formula to be the set of infinite words over $\Sigma = 2^{\mathsf{AP}}$ satisfying the formula over some set of atomic propositions AP. We call a DBA equivalent to an LTL formula if their languages are the same.

3   Minimising Deterministic Büchi Automata

Minimising deterministic Büchi automata is different from minimising deterministic automata over finite words: while for the latter, there exists a suitable polynomial algorithm, which is based on merging states with the same *language*, the same idea cannot be exploited for Büchi automata. The left part of Figure 1 shows a Büchi automaton over the alphabet $\Sigma = 2^{\{a,b\}}$ equivalent to the LTL formula $(\mathsf{GF}a) \wedge (\mathsf{GF}b)$. This DBA has four states, whereas the smallest Büchi automata equivalent to this formula have only three states. One such DBA is depicted in the right part of Figure 1. It is by no means obvious how to *restructure* the left automaton in order to obtain such a smaller one.

This example shows that we cannot only rely on language equivalence for minimising deterministic Büchi automata. We thus propose a different approach here. Assume that some n-state *reference* automaton $\mathcal{A}' = (Q', \Sigma, \delta', q'_0, F')$ is given. We use a SAT solver to consider all possible $n-1$-state *candidate automata* $\mathcal{A} = (Q, \Sigma, \delta, q_0, F)$ and encode the equivalence check of the languages of $\mathcal{A}$ and $\mathcal{A}'$ in clausal form. While such a check is PSPACE-complete for non-deterministic Büchi automata, it can be performed in polynomial time for deterministic Büchi

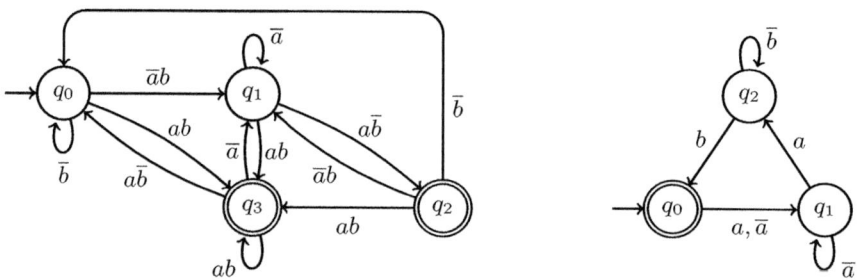

Fig. 1. A non-minimal DBA (left) and a minimal DBA (right) equivalent to the LTL formula $(\mathsf{GF}a) \wedge (\mathsf{GF}b)$ over the atomic proposition set $\mathsf{AP} = \{a, b\}$. Both DBA have a total transition relation. Accepting states are doubly-circled.

automata and can also be efficiently encoded into a SAT instance. By repeatedly applying this reduction technique until the resulting SAT instance becomes unsatisfiable, we obtain an automaton of minimal size.

For encoding the equivalence check, we use the observation that we can deduce from the *product* of $\mathcal{A}$ and $\mathcal{A}'$ if the two deterministic Büchi automata have the same language. More precisely, we build the graph $G = \langle V, E \rangle$ with the set of vertices $V = Q \times Q'$ and edges $E = \{((q_1, q_1'), (q_2, q_2')) \mid \exists s \in \Sigma : \delta(q_1, s) = q_2 \wedge \delta'(q_1', s) = q_2'\}$. If there is some loop $(v_0, v_0')(v_1, v_1') \ldots (v_k, v_k')$ in $\langle V, E \rangle$ with (v_0, v_0') being reachable from (q_0, q_0') such that on it, accepting states are only visited for $\mathcal{A}$, but not for $\mathcal{A}'$, i.e., $\{v_0, \ldots, v_k\} \cap F \neq \emptyset$ and $\{v_0', \ldots, v_k'\} \cap F' = \emptyset$, then there exists some $w \in \mathcal{L}(\mathcal{A})$ such that $w \notin \mathcal{L}(\mathcal{A}')$, so $\mathcal{A}$ and $\mathcal{A}'$ are inequivalent. Dually, if $\{v_0, \ldots, v_k\} \cap F = \emptyset$ and $\{v_0', \ldots, v_k'\} \cap F' \neq \emptyset$, then there exists a word $w \notin \mathcal{L}(\mathcal{A})$ such that $w \in \mathcal{L}(\mathcal{A}')$. If no such loops can be found, $\mathcal{A}$ and $\mathcal{A}'$ are equivalent.

3.1 Encoding as a SAT Problem

Using the observation stated above, we can build a SAT problem instance for solving the DBA state reduction problem. For notational convenience, in the following, primed state variables always refer to states in the reference automaton, whereas unprimed state variables refer to the candidate automaton. We use the following set of variables:

$$\{\langle q \rangle_F \mid q \in Q\} \cup \{\langle q_1, s, q_2 \rangle_\delta \mid q_1, q_2 \in Q, s \in \Sigma\} \cup \{\langle q, q' \rangle_G \mid q \in Q, q' \in Q'\}$$
$$\cup \{\langle q_1, q_1', q_2, q_2' \rangle_X \mid q_1, q_2 \in Q, q_1', q_2' \in Q', X \in \{N, A\}\}$$

The SAT clauses are defined as follows:

$$\bigwedge_{q_1 \in Q, s \in \Sigma} \bigvee_{q_2 \in Q} \langle q_1, s, q_2 \rangle_\delta \tag{1}$$

$$\wedge \bigwedge_{q_1, q_2 \in Q, q' \in Q', s \in \Sigma} \langle q_1, q' \rangle_G \wedge \langle q_1, s, q_2 \rangle_\delta \Rightarrow \langle q_2, \delta'(q', s) \rangle_G \tag{2}$$

$$\wedge \bigwedge_{\substack{q_1, q_2, q_3 \in Q, q_1', q_2' \in Q', s \in \Sigma, \\ \delta'(q_2', s) \notin F', (q_1' \neq \delta(q_2', s)) \vee (q_1 \neq q_3)}} \Rightarrow \begin{array}{c} \langle q_1, q_1', q_2, q_2' \rangle_N \wedge \langle q_2, s, q_3 \rangle_\delta \\ \langle q_1, q_1', q_3, \delta'(q_2', s) \rangle_N \end{array} \tag{3}$$

$$\wedge \bigwedge_{\substack{q_1, q_2 \in Q, q_1', q_2' \in Q', s \in \Sigma, \\ q_1' \notin F', q_1' = \delta(q_2', s))}} \langle q_1, q_1', q_2, q_2' \rangle_N \wedge \langle q_2, s, q_1 \rangle_\delta \Rightarrow \neg \langle q_1 \rangle_F \tag{4}$$

$$\wedge \bigwedge_{\substack{q_1, q_2, q_3 \in Q, q_1', q_2' \in Q', s \in \Sigma, \\ q_1' \in F', q_1' \neq \delta(q_2', s)) \vee q_1 \neq q_3}} \begin{array}{c} \langle q_1, q_1', q_2, q_2' \rangle_A \wedge \langle q_2, s, q_3 \rangle_\delta \wedge \\ \neg \langle q_3 \rangle_F \Rightarrow \langle q_1, q_1', q_3, \delta'(q_2', s) \rangle_A \end{array} \tag{5}$$

$$\wedge \bigwedge_{\substack{q_1, q_2 \in Q, q_1', q_2' \in Q', s \in \Sigma, \\ q_1' \in F', q_1' = \delta(q_2', s))}} \langle q_1, q_1', q_2, q_2' \rangle_A \wedge \langle q_2, s, q_1 \rangle_\delta \Rightarrow \langle q_1 \rangle_F \tag{6}$$

$$\wedge \langle q_0, q_0' \rangle_G \wedge \bigwedge_{q \in Q, q' \in Q} (\langle q, q' \rangle_G \Rightarrow \langle q, q', q, q' \rangle_N) \wedge (\langle q, q' \rangle_G \Rightarrow \langle q, q', q, q' \rangle_A) \quad (7)$$

The variables $\langle \cdot \rangle_F$ represent whether a state is accepting. The transition function of the candidate automaton is defined in the variables $\langle \cdot \rangle_\delta$. The clauses (1) make sure that the transition function is total. For the vertices in G that are reachable from (q_0, q_0'), the first part of (7) and (2) enforce that the respective variables in $\langle \cdot \rangle_G$ are set to **true**. In particular, the first conjunct of (7) makes sure that (q_0, q_0') is defined as being reachable and (2) forces successors of reachable vertices in G to be marked as being reachable as well.

If there is path from some reachable vertex $(q_1, q_1') \in V$ to some vertex $(q_2, q_2') \in V$ in G such that no accepting state of $\mathcal{A}'$ is visited along the path, then $\langle q_1, q_1', q_2, q_2' \rangle_N$ indicates this fact. This is made sure by (3) in conjunction with (7). We use these path witness variables for detecting the loops in G that are non-accepting for $\mathcal{A}'$. The clauses (4) state that the states of $\mathcal{A}$ along such loops then also have to be non-accepting.

Dually, $\langle q_1, q_1', q_2, q_2' \rangle_A$ is set to true if there exists a path from some reachable vertex $(q_1, q_1') \in V$ to some vertex $(q_2, q_2') \in V$ in G such that no accepting state of $\mathcal{A}$ is visited in between. This is assured by the clauses (5) and (7). We add the clauses (6) to make sure that such a path may not form a loop if one of its $\mathcal{A}'$-states is accepting.

Note that there are no clauses enforcing that not too many variables in $\langle \cdot \rangle_G$, $\langle \cdot \rangle_\delta$, $\langle \cdot \rangle_N$ or $\langle \cdot \rangle_A$ are set to **true**. This is not necessary, as this only makes finding a satisfying assignment harder, but never results in false-positives. If a variable valuation satisfying all constraints has some state for which there is more than one transition possible for some input symbol, then the encoding makes sure that picking any of the possible successors always results in a correct DBA. Apart from the clauses (1), all conjuncts are Horn clauses (if we negate all values of $\langle \cdot \rangle_F$). While the generated instance is relatively large (of size $O(|Q'|^4)$), the fact that most clauses are of Horn type simplifies solving such SAT instances.

For speeding up the SAT solving process, symmetry breaking clauses [16] can also be added. For simplicity, we only break symmetry partially in our experimental evaluation. In particular, for $Q = \{q_1, \ldots, q_{|Q|}\}$ and $\Sigma = \{s_1, \ldots, s_{|\Sigma|}\}$, we add the following conjuncts that encode some relaxed form of lexicographical minimality of the candidate automaton over the automata whose graphs are isomorphic to the candidate solution:

$$\bigwedge_{1 \leq i < |Q|, \, i+1 < j \leq |Q|, \, (i-1) \cdot |Q| + j + 2 \leq k \leq |\Sigma|} \neg \langle q_i, s_k, q_j \rangle_\delta$$

4   Experimental Evaluation[1]

We ran a prototype implementation of our technique on a couple of LTL formulas that are typical for model checking and synthesis. The upper part of Table 1

---

[1] Details and a downloadable implementation of the approach can be found at http://react.cs.uni-saarland.de/tools/dbaminimizer

Table 1. Experimental results of our DBA minimisation technique

| LTL specification | # States | | First instance | | Total |
	From	To	# Vars	# Clauses	time
$F(q \wedge X(pUr))$	3	3	94	697	0.01 s
$pUqUr \vee qUrUp \vee rUpUq$	3	3	112	699	0.01 s
$F(p \wedge X(q \wedge XFr))$	4	4	279	3785	0.01 s
$G(p \rightarrow qUr)$	5	3	852	13508	0.03 s
$pU(q \wedge X(rUs))$	5	5	780	26972	0.04 s
$F(p \wedge XF(q \wedge XF(r \wedge XFs)))$	5	5	780	26972	0.01 s
$pU(q \wedge X(r \wedge F(s \wedge XF(u$ $\wedge XF(v \wedge XFw)))))$	9	9	14752	5383278	30.36 s
$GFp \wedge GFq \wedge GFr \wedge GFs \wedge GFu$	14	6	51467	13856026	63.03 s
$GFa \vee GFb \vee GFc$	2	2	19	60	0.01 s
$G(a \rightarrow Fb)$	4	2	339	1907	0.03 s
$G(aUbU\neg aU\neg b)$	4	2	339	1907	0.03 s
$(Ga \rightarrow Fb) \wedge (G\neg a \rightarrow F\neg b)$	4	4	291	1904	0.01 s
$G\neg c \wedge G(a \rightarrow Fb) \wedge G(b \rightarrow Fc)$	5	2	852	13508	0.04 s
$G(a \rightarrow Fb) \wedge Gc$	5	3	852	13508	0.03 s
$GF(a \rightarrow XXXb)$	7	2	3720	43457	0.08 s
$G(a \rightarrow Fb) \wedge G(\neg a \rightarrow F\neg b)$	8	4	5635	89511	0.14 s
$GF(a \leftrightarrow XXb)$	9	6	8760	168358	1.57 s
$G(a \rightarrow Fb) \wedge G(b \rightarrow Fc)$	10	5	14247	589925	0.98 s
$G(a \rightarrow XXXb)$	10	9	16623	295076	3.71 s
$G(a \rightarrow Fb) \wedge G(c \rightarrow Fd)$	15	6	72660	9926065	23.96 s
$GF(a \leftrightarrow XXXb)$	17	15	116912	4752970	3617.3 s*

contains results for the 8 out of 12 LTL formulas stated in [7] that are representable as DBAs. In the lower half of the table, we give results for some typical synthesis specification parts and added some more complex formulas to allow for a more meaningful evaluation of our approach. Note that in both cases, the formulas occurring are mostly rather small, as in practice larger specifications are usually split up such that their conjuncts can be translated separately and composed to an overall automaton afterwards.

We used the tool ltl2dstar v.0.5.1 [10] in conjunction with ltl2ba v.1.1 [9] to obtain initial non-optimised deterministic Rabin automata equivalent to the input formulas given in the first column of the table. By applying the algorithm described in [11], we converted the deterministic Rabin automaton to a deterministic Büchi automaton whenever this is possible (and aborted otherwise). The ltl2dstar tool applies bisimulation quotienting, so the automata obtained are already heuristically optimised.

The number of states of these automata is given in the second column of Table 1. Columns four and five state the sizes of the reduction problems of these initial DBAs. The reduction process is repeated until no further improvements are possible. The resulting number of states in the minimised DBAs is shown in the third column. Finally, the total computation time of the SAT solver picosat v.913

[3] observed on a computer with an Intel Core 2 Duo 1.86GHz processor for all reduction steps is given in the last column. We restricted the SAT solving time to one hour. Exceeding of this time bound is denoted by a star. Consequently, in such cases, the resulting DBA is not guaranteed to be minimal. The computation times for obtaining the initial DBA are negligible (< 0.05 seconds in all cases) and thus not added to the total time value. Furthermore, the computation of the SAT instances from the automata has also not been taken into account as most time was spent on writing the SAT instance to disk here, which can be circumvented by a future tighter integration with the SAT solver.

The table shows that except for one instance, the minimisation problem was always solved quickly. Thus, our technique is well-suited for being used as an optimisation step for the applications discussed in this paper. As the problem definition inherently induces a lot of symmetries in the SAT instance, we conjecture that future advancements in dynamic symmetry breaking [16] will allow tackling even bigger problem instances.

References

1. Baier, C., Katoen, J.P.: Principles of Model Checking. MIT Press, Cambridge (2008)
2. Barrett, C., Sebastiani, R., Seshia, S., Tinelli, C.: Satisfiability Modulo Theories. In: [5], pp. 825–885.
3. Biere, A.: Picosat essentials. JSAT 4(2-4), 75–97 (2008)
4. Biere, A.: Bounded Model Checking. In: [5], pp. 457–474
5. Biere, A., Heule, M., van Maaren, H., Walsh, T. (eds.): Handbook of Satisfiability. IOS Press, Amsterdam (2009)
6. Bloem, R., Galler, S., Jobstmann, B., Piterman, N., Pnueli, A., Weiglhofer, M.: Specify, compile, run: Hardware from PSL. Electr. Notes Theor. Comput. Sci. 190(4), 3–16 (2007)
7. Etessami, K., Holzmann, G.J.: Optimizing Büchi automata. In: Palamidessi, C. (ed.) CONCUR 2000. LNCS, vol. 1877, pp. 153–167. Springer, Heidelberg (2000)
8. Etessami, K., Wilke, T., Schuller, R.A.: Fair simulation relations, parity games, and state space reduction for Büchi automata. SIAM J. Comput. 34(5), 1159–1175 (2005)
9. Gastin, P., Oddoux, D.: Fast LTL to Büchi automata translation. In: Berry, G., Comon, H., Finkel, A. (eds.) CAV 2001. LNCS, vol. 2102, pp. 53–65. Springer, Heidelberg (2001)
10. Klein, J., Baier, C.: Experiments with deterministic ω-automata for formulas of linear temporal logic. Theor. Comput. Sci. 363(2), 182–195 (2006)
11. Krishnan, S.C., Puri, A., Brayton, R.K.: Deterministic ω automata vis-a-vis deterministic Buchi automata. In: Du, D.-Z., Zhang, X.-S. (eds.) ISAAC 1994. LNCS, vol. 834, pp. 378–386. Springer, Heidelberg (1994)
12. Kupferman, O., Morgenstern, G., Murano, A.: Typeness for omega-regular automata. Int. J. Found. Comput. Sci. 17(4), 869–884 (2006)
13. Löding, C.: Efficient minimization of deterministic weak omega-automata. Inf. Process. Lett. 79(3), 105–109 (2001)
14. Piterman, N., Pnueli, A., Sa'ar, Y.: Synthesis of reactive(1) designs. In: Emerson, E.A., Namjoshi, K.S. (eds.) VMCAI 2006. LNCS, vol. 3855, pp. 364–380. Springer, Heidelberg (2005)
15. Pnueli, A., Rosner, R.: On the synthesis of a reactive module. In: POPL, pp. 179–190 (1989)
16. Sakallah, K.A.: Symmetry and Satisfiability. In: [5], pp. 289–338 (2009)

Exploiting Circuit Representations in QBF Solving

Alexandra Goultiaeva and Fahiem Bacchus

Department of Computer Science
University of Toronto
{alexia,fbacchus}@cs.toronto.edu

Abstract. Previous work has shown that circuit representations can be exploited in QBF solvers to obtain useful performance improvements. In this paper we examine some additional techniques for exploiting a circuit representations. We discuss the techniques of propagating a dual set of values through the circuit, conversion from simple negation normal form to a more optimized circuit representation, and adding phase memorization during search. We have implemented these techniques in a new QBF solver called CirQit2 and evaluated their impact experimentally. The solver has also displayed superior performance in the non-prenex non-CNF track of the QBFEval'10 competition.

1 Introduction

Quantified Boolean Formulas (QBF) are a PSPACE-complete extension of satisfiability (SAT) in which the propositional variables can be either universally or existentially quantified. The addition of quantifiers, and the arbitrary nesting of quantifiers, provides considerable additional representational power: QBFs can compactly represent a much wider range of problems than SAT. This can make QBF more effective than SAT for representing and solving some problems [1].

While traditionally QBF solvers have represented their input in prenex conjunctive normal form (CNF), some recent work has explored non-CNF solvers. For example, in [2] a QBF solver utilizing a circuit representation was presented and shown to have some advantages over CNF based solvers.

In this paper we present some additional ways of exploiting a circuit representation to obtain better performance. We present an overview of dual propagation (explored more fully in [3]) which provides superior detection and learning of solutions; some techniques for converting negation normal form to a more optimized circuit representation; and extend the phase memorization technique utilized in SAT solvers [4] to QBF. We have implemented all of these techniques as an extension of the CirQit solver [2], and we present empirical evidence supporting their usefulness.

2 Background

A QBF has the form $Q.\phi$, where ϕ is an arbitrary propositional formula and Q is a sequence of quantified variables ($\forall x$ or $\exists x$) one for each variable in ϕ (i.e., the formula

O. Strichman and S. Szeider (Eds.): SAT 2010, LNCS 6175, pp. 333–339, 2010.

has no free variables).[1] The truth value of a QBF is defined recursively: $\exists x Q.\phi$ is true iff there is at least one value v of x for which $Q.\phi|_{x=v}$ is true, and $\forall x Q.\phi$ is true iff $Q.\phi|_{x=v}$ is true for both values v of x.

Typically QBF solvers represent the body of the QBF, ϕ, in CNF. However, ϕ can be represented in other ways. In circuit-based solvers, e.g., the CirQit solver of [2], ϕ is represented as a **logical circuit** consisting of AND, OR and NOT gates along with lines running from gate outputs to the inputs of other gates. The quantified variables of Q are the inputs to the circuit and the output line of each gate is labeled with a new variable. Formally, these variables can be treated as new existentially quantified auxiliary variables scoped by all of the input variables with a path to them in the circuit.

The CirQit solver explores different settings of the input variables during a backtracking search and propagates **primal values** through the circuit to determine logical consequences of these settings. During search the aim is to determine when the QBF body ϕ is satisfied. This occurs when the circuit inputs force the circuit output to be true. Hence, CirQit initially sets the circuit output to 1 and propagates this **primal** value back through the circuit. Whenever a circuit line is forced to be both 0 (false) and 1 (true) (by propagation from the inputs and from the output's primal value) a contradiction is detected: the circuit cannot be satisfied under the current input settings. Similarly, a contradiction is detected whenever the value of a universally quantified input is forced.

As shown in [2] clausal reasons for each forced line can be extracted from the circuit, and clause learning can be performed when conflicts are detected. Further, when ϕ is satisfied **cubes** can be extracted by finding a subset of the input lines sufficient to propagate 1 to the circuit output. Finally, the circuit representation naturally supports don't care reasoning. In particular, certain lines of the circuit, can be marked as don't care during a don't care propagation process. Input lines that become don't care do not need to be branched on during search, and do not need to appear in the extracted cubes. This can yield better (smaller) cubes than the technique employed in CNF solvers.[2]

3   New Techniques for Exploiting the Circuit Representation

3.1   Dual Propagation

While clauses are used in QBF solvers to learn settings that falsify the QBF formula, cubes are used to learn settings that satisfy the formula. However, cubes are usually not as effective as clauses in improving the efficiency of QBF solvers. In particular, the solver starts off with no cubes, and the cubes it learns initally are usually quite large and thus only useful quite deep in the search tree. On the other hand, the solver starts off with many short clauses in its input (in the circuit representation these clauses are implicit in the logical relationships between the gate inputs and its output), and even initially learnt clauses can be quite short.

---

[1] This is the prenex form where the quantifiers precede the formula body. Non-prenex representations allow quantifiers to appear in front of any sub-formula of the body. We do not address the possible advantages of non-prenex representations in this paper.

[2] In CNF a cube is extracted by finding a set of true literals sufficient to satisfy every clause. Even clauses from the don't care part of the circuit must be satisfied.

In [3] a new technique is presented that allows the identical technique of clauses and clause learning to be used to detect both satisfying and falsifying input settings. This new technique is called **dual propagation**, and we present a brief overview of its main ideas here, leaving the details and comparison with prior methods to [3].

Consider the negation of the QBF being solved $\neg Q.\phi$. This formula is false iff the original QBF $Q.\phi$ is true. Taking the negation in we obtain $(\neg Q).(\neg\phi)$, where $\neg Q$ is the same as Q except that its quantifiers are flipped. For $\neg\phi$ we can exploit the circuit representation. This formula can be represented by the same circuit used to represent ϕ, C_ϕ, by simply passing the output of C_ϕ through a NOT gate.

If we want to solve $\neg Q.\phi$ with a circuit based solver we would take the circuit NOT(C_ϕ) and set its output to 1/true. The 1 would propagate back through the final NOT gate, and set the output of C_ϕ to 0. So we see that the final NOT gate can be discarded; it suffices to set the output of C_ϕ to 0. Now the solver would search various settings of the circuit input lines. These are identical to the input lines of C_ϕ but have reversed quantifiers. Propagation and clause learning operate just as before.

Conflicts discovered while solving $\neg Q.\phi$ are actually solutions for $Q.\phi$; if $\neg Q.\phi$ cannot be made true under the current input settings, $Q.\phi$ cannot be made false. Propagation on $\neg Q.\phi$ can force variables that in $Q.\phi$ are universal. This indicates that in $Q.\phi$ the other value is guaranteed to lead to a solution and need not be explored. Similarly, clause learning on $\neg Q.\phi$ yields clauses that if falsified indicate that $Q.\phi$ is satisfied. Thus, propagation and clause learning on $\neg Q.\phi$ detects settings that satisfy $Q.\phi$ using the same techniques used on $Q.\phi$ to detect settings that falsify $Q.\phi$.

The circuit representation can be exploited to implement this idea by simply propagating two sets of values through the circuit: the original **primal** values generated by setting the circuit output to 1, and a new set of **dual** values generated by setting the circuit output to 0. Since both $\neg Q.\phi$ and $Q.\phi$ have the same input lines (with flipped quantifiers) the input lines always have identical primal and dual values. Thus values can be transferred between the primal and dual channels via the input lines.

As before, backtracking search sets input lines, and both primal and dual values are propagated through the circuit. Thus propagation might set either or both values of the auxiliary variables. Contradictions are detected from both the primal and dual values, and in either case a clause is learnt and the solver backtracks. Primal and dual clauses are put in separate databases: unit propagating primal clauses forces primal values while unit propagating dual clauses forces dual values.

It can be seen that a primal conflict causes backtrack and an existential of $Q.\phi$ to be forced (the opposite value must falsify $Q.\phi$), while a dual conflict causes backtrack and a universal of $Q.\phi$ to be forced (the opposite value must truthify $Q.\phi$). Furthermore, the solver always encounters either a primal conflict or a dual conflict along each path it explores—once a sufficient number of input lines have been set either a 0 or a 1 must be propagated to the circuit output causing a conflict with the other value. The solver terminates on learning either an empty primal or an empty dual clause indicating that the input QBF is false or true respectively. Finally, don't care propagation can be extended to work with dual propagation.

3.2 Better Circuits from NNF

Negation normal form (NNF) has recently been adopted as the standard non-clausal input format for the QBF evaluation. However, because NNF involves pushing all negations down to the level of the propositional variables, it can obscure structure in the input formula. For example, consider the propositional formula $(c \vee F(\boldsymbol{x})) \wedge (d \vee \neg F(\boldsymbol{x}))$, where $F(\boldsymbol{x})$ is some sub-formula over the variables $\boldsymbol{x}$. In a circuit representation a single sub-circuit, $C_{F(\boldsymbol{x})}$, could be used to represent $F(x)$, its output feed into the gate $\text{OR}(c, F(\boldsymbol{x}))$, and the negation of its output feed into the gate $\text{OR}(d, \text{NOT}(F(\boldsymbol{x})))$. Initially, if the formula must be true, the output of both OR gates is forced to 1. Then, e.g., if during search c was set to 0, propagation would set the output line of $C_{F(\boldsymbol{x})}$ to 1, the output line of $\text{NOT}(F(\boldsymbol{x}))$ to 0, and thus force d to 1. If this formula was first converted to NNF, then a separate sub-circuit, $C_{\neg F(\boldsymbol{x})}$, would have to be constructed for $\neg F(\boldsymbol{x})$ which would share only inputs (and negated inputs) with $C_{F(\boldsymbol{x})}$. Depending on the complexity of $C_{F(\boldsymbol{x})}$, it is likely that forcing the output of $C_{F(\boldsymbol{x})}$ would have no effect on the output of $C_{\neg F(\boldsymbol{x})}$—e.g., if the output of $C_{F(\boldsymbol{x})}$ failed to propagate down to any of its inputs.

Here we suggest a technique for recovering structure that might be lost in an NNF representation. We define an inductive relation of structural equivalence $\overset{i}{\equiv}$ and structural negated equivalence $\overset{ni}{\equiv}$ between two propositional formulas represented in NNF:

- For every literal l, $l \overset{i}{\equiv} l$ and $l \overset{ni}{\equiv} \neg l$
- $G_1 \circ \cdots \circ G_n \overset{i}{\equiv} F_1 \circ \cdots \circ F_n$ for $\circ \in \{\wedge, \vee\}$ if $G_i \overset{i}{\equiv} F_{f(i)}$ for every $i \in \{1, \ldots, n\}$ under some bijection $f : \{1, \ldots, n\} \to \{1, \ldots, n\}$.
- $G_1 \circ \cdots \circ G_n \overset{ni}{\equiv} F_1 \bullet \cdots \bullet F_n$ for $\{\circ = \wedge, \bullet = \vee\}$ or $\{\circ = \vee, \bullet = \wedge\}$ if $G_i \overset{ni}{\equiv} F_{f(i)}$ for every $i \in \{1, \ldots, n\}$, under some bijection $f : \{1, \ldots, n\} \to \{1, \ldots, n\}$.

It can proved by a simple induction that if $G \overset{i}{\equiv} F$ then G and F represent the same boolean function, and that if $G \overset{ni}{\equiv} F$ then G and $\neg F$ represent the same boolean function. A simple example of this definition is that $(a \wedge (b \vee c)) \overset{ni}{\equiv} (\neg a \vee (\neg b \wedge \neg c))$.

To simplify an NNF we first generate for every subformula F a function identifier (id) $f_{id}(F)$ and a negated function id $nf_{id}(F)$ so that for any two subformulas G and F, $G \overset{i}{\equiv} F$ iff $f_{id}(G) = f_{id}(F)$, and $G \overset{ni}{\equiv} F$ iff $f_{id}(G) = nf_{id}(F)$.

This accomplished by using hashing techniques to identify structurally equivalent sub-formulas (similar to how BDDs are constructed). Ids for the subformulas are assigned bottom up so that a subformula obtains an id only after its children subformulas have obtained theirs. For each variable v unique ids are created for v and for $\neg v$. In addition we set $nf_{id}(v) = f_{id}(\neg v)$ and $nf_{id}(\neg v) = f_{id}(v)$. Each function id assigned is kept in a hash table. The subformula $F_1 \circ \cdots \circ F_k$ is assigned an id by first sorting its children F_i by their ids. Then an id is computed that is a function of the operator $\circ$ ($\vee$ or $\wedge$) and the ids of the sorted children. Similarly a negated id is computed by repeating the computation using the dual operator of $\circ$ and the (already computed) negated ids of the children (resorted by their negated ids). Both ids are stored in the hash table. It is not hard to show that this procedure generates ids that satisfy the conditions stated above, i.e., $G \overset{i}{\equiv} F$ iff $f_{id}(G) = f_{id}(F)$, and $G \overset{ni}{\equiv} F$ iff $f_{id}(G) = nf_{id}(F)$.

Using the subformula ids an optimized circuit can now be constructed to represent the NNF. Again we build up the circuit bottom up, first creating input lines corresponding to $f_{id}(v)$ and $nf_{id}(\neg v)$ for each variable. To construct the circuit for the subformula $F = F_1 \circ \cdots \circ F_k$ we first check to see if subcircuits for $f_{id}(F)$ or $nf_{id}(F)$ have already been constructed. If so we simply reuse the output line (or output line run through a NOT gate) as the line representing F. Otherwise we create a new gate representing $\circ$ whose inputs are the outputs of the subcircuits corresponding to the F_i, and mark this as the circuit that has already been constructed for $f_{id}(F)$. This allows the subcircuit constructed for F to be reused to represent future subformulas.

This process creates a more compact circuit. In particular, the circuit needs only one sub-circuit (and perhaps an additional NOT gate) to represent an entire set of sub-formulas $\{F_1, \ldots, F_k\}$ of the NNF with either $F_i \overset{i}{\equiv} F_j$ or $F_i \overset{ni}{\equiv} F_j$ ($1 \leq i, j \leq k$).

3.3 Simplifying the NNF

Using the ids constructed as described above we can perform some further simplifica-tions of the NNF. This utilizes two easy to verify logical rules: $\phi \wedge F(\phi) \equiv \phi \wedge F(\text{TRUE})$ and $\phi \vee F(\phi) \equiv \phi \vee F(\text{FALSE})$, where ϕ is any propositional formula.

The algorithm is simple and is applied before the circuit representation is generated: using the f_{id} and nf_{id} identifiers, any repeated occurrences of a subformula in its sibling subformula can be replaced with a constant according to the above logical rules. The NNF is then simplified, function ids regenerated, and the rules are applied again. This process is repeated until no further simplifications are possible, after which the circuit representation can be generated.

3.4 Phase Memorization

Phase memorization in SAT involves setting variables to their previously set values on the new descent after backtrack [4]. The main idea rests on the intuition that often a problem decomposes into different subproblems. Reusing the previous values after backtrack allows the solver to retain some of the work done on other subproblems.

In QBF, we have two types of backtrack—success and failure. Also, universal and existential variables play very different roles in QBF. We consider this interplay in light of the above idea and in light of the now completely symmetric processes of failure and success backtracking under dual propagation.

In a conjunction of different subproblems, the universal variables in one subproblem cannot affect the solution of another one—all possible settings need to be checked any-way. This leads to the idea that remembering existential variables might be beneficial after a conflict is generated on one of the subproblems, but that the universal settings are irrelevant. Symmetrically, if a solution is found, only the settings of the universal variables are important.

Based on this intuition, we implemented the following phase memorization schema: we restore phase values only for those variables that are existential with respect to the primal or dual conflict that occurs. This means that phases are restored for the exis-tential variables when backtracking from a conflict, and for universal variables when backtracking from a solution.

4 Experimental Results

We have modified the solver CirQit2 [3], which includes dual propagation. The new solver, CirQit2.1, is slightly more optimized, reads NNF input and includes all techniques discussed here. The experiments were run on all the non-Prenex, non-CNF benchmarks currently available from QBFLIB [5]. All tests were run on a 2.8GHz machine with 12GB of RAM under a time limit of 1200 CPU seconds per instance.

Table 1 shows the comparison between CirQit2.1, CirQit2 [3], CirQit [2] and some state-of-the-art CNF-based QBF solvers: quantor (version 3.0, with the recommended picosat back end) [6], Qube (version 6.5) [7], nenofex [8] and depqbf [9]—the latter two are the versions submitted to the main track of QBFEVAL'10. The non-prenex non-CNF instances are converted to prenex CNF using a conversion tool available on the QBFEval site. Internally, CirQit converts input instances to prenex form using a naive algorithm that places the variables in the same order in which they are encountered.

On the domains from the set *BMC_QBF_1.0* ("assertion", "consistency" and "possibility"), the bottom-up solver quantor outperforms CirQit2.1, as well as any other search-based solver. With this exception, CirQit2.1 proves to be superior. The drastic difference between CirQit and CirQit2 shows the effectiveness of dual propagation, while the improvement of CitQit2.1 shows the effectiveness of the other techniques. Each technique contributes to the improvement. Turning off any one of them reduces the performance of CitQit2.1. Without generating better circuits, it can solve only 344 problems (out of 492); without simplification, 340; without phase memorization, 344. Overall the preprocessing techniques reduced the number of variables by 0.99%, but only because they were less useful on the more numerous benchmark families.

Table 1. Comparison between CirQit2.1 and other state-of-the-art CNF-based solvers. The largest number of instances solved is shown in **bold**, with ties broken by the time taken to solve those instances. Percent decrease in the number of variables (from CirQit2.1 to CirQit2) is also shown.

	CirQit2.1			CirQit2		CirQit		quantor		Qube6.5		nenofex		depqbf	
	Solved	Time	% Dec	Sol.	Time	Sol.	Time	Sol.	Time	Sol.	Time	Sol.	Time	Sol.	Time
Seidl (150)	**150**	43	12.63	150	318	147	2,281	42	3,272	149	2,485	82	1,160	150	557
assertion (120)	48	16,166	0.67	40	14,503	3	1	**119**	8,736	6	1,180	12	4,170	24	145
consistency (10)	7	2,562	0.51	4	1,283	0	0	**10**	720	0	0	1	306	0	0
counter (45)	**43**	1,368	7.77	40	492	39	1,315	28	414	31	540	29	1,727	31	70
dme (11)	**11**	11	27.59	10	5	10	15	0	0	7	88	8	94	11	901
possibility (120)	57	17,535	0.50	45	16,121	10	1,707	**111**	7,976	14	4,713	12	4,037	10	143
ring (20)	**20**	40	24.31	20	53	15	60	11	479	16	189	11	4	13	243
semaphore (16)	**16**	2	39.99	16	3	16	7	16	12	16	361	16	1,193	16	39
Total (492)	**352**	37,728	0.99	325	32,779	240	5,389	337	21,613	239	9,557	171	12,692	255	2,098

References

1. Mangassarian, H., Veneris, A.G., Safarpour, S., Benedetti, M., Smith, D.: A performance-driven QBF-based iterative logic array representation with applications to verification, debug and test. In: International Conference on Computer-Aided Design, pp. 240–245 (2007)

2. Goultiaeva, A., Iverson, V., Bacchus, F.: Beyond CNF: A circuit-based QBF solver. In: Kullmann, O. (ed.) SAT 2009. LNCS, vol. 5584, pp. 412–426. Springer, Heidelberg (2009)
3. Goultiaeva, A., Bacchus, F.: Exploiting QBF duality on a circuit representation. In: Proceedings of the AAAI National Conference (2010) (accepted for publication)
4. Pipatsrisawat, K., Darwiche, A.: A lightweight component caching scheme for satisfiability solvers. In: Marques-Silva, J., Sakallah, K.A. (eds.) SAT 2007. LNCS, vol. 4501, pp. 294–299. Springer, Heidelberg (2007)
5. Giunchiglia, E., Narizzano, M., Tacchella, A.: Quantified Boolean Formulas satisfiability library (QBFLIB) (2001), www.qbflib.org
6. Biere, A.: Resolve and expand. In: Hoos, H., Mitchell, D.G. (eds.) SAT 2004. LNCS, vol. 3542, pp. 238–246. Springer, Heidelberg (2005)
7. Giunchiglia, E., Narizzano, M., Tacchella, A.: QUBE: A system for deciding Quantified Boolean Formulas satisfiability. In: Proceedings of the International Joint Conference on Automated Reasoning, pp. 364–369 (2001)
8. Lonsing, F., Biere, A.: Nenofex: Expanding NNF for QBF solving. In: Kleine Büning, H., Zhao, X. (eds.) SAT 2008. LNCS, vol. 4996, pp. 196–210. Springer, Heidelberg (2008)
9. Lonsing, F., Biere, A.: A compact representation for syntactic dependencies in QBFs. In: Kullmann, O. (ed.) SAT 2009. LNCS, vol. 5584, pp. 398–411. Springer, Heidelberg (2009)

Reconstructing Solutions after Blocked Clause Elimination[*]

Matti Järvisalo[1] and Armin Biere[2]

[1] Department of Computer Science, University of Helsinki, Finland
[2] Institute for Formal Models and Verification, Johannes Kepler University Linz, Austria

Abstract. Preprocessing has proven important in enabling efficient Boolean satisfiability (SAT) solving. For many real application scenarios of SAT it is important to be able to extract a full satisfying assignment for original SAT instances from a satisfying assignment for the instances after preprocessing. We show how such full solutions can be efficiently reconstructed from solutions to the conjunctive normal form (CNF) formulas resulting from applying a combination of various CNF preprocessing techniques implemented in the PrecoSAT solver—especially, blocked clause elimination combined with SatElite-style variable elimination and equivalence reasoning.

1   Introduction

CNF-level preprocessing has proven important in enabling efficient SAT solving. This is highlighted for instance by PrecoSAT[1]—one of the most successful SAT solvers in the 2009 SAT Competition—that applies a combination of different preprocessing techniques both before and during search. On the other hand, for many real applications scenarios it is important to be able to extract a full satisfying assignment for the original instances from satisfying assignments for preprocessed instances. However, CNF-level preprocessing/simplification techniques, such as SatElite-style variable elimination [1] and blocked clause elimination [2], often preserve only satisfiability, not the set of satisfying assignments. Especially, reconstruction of an original solution becomes non-straightforward when applying combinations of preprocessing techniques.

In this paper we show how such full satisfying assignments can be efficiently reconstructed from solutions to the CNFs resulting from applying combinations of various preprocessing techniques. Especially, we concentrate on the non-trivial case of combining blocked clause elimination [2] (BCE)—which has proven surprisingly powerful, being able to achieve the same level of simplification as the Plaisted-Greenbaum polarity-based CNF encoding and a combination of specific circuit-level simplification techniques— with SatElite-style variable elimination and equivalence reasoning [3,4,5]. We explain how solution reconstruction is done in practice in PrecoSAT, and formally justify the correctness of this process. The presented reconstruction techniques are both time and space wise linear, and hence have no real overhead w.r.t. solving.

---

[*] First author financially supported by Academy of Finland (grant 132812).
[1] See http://fmv.jku.at/precosat

O. Strichman and S. Szeider (Eds.): SAT 2010, LNCS 6175, pp. 340–345, 2010.
© Springer-Verlag Berlin Heidelberg 2010

2   Preliminaries

CNF.   For a Boolean variable x, there are two *literals*, the positive literal, denoted by x, and the negative literal, denoted by $\bar{x}$, the *negation of* x. A *clause* is a disjunction of distinct literals and a CNF formula is a conjunction of clauses. When convenient, a clause is seen as a finite set of literals and a CNF formula as a finite set of clauses. A clause is a *tautology* if it contains both x and $\bar{x}$ for some variable x. A truth assignment for a CNF formula F is a function τ that maps variables in F to $\{\mathbf{t}, \mathbf{f}\}$. If $\tau(x) = v$, then $\tau(\bar{x}) = \neg v$, where $\neg\mathbf{t} = \mathbf{f}$ and $\neg\mathbf{f} = \mathbf{t}$. A clause is satisfied by τ if it contains at least one literal l such that $\tau(l) = \mathbf{t}$. An assignment τ *satisfies* F if it satisfies every clause in F. Finally, given an assignment τ, let τ_x (resp., $\tau_{\bar{x}}$) denote the assignment for which $\tau_x(x) = \mathbf{t}$ (resp., $\tau_{\bar{x}}(x) = \mathbf{f}$) and which otherwise is identical to τ.

Resolution.   The resolution rule states that, given two clauses $C_1 = \{x, a_1, \dots, a_n\}$ and $C_2 = \{\bar{x}, b_2, \dots, b_m\}$, the implied clause $C = \{a_1, \dots, a_n, b_1, \dots, b_m\}$, called the *resolvent* of C_1 and C_2, can be inferred by *resolving* on the variable x. We write $C = C_1 \otimes_x C_2$. A sequence of clauses $(C_0, C_1, \dots, C_n)$ is a resolution derivation of the clause C from a CNF formula F if (i) $C_n = C$, and (ii) each C_i, where $0 \leq i < n$, is either a clause in F (in this case C_i is called an *input clause*), or C_i is the resolvent of two clauses C_j and C_k, where $j, k < i$. We denote by $F \vdash C$ the fact that there is a resolution derivation of the clause C from the CNF formula F. A well-known refinement of resolution is tree-like resolution, where derivations have to be representable as trees.

Variable Elimination as SatElite-style Preprocessing.   Following the Davis-Putnam procedure [6] (DP), a preprocessing technique VE, referred to as *variable elimination by clause distribution* in [1], can be defined. For a CNF formula F, let $S_x \subseteq F$ and $S_{\bar{x}} \subseteq F$ consist of all the clauses in F that contain the literal x and $\bar{x}$, respectively. The elimination of a variable x in the whole CNF can be computed by pair-wise resolving each clause in S_x with every clause in $S_{\bar{x}}$. Formally, the resolution operator $\otimes$ can be lifted to sets of clauses:

$$S_x \otimes_x S_{\bar{x}} = \{C_1 \otimes_x C_2 \mid C_1 \in S_x, C_2 \in S_{\bar{x}}, \text{ and } C_1 \otimes_x C_2 \text{ is not a tautology}\}.$$

Now, replacing the original clauses in $S_x \cup S_{\bar{x}}$ with the set $S = S_x \otimes_x S_{\bar{x}}$ of non-tautological resolvents gives the CNF $(F \setminus (S_x \cup S_{\bar{x}})) \cup S$ which is satisfiability-equivalent to F. Since DP is a complete proof procedure for CNFs, with exponential worst-case space complexity, for practical applications as a preprocessing technique, variable elimination needs to be bounded; e.g., SatElite eliminates a considered variable only when the resulting CNF formula $(F \setminus (S_x \cup S_{\bar{x}})) \cup S$ will not contain more clauses as the original formula F. For the following, let $\text{VE}(F, x)$ denote the result of applying variable elimination to F w.r.t x.

Blocked Clause Elimination (BCE) is a satisfiability-preserving CNF preprocessing techniques which removes so called *blocked clauses* [7] from CNF formulas.

Definition 1.   *A literal l in a clause C of a CNF F blocks C w.r.t. F if for every clause $C' \in F$ with $\bar{l} \in C'$, the resolvent $(C \setminus \{l\}) \cup (C' \setminus \{\bar{l}\})$ obtained from resolving C and C' on l is a tautology.*

With respect to a fixed CNF and its clauses we have:

Definition 2. *A clause is* blocked *if it has a literal that blocks it.*

Example 1. Consider the formula $F_{\text{blocked}} = (a \vee b) \wedge (a \vee \bar{b} \vee \bar{c}) \wedge (\bar{a} \vee c)$. Only the first clause of F_{blocked} is not blocked. The second clause contains two blocked literals: a and $\bar{c}$. Also literal c in the last clause is blocked. Notice that after removing either $(a \vee \bar{b} \vee \bar{c})$ or $(\bar{a} \vee c)$, the clause $(a \vee b)$ becomes blocked. This is actually an extreme case in which BCE can remove all clauses of a formula, resulting in a trivially satisfiable formula. $\square$

In the example, notice that although BCE alone can show that the original formula is satisfiable, a solution to the original CNF is not directly available.

Recent work [2] shows that, although a simple technique, BCE is surprisingly powerful. For example, without any circuit-level information, on the standard Tseitin CNF encoding BCE can achieve at least the same level of simplification as the Plaisted-Greenbaum polarity-based CNF encoding and a combination of specific circuit-level simplification techniques. Moreover, as shown in [2], BCE and SatElite-style variable elimination are to some extend orthogonal preprocessing techniques, which justifies combining these techniques for even more effective preprocessing. Notice also that, in contrast to variable elimination, BCE has a unique fixpoint for any CNF formula, i.e., BCE is confluent. This is due to the following.

Proposition 1 ([7]). *Given a CNF formula F, let clause $C \in F$ be blocked with respect to F. Any clause $C' \in F$, where $C' \neq C$, that is blocked with respect to F is also blocked with respect to $F \setminus \{C\}$.*

Exploiting Equivalent Literals.   For two literals l_1 and l_2, let $l_1 \equiv l_2$ denote the CNF formula $\{\{l_1, \bar{l}_2\}, \{\bar{l}_1, l_2\}\}$. For a given CNF formula F, if $F \vdash l_1 \equiv l_2$, the equivalent literals l_1 and l_2 can be exploited by the equivalence reduction in which all occurrences of l_2 are substituted by l_1 (or vice versa), eliminating the variable of l_2 (or l_1). For example, *hyper binary resolution*, in which the clause $\{l, l'\}$ can be derived in one step from the clauses $\{l, l_1, \ldots, l_n\}$ and $\{\bar{l}_i, l'\}$ where $1 \leq i \leq n$, can be used to derive new binary clauses [3,4,5].

For detecting and exploiting equivalent literals in preprocessing/simplification, PrecoSAT implements a lazy version of hyper binary resolution. It also finds equivalent literals during failed literal probing. Equivalences are represented with a union find data structure. During garbage collection, not only top-level satisfied clauses are removed but all their literals are mapped to their representatives of the union find data structure. This essentially removes equivalent literals from the CNF; afterwards, only the representatives remain.

3   Solution Reconstruction for Individual Techniques

In this section we describe how to reconstruct solutions for each of the considered preprocessing techniques separately. We start with variable elimination for which reconstruction can be seen as part of the completeness proof of DP.

Proposition 2. *Let τ be a satisfying assignment for* $\mathrm{VE}(F, x)$. *Either τ_x or $\tau_{\bar{x}}$ satisfies* $S_x \cup S_{\bar{x}}$, *and, the one that does, also satisfies* $F = \mathrm{VE}(F, x) \cup (S_x \cup S_{\bar{x}})$.

To reconstruct a solution after <u>VE</u> has been applied repeatedly for the variables $x_1, \ldots,$ x_m, it is enough to save (remember) the clauses $(S_{x_1} \cup S_{\bar{x}_1}), \ldots, (S_{x_m} \cup S_{\bar{x}_m})$. Assume that τ satisfies $\mathrm{VE}(\cdots \mathrm{VE}(\mathrm{VE}(F, x_1), x_2) \cdots, x_m)$. Let $\tau^{m+1} = \tau$, and, iteratively from $i = m$ to 1, define τ^i as the one of $\tau_{x_i}^{i+1}$ and $\tau_{\bar{x}_i}^{i+1}$ which satisfies $(S_{x_m} \cup S_{\bar{x}_m})$. Proposition 2 guarantees that τ^1 is a satisfying assignment for the original formula F.

If the application only requires to reconstruct one solution, then in practice[2] it is enough to only save either S_{x_i} or $S_{\bar{x}_i}$. W.l.o.g. assume S_{x_i} is saved. Then, if $\tau_{\bar{x}_i}^{i+1}$ satisfies the saved S_{x_i}, we pick $\tau^i = \tau_{\bar{x}_i}^{i+1}$, since this truth assignment obviously satisfies $S_{\bar{x}_i}$ as well. Otherwise x_i is forced to be **t** and we must set $\tau^i = \tau_{x_i}^{i+1}$. This case occurs if and only if there is a clause in S_{x_i} for which τ^{i+1} assigns all literals except x_i to **f**.

In an actual implementation only the smaller of the two sets is saved. Thus this technique is also efficient in the case where VE is used for pure literal elimination as discussed in [2]. In addition to plain VE, it also works for functional substitution [1] as in the SatElite preprocessor. The only difference between VE and functional substitution is that the latter removes some redundant clauses from $S_x \otimes_x S_{\bar{x}}$ while maintaining the set of satisfying assignments.

<u>Equivalent literals</u> are substituted by their representatives during preprocessing. Clauses used to derive equivalent literals become trivial and are removed during garbage collection. However, the relation between original literals and their representatives is maintained. If a satisfying assignment for the remaining clauses is found, the truth value of a substituted literal is defined to be the value of its representative. This extends the satisfying assignment for the remaining clauses to a satisfying assignment for the original formula.

Finally, consider <u>BCE</u>. In analogy to the case of VE, the proof [7] which shows that removal of a blocked clause does not turn an unsatisfiable formula into a satisfiable formula, gives us grounds to reconstruct solutions for BCE.

Proposition 3. *Assume that literal l blocks C w.r.t. F. Let τ be a satisfying assignment for $F \setminus \{C\}$. If τ does not satisfy C, then τ_l satisfies both $F \setminus \{C\}$ and C and thus F.*

In practice it is enough to save all removed blocked clauses $C_1, \ldots, C_m$ together with their blocking literals $l_1, \ldots, l_m$.[3] Let τ^m be a satisfying assignment for F_m, where $F_i = F \setminus \cup_{j=1}^i \{C_j\}$ for $i = 1 \ldots m$ and $F_0 = F$. If τ^i satisfies C_i, we pick $\tau^{i-1} = \tau^i$, and otherwise $\tau^{i-1} = \tau_{l_i}^i$. Using Proposition 3, one can show by induction that τ^i satisfies F_i, and thus τ^0 is a satisfying assignment for F.

4   Combined Solution Reconstruction

First, BCE and VE can be combined by saving clauses for reconstructing solutions after BCE resp. VE on the same reconstruction stack. Reconstruction works in reverse order

---

[2] By private communication with Niklas Sörensson.

[3] A space efficient way to save this information is to maintain l_i as the first literal in the saved clause C_i. This also allows to keep track of eliminated variables in VE.

in which these clauses have been saved. This also works nicely if BCE is applied on-the-fly during VE: while counting the non-trivial resolvents of $S_x \otimes_x S_{\bar{x}}$ to determine whether VE is applied to x, it may occur that a clause $C \in (S_x \cup S_{\bar{x}})$ has only trivial resolvents w.r.t. x, even though the overall number of non-trivial resolvents exceeds $|S_x \cup S_{\bar{x}}|$, which prevents x from being eliminated. Yet C can be removed as a blocked clause and is saved on the reconstruction stack.

In order to combine VE and equivalent reasoning it is enough to make sure that VE is only attempted after all equivalent literals have been first substituted. Enforcing this order of using equivalent literal reasoning and VE makes sure that variables eliminated with VE are always representatives and the only remaining variables of their equivalence class. Eliminating a representative through VE will eliminate its whole equivalence class, and after this it is not possible that further equivalent literals could be added to the equivalence class of an eliminated variable.

When combining BCE with equivalent literal reasoning, however, the situation is different: at some point after removing a blocked clause C, a literal l which blocked C may become equivalent to another literal and may even become a representative of its equivalence class. On the other hand, one may be forced to flip the value of l during solution reconstruction since BCE removed C (recall Sect. 3). Hence the values of all the literals in the equivalence class should be flipped, which appears not to be sound since this could make some other clause unsatisfied. However, as we show in the next section, the value of l will never have to be flipped in such a situation.

5   Equivalent Literals and Blocked Clause Satisfiability

Equivalent literals detected and applied in simplifying a CNF after removing blocked clauses cannot make the removed blocked clauses to not to be satisfied under a satisfying assignment for the rest of the formula.

Theorem 1. *Assume a CNF formula F, a clause $C \in F$ which is blocked for $l \in C$ w.r.t. F, and a literal l'. If $F \setminus \{C\} \vdash l \equiv l'$, then $(F \setminus \{C\}) \cup (l \equiv l') \models C$.*

In other words, any satisfying assignment for $(F \setminus \{C\}) \cup (l \equiv l')$ also satisfies the blocked clause C. This means that binary equivalences detected during preprocessing can be exploited when applying BCE, at the same time guaranteeing all the blocked clauses removed by BCE will be satisfied by any satisfying assignment for the resulting preprocessed CNF formula. Notice that this lemma is independent of the techniques used for deriving the clauses in $l \equiv l'$.

Proof (of Theorem 1). Assume a CNF formula F, a clause $C = \{l, l_1, \ldots, l_k\} \in F$ which is blocked for $l \in C$ w.r.t. F. Denote by $B \subset F$ the set of clauses which contain the literal $\bar{l}$. Hence each clause in B contains at least one of the literals $\bar{l}_1, \ldots, \bar{l}_k$. Assume that $F \setminus \{C\} \vdash l \equiv l'$ for some literal l', and hence there is a resolution derivation of $\{l, \bar{l'}\}$ and $\{\bar{l}, l'\}$ from $F \setminus \{C\}$.

If F is unsatisfiable, $F \setminus \{C\}$ is also unsatisfiable since C is blocked, and hence trivially $(F\setminus\{C\})\cup(l \equiv l') \models C$. Now consider the case that F and (thus) also $F\setminus\{C\}$ and $(F \setminus \{C\}) \cup (l \equiv l')$ are satisfiable. Take an arbitrary satisfying assignment τ for $(F \setminus \{C\}) \cup (l \equiv l')$. We will show that any such τ also satisfies C.

The case in which $\tau(l) = \mathbf{t}$ (that is, τ satisfies l) is trivial. Now assume $\tau(l) = \mathbf{f}$. Then $\tau(l') = \mathbf{f}$ since τ satisfies $l \equiv l'$. Consider an arbitrary resolution derivation $\pi = (C_1, \ldots, C_m)$ of $C_m = \{\bar{l}, l'\}$ from $F \setminus \{C\}$. Assume w.l.o.g. that π is tree-like. We claim that there is an input clause $C' = \{\bar{l}, l'_1, \ldots, l'_k\} \in B$ in π such that $\tau(l'_i) = \mathbf{f}$ for all i. Since $C' \in B$, it then follows that one of the l'_is is one of the literals $\bar{l}_1, \ldots, \bar{l}_k$, and hence τ satisfies C (recall that $C = \{l, l_1, \ldots, l_k\}$).

To prove the claim, we show that there is a path $P_1, \ldots, P_n$ of clauses in π (seen as a tree) from the root of the tree ($P_1 = C_m$) to a leaf (P_n is an input clause of π), such that each clause P_i on the path contains $\bar{l}$ and τ assigns all literals in P_i except $\bar{l}$ to $\mathbf{f}$.

First notice that for $P_1 = C_m$ we know that $\tau(\bar{l}) = \mathbf{t}$ and $\tau(l') = \mathbf{f}$. Now assume that $P_i = \{\bar{l}\} \cup D$, where D is a set of literals such that τ assigns every literal in D to $\mathbf{f}$, was directly derived from clauses C_a and C_b in π resolving on the variable x. Notice that at least one of C_a and C_b must contain $\bar{l}$. First consider the case that C_a contains $\bar{l}$ and C_b does not. Since τ assigns all literals in D to $\mathbf{f}$, τ must satisfy the literal for x in C_b. (Otherwise τ does not satisfy C_b which would imply that τ does not satisfy an input clause in π and hence τ cannot be a satisfying truth assignment for $(F \setminus \{C\}) \cup (l \equiv l')$, in contradiction to our assumption.) Hence τ assigns all literals in C_a apart from $\bar{l}$ to $\mathbf{f}$. In this case let $P_{i+1} = C_a$. The case that C_b contains $\bar{l}$ and C_a does not is identical.

Now consider the case that both C_a and C_b contain $\bar{l}$. Since τ assigns a unique truth value to x, τ assigns all literals in either C_a or C_b apart from $\bar{l}$ to $\mathbf{f}$. In this case let P_{i+1} be this particular clause. $\qquad\square$

6   Conclusions

We showed how and why—in theory and in practice—full solutions to CNF formulas can be reconstructed from solutions to the CNF after applying both individual and combinations of preprocessing techniques, including blocked clause elimination, SatElite-style variable elimination and equivalence reasoning.

References

1. Eén, N., Biere, A.: Effective preprocessing in SAT through variable and clause elimination. In: Bacchus, F., Walsh, T. (eds.) SAT 2005. LNCS, vol. 3569, pp. 61–75. Springer, Heidelberg (2005)
2. Järvisalo, M., Biere, A., Heule, M.: Blocked clause elimination. In: Esparza, J., Majumdar, R. (eds.) TACAS 2010. LNCS, vol. 6015, pp. 129–144. Springer, Heidelberg (2010)
3. Bacchus, F.: Enhancing Davis Putnam with extended binary clause reasoning. In: Proc. AAAI 2002, pp. 613–619. AAAI Press, Menlo Park (2002)
4. Bacchus, F., Winter, J.: Effective preprocessing with hyper-resolution and equality reduction. In: Giunchiglia, E., Tacchella, A. (eds.) SAT 2003. LNCS, vol. 2919, pp. 341–355. Springer, Heidelberg (2004)
5. Gershman, R., Strichman, O.: Cost-effective hyper-resolution for preprocessing CNF formulas. In: Bacchus, F., Walsh, T. (eds.) SAT 2005. LNCS, vol. 3569, pp. 423–429. Springer, Heidelberg (2005)
6. Davis, M., Putnam, H.: A computing procedure for quantification theory. Journal of the ACM 7(3), 201–215 (1960)
7. Kullmann, O.: On a generalization of extended resolution. Discrete Applied Mathematics 96-97, 149–176 (1999)

An Empirical Study of Optimal Noise and Runtime Distributions in Local Search[*]

Lukas Kroc[1,2], Ashish Sabharwal[1], and Bart Selman[1]

[1] Dept. of Computer Science, Cornell University, Ithaca NY 14853-7501, USA
[2] CCS-3, Los Alamos National Lab, Los Alamos, NM 87545, USA
{kroc,sabhar,selman}@cs.cornell.edu

Abstract. This paper presents a detailed empirical study of local search for Boolean satisfiability (SAT), highlighting several interesting properties, some of which were previously unknown or had only anecdotal evidence. Specifically, we study hard random 3-CNF formulas and provide surprisingly simple analytical fits for the optimal (static) noise level and the runtime at optimal noise, as a function of the clause-to-variable ratio. We also demonstrate, for the first time for local search, a power-law decay in the tail of the runtime distribution in the low noise regime. Finally, we discuss a Markov Chain model capturing this intriguing feature.

Designing, understanding, and improving, local search methods for constraint reasoning, and in particular for Boolean satisfiability (SAT), has been the focus of hundreds of research papers since the 1990's and even earlier. For SAT, techniques such as greedy local search, tabu search, solution guided search, focused random walk, and reactive or adaptive search have led to much success. Specifically, Walksat [7] stands out as one of the initial solvers that introduced many of the key ideas in use today and, is still competitive with the state of the art.

While many attempts have been made to understand the behavior of local search methods in terms of local minima, exploring "plateaus", the exploration vs. exploitation tradeoff, etc., our formal understanding is limited mostly to relatively simple variants of local search, such as a pure greedy search, a pure random walk, or a combination of the two. This is not surprising as the techniques employed by Walksat and other state-of-the-art local search solvers are too complex to allow a formal analysis in terms of, for example, a traditional Markov Chain. At the same time, there is a wealth of information available from observations of the behavior of local search methods on a variety of domains, most notably for random 3-SAT. There is either formal or anecdotal evidence of various features, such as Walksat scaling linearly at optimal noise but exponentially at sub-optimal noise, and suggestions that the runtime distribution of local search on a single random instance has an exponentially decaying tail. This work provides convincing empirical evidence in favor of, or even against, such

---

[*] Supported by NSF (Expeditions in Computing award for Computational Sustainability, 0832782; IIS grant 0514429) & AFOSR (IISI, grant FA9550-04-1-0151). The authors thank Yahoo! for generously providing access to their M45 compute cloud.

O. Strichman and S. Szeider (Eds.): SAT 2010, LNCS 6175, pp. 346–351, 2010.

anecdotal insights and observations. We study the behavior of Walksat on hard, large, random 3-CNF formulas and investigate its time complexity in relation to the *clause-to-variable ratio* α and the (static) *noise level*—both of which Walksat is highly sensitive to. Unlike previous studies, our conclusions are based on very large formulas and are thus free of "small N effects". This might explain the difference between our conclusions and those of, e.g., Hoos and Stützle [4].

While many new local search SAT solvers are based on "adaptive" or "dynamic" noise, these solvers are apparently unable to settle on the optimal noise setting for hard random 3-CNF formulas, doing much worse than optimal static noise. E.g., we found that the SAT Competition 2009 winners in the satisfiable Random category, TNM and gnovelty+2, were slower than Walksat at optimal noise by a factor of roughly 4x for N=10,000 variable formulas with $\alpha = 4.2$, 13x for N=20,000, 31x for N=30,000, 54x for N=40,000, and 785x for N=50,000. This also shows that, unlike Walksat, these adaptive noise solvers scale super-linearly in this domain, justifying the interest in our study of static noise.

Our first result is a surprisingly simple step-linear analytical fit for the value of the *optimal noise* as a function of α, and an equally simple analytical expression for the mean running time of Walksat (measured as the number of flips) at this optimal noise. This fit as well as our data exhibit linear scaling with N for α close to the phase transition region for 3-SAT. Second, we study the runtime distribution of Walksat on single instances and find first clear evidence of *power-law decay* in the probability of failure in T flips in the tail of the distribution. Power-law decays and heavy-tailed runtime distributions [3] have been one of the key observations for DPLL-style systematic search solvers and have led to methodologies such as rapid restarts and algorithm portfolios. This phenomenon, however, is usually not associated with local search. We show that after a (relatively long) "flat" region, the probability of failure decays exponentially in the high noise regime but as a power-law in the low noise regime. Third, we show that as Walksat proceeds, the number of unsatisfied clauses exhibits an interesting *gradual decay* that happens only at near-optimal noise.

A model that captures such features and is yet simple to describe and simulate can be a very useful tool for understanding and exploiting the tradeoffs inherent in local search. We therefore propose a preliminary Markov Chain model capturing, e.g., exponential scaling with N and power-law decay at low noise.

The kind of empirical study pursued here requires a significant computational power (e.g., 100,000 runs for some low noise levels to observe a clear trend). We used Yahoo!'s Apache Hadoop based M45 cloud computing platform with the net computational effort being equivalent to around 14 years of single CPU time.

We assume basic familiarity with SAT, CNF formulas, and local search solvers such as Walksat. N and M will denote the number of variables and clauses, resp., of a CNF formula F, with $\alpha = M/N$. A random 3-CNF formula is created by choosing, with repetition, M clauses of size 3 each uniformly at random. Ignoring the details of the "freebie move", the noise parameter $n \in [0, 1]$ (or in $[0\%, 100\%]$) of Walksat is essentially the probability with which it makes a (possibly uphill) random walk move, instead of a greedily chosen downhill or plateau move.

1   Local Search in SAT: Empirical Findings

1.1   Analytical Expressions for Optimal Noise and Mean Runtime

It has been observed that Walksat behaves quite predictably on large random formulas, in terms of the average (or median) running time or the noise levels that perform the best. For example, Seitz et al. [6] have given evidence that Walksat scales linearly when a noise value of 0.57 is used for formulas with $\alpha \approx 4.2$.

The left pane of Fig. 1 shows how the optimal noise $n*$ (y-axis) changes as α (x-axis) increases. For this experiment, we considered values of α in the range $[1.5, 4.2]$ and random 3-CNF formulas with N ranging from 100,000 to 400,000. For each (α, N) pair, we considered 10 formulas (the variation amongst formulas is not much at such large values of N) and did a binary search to estimate $n*$ for each formula up to a granularity of 0.1%. $n*$ values for these formulas, as one might expect, do *not* depend on N. The resulting average $n*$ and its standard deviation are plotted in the left pane of Fig. 1. We see three clear linear regimes, which can be fitted well with the following step-linear model, shown as the solid black line in the figure:

$$\text{OptNoise } (\%) \approx \begin{cases} -7.617 + 14.588\alpha & \text{for } \alpha \in [1.5, 3.0] \\ -2.558 + 12.347\alpha & \text{for } \alpha \in (3.0, 3.85) \\ -77.53 + 31.970\alpha & \text{for } \alpha \in [3.85, 4.2] \end{cases} \tag{1}$$

Interestingly, the transition points between the three regimes correspond to values of α that have been studied before. The first transition point corresponds to the threshold of $c_3 \approx 3.003$ which was proven by Frieze and Suen [2] to be the precise point up till which a simple "generalized unit clause" rule, GUC, almost surely solves random 3-SAT instances. The second transition point corresponds to the threshold of ≈ 3.9 up till which the purely greedy version of Walksat, namely GSAT, has been empirically seen to be successful (also the point at which the solution space structure is believed to change drastically [5]).

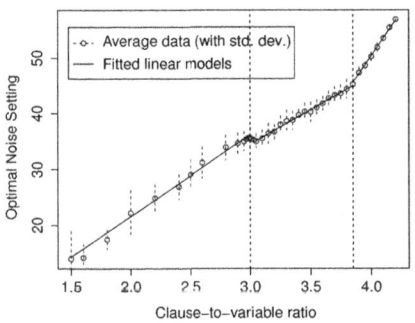
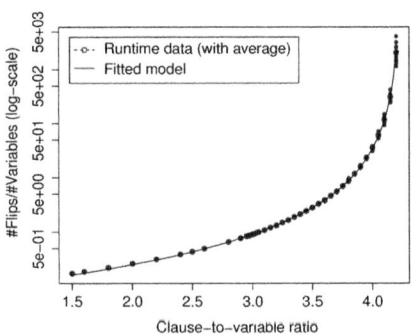

Fig. 1. Left: Linear model fitted to the optimal noise data (y-axis). Right: Model fitted to the mean runtime of Walksat (y-axis) at optimal noise. Both plotted as functions of clause-to-variable ratios (x-axis).

The right pane of Fig. 1 shows similar data for the average running time of Walksat at the optimal noise predicted by Eq. (1), for $\alpha \in [1.5, 4.2]$. In order to account for different formula sizes, we scaled the runtimes by dividing by N and found that this normalized runtime at optimal noise is indeed independent of N, implying linear scaling with N (as generally believed). We fit this curve with a model, depicted as a solid black line. For $\alpha < 4.2355$, our fitted model captures the linear scaling of the mean runtime with N at optimal noise. Specifically:

$$\frac{\text{numflips at OptNoise}}{N} \approx \frac{1.4358}{(4.2355 - \alpha)^{2.1894}} \tag{2}$$

1.2 Runtime Distribution: Exponential or Power-Law Decay

Fix α to 4.2 and consider the following question: *if we take a single formula F and perform several runs of Walksat, what is the distribution of the runtime needed to find a solution?* We measure this as the *probability of failure* as a function of the number of flips, i.e., what fraction of the runs fail to find a solution in T flips. The result for various noise levels for a 100,000 variable formula with $\alpha = 4.2$ is shown in Fig. 2, in both log-linear and log-log scales. For each low and high noise level, we performed 110,000 and 10,000 runs, resp., of Walksat with a cutoff of 100B (100×10^9) and 4B flips, resp. The median runtime to solve these formulas ranges between 140M to 700M flips at different noise levels.

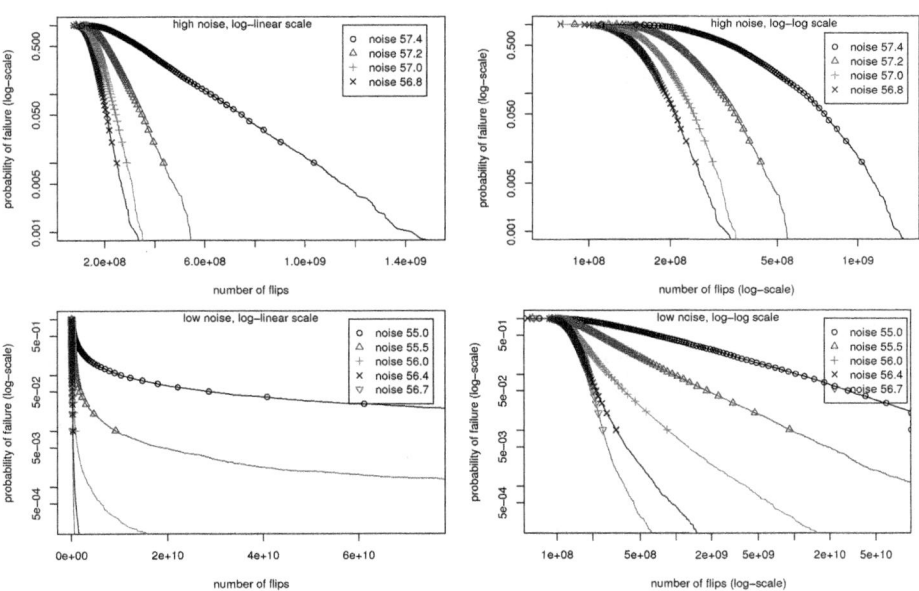

Fig. 2. Runtime distribution of Walksat on a large 3-CNF formula (100k variables, 420k clauses), with probability of failure on y-axis. Top: *high noise regime with exponential decay* (straight line in log-linear plot at top-left). Bottom: *low noise regime with power-law decay* (straight line in log-log plot at bottom-right).

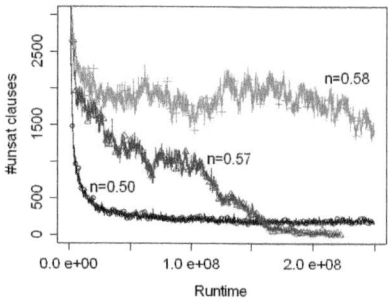
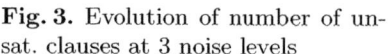

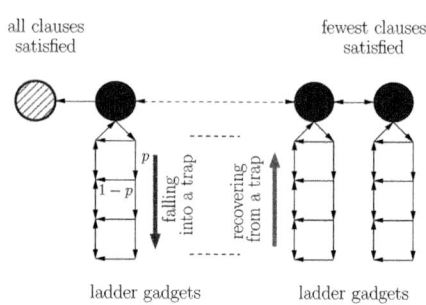

Fig. 3. Evolution of number of unsat. clauses at 3 noise levels

Fig. 4. A Markov Chain model for local search in SAT

The figure shows a clear qualitative distinction between the runtime distributions at high noise levels ($n > 0.567$, top row) and at low noise levels (bottom row). In particular, the log-linear plot in the top-left shows that the probability of failure for high noise decays exponentially with the number of flips: for large enough T, $\Pr[\text{failure after } T \text{ flips}] \approx \exp(-cT)$ for some constant c. On the other hand, the bottom-left plot, also in log-linear scale, shows that the decay is clearly slower than exponential for noise levels below the optimal. The bottom-right plot, showing relatively straight lines in the log-log scale, demonstrates that for low noise levels, the decay rate is very close to being a power-law (at the tail): for large enough T, $\Pr[\text{failure after } T \text{ flips}] \approx T^{-c'}$ for some $c' > 0$.

1.3   Evolution of the Number of Violated Clauses

Fig. 3 shows how the number of violated clauses, denoted $\hat{M}$, progresses when `Walksat` is run in a formula with 100,000 variables with $\alpha = 4.2$ at noise levels 0.50 (too low), 0.57 (optimal), and 0.58 (too high). For all noise levels considered, there is a very steep initial descent until $\hat{M}$ is under 2,500, or 2.5% of the original clauses. After this, the behavior changes drastically depending on the noise level. For optimal noise n_*, $\hat{M}$ shows a *gradually decreasing, almost linear, trend* (with small random variations, as one would expect) until a solution is found, depicted by the middle curve in red. For high noise, $\hat{M}$ decreases rather fast in the beginning but then settles at a non-zero value that is higher the higher the noise is (the highest curve, in green). For low noise, $\hat{M}$ decreases very rapidly and settles at a relatively low but non-zero value (the lowest curve, in blue).

2   Markov Chain Model Capturing Power-Law Decay

We briefly sketch a Markov Chain (MC) model that may shed light, at least qualitatively, on exponential scaling and power-law decay. The model has two parts (see Fig. 4). The first (top, horizontal) part is a linear MC with states corresponding to truth assignments that satisfy the same fraction of clauses of F, with the leftmost state encapsulating all solutions. Second, hanging from each state in the top chain is a "trap gadget", which captures the behavior of Walksat when it "gets lost" exploring parts of the search space without any solutions.

The probability of moving left or right in the top chain is determined by combining two effects. First, we assume that in the very high noise setting, the search will prefer to choose a neighboring state s with probability proportional to the number of truth assignments in s. Assuming a roughly binomial distribution of the number of assignments satisfying k clauses, the chain will then have a tendency to drift towards the middle, away from the solutions. Second, for relatively low noise settings, the search will be more focused and will choose among *immediately better* neighbors, i.e., either left or down into one of the trap gadgets. These two forces will balance in different ways for different noise settings. The trap gadget can itself be represented as an MC: two vertical linear chains, connected horizontally at each level (a "ladder", see Fig. 4). Entering this chain would first allow the search to either go down along the right side of the ladder (with probability p at each step), or across to the left side (probability $1 - p$). This part models Walksat's decision to either keep searching deeper in the trap (downward), or to try to escape (across). The left side of the ladder models the effort to escape a local minimum. If the search descends to level k of the trap before making the move across, it may become exponentially difficult for it (in k, say b^k steps) to escape, again depending on the noise; this b^k behavior is itself easily modeled by an MC biased downward while the search tries to go upward. Thus, once in a trap, the search will, with probability p^k, descend to depth k before initiating an escape move, and then spend $\approx b^k$ steps to get back to the top, yielding $\sum_k p^k \cdot b^k$ steps in expectation—and resulting in a power-law distribution, similar to previous analysis [1]. Thus, the cumulative runtime of the MC will be a sum of power-law distributions, in agreement with our data. On the other hand, for very high noise settings, the search will be successfully able to avoid traps, but will also be attracted towards the middle of the top chain, thus taking, in expectation, exponentially long to reach the left end (a solution).

References

[1] Chen, H., Gomes, C., Selman, B.: Formal models of heavy-tailed behavior in combinatorial search. In: Walsh, T. (ed.) CP 2001. LNCS, vol. 2239, p. 408. Springer, Heidelberg (2001)
[2] Frieze, A., Suen, S.: Analysis of two simple heuristics on a random instance of k-SAT. J. Algorithms 20(2), 312–355 (1996)
[3] Gomes, C.P., Selman, B., Kautz, H.: Boosting combinatorial search through randomization. In: 15th AAAI, Madison, WI, July 1998, pp. 431–437 (1998)
[4] Hoos, H.H., Stützle, T.: Towards a characterization of the behavior of stochastic local search algorithms for SAT. AI J. 112(1-2), 213–232 (1999)
[5] Mézard, M., Mora, T., Zecchina, R.: Clustering of solutions in the random satisfiability problem. Phy. Rev. Lett. 94, 197–205 (2005)
[6] Seitz, S., Alava, M., Orponen, P.: Focused local search for random 3-satisfiability. J. Stat. Mech., P06006 (2005)
[7] Selman, B., Kautz, H., Cohen, B.: Local search strategies for satisfiability testing. In: Johnson, D.S., Trick, M.A. (eds.) Cliques, Coloring and Satisfiability: the Second DIMACS Implementation Challenge. DIMACS Series in DMTCS, vol. 26, pp. 521–532. Amer. Math. Soc., Providence (1996)

Green-Tao Numbers and SAT

Oliver Kullmann

Computer Science Department
Swansea University
O.Kullmann@Swansea.ac.uk
http://cs.swan.ac.uk/~csoliver

Abstract. We aim at using the problems from exact Ramsey theory, concerned with computing Ramsey-type numbers, as a rich source of test problems for SAT solving, targeting especially hard problems. Particularly we consider the links between Ramsey theory in the integers, based on van der Waerden's theorem, and (boolean, CNF) SAT solving. Based on Green-Tao's theorem ("the primes contain arbitrarily long arithmetic progressions") we introduce the *Green-Tao numbers* $\mathrm{grt}_m(k_1, \ldots, k_m)$, which in a sense combine the strict structure of van der Waerden problems with the quasi-randomness of the distribution of prime numbers. In general the problem sizes become quickly infeasible here, but we show that for *transversal extensions* these numbers only grow linearly, thus having a method at hand to produce more problem instances of feasible sizes. Using standard SAT solvers (look-ahead, conflict-driven, and local search) we determine the basic Green-Tao numbers. It turns out that already for this single case of a Ramsey-type problem, when considering the best-performing solvers a wide variety of solver types is covered. This is different to van der Waerden problems, where apparently only simple look-ahead solvers succeed (regarding complete methods). For $m > 2$ the problems are non-boolean, and we introduce the *generic translation scheme* for translating non-boolean clause-sets to boolean clause-set. This general method offers an infinite variety of translations ("encodings") and covers the known methods. In most cases the special instance called *nested translation* proved to be far superior over its competitors (including the direct translation).

1 Introduction

The applicability of SAT solvers has made tremendous progress over the last 15 years; see the recent handbook [3]. We are concerned here with solving (concrete) combinatorial problems (see [24] for an overview). Especially we are concerned with the computation of van-der-Waerden-like numbers, which is about colouring hypergraphs of arithmetic progressions; see [18] for the underlying report.

An *arithmetic progression* of size $k \in \mathbb{N}_0$ in $\mathbb{N}$ is a set $P \subset \mathbb{N}$ of size k such that after ordering (in the natural order), two neighbours always have the same distance. So the arithmetic progressions of size $k > 1$ are the sets of the form $P = \{a + i \cdot d : i \in \{0, \ldots, k-1\}\}$ for $a, d \in \mathbb{N}$. Van der Waerden's Theorem ([23])

O. Strichman and S. Szeider (Eds.): SAT 2010, LNCS 6175, pp. 352–362, 2010.
© Springer-Verlag Berlin Heidelberg 2010

shows that for every progression size $k \in \mathbb{N}$ and every number $m \in \mathbb{N}$ of parts there exists some $n_0 \in \mathbb{N}$ such that for $n \geq n_0$, every partitioning of $\{1, \ldots, n\}$ into m parts has some part which contains an arithmetic progression of size k. The smallest such n_0 is denoted by $\mathbf{vdw}_m(k)$, and is called a *vdW-number*. The subfield of Ramsey theory concerned with van der Waerden's theorem is for over 70 years now an active field of mathematics and combinatorics; for an elementary introduction see [19].

We are concerned here with *exact Ramsey theory*, that is, computing vdW-like numbers if possible, or otherwise producing (concrete) lower bounds. [5] introduced the application of SAT for computing vdW-numbers, showing that all known vdW-numbers (at that time) were rather easily computable with SAT solvers. With [11] yet SAT had its biggest success, computing the new (major) vdW-number $vdw_2(6) = 1132$ (mentioned in [19] as a difficult research problem). See [1,2,8] for the current state-of-the-art. And in the underlying report [18] we made an effort at a systematic representation.

We introduce *Green-Tao numbers* $grt_m(k)$ ("GT-numbers"; see Definition 3), which are defined as the vdW-numbers but using the first n *prime numbers* instead of the first n natural numbers. The existence of these numbers is given by the celebrated Green-Tao Theorem ([7]). We are concerned here also with the "mixed" GT-numbers $grt_m(k_1, \ldots, k_m)$ (with $grt_m(k) = grt_m(k, \ldots, k)$). In Theorem 4 we show that *transversal extension GT-numbers*, which are of the form $grt_{m+p}(2, \ldots, 2, k_1, \ldots, k_p)$, grow only linearly in m. In the remainder of the article we are concerned with computing "all feasible" GT-numbers (computable within up to, say, a week by a single processor with the best SAT method).

For binary parameter tuples ($m = 2$ above) the problem of computing vdW- or GT-numbers has a canonical translation to (boolean) SAT problems. For $m > 2$ we still have a canonical translation into *non-boolean* SAT problems, as is the case in general for hypergraph colouring problems (see [16]), but for using standard (boolean) SAT solvers the problem of a boolean translation arises. In Section 3 we introduce the (general) *generic translation scheme*, with seven natural instances, amongst them the well-known direct and logarithmic translations. As it turns out, in nearly all cases for all solver types the *weak nested translation* (introduced in [14]) performed far best, with the only exception that for relatively large numbers of colours the logarithmic translation was better.

For this (initial) phase of investigations into GT-numbers we only used "off-the-shelves" SAT solvers, establishing the "ground level" by providing the best solvers for the various parameter ranges. For over one year on average 10 processors were running, with a lot of manual interaction and adjustment to find the right solvers and translations, and to set the parameters (most basic the number of vertices), establishing the basic Green-Tao numbers. All generators and the details of the computations are available in the open-source research platform OKlibrary (see [13]).[1] See Section 4 for the results of these computations. We conclude this article by a discussion of research directions in Section 5.

---

[1] http://www.ok-sat-library.org

2   The Theorem of Green-Tao, and Green-Tao Numbers

We use $\mathbb{N}_0 = \mathbb{Z}_{\geq 0}$ and $\mathbb{N} = \mathbb{N}_0 \setminus \{0\}$. A *finite hypergraph* G is a pair $G = (V, E)$ where V is a finite set and $E \subseteq \mathbb{P}(V)$ (that is, E is set of subsets of V); we use $V(G) := V$ and $E(G) := E$. An *m-colouring* of a hypergraph G is a map $f : V(G) \to \{1, \ldots, m\}$ such that no hyperedge is monochromatic, that is, for every $H \in E(G)$ there are $v, w \in H$ with $f(v) \neq f(w)$. Regarding (boolean) clause-sets, complementation of boolean variables v is denoted by $\overline{v}$, (boolean) clauses are finite and clash-free sets of (boolean) literals, and (boolean) clause-sets are finite sets of (boolean) clauses. The numbers $\text{vdw}_m(k)$ introduced in Section 1 are "diagonal vdW-numbers", while we consider also the "non-diagonal" or *mixed vdW-numbers*, which are defined as follows.

Definition 1. *A **parameter tuple** is an element of $\mathbb{N}^m_{\geq 2}$ for some $m \in \mathbb{N}$ which is monotonically non-decreasing (that is, sorted in non-decreasing order). For a parameter tuple $(k_1, \ldots, k_m)$ the **vdW-number** $\text{vdw}_m(k_1, \ldots, k_m)$ is the smallest $n_0 \in \mathbb{N}$ such that for every $n \geq n_0$ and every $f : \{1, \ldots, n\} \to \{1, \ldots, m\}$ there exists some "colour" (or "part") $i \in \{1, \ldots, m\}$ such that $f^{-1}(i)$ contains an arithmetic progression of size k_i.*

Note that $\text{vdw}_m(k) := \text{vdw}_m(k, \ldots, k)$ is the smallest n such that the hypergraph $\text{ap}(k, n)$ is not m-colourable, where $V(\text{ap}(k, n)) = \{1, \ldots, n\}$, while $E(\text{ap}(k, n))$ is the set of arithmetic progressions of size k contained in $\{1, \ldots, n\}$. In [18] we have compiled an up-to-date collection of all van-der-Waerden numbers, plus various new numbers, observations and conjectures, using the following organisation of the parameter space.

Definition 2. *A parameter tuple is **trivial** if all entries are equal to 2, otherwise it is **non-trivial**. A **simple** parameter tuple has length 1, otherwise it is **non-simple**. A parameter tuple is a **core tuple** if it is non-simple and if all entries are greater than or equal to 3. A parameter tuple t is a **(transversal) extension** of a parameter tuple t' if t can be obtained from t' by adding entries equal to 2 to the front of t'. A transversal extension of a simple parameter tuple is called an **extended simple tuple** or a **transversal tuple**, while an transversal extension of a core tuple is called an **extended core tuple**. Finally a parameter tuple is **diagonal**, if it is constant (all entries are equal), while otherwise it is **non-diagonal** or **mixed**. Accordingly we speak of **trivial vdW-numbers**, **simple vdW-numbers**, **core vdW-numbers**, **transversal vdW-numbers**, **extended core vdW-numbers**, and **diagonal vdW-numbers**.*

The trivial vdW-numbers are $\text{vdw}_m(2) = m + 1$, while the simple vdW-numbers are given by $\text{vdw}_1(k) = k$. We turn to a major strengthening of van der Waerden's theorem. Now the hypergraphs are given as $\text{ap}_{\text{pr}}(k, n)$ for fixed $k \in \mathbb{N}$, where the vertex set of $\text{ap}_{\text{pr}}(k, n)$ is the set of the first n prime numbers, while the hyperedges are the arithmetic progressions of size k (within the first n prime numbers). As before, every set of prime numbers having at most two elements is an arithmetic progression, but now the first arithmetic progression of size 3 is $\{3, 5, 7\}$, and the first arithmetic progression of size 4 is $\{5, 11, 17, 23\}$. Until 2004 it was even unknown whether the primes contain arbitrarily long arithmetic progressions, and only with [7] the much stronger property, that the relative size of independent sets in $\mathrm{ap}_{\mathrm{pr}}(k, n)$ (that is, sets of prime numbers not containing arithmetic progressions of length k) tend to 0 (for fixed k). In analogy to Definition 1 we define *Green-Tao numbers* ("GT-numbers").

Definition 3. *For a parameter tuple $(k_1, \ldots, k_m)$ let the **Green-Tao number** $\mathrm{grt}_m(k_1, \ldots, k_m)$ be defined as the smallest $n_0 \in \mathbb{N}$ such that for every $n \geq n_0$ and every $f : \{p_1, \ldots, p_n\} \to \{1, \ldots, m\}$, where $p_1, \ldots, p_n$ are the first n prime numbers, there exists some $i \in \{1, \ldots, m\}$ such that $f^{-1}(i)$ contains an arithmetic progression of size k_i. According to Definition 2 we speak of **trivial GT-numbers**, **simple GT-numbers**, **core GT-numbers**, **transversal GT-numbers**, **extended core GT-numbers**, and **diagonal GT-numbers**.*

Theorem 10 in [18] and Green-Tao's theorem ([7]) yields that extended GT-numbers grow linearly. Let $\mathrm{cr}_{\mathrm{ap}}^{\mathrm{pr}}(t, q)$ for parameter tuples t and $q \in \mathbb{R}_{>0}$ be the smallest $n \in \mathbb{N}$ such that for every set P of primes in $\{1, \ldots, n\}$ not containing an arithmetic progression of length k holds $\frac{|P|}{n} < q$.

Theorem 4. *For a parameter tuple t of length $l \in \mathbb{N}$, for $m \in \mathbb{N}_0$ and for $s \in \mathbb{R}_{>1}$ we have $\mathrm{grt}_{m+l}((2, \ldots, 2); t) \leq \max(s \cdot m + 1, \mathrm{cr}_{\mathrm{ap}}^{\mathrm{pr}}(t, 1 - \frac{1}{s}))$.*

3   The Generic Translation Scheme from Non-boolean Clause-Sets to Boolean Clause-Sets

GT-problems of the form "$\mathrm{grt}_2(k_1, k_2) > n$?" have a natural formulation as (boolean) SAT problems by just excluding the arithmetic progressions of sizes k_1 and k_2, e.g. the problem "$\mathrm{grt}_2(2, 3) > 4$?" yields the (satisfiable) clause-set $\{\{2, 3\}, \{2, 5\}, \{2, 7\}, \{3, 5\}, \{3, 7\}, \{5, 7\}, \{-3, -5, -7\}\}$ over the variable-set $\{2, 3, 5, 7\}$ (thus the answer is "yes"). A natural translation for arbitrary m is given when using *generalised clause-sets* as systematically studied in [14,16,17], which allow variables v with finite domains D_v and literals of the form "$v \neq \varepsilon$" for values $\varepsilon \in D_v$. The problem of colouring a hypergraph G with m colours is naturally translated into a SAT problem for generalised clause-sets via using m clauses for every hyperedge $H \in E(G)$, namely for every value $\varepsilon \in \{1, \ldots, m\}$ the clause $\{v \neq \varepsilon : v \in H\}$, stating that not all vertices in H can have value ε (note that the vertices of G are used as variables with (uniform) domain $\{1, \ldots, m\}$). Accordingly we arrive at the natural generalisation $\mathrm{F}_{k_1, \ldots, k_m}^{\mathrm{GT}}(n)$ of the boolean formulation, using as variables the first n prime numbers, each with domain $\{1, \ldots, m\}$, where the clauses are obtained from the hyperedges of $\mathrm{ap}_{\mathrm{pr}}(k_i, n)$ for $i \in \{1, \ldots, m\}$ by using literals "$v \neq i$".

As a running example consider $m = 3$, $k_1 = k_2 = k_3 = 3$ and $n = 5$. We remark that we have $\mathrm{grt}_3(3) = 137$, as can be seen in Section 4. Only one hypergraph needs to be considered here (since all k_i-values coincide), namely

$\mathrm{ap_{pr}}(3,5) = (\{2,3,5,7,11\}, \{\{3,5,7\}, \{3,7,11\}\})$. Now the (non-boolean) clause-set $\mathrm{F}^{\mathrm{GT}}_{3,3,3}(5)$ uses the five (formal[2]) variables $2,3,5,7,11$, each with domain $\{1,2,3\}$, while we have $3 \cdot 2 = 6$ clauses (each of length 3), namely the clauses $\{(3,i),(5,i),(7,i)\}, \{(3,i),(7,i),(11,i)\}$ for $i \in \{1,2,3\}$.

In [14] the *nested translation* from generalised clause-sets to boolean clause-sets was introduced, while the generalisation to the *generic translation scheme* is outlined in [17]. Given a generalised clause-set F, for every variable an (arbitrary) unsatisfiable boolean clause-set $T(v)$ is chosen, such that for different variables these clause-sets are variable-disjoint. Furthermore for every value $\varepsilon \in D_v$ a necessary clause $\gamma_v(\varepsilon) \in T(v)$ is chosen (that is, $T(v) \setminus \{\gamma_v(\varepsilon)\}$ is satisfiable), such that to different values different clauses are assigned. Now the translation $T_\gamma(F)$ of F under T and γ replaces for every clause $C \in F$ the (non-boolean) literals $v \neq \varepsilon$ by the (boolean) literals in clause $\gamma_v(\varepsilon)$, and adds for every variable $v \in \mathrm{var}(F)$ the clauses of the (boolean) clause-set $T(v) \setminus \{\gamma_v(\varepsilon) : \varepsilon \in D_v\}$. The clauses $\gamma_v(\varepsilon)$ are called the *main clauses* of $T(v)$, while the other clauses of $T(v)$ constitute the *remainder*.

Lemma 5. $T_\gamma(F)$ *is satisfiability-equivalent to* F.

Proof. If φ is a satisfying assignment for F, then for every variable $v \in \mathrm{var}(\varphi)$ choose a satisfying assignment ψ_v of $T(v) \setminus \{\gamma_v(\varphi(v))\}$, and the union of these (compatible) assignments ψ_v yields a satisfying assignment for $T_\gamma(F)$ (here it is used that for $\varepsilon \in D_v \setminus \{\varphi(v)\}$ we have $\gamma_v(\varepsilon) \neq \gamma_v(\varphi(v))$). If on the other hand ψ is a satisfying (total) assignment for $T_\gamma(F)$, then for every clause-set $T(v)$ there exists some $\varepsilon_v \in D_v$ such that the clause $\gamma_v(\varepsilon)$ is falsified by ψ; now the assignment $v \mapsto \varepsilon_v$ satisfies F. □

The seven instances of the generic scheme used in this paper, where the domain of variable v is $\{1, \ldots, m\}$, and where the boolean variables are v_i for appropriate indices i, are as follows:

$T(v) = D_m := \{\{v_1\}, \ldots, \{v_m\}, \{\overline{v_1}, \ldots, \overline{v_m}\}\}$ with m variables is used for the *weak direct translation*, where $\gamma_v(i) := \{v_i\}$. D_m is a marginal minimally unsatisfiable clause-set[3] with deficiency 1 (that is, with $m + 1$ clauses). The *strong direct translation* uses $T(v) = D'_m := D_m \cup \{\{v_i, v_j\} : 1 \leq i < j \leq m\}$ and the same γ_v.[4]

---

[2] Note that variable 2 does not occur here; it occurs only for $k_i = 2$, and one could ignore it in general, however then we always had to use the offset 1 when comparing with prime number tables.

[3] See [10] for an overview on minimally unsatisfiable clause-sets.

[4] In [21] the "strong direct translation" is called "direct encoding", starting from arbitrary CSP-problems (while we start from generalised clause-sets). We prefer to distinguish between "encodings", which are about variables and the mapping of assignments, and "translations", which concern the whole process, and which can use quite different but semantically equivalent clause-sets for example. For the direct translation it seems that always the strong form is better, but this is not the case for other translations, and so we explicitly distinguish between "weak" and "strong".

The *weak reduced translation* uses $m - 1$ variables with $T(v) = D_{m-1}$ and an arbitrary bijection γ_v (note that D_{m-1} has m clauses), while the *strong reduced translation* uses the same γ_v and $T(v) = D'_{m-1}$. Different from the direct translations, here γ_v plays a role now, namely the question is to which value one associates the long clause $\{\overline{v_1}, \ldots, \overline{v_{m-1}}\}$, and so we have m (essentially) different choices. Note that clause-set D_{m-1} can be obtained from D_m by DP-reduction for variable v_m (replacing all clauses containing variable v_m by their resolvents on v_m), and accordingly from a clause-set translated by the (weak/strong) direct translation we obtain the clause-set translated by the (weak/strong) reduced translation by performing DP-reduction on all such variables v_m (using that the remainder-clauses are just used as they are, without additional literals in them).

The *weak nested translation* uses $m - 1$ variables and $T(v) = H_{m-1}$, where

$$H_m := \left\{ \{v_1\}, \{\overline{v_1}, v_2\}, \ldots, \{\overline{v_1}, \ldots, \overline{v_{m-1}}, v_m\}, \{\overline{v_1}, \ldots, \overline{v_m}\} \right\},$$

using some arbitrary bijection γ_v (note that H_m has deficiency 1, and thus H_{m-1} has m clauses). H_m is up to isomorphism the unique saturated minimally unsatisfiable Horn clause-set with m variables, and in fact is a saturation of the minimally unsatisfiable clause-set D_m (see [10]). The *strong nested translation* uses the same γ_v, and, similar to the strong direct translation, $T(v) = H'_{m-1} := H_{m-1} \cup \{\{v_i, v_j\} : 1 \leq i < j \leq m - 1\}$. For both forms now we have $m!/2$ (essentially) different choices for γ_v (note that only the two clauses of length m in H_m can be mapped to each other by an isomorphism of H_m). The motivation for the introduction of the weak nested translation in [14,17] was that first the number of clauses is not changed by the translation, that is, $T(v)$ is minimally unsatisfiable (also D_{m-1} fulfils this), and second that $T(v)$ is a hitting clause-set, that is, every pair of different clauses clashes in at least one variable. These two requirements ensure that the conflict structure of the original (non-boolean) clause-set is preserved by the (boolean) translation. Instead of using H_{m-1} one could actually use any unsatisfiable hitting clause-set with m clauses here.

The *simple logarithmic translation*[5] considers the smallest natural number p with $2^p \geq m$, and sets $T(v) = A_p$, where A_p consists of all 2^p full clauses over variables $v_1, \ldots, v_p$, while γ_v is an arbitrary injection.[6]

With the exception of the direct translation, which is fully symmetric in the clauses $\gamma_v(\varepsilon)$, one has to decide about the choice γ_v of necessary clauses. With the exception of the simple logarithmic translation this is the choice of a suitable bijection, i.e., a question of ordering the values of the variables. In this initial study we have chosen a "standard ordering", with the aim of minimising

---

[5] Called the "log encoding" (for CSP-problems) in [21].

[6] If $2^p = m$, then there is (essentially) only one choice for γ_v, however otherwise the situation is more complicated, and also resolutions are possible between the remaining clauses, shortening these clauses, and these shortened clauses can be used to shorten the main clauses. Therefore we speak of the "simple" translation, and further investigations are needed to find stronger schemes when $m < 2^p$.

the size of the clause-set, by simply assigning the larger clauses to the larger k-values (since the larger the size of arithmetic progressions the fewer there are). Considering our running example $F^{\mathrm{GT}}_{3,3,3}(5)$ we obtain the following 7 translations:

For the direct encoding we get $5 \cdot 3 = 15$ boolean variables $v_{p,i}$, $p \in \{2,3,5,7\}$ and $i \in \{1,2,3\}$. The clause $\{(3,i),(5,i),(7,i)\}$ is replaced by $\{v_{3,i}, v_{5,i}, v_{7,i}\}$ for $i \in \{1,2,3\}$, while clause $\{(3,i),(7,i),(11,i)\}$ is replaced by $\{v_{3,i}, v_{7,i}, v_{11,i}\}$. For the weak translation we have the 5 additional clauses $\{\overline{v_{p,1}}, \overline{v_{p,2}}, \overline{v_{p,3}}\}$ for $p \in \{2,3,5,7,11\}$, while for the strong translation additionally we have the $5 \cdot \binom{3}{2} = 15$ clauses $\{v_{p,i}, v_{p,j}\}$, $p \in \{2,3,5,7,11\}$ and $i,j \in \{1,2,3\}$, $i < j$.

The reduced encoding has $5 \cdot 2 = 10$ boolean variables $v_{p,i}$, $p \in \{2,3,5,7,11\}$, $i \in \{1,2\}$. The clause $\{(3,i),(5,i),(7,i)\}$ is replaced by $\{v_{3,i}, v_{5,i}, v_{7,i}\}$ for $i \in \{1,2\}$ resp. by $\{\overline{v_{3,1}}, \overline{v_{3,2}}, \overline{v_{5,1}}, \overline{v_{5,2}}, \overline{v_{7,1}}, \overline{v_{7,2}}\}$ for $i = 3$, while $\{(3,i),(7,i),(11,i)\}$ is replaced by $\{v_{3,i}, v_{7,i}, v_{11,i}\}$ for $i \in \{1,2\}$ resp. by $\{\overline{v_{3,1}}, \overline{v_{3,2}}, \overline{v_{7,1}}, \overline{v_{7,2}}, \overline{v_{11,1}}, \overline{v_{11,2}}\}$ for $i = 3$. For the weak translation there are no additional clauses, while for the strong translation we have $5 \cdot \binom{2}{2} = 5$ additional binary clauses $\{v_{p,1}, v_{p,2}\}$ for $p \in \{2,3,5,7,11\}$. Note that due to our standardisation scheme the long replacement-clause is uniformly used for $i = 3$, while actually for each of the five (non-boolean) variables $2,3,5,7,11$ one could use a different $i \in \{1,2,3\}$.

The nested encoding has the same variables. $\{(3,i),(5,i),(7,i)\}$ is resp. replaced by $\{v_{3,1}, v_{5,1}, v_{7,1}\}$, $\{\overline{v_{3,1}}, v_{3,2}, \overline{v_{5,1}}, v_{5,2}, \overline{v_{7,1}}, v_{7,2}\}$, $\{\overline{v_{3,1}}, \overline{v_{3,2}}, \overline{v_{5,1}}, \overline{v_{5,2}}, \overline{v_{7,1}}, \overline{v_{7,2}}\}$, while $\{(3,i),(7,i),(11,i)\}$ for $i = 1,2,3$ is replaced by resp. $\{v_{3,1}, v_{7,1}, v_{11,1}\}$, $\{\overline{v_{3,1}}, v_{3,2}, \overline{v_{7,1}}, v_{7,2}, \overline{v_{11,1}}, v_{11,2}\}$, $\{\overline{v_{3,1}}, \overline{v_{3,2}}, \overline{v_{7,1}}, \overline{v_{7,2}}, \overline{v_{11,1}}, \overline{v_{11,2}}\}$. The weak translation has no additional clauses, while for the strong translation we have $5 \cdot \binom{2}{2} = 5$ additional binary clauses $\{v_{p,1}, v_{p,2}\}$ for $p \in \{2,3,5,7,11\}$. Note (again) that due to our standardisation scheme the order of the three replacement-clauses is fixed for each variable, while for each variable one could use one of the $3! = 6$ possible orders.

Finally, for the logarithmic encoding we get (again, but here this is just an exception) $5 \cdot 2 = 10$ boolean variables $v_{p,i}$, $p \in \{2,3,5,7,11\}$, $i \in \{1,2\}$. We use the order $A_2 = \{ \{v_1, v_2\}, \{\overline{v_1}, v_2\}, \{\overline{v_1}, \overline{v_2}\}, \{v_1, \overline{v_2}\} \}$, where the first three clauses are used for the values $i = 1,2,3$. Then the clause $\{(3,i),(5,i),(7,i)\}$ is replaced for $i = 1,2,3$ by $\{v_{3,1}, v_{3,2}, v_{5,1}, v_{5,2}, v_{7,1}, v_{7,2}\}$, $\{\overline{v_{3,1}}, v_{3,2}, \overline{v_{5,1}}, v_{5,2}, \overline{v_{7,1}}, v_{7,2}\}$, $\{\overline{v_{3,1}}, \overline{v_{3,2}}, \overline{v_{5,1}}, \overline{v_{5,2}}, \overline{v_{7,1}}, \overline{v_{7,2}}\}$ respectively, and $\{(3,i),(7,i),(11,i)\}$ is replaced respectively by $\{v_{3,1}, v_{3,2}, v_{7,1}, v_{7,2}, v_{11,1}, v_{11,2}\}$, $\{\overline{v_{3,1}}, v_{3,2}, \overline{v_{7,1}}, v_{7,2}, \overline{v_{11,1}}, v_{11,2}\}$, $\{\overline{v_{3,1}}, \overline{v_{3,2}}, \overline{v_{7,1}}, \overline{v_{7,2}}, \overline{v_{11,1}}, v_{11,2}\}$. Additionally we have the 5 clauses $\{v_{p,1}, \overline{v_{p,2}}\}$.

Somewhat surprisingly, in many cases considered in this paper the weak nested translation turned out to be best (from the above 7 translations considered), for all three types of solvers, look-ahead, conflict-driven and local-search solvers (where for the latter an appropriate algorithm has to be chosen). Only for larger number of colours is the logarithmic translation superior (for local search, with various best algorithms; complete solvers were not successful on any of these instances (with larger number of colours)). In all cases the weak nested translation was superior over the direct translation (weak or strong, for all solver types).

4   Computing Green-Tao Numbers

For trivial GT-numbers as with vdW-numbers we have $\mathrm{grt}_m(2) = m+1$. However the simple GT-numbers are non-trivial: $\mathrm{grt}_1(k)$ is the smallest n such that the first n prime numbers contain an arithmetic progression of size k. Only the values for $2 \leq k \leq 21$ are known, given by the sequence $2, 4, 9, 10, 37, 155, 263, 289, 316,$ $21'966, 23'060, 58'464, 2'253'121, 9'686'320, 11'015'837, 227'225'515, 755'752'809,$ $3'466'256'932, 22'009'064'470, 220'525'414'079$. It seems likely that consideration of GT-numbers for core tuples involving $k \geq 11$ is infeasible (since the first 21966 prime numbers need to be considered just to see the *first* progression of size 11).

Solvers used are the algorithms from the Ubcsat local-search suite ([22]), minisat2 ([6]) for conflict-driven solvers (on our instances either minisat2 was superior or not much worse than all other publicly available conflict-driven solvers, and thus it seems that the optimisations applied to minisat2 in other solvers don't improve performance on our instances), and OKsolver-2002 ([12]), march_pl ([9]) and satz ([20]) for look-ahead solvers. In one (largest) case survey propagation ([4]) was successful (with 708206 clauses of length 5). If not stated otherwise, for all non-boolean cases the weak (standard) nested translation is best (considering complete and incomplete solvers), and if not otherwise stated, for lower bounds rnovelty+ is best. A lower bound stated as "$\geq n$" means that we conjecture that actually equality holds.

We were able to compute five core GT-numbers, for 3 core numbers we have reasonable conjectures, and for 9 core numbers we have hopefully not unreasonable lower bounds. Furthermore we were able to compute 12 extended core GT-numbers, while for 16 cases we have conjectures. Transversal GT-numbers are presented in [18]. 4 binary core GT-numbers $\mathrm{grt}_2(a, b)$ have been computed:

a \\ b	3	4	5	6	7
3	23	79	528	≥ 2072	> 13800
4	-	512	> 4231		
5	-	-	≥ 34309		

For $(5, 5)$ we experienced the only case where survey propagation was successful (converging for $n < 34309$, diverging for $n \geq 34309$). For the other lower bounds adaptnovelty+ is best. OKsolver-2002 is best for $(4, 4)$, while for $(3, 5)$ minisat2 is best, followed by march_pl. One ternary core GT-number $\mathrm{grt}_3(a, b, c)$ has been computed:

a, b \\ c	3	4	5
3, 3	137	≥ 434	> 1989
3, 4	-	> 1662	> 8300
4, 4	-	> 5044	

For $(3, 3, 3)$ the logarithmic translation performed best, with minisat2 fastest, followed by OKsolver-2002. For $(3, 4, 5)$ rnovelty performed best. No core GT-number $\mathrm{grt}_4(a, b, c, d)$ could be computed:

a,b,c d	3	4
$3,3,3$	> 384	> 1052
$3,3,4$	-	> 2750

Extending $(3,3)$ by m 2's, i.e., the numbers $grt_{m+2}(2,\ldots,2,3,3)$:

m	0	1	2	3	4	5	6	7	8	9	10	11	12	13	14
	23	31	39	41	47	53	55	≥ 60	≥ 62	≥ 67	≥ 71	≥ 73	≥ 82	≥ 83	≥ 86

minisat2 is the best complete solver here (also for the other (complete) cases below). For the lower bounds the logarithmic translation is best, with rsaps except for $m = 13$ where walksat-tabu without null-flips is best. Extending $(3,k)$ for $k \geq 4$ by m 2's, i.e., the numbers $grt_{m+2}(2,\ldots,2,3,k)$:

k	m	0	1	2	3	4	5	6
4		79	117	120	128	136	≥ 142	≥ 151
5		528	581	≥ 582	≥ 610			

For $k = 5$, $m = 2$ saps is best, and for $m = 3$ walksat. For $k = 4$, $m = 6$ walksat-tabu with the logarithmic translation is best. Extending $(4,k)$ by m 2's, i.e., the numbers $grt_{m+2}(2,\ldots,2,4,k)$:

k	m	0	1	2
4		512	≥ 553	> 588

sapsnr is best (for the lower bounds). Extending $(3,3,k)$ by m 2's, i.e., the numbers $grt_{m+3}(2,\ldots,2,3,3,k)$:

k	m	0	1	2
3		137	151	≥ 154
4		≥ 434	≥ 453	> 471

5 Open Problems and Outlook

Regarding the generic translation scheme, further extensive experimentation is needed w.r.t. the problem of ordering the values and of mixing translation schemes (recall that every variable can be treated on its own). Also further instances of the generic scheme need to considered, starting with refining the logarithmic translation when the number of values is not a power of 2. Of course, finally some form of understanding needs to be established, and we hope that the generic scheme offers a suitable environment for such considerations.

The translation of transversal extension problems into boolean SAT problems can use cardinality constraints, and this needs to be explored systematically. This includes the special case of transversal extensions of simple tuples, which is basically the hypergraph transversal problem (for these special hypergraphs). See [18] for more information.

A fundamental problem is to improve performance on *unsatisfiable* instances (of complete solvers). The most promising general approach seems to us to systematically study the optimisation of heuristics as outlined in [15]. Investigating the tree-resolution and full-resolution complexity of these instances should be of great interest; we noticed that especially with the `OKsolver-2002` the search trees show remarkable regularities (of a number-theoretical touch, in a kind of "fractal" way). Exploiting the monotone nature of the hypergraph sequences of vdW- or GT-hypergraphs seems also necessary to reach the next level of vdW- or GT-numbers (regarding core parameter tuples), where some first (sporadic) methods one finds in [11].

In general, it seems to us that instances from Ramsey theory, like vdW-instances or GT-instances as considered in this paper, or like the Ramsey-instances (and there are many other families), provide very good benchmarks for SAT solvers, combining the power of systematic creation as for random instances with various types of "structures", where the interplay between these structures and SAT solving should be of great interest and potential.

References

1. Ahmed, T.: Some new van der Waerden numbers and some van der Waerden-type numbers. INTEGERS: Electronic Journal of Combinatorial Number Theory 9, 65–76 (2009)
2. Ahmed, T.: Two new van der Waerden numbers: w(2;3,17) and w(2;3,18). To appear in INTEGERS: Electronic Journal of Combinatorial Number Theory (2010)
3. Biere, A., Heule, M.J.H., van Maaren, H., Walsh, T. (eds.): Handbook of Satisfiability. Frontiers in Artificial Intelligence and Applications, vol. 185. IOS Press, Amsterdam (2009)
4. Braunstein, A., Mézard, M., Zecchina, R.: Survey propagation: An algorithm for satisfiability. Random Structures and Algorithms 27(2), 201–226 (2005)
5. Dransfield, M.R., Liu, L., Marek, V.W., Truszczyński, M.: Satisfiability and computing van der Waerden numbers. The Electronic Journal of Combinatorics 11(#R41) (2004)
6. Eén, N., Sörensson, N.: An extensible SAT-solver. In: Giunchiglia, E., Tacchella, A. (eds.) SAT 2003. LNCS, vol. 2919, pp. 502–518. Springer, Heidelberg (2004)
7. Green, B., Tao, T.: The primes contain arbitrarily long arithmetic progressions. Annals of Mathematics 167(2), 481–547 (2008)
8. Herwig, P.R., Heule, M.J.H., van Lambalgen, P.M., van Maaren, H.: A new method to construct lower bounds for van der Waerden numbers. The Electronic Journal of Combinatorics 14(#R6) (2007)
9. Heule, M.J.H.: SMART solving: Tools and techniques for satisfiability solvers. PhD thesis, Technische Universiteit Delft, ISBN 978-90-9022877-8 (2008)
10. Büning, H.K., Kullmann, O.: Minimal unsatisfiability and autarkies. In: Biere, et al. (eds.) [3], ch. 11, pp. 339–401.
11. Kouril, M., Paul, J.L.: The van der Waerden number $W(2,6)$ is 1132. Experimental Mathematics 17(1), 53–61 (2008)
12. Kullmann, O.: Investigating the behaviour of a SAT solver on random formulas. Technical Report CSR 23-2002, Swansea University, Computer Science Report Series, 119 pages (October 2002), http://www-compsci.swan.ac.uk/reports/2002.html
13. Kullmann, O.: The OK library: Introducing a holistic research platform for (generalised) SAT solving. Studies in Logic 2(1), 20–53 (2009)

14. Kullmann, O.: Constraint satisfaction problems in clausal form: Autarkies and minimal unsatisfiability. Technical Report TR 07-055, version 02, Electronic Colloquium on Computational Complexity (ECCC) (January 2009)
15. Kullmann, O.: Fundaments of branching heuristics. In: Biere, et al. (eds.) [3], ch. 7, pp. 205–244
16. Kullmann, O.: Constraint satisfaction problems in clausal form I: Autarkies and deficiency. In: Fundamenta Informaticae (to appear, 2010)
17. Kullmann, O.: Constraint satisfaction problems in clausal form II: Minimal unsatisfiability and conflict structure. In: Fundamenta Informaticae (to appear, 2010)
18. Kullmann, O.: Exact Ramsey theory: Green-Tao numbers and SAT. Technical Report arXiv:1004.0653v2 [cs.DM], arXiv (April 2010)
19. Landman, B.M., Robertson, A.: Ramsey Theory on the Integers. Student mathematical library, vol. 24. American Mathematical Society, Providence (2003) ISBN 0-8218-3199-2
20. Li, C.M.: A constraint-based approach to narrow search trees for satisfiability. Information Processing Letters 71(2), 75–80 (1999)
21. Prestwich, S.: CNF encodings. In: Biere, et al. (eds.) [3], ch. 2, pp. 75–97
22. Tompkins, D.A.D., Hoos, H.H.: UBCSAT: An implementation and experimentation environment for SLS algorithms for SAT and MAX-SAT. In: Hoos, H. H., Mitchell, D.G. (eds.) SAT 2004. LNCS, vol. 3542, pp. 306–320. Springer, Heidelberg (2005)
23. van der Waerden, B.L.: Beweis einer Baudetschen Vermutung. Nieuw Archief voor Wiskunde 15, 212–216 (1927)
24. Zhang, H.: Combinatorial designs by SAT solvers. In: Biere, et al. (eds.) [3], ch. 17, pp. 533–568

Exact MinSAT Solving[*]

Chu Min Li[1,2], Felip Manyà[3], Zhe Quan[2], and Zhu Zhu[2]

[1] Hunan Normal University, Changsha, China
[2] MIS, Université de Picardie Jules Verne, Amiens, France
[3] Artificial Intelligence Research Institute (IIIA-CSIC), Bellaterra, Spain

Abstract. We present an original approach to exact MinSAT solving based on solving MinSAT using MaxSAT encodings and MaxSAT solvers, and provide empirical evidence that our generic approach is competitive.

1 Introduction

MinSAT is the problem of finding a truth assignment that minimizes the number of satisfied clauses in a CNF formula. Despite that MaxSAT has focused the interest of the SAT community in recent years [LM09], we believe that it is worth studying MinSAT and, in particular, to devise fast exact MinSAT solving techniques. It has both theoretical and practical interest: From the theoretical point of view, we highlight the existing work on approximation algorithms for MinSAT (see [MR96] and the references therein); and from the practical point of view, we emphasize its applicability in areas such as Bioinformatics [GKZ05].

Since there exists no exact MinSAT solver similar to the modern branch-and-bound solvers developed for MaxSAT, we believe that it is worth exploring solving MinSAT using a generic problem solving approach. Our proposal relies on defining efficient and original encodings from MinSAT into MaxSAT, solving the resulting MaxSAT encodings with a modern MaxSAT solver, and then derive a MinSAT optimal solution from a MaxSAT optimal solution. To be more precise, we define three encodings. The first one is a straightforward reduction of MinSAT into Partial MaxSAT. The other two encodings are more involved because they first reduce MinSAT to find an optimal MaxClique solution in a graph that represents the interactions among the clauses of the MinSAT instance, and reduce MaxClique to Partial MaxSAT. The difference between both encodings is that one uses the usual encoding from MaxClique into Partial MaxSAT, while the other uses a novel encoding which is based on identifying a minimum clique partition. Moreover, we performed an empirical investigation that shows that our approach is competitive.

The paper is structured as follows. In Section 2 we define three encoding from Min-SAT into Partial MaxSAT. In Section 3 we report on the empirical evaluation we have conducted in order to evaluate our approach to MinSAT solving. We refer to [LM09] for the basic definitions of MaxSAT, and to [MR96] for the used definitions of graphs.

---

[*] Research supported by Generalitat de Catalunya (2009-SGR-1434), and *Ministerio de Ciencia e Innovación* (CONSOLIDER CSD2007-0022, INGENIO 2010,TIN2007-68005-C04-04, Acción Integrada HA2008-0017). Acknowledgements: The MinSAT problem was originally asked by Laurent Simon to the first author.

O. Strichman and S. Szeider (Eds.): SAT 2010, LNCS 6175, pp. 363–368, 2010.

2   Encodings

Definition 1. *Given a MinSAT instance I consisting of the clause set $C_I = \{C_1, \ldots, C_m\}$ and variable set X_I, the direct MaxSAT encoding of I is defined as follows: (i) The set of propositional variables is $X_I \cup \{c_1, \ldots, c_m\}$, where $\{c_1, \ldots, c_m\}$ is a set of auxiliary variables; (ii) for every clause $C_i \in C_I$, the hard clause $c_i \leftrightarrow C_i$ is added; and (iii) for every auxiliary variable c_i, the unit soft clause $\neg c_i$ is added.*

The hard part establishes that c_i is true iff C_i is satisfied. In the soft part, the number of unsatisfied clauses in C_I is maximized. Therefore, if the auxiliary variables set to true in an optimal solution of the MaxSAT encoding are $\{c_{i_1}, \ldots, c_{i_k}\}$, then $\{C_{i_1}, \ldots, C_{i_k}\}$ is a minimum set of satisfied clauses in the MinSAT instance I.

Once we have defined the encoding based on reducing MinSAT to Partial MaxSAT, we define two encodings based on reducing MinSAT to MaxClique, and then Max-Clique to Partial MaxSAT. First, we introduce the concept of auxiliary graph:

Let I be a MinSAT instance consisting of the clause set C_I and variable set X_I. The *auxiliary graph* $G_I(V_I, E_I)$ corresponding to I is constructed as follows: the vertex set V_I is in one-to-one correspondence with the clause set C_I; in the sequel, vertex c_i corresponds to clause C_i. For any two vertices c_i and c_j in V_I, the edge $\{c_i, c_j\}$ is in E_I iff the corresponding clauses C_i and C_j are such that there is a variable $x \in X_I$ that appears in uncomplemented form in C_i and in complemented form in C_j, or vice versa. The *complement of an auxiliary graph* G_I is a graph $\overline{G_I}$ on the same vertices such that two vertices of $\overline{G_I}$ are adjacent iff they are not adjacent in G_I.

It was proved in [MR96] that the number of clauses of a MinSAT instance I satisfied by an optimal assignment is equal to the cardinality of a minimum vertex cover for the auxiliary graph G_I. On the other hand, the set of vertices not belonging to a maximum clique in the complement graph $\overline{G_I}$ are a minimum vertex cover of G_I. This follows from the fact that, for any graph $G(V, E)$, $V' \subseteq V$ is a vertex cover iff $V - V'$ in a clique in the complement graph of G. Therefore, the number of clauses of a MinSAT instance I satisfied by an optimal assignment is equal to the total number of vertices minus the cardinality of a maximum clique in $\overline{G_I}$.

Moreover, an optimal assignment for the MinSAT instance I can be derived from a minimum vertex cover for G_I as follows [MR96]: The variables occurring in clauses corresponding to vertices not belonging to the minimum vertex cover must be assigned in such a way that these clauses are set to false, which is possible because, by construction of G_I, these clauses do not contain both x and $\bar{x}$ for any $x \in X_I$; the rest of variables are assigned to an arbitrary value. Therefore, we can derive an optimal assignment for the MinSAT instance I from a maximum clique of $\overline{G_I}$: The variables occurring in clauses corresponding to vertices belonging to the maximum clique must be assigned in such a way that these clauses are set to false; the rest of variables are assigned to an arbitrary value. Therefore, in order to find an optimal assignment for the MinSAT instance I, we can search for a maximum clique in $\overline{G_I}$. The MaxClique problem for $\overline{G_I}$ can be naturally encoded into a Partial MaxSAT problem as follows.

Definition 2. *Given a MinSAT instance I consisting of the clause set $C_I = \{C_1, \ldots, C_m\}$, the MaxClique-based MaxSAT encoding of I is defined as follows: (i) The set of propositional variables is $\{c_1, \ldots, c_m\}$; (ii) for every two clauses C_i, C_j in C_I such* *that C_i contains an occurrence of a literal l and C_j contains an occurrence of $\neg l$, the hard clause $\neg c_i \vee \neg c_j$ is added; and (iii) for every propositional variable c_i, the unit soft clause c_i is added.*

Observe that there is a hard clause for every two non-adjacent vertices in $\overline{G_I}$ encoding that the two vertices cannot be in the same clique, and a soft unit clause for every vertex in $\overline{G_I}$. The set of propositional variables evaluated to true in an optimal assignment of the resulting Partial MaxSAT problem forms a maximum clique in $\overline{G_I}$.

The MaxClique-based MaxSAT encoding can be improved by reducing the number of soft clauses by taking into account the following fact: If the auxiliary graph of a MinSAT instance contains a clique $C = \{c_{i_1}, \ldots, c_{i_k}\}$, then the hard part contains the clauses $\neg c_{i_j} \vee \neg c_{i_k}$ for all i, j such that $1 \leq i < j \leq k$. Observe that these clauses encode the following at-most-one condition: There is at most one literal in $\{c_{i_1}, \ldots, c_{i_k}\}$ that evaluates to true. In other words, they encode that any feasible solution assigns to true at most one variable of the subset of variables $\{c_{i_1}, \ldots, c_{i_k}\}$. Therefore, we can replace the k unit soft clauses $c_{i_1}, \ldots, c_{i_k}$ with the soft clause $c_{i_1} \vee \cdots \vee c_{i_k}$ because the number of satisfied clauses is the same in both encodings for any feasible assignment.

Definition 3. *Let I be a MinSAT instance, let $G_I(V_I, E_I)$ be the auxiliary graph of I, and let $V_1, \ldots, V_k$ be a clique partition of G_I. The improved MaxClique-based MaxSAT encoding of I is obtained from the MaxClique-based MaxSAT encoding of Definition 2 by replacing, for every clique $V_i = \{c_{i_1}, \ldots, c_{i_k}\}$ in the partition, the soft unit clauses $c_{i_1}, \ldots, c_{i_k}$ with the soft unit clause $c_{i_1} \vee \cdots \vee c_{i_k}$.*

Example 1. Let I be the MinSAT instance $\{C_1, C_2, C_3, C_4, C_5\}$, where $C_1 = a \vee b$, $C_2 = a \vee \neg b$, $C_3 = \neg a \vee b$, $C_4 = \neg a \vee \neg b$, and $C_5 = a \vee c$. The direct MaxSAT encoding of I is formed by the hard clauses $c_1 \leftrightarrow a \vee b$, $c_2 \leftrightarrow a \vee \neg b$, $c_3 \leftrightarrow \neg a \vee b$, $c_4 \leftrightarrow \neg a \vee \neg b$, $c_5 \leftrightarrow a \vee c$, and the soft clauses $\neg c_1, \neg c_2, \neg c_3, \neg c_4, \neg c_5$. The MaxClique-based MaxSAT encoding of I is formed by the hard clauses $\neg c_1 \vee \neg c_2, \neg c_1 \vee \neg c_3, \neg c_1 \vee \neg c_4, \neg c_2 \vee \neg c_3, \neg c_2 \vee \neg c_4, \neg c_3 \vee \neg c_4, \neg c_3 \vee \neg c_5, \neg c_4 \vee \neg c_5$, and the soft unit clauses c_1, c_2, c_3, c_4, c_5. Since $\{\{c_1, c_2, c_3, c_4\}, \{c_5\}\}$ is a clique partition of the auxiliary graph of I, an improved MaxClique-based MaxSAT encoding of I is obtained by replacing the soft clauses of the previous encoding with the following soft clauses: $c_1 \vee c_2 \vee c_3 \vee c_4, c_5$.

Since finding a minimum clique partition is NP-hard, we propose to apply a heuristic clique partition method for deriving the improved encoding. The heuristic used in our empirical investigation is Algorithm 1. Given a graph G and a list P of disjoint cliques in G, where P is initially empty, let #c(v) denote the number of cliques in P in which vertex v can be inserted to enlarge them. Algorithm 1 selects the vertex v with the minimum #c(v), breaking ties in favor of the vertex with the smallest degree (i.e., with the minimum number of adjacent vertices). Let C be a clique in P, and let v_i and v_j be two non-adjacent vertices that can be respectively inserted into C to get a larger clique, observe that the insertion of v_i into C prevents v_j from being inserted into C, and vice versa. The intuition behind the selection of v in Algorithm 1 is that the most constrained vertex (i.e., with the fewest possibilities in P) is inserted first, avoiding to create a new clique for this vertex after inserting other vertices. The selected v is then inserted into a clique in P to enlarge this clique. If this clique does not exist, a new clique is created and v is inserted there.

Algorithm 1. cliquePartition(G)

Input: A graph $G=(V, E)$
Output: A clique partition of G

1 **begin**
2     $P \leftarrow \emptyset$;
3   **while** G *is not empty* **do**
4   For each vertex v in G, compute the number #c of cliques in P in which v is adjacent to all vertices;
5        $v \leftarrow$ the vertex of G with the minimum #c, breaking ties in favor of the vertex with the smallest degree;
6   remove v from G;
7   **if** *there is a clique C in P in which v is adjacent to all vertices* **then**
8   insert v into C;
9   **else**
10   create a new clique C;
11   insert v into C;
12           $P \leftarrow P \cup \{C\}$;
13   return P;
14 **end**

---

3   Experimental Results

We conducted an empirical comparison of the three proposed encodings on several MaxSAT solvers, and also compared our results with the results obtained by solving MinSAT with two of the best performing state-of-the-art exact MaxClique solvers.

The MaxSAT solvers used in our empirical investigation are the versions that that participated in the 2009 MaxSAT Evaluation of the following solvers: WBO [MSP09], MaxSatz [LMP07][1], PM2 [ABL09], and SAT4J-Maxsat[2]. The MaxClique solvers used in our empirical investigation are: MaxCliqueDyn [KJ07][3], and Cliquer [Ost02][4].

The MaxClique solvers MaxCliqueDyn and Cliquer solve a MinSAT instance I by computing a maximum clique in the complement auxiliary graph $\overline{G_I}$. The MaxSAT solvers MaxSatz, WBO, PM2, and SAT4J-Maxsat solve I using the three encodings of MinSAT into Partial MaxSAT defined in the previous sections.

As benchmarks, we used randomly generated Min-2-SAT and Min-3-SAT instances. The number of variables in the instances ranged from 40 to 100, and the clause-to-variable ratios solved for Min-2-SAT were 1 and 3, and the clause-to-variable (C/V)

---

[1] We introduced a small modification in the heuristic used to select the variables to which failed literal detection is applied in the lower bound computation: instead of applying it to all the variables having both at least two negative and two positive occurrences in binary clauses, it is now applied to all the variables having at least one positive occurrence and at least two negative occurrences in binary clauses, and to all the variables having at least one negative occurrence and at least two positive occurrences in binary clauses.

[2] http://www.sat4j.org/

[3] We got the source code of this solver from D. Janezic in January 2010.

[4] We used the last publicly available version of this solver: http://users.tkk.fi/pat/cliquer.html

Table 1. Number of solved instances and mean time (seconds) of MaxSatz, MaxCliqueDyn (Dyn), Cliquer (Clq), PM2, WBO and SAT4J-Maxsat on random Min-2-SAT

instance		MaxSatz			Dyn	Clq	PM2			WBO			sat4j-maxsat		
#var	C/V	E1	E2	E3			E1	E2	E3	E1	E2	E3	E1	E2	E3
90	1.0	0.58	9.38	**0.00**	0.02	**0.00**	0.01	**0.00**	**0.00**	0.02	0.01	**0.00**	2710	1995	0.08
		(50)	*(50)*	***(50)***	*(50)*	***(50)***	*(50)*	***(50)***	***(50)***	*(50)*	*(50)*	***(50)***	*(24)*	*(45)*	*(50)*
100	1.0	0.26	27.34	**0.00**	0.02	**0.00**	0.01	0.01	**0.00**	0.02	0.01	**0.00**	2596	6358	0.11
		(50)	*(50)*	***(50)***	*(50)*	***(50)***	*(50)*	*(50)*	***(50)***	*(50)*	*(50)*	***(50)***	*(14)*	*(12)*	*(50)*
40	3.0	0.12	52.95	0.01	0.03	0.01	0.19	0.51	**0.00**	0.07	0.52	**0.00**	9904	-	0.20
		(50)	*(50)*	*(50)*	*(50)*	*(50)*	*(50)*	*(50)*	***(50)***	*(50)*	*(50)*	***(50)***	*(4)*	*(0)*	*(50)*
50	3.0	0.89	2053	0.01	0.04	1.23	5.25	7.89	**0.00**	42.35	51.12	0.01	-	-	0.43
		(50)	*(48)*	*(50)*	*(50)*	*(50)*	*(50)*	*(50)*	***(50)***	*(50)*	*(50)*	*(50)*	*(0)*	*(0)*	*(50)*
60	3.0	7.42	-	0.02	0.07	15.88	49.35	79.77	**0.01**	207.6	93.17	0.20	-	-	2.41
		(50)	*(0)*	*(50)*	*(50)*	*(50)*	*(50)*	*(50)*	***(50)***	*(50)*	*(50)*	*(50)*	*(0)*	*(0)*	*(50)*
70	3.0	43.25	-	0.03	0.13	68.01	144.3	135.8	**0.01**	225.1	143.9	0.08	-	-	2.42
		(50)	*(0)*	*(50)*	*(50)*	*(50)*	*(50)*	*(50)*	***(50)***	*(50)*	*(50)*	*(50)*	*(0)*	*(0)*	*(50)*
80	3.0	201.2	-	0.05	0.25	302.4	199.9	231.8	**0.02**	416.8	432.9	0.70	-	-	3.35
		(50)	*(0)*	*(50)*	*(50)*	*(50)*	*(50)*	*(50)*	***(50)***	*(50)*	*(50)*	*(50)*	*(0)*	*(0)*	*(50)*
90	3.0	1355	-	0.08	0.91	895.5	353.1	330.3	**0.05**	558.8	295.1	1.37	-	-	5.38
		(49)	*(0)*	*(50)*	*(50)*	*(42)*	*(50)*	*(50)*	***(50)***	*(50)*	*(50)*	*(50)*	*(0)*	*(0)*	*(50)*
100	3.0	3258	-	0.11	1.51	1185	256.7	455.3	**0.09**	548.5	665.9	29.27	-	-	96.91
		(46)	*(0)*	*(50)*	*(50)*	*(44)*	*(10)*	*(50)*	***(50)***	*(50)*	*(50)*	*(50)*	*(0)*	*(0)*	*(50)*

ratios solved for Min-3-SAT were 4.25 and 5. Experiments were performed on a Macpro with a 2.8Ghz intel Xeon processor and 4 Gb memory with MAC OS X 10.5. The cut-off time was set to 3 hours.

Table 1 and Table 2 contain the experimental results obtained when solving the Min-2-SAT and Min-3-SAT instances with the MaxSAT and MaxClique solvers. The tables show, for each solver, the number of instances solved within 3 hours (between brackets) in a set of 50 instances at each point, and the mean time needed to solve these solved instances. Table 1 shows results just for 90-variable and 100-variable instances when $C/V = 1$ because they can be easily solved and we have no space. The MaxSAT solvers not included in Table 2 were far from being competitive on Min-3-SAT instances. The three encodings are denoted by $E1$ (Encoding 1), $E2$ (Encoding 2), and $E3$ (Encoding 3).

The experimental results show that the direct encoding of MinSAT into Partial MaxSAT (Encoding 1) is better than the MaxClique based encoding (Encoding 2) for all the MaxSAT solvers. However, when the MaxClique-based encoding is improved using a good clique partition of G_I, then the improved MaxClique-based encoding (Encoding 3) is by far the best performing encoding for all the MaxSAT solvers. More importantly, Encoding 3 even makes MaxSatz significantly better for computing a maximum clique in $\overline{G_I}$ than the state-of-the-art specific MaxClique solvers MaxCliqueDyn and Cliquer.

Table 2. Number of solved instances and mean time (seconds) of MaxSatz, MaxCliqueDyn (Dyn in the table) and Cliquer on random Min-3-SAT

instance		MaxSatz			Dyn	Cliquer
#var	C/V	E1	E2	E3		
40	4.25	4.67(50)	992.5(50)	0.28(50)	**0.12(50)**	15.94(50)
50	4.25	75.6(50)	- (0)	1.57(50)	**0.98(50)**	945.0(49)
60	4.25	1153(50)	- (0)	**8.31(50)**	9.94(50)	5385(10)
70	4.25	5989(5)	- (0)	**42.77(50)**	106.4(50)	- (0)
80	4.25	- (0)	- (0)	**186.3(50)**	917.4(50)	- (0)
90	4.25	- (0)	- (0)	**760.4(50)**	4453(41)	- (0)
100	4.25	- (0)	- (0)	**2819(26)**	- (0)	- (0)
40	5.00	10.87(50)	5693(48)	0.80(50)	**0.30(50)**	63.98(50)
50	5.00	226.8(50)	- (0)	5.24(50)	**3.97(50)**	3035(35)
60	5.00	3803(48)	- (0)	**39.3(50)**	60.14(50)	- (0)
70	5.00	- (0)	- (0)	**243.4(50)**	735.2(50)	- (0)
80	5.00	- (0)	- (0)	**1512(50)**	6355(41)	- (0)
90	5.00	- (0)	- (0)	**5167(39)**	- (0)	- (0)
100	5.00	- (0)	- (0)	- (0)	- (0)	- (0)

References

[ABL09]   Ansótegui, C., Bonet, M.L., Levy, J.: Solving (weighted) partial MaxSAT through satisfiability testing. In: Kullmann, O. (ed.) SAT 2009. LNCS, vol. 5584, pp. 427–440. Springer, Heidelberg (2009)

[GKZ05]   Goldstein, A., Kolman, P., Zheng, J.: Minimum common string partition problem: Hardness and approximations. Electr. J. Comb. 12 (2005)

[KJ07]   Konc, J., Janezic, D.: An improved branch and bound algorithm for the maximum clique problem. Communications in Mathematical and in Computer Chemistry 58, 569–590 (2007)

[LM09]   Li, C.M., Manyà, F.: Max-SAT, hard and soft constraints. In: Biere, A., van Maaren, H., Walsh, T. (eds.) Handbook of Satisfiability, pp. 613–631. IOS Press, Amsterdam (2009)

[LMP07]   Li, C.M., Manyà, F., Planes, J.: New inference rules for Max-SAT. Journal of Artificial Intelligence Research 30, 321–359 (2007)

[MR96]   Marathe, M.V., Ravi, S.S.: On approximation algorithms for the minimum satisfiability problem. Information Processing Letters 58, 23–29 (1996)

[MSP09]   Manquinho, V.M., Silva, J.P.M., Planes, J.: Algorithms for weighted boolean optimization. In: Kullmann, O. (ed.) SAT 2009. LNCS, vol. 5584, pp. 495–508. Springer, Heidelberg (2009)

[Ost02]   Ostergard, P.R.J.: A fast algorithm for the maximum clique problem. Discrete Applied Mathematics 120, 197–207 (2002)

Uniquely Satisfiable k-SAT Instances with Almost Minimal Occurrences of Each Variable

William Matthews and Ramamohan Paturi[*]

University of California, San Diego

Abstract. Let (k, s)-SAT refer the family of satisfiability problems restricted to CNF formulas with exactly k distinct literals per clause and at most s occurrences of each variable. Kratochvíl, Savický and Tuza [6] show that there exists a function $f(k)$ such that for all $s \leq f(k)$, all (k, s)-SAT instances are satisfiable whereas for $k \geq 3$ and $s > f(k)$, (k, s)-SAT is NP-complete. We define a new function $u(k)$ as the minimum s such that uniquely satisfiable (k, s)-SAT formulas exist. We show that for $k \geq 3$, unique solutions and NP-hardness occur at almost the same value of s: $f(k) \leq u(k) \leq f(k) + 2$.

We also give a parsimonious reduction from SAT to (k, s)-SAT for any $k \geq 3$ and $s \geq f(k) + 2$. When combined with the Valiant–Vazirani Theorem [8], this gives a randomized polynomial time reduction from SAT to UNIQUE-(k, s)-SAT.

1 Introduction

Let (k, s)-SAT refer to the family of satisfiability problems restricted to CNF formulas with exactly k distinct literals per clause and at most s occurrences of each variable. Since $(2, s)$-SAT is in P for all s, we restrict our attention to $k \geq 3$.

Tovey [7] first observed that $(3, 3)$-SAT was trivial since every instance is satisfiable, and showed that $(3, 4)$-SAT was NP-hard. This was generalized to larger k by Kratochvíl, Savický and Tuza [6] who showed that for each $k \geq 4$ there exists a threshold $f(k)$ such that for all $s \leq f(k)$, (k, s)-SAT is trivial whereas for all $s > f(k)$, (k, s)-SAT is NP-hard.

Using Hall's Theorem, Tovey [7] showed that every (k, k)-SAT instance is satisfiable, giving the first lower bound $f(k) \geq k$. This was improved by Kratochvíl, Savický and Tuza [6] who used the Lovász local lemma to show that all $(k, \lfloor 2^k/ek \rfloor)$-SAT instances are satisfied by random assignments with positive probability, implying $f(k) \geq \lfloor 2^k/ek \rfloor$.

Trivially, $f(k) < 2^k$ since enumerating all 2^k possible clauses for k variables gives an unsatisfiable formula. Kratochvíl, Savický and Tuza [6] proved that $f(k + 1) \leq 2f(k) + 1$. Combined with the fact that $f(3) = 3$, we get $f(k)$

[*] This research is supported by NSF grant CCF-0947262 from the Division of Computing and Communication Foundations. Any opinions, findings and conclusions or recommendations expressed in this material are those of the authors and do not necessarily reflect the views of the National Science Foundation.

O. Strichman and S. Szeider (Eds.): SAT 2010, LNCS 6175, pp. 369–374, 2010.
© Springer-Verlag Berlin Heidelberg 2010

$\leq 2^{k-1} - 1$ (this may be improved slightly by using a base case of $f(k)$ for larger k). Subsequently, this has been improved to $f(k) = \Theta\left(2^k/k\right)$ [4]. However the exact value of $f(k)$, or even whether $f(k)$ is computable, remains unknown.

Valiant and Vazirani [8] showed that deciding whether a SAT formula has zero or one solution is essentially as hard as SAT in general. In particular, they prove the following theorem:

Theorem 1 (Valiant–Vazirani Theorem [8]). *There exists a randomized polynomial time reduction from SAT to UNIQUE-SAT.*

By the standard parsimonious reduction from SAT to k-SAT, the Valiant–Vazirani Theorem implies the same hardness for UNIQUE-k-SAT. However, what happens when the number of occurrences of each variable is also limited? Specifically, what can be said about UNIQUE-(k, s)-SAT for various values of s?

We give a parsimonious reduction from 3-SAT to (k, s)-SAT, for any $k \geq 3$ and $s \geq f(k) + 2$. Thus, UNIQUE-$(k, f(k) + 2)$-SAT is as hard as UNIQUE-SAT. In contrast, UNIQUE-$(k, f(k))$-SAT is trivial since every formula is satisfiable.

Calabro et al. [3] give additional evidence that UNIQUE-k-SAT is no easier than k-SAT, not just for polynomial time algorithms (as shown by Valiant and Vazirani), but for super-polynomial time algorithms. They show that if UNIQUE-3-SAT is in randomized subexponential time $(\cap_{\epsilon > 0} \mathrm{RTIME}(2^{\epsilon n}))$, then so is k-SAT for all $k \geq 3$. Our parsimonious reduction from 3-SAT to (k, s)-SAT combined with their result implies that if UNIQUE-(k, s)-SAT is in randomized subexponential time for some $k \geq 3$ and $s \geq f(k) + 2$, then so is k'-SAT for all $k' \geq 3$. We omit the details which follow fairly straightforwardly from [3,5].

A key component in our reduction is a construction of uniquely satisfiable $(k, s + 1)$-SAT formulas from unsatisfiable (k, s)-SAT formulas. Starting with unsatisfiable $(k, f(k) + 1)$-SAT formulas allows us work with uniquely satisfiable formulas with almost the minimum number of occurrences of each variable, and also argue about the transition where uniquely satisfiable formulas first occur. Since the smallest s we argue about for each k is $f(k) + 2$, the questions of whether there exists a uniquely satisfiable $(k, f(k) + 1)$-SAT formula and the complexity of UNIQUE-$(k, f(k) + 1)$-SAT remain open.

Since our reduction requires the existence of an unsatisfiable (k, s)-SAT formula, we require that k and $s > f(k)$ be constants. In this case we know there exists an unsatisfiable formula of constant size. If we could give an upper bound on the size of this formula in terms of k and s it would imply that $f(k)$ was computable, which would be an independently interesting result.

Let $F_k(n, m)$ refer to a random k-SAT formula with n variables and m clauses. Just as there is a transition in (k, s)-SAT as s increases from trivial to NP-hard, there is a similar transition in $F_k(n, rn)$ as r increases from satisfiable with high probability to unsatisfiable w.h.p. (See [1] and its references). It is conjectured that for each k the transition occurs at a sharp threshold r_k. Achiloptas and Ricci-Tersenghi [2] show that for sufficiently large k and $r < r_k$, w.h.p., $F_k(n, rn)$ has exponentially many, widely separated, small clusters of solutions. In some ways, small, widely separated clusters of solutions are similar to unique solutions. In both cases, they seem to be some of the hardest instances for algorithms. While we don't consider random SAT formulas in this paper, we view the similarities as additional motivation.

2   Definitions and Results

Definition 1. *We let SAT refer to the satisfiability problem restricted to formulas in conjunctive normal form; k-SAT to SAT restricted to formulas with exactly k literals per clause; and (k, s)-SAT to k-SAT restricted to formulas where each variable occurs in at most s clauses.*

Definition 2. *We let UNIQUE-SAT refer to the promise problem of deciding whether a SAT formula is unsatisfiable or has a unique satisfying assignment. UNIQUE-k-SAT and UNIQUE-(k, s)-SAT are defined similarly.*

Definition 3 (Valiant and Vazirani [8]). *A randomized polynomial time reduction M from a problem A to a problem B is a randomized polynomial time Turing machine such that for all $x \notin A$ we are guaranteed that $M(x) \notin B$, and for all $x \in A$ we get $M(x) \in B$ with probability at least $1/poly(|x|)$.*

Definition 4. *In the context of SAT, a reduction M is said to be* parsimonious *if the formulas x and $M(x)$ have the same number of satisfying assignments.*

In particular, parsimonious reductions preserve the existence of unique satisfying assignments.

Definition 5 (Kratochvíl, Savický and Tuza [6]). *For each $k \geq 3$, $f(k)$ is defined as the largest value of s such that all (k, s)-SAT instances are satisfiable.*

Equivalently, we may think of $f(k) + 1$ as the smallest value of s such that there exist unsatisfiable (k, s)-SAT instances.

Definition 6. *For each $k \geq 3$, we define $u(k)$ as the smallest value of s such that there exist (k, s)-SAT instances with exactly one satisfying assignment.*

Theorem 2. *For all $k \geq 3$, $f(k) \leq u(k) \leq f(k) + 2$.*

This theorem follows directly from the following two lemmas:

Lemma 1. *For all $k \geq 3$ and $s \geq u(k)$, there exist unsatisfiable $(k, s + 1)$-SAT instances.*

Lemma 2. *For all $k \geq 3$ and $s \geq f(k) + 1$, there exist uniquely satisfiable $(k, s + 1)$-SAT instances.*

To prove that $(k, f(k) + 1)$-SAT is NP-hard, Kratochvíl, Savický and Tuza [6] give a reduction from k-SAT to (k, s)-SAT for any $s > f(k)$. Combining their proof with Lemma 2 we get the following lemma:

Lemma 3. *For any constants $k \geq 3$ and $s \geq f(k) + 2$, there is a parsimonious polynomial time reduction from 3-SAT to (k, s)-SAT.*

Composing the Valiant–Vazirani Theorem (Theorem 1), the standard parsimonious reduction from SAT to 3-SAT, and Lemma 3, we get the following:

Corollary 1. *For any constants $k \geq 3$ and $s \geq f(k) + 2$, there is a randomized polynomial time reduction from SAT to UNIQUE-(k, s)-SAT.*

3   Proofs

Proof (Lemma 1). Since $s \geq u(k)$, there exists a uniquely satisfiable (k, s)-SAT formula F. Add a single clause to F which is violated by the unique satisfying assignment. We add at most 1 occurrence of each variable, so this gives an unsatisfiable $(k, s + 1)$-SAT formula.   □

To prove Lemma 2, we will construct a uniquely satisfiable $(k, s + 1)$-SAT formula in a sequence of steps from an unsatisfiable (k, s)-SAT formula. We classify variables in each of these formulas as either *forced* or *unforced*. If every satisfying assignment for a formula sets a variable to the same value, we say that the variable is forced. Otherwise, we say that the variable is unforced. We will be particularly interested in forced variables that occur exactly once in the formula. Without loss of generality, we will always assume that forced variable must be set to false in all satisfying assignments (otherwise replace every occurrence of the variable with its negation). Note that uniquely satisfiable formulas are equivalent to formulas where every variable is forced.

Our construction can be broken down into 3 steps formalized by the following lemmas: Lemma 4 constructs a formula with a few forced variables. Lemma 5 increases the number of forced variables without increasing the number of unforced variables. Lemma 6 uses the newly created forced variables to force all of the unforced variables.

Lemma 4. *We can transform an unsatisfiable (k, s)-SAT formula into a satisfiable (k, s)-SAT formula with k forced variables that only occur once.*

Lemma 5. *We can transform a (k, s)-SAT formula with n unforced variables and $t \geq k - 1$ forced variables that only occur once (and possibly other variables that are forced but occur more than once) into a (k, s)-SAT formula with n unforced variables and $t + (s - k) > t$ forced variables that only occur once.*

Lemma 6. *We can transform a (k, s)-SAT formula with n unforced variables and at least $n + k$ forced variables that only occur once into a $(k, s + 1)$-SAT formula where every variable is forced.*

Proof (Lemma 4). Let F be a minimal unsatisfiable (k, s)-SAT formula. (A formula is minimally unsatisfiable if removing any clause would make it satisfiable.) Transform F by renaming variables and replacing variables with their negations so that F can be written as $(x_1 \vee x_2 \vee \cdots \vee x_k) \wedge G$, where G is satisfied by the all-false assignment. Within G, the variables $x_1, \ldots, x_k$ each occur at most $s - 1$

times and are forced to false (any satisfying assignment that didn't set them to false would also satisfy F).

Let $G^{(1)}, \ldots, G^{(k-1)}$ be $k-1$ disjoint copies of G. Let $x_1^{(i)}, \ldots, x_k^{(i)}$ denote the copy of $x_1, \ldots, x_k$ occurring in $G^{(i)}$. Return the formula $G^{(1)} \wedge \cdots \wedge G^{(k-1)} \wedge H$, where $H = \bigwedge_{i=1}^{k}(x_i^{(1)} \vee \cdots \vee x_i^{(k-1)} \vee \overline{y_i})$ and $y_1, \ldots, y_k$ are fresh variables. Each variable y_i occurs in exactly 1 clause and must be set to false to satisfy that clause since all of the other variables in the clause are already forced. $\square$

Proof (Lemma 5). Let G be a (k, s)-SAT formula with n unforced variables and $t \geq k - 1$ forced variables that only occur once. Let $y_1, \ldots, y_{k-1}$ denote $k-1$ of these forced variables. Let $H = \bigwedge_{i=1}^{s-1}(y_1 \vee \cdots \vee y_{k-1} \vee \overline{z_i})$, where $z_1, \ldots z_{s-1}$ are fresh variables. Return the formula $G \wedge H$. Each of the variables $y_1, \ldots, y_{k-1}$ is still forced, but now each occurs s times. In their place, we have $s - 1$ new forced variables $z_1, \ldots, z_{s-1}$ which each only occur once, for a total of $t + (s - k)$ such variables. Whether other variables are forced remains unchanged. Note that $s > f(k) \geq k$ since unsatisfiable (k, s)-SAT instances exist. $\square$

Proof (Lemma 6). Let F be a (k, s)-SAT formula with n unforced variables and $n + k$ forced variables that only occur once. Let $x_1, \ldots, x_n$ denote the unforced variables. Let $y_1, \ldots, y_{n+k}$ denote the forced variables that only occur once. Let $m = \lceil \frac{n}{k-1} \rceil$. Arbitrarily partition the variables $x_1, \ldots, x_n$ into m sets $X_1, \ldots, X_m$ of size $k - 1$. Add new variables as needed so that every set contains exactly $k - 1$ variables. Arbitrarily partition the variables $y_1, \ldots, y_{(k-1)m}$ into m sets $Y_1, \ldots, Y_m$ of size $k - 1$.

For each $1 \leq i \leq m$, we will construct a formula H_i using the variables in sets X_i and Y_i. For simplicity, let $X_i = \{x_1, \ldots, x_{k-1}\}$ and $Y_i = \{y_1, \ldots, y_{k-1}\}$. For each i, let $H_i = \bigwedge_{j=1}^{k-1}(y_1 \vee \cdots \vee y_{k-1} \vee \overline{x_j})$.

Return the formula $F \wedge H_1 \wedge \cdots \wedge H_m$. Each H_i uses the variables in Y_i to force the variables in X_i. Since each variable in Y_i is forced, the variables in X_i must be false to satisfy the clauses in H_i. This adds $k - 1$ occurrences for each variable in Y_i and one occurrence for each variable in X_i. Each variable in Y_i now occurs $k < s$ times and each variable in X_i now occurs at most $s + 1$ times. $\square$

Proof (Lemma 2). Since $s \geq f(k) + 1$, there exists an unsatisfiable (k, s)-SAT formula F. Use Lemma 4 to construct a (k, s)-SAT formula G with k forced variables that only occur once. Let n denote the number of unforced variables in G. Use Lemma 5 sufficiently many times starting with G to get a formula H with at least $n + k$ forced variables that only occur once. Note that H still contains only n unforced variables. Using Lemma 6 on H gives a uniquely satisfiable $(k, s + 1)$-SAT formula. $\square$

By repeating Lemma 5 sufficiently many additional time before using Lemma 6, we get the following corollary:

Corollary 2. *For any constants $k \geq 3$ and $s \geq f(k) + 2$, and any $m \geq 0$, we can construct a uniquely satisfiable (k, s)-SAT formula with at least m forced variables that only occur once in time polynomial in m.*

The following proof of Lemma 3 is the same as the reduction given by Kratochvíl, Savický and Tuza [6] to prove that $(k, f(k) + 1)$-SAT is NP-hard with one exception. We use Corollary 2 to supply forced variables whereas they used a $(k, f(k) + 1)$-SAT formula with potentially many satisfying assignments.

Proof (Lemma 3). For any $k \geq 3$ and $s \geq f(k)+2$, we transform a 3-SAT formula F parsimoniously into a (k, s)-SAT formula in 2 steps:

First, we reduce the number of occurrences of each variable to at most s, which introduces additional 2-variable clauses. For each variable x occurring $t > s$ times, replace each occurrence of x with a new variable x_i, $1 \leq i \leq t$. Add clauses $(\overline{x_i} \vee x_{i+1})$ for $1 \leq i \leq t - 1$, and $(\overline{x_t} \vee x_1)$. These clauses ensure that in any satisfying assignment all the variables x_i are assigned the same value. Thus, we maintain exactly the same number of satisfying assignments. Each of these new variables occurs exactly $3 < s$ times. Let G denote the resulting formula, and m the number of clauses in G.

Second, we pad each clause with forced variables so that all clauses contain exactly k variables. Using Corollary 2, there exists a (k, s)-SAT formula H with at least mk forced variables that only occur once. For each clause c of length $\ell < k$ in G, replace c with $(c \vee y_1 \vee \cdots \vee y_{k-\ell})$, where $y_1, \ldots, y_{k-\ell}$ are arbitrary forced variables from H occurring fewer than s times. Let G' denote the result of these replacements. Return the formula $G' \wedge H$. Since the only satisfying assignment to H sets all variables to false, the padded clauses in G' are satisfied by exactly the same assignments that satisfy G. Thus, $G' \wedge H$ has exactly the same number of satisfying assignments as G. □

References

1. Achlioptas, D., Peres, Y.: The threshold for random k-SAT is $2^k \ln 2 - O(k)$. J. AMS 17, 947–973 (2004)
2. Achlioptas, D., Ricci-Tersenghi, F.: On the solution-space geometry of random constraint satisfaction problems. In: Kleinberg, J.M. (ed.) STOC, pp. 130–139. ACM, New York (2006)
3. Calabro, C., Impagliazzo, R., Kabanets, V., Paturi, R.: The complexity of unique k-SAT: An isolation lemma for k-CNFs. J. Comput. Syst. Sci. 74(3), 386–393 (2008)
4. Gebauer, H.: Disproof of the neighborhood conjecture with implications to SAT. In: Fiat, A., Sanders, P. (eds.) ESA 2009. LNCS, vol. 5757, pp. 764–775. Springer, Heidelberg (2009)
5. Impagliazzo, R., Paturi, R., Zane, F.: Which problems have strongly exponential complexity? J. Comput. Syst. Sci. 63(4), 512–530 (2001)
6. Kratochvíl, J., Savický, P., Tuza, Z.: One more occurrence of variables makes satisfiability jump from trivial to NP-complete. SIAM Journal on Computing 22(1), 203–210 (1993)
7. Tovey, C.A.: A simplified NP-complete satisfiability problem. Discrete Applied Mathematics 8(1), 85–89 (1984)
8. Valiant, L.G., Vazirani, V.V.: NP is as easy as detecting unique solutions. Theor. Comput. Sci. 47(1), 85–93 (1986)

Assignment Stack Shrinking

Alexander Nadel[1] and Vadim Ryvchin[1,2]

[1] Intel Corporation, P.O. Box 1659, Haifa 31015 Israel
{alexander.nadel,vadim.ryvchin}@intel.com
[2] Information Systems Engineering, IE, Technion, Haifa, Israel

Abstract. Assignment stack shrinking is a technique that is intended
to speed up the performance of modern complete SAT solvers. Shrinking
was shown to be efficient in SAT'04 competition winners Jerusat and
Chaff. However, existing studies lack the details of the shrinking algorithm. In addition, shrinking's performance was not tested in conjunction
with the most modern techniques. This paper provides a detailed description of the shrinking algorithm and proposes two new heursitics for it.
We show that using shrinking is critical for solving well-known industrial
benchmark families with the latest versions of Minisat and Eureka.

1   Introduction

Modern SAT solvers are known to be extremely efficient on many industrial problems which may comprise up to millions of variables and clauses. Among the key
features that enable the solvers to be so efficient, despite the apparent difficulty
of solving huge instances of NP-complete problems, are *dynamic behavior* and
search locality, that is, the ability to maintain the set of assigned variables and
recorded clauses relevant to the currently explored space. This effect is achieved
by applying various techniques, such as the VSIDS decision heuristic [1] (which
gives preference to variables that participated in recent conflict clause derivations) and local restarts (such as [2]). Another important feature of modern
SAT solvers is that they tend to pick *interrelated variables*, that is, variables
whose joint assignment increases the chances of quickly reaching conflicts in unsatisfiable branches and satisfying clauses in satisfiable branches. Clause-based
heuristics (such as CBH [3]), which prefer to pick variables from the same clause,
increase the interrelation of the assigned variables.

Assignment stack shrinking (or, simply, shrinking) is a technique that seeks to
boost the performance of modern SAT solvers by making their behavior more local
and dynamic, as well as by improving the interrelation of the assigned variables.

Shrinking was introduced in [4] and implemented in the Jerusat SAT solver.
After a conflict, Jerusat applies shrinking if its *shrinking condition* is satisfied.
The shrinking condition of Jerusat is satisfied if the conflict clause contains no
more than one variable from each decision level. The solver then sorts the conflict
clause literals according to its *sorting scheme*. The sorting scheme of Jerusat
sorts the clause by decision level from lowest to highest. Afterwards Jerusat
backtracks to the *shrinking backtrack level*. The shrinking backtrack level for

O. Strichman and S. Szeider (Eds.): SAT 2010, LNCS 6175, pp. 375–381, 2010.
© Springer-Verlag Berlin Heidelberg 2010

Jerusat is the highest possible decision level where all the literals of the conflict clause become unassigned. Jerusat then guides the decision heuristic to select the literals of the conflict clause according to the sorted order and assign them the value false, whenever possible. As usual, Boolean Constraint Propagation (BCP) follows each assignment.

One can pick out three important components of the shrinking algorithm that can be tuned heuristically: the shrinking condition, the sorting scheme, and the determination of the shrinking backtrack level. Shrinking was implemented in the 2004 version of the Chaff SAT solver [5] with important modifications in each one of these components, as described below.

2   Algorithmic Details and New Heuristics

Chaff had two versions: *zchaff.2004.5.13* and *zchaff_rand*. We concentrate on *zchaff_rand*'s version of shrinking, since it was shown to be more useful in [5], and also performed better in the SAT'04 competition [6]. Suppose Chaff encounters a conflict. Chaff considers applying shrinking if the length of the conflict clause exceeds a certain threshold x. The clause is sorted according to decision levels. The algorithm finds the lowest decision level that is less than the next higher decision level by at least 2. (If no such decision level is found, shrinking is not performed.) The algorithm backtracks to this decision level, and the decision strategy starts reassigning the value false to the unassigned literals of the conflict clause, whenever possible. Chaff reassigns the variables in the reverse order, that is, in descending order of decision levels, since this sorting scheme was found to perform slightly better than Jerusat's in [5]. The threshold value x for applying shrinking is set dynamically using some measured statistics. More specifically, Algorithm 1 is used in Chaff for adjusting x after every y conflicts. Chaff measures the mean and standard deviation of the lengths of the recently learned conflict clauses and tries to adjust x to keep it at a value greater than the mean. The threshold on the number of conflicts y is 600 for Chaff.

Chaff's shrinking algorithm was implemented in Intel's SAT solver Eureka with two minor differences: (1) The threshold on the number of conflicts y is 2000; (2) Eureka forbids performing shrinking for two conflicts in a row.

An important detail for understanding the reasons for the efficiency of shrinking is that a conflict clause is recorded even when shrinking is applied. Hence the solver always explores a different subspace after performing shrinking. Previous works [4, 5, 7] claimed that a "similar" conflict must follow an application of shrinking, on the assumption that a conflict clause is not recorded when shrinking is applied, but this claim does not fit the actual way shrinking is implemented in Jerusat, Chaff, and Eureka.

Applying shrinking contributes to search locality and makes the solver more dynamic, since the set of assigned variables becomes more relevant to the recently explored search space as irrelevant variables become unassigned. Also, since the variables on the assignment stack are precisely those that appeared in recent conflict clauses, conflict clauses are more likely to share common interrelated

Algorithm 1. Adjust Threshold for Shrinking (Threshold for shrinking x, Threshold for number of learned clauses y)

Require: x is initialized with the value 95 at the beginning of SAT solving.
  ($mean, stdev$) := mean and standard deviation of last y learned clause lengths
  $center := mean + 0.5 * stdev$; $ulimit := mean + stdev$
  if $x \geq center$ **then**
    $x := x - 5$
  end if
  if $x < center$ **then**
    $x := x + 5$
  end if
  if $x > ulimit$ **then**
    $x := ulimit$
  end if
  if $x < 5$ **then**
    $x := 5$
  end if
  return x

---

variables. Shrinking often reduced the average length of learned conflict clauses and led to faster solving times, especially for the microprocessor verification benchmarks in Chaff [5].

We propose two new heuristics for shrinking. First, we propose generalizing the shrinking condition of Jerusat. We count the number of decision levels associated with a conflict clause's variables and perform shrinking if this number is greater than a threshold x. The threshold is calculated exactly like the conflict clause size threshold in Chaff in Algorithm 1, using the number of decision levels in the clauses instead of their lengths. We dub our proposal the *decision-level-based shrinking condition*. Interestingly, Jerusat's shrinking condition and its proposed generalization correspond to the recent observation that a "good" clause should contain as few decision levels as possible [8]. The clause deletion scheme of SAT'09 competition winner Glucose is based on this observation. Second, we propose using a new sorting scheme, called *activity ordering*. Our scheme sorts the variables of the conflict clause according to VSIDS's scores, from highest to lowest. Our proposal is intended to make the solver even more dynamic, since it reorders the relevant variables according to their contribution to the derivation of recent conflict clauses.

3   Experimental Results and Discussion

We used Eureka and Minisat for our experiments. Minisat was enhanced by a restart strategy that was found to be optimal for this solver in [2]. We used eight publicly available benchmark families: sat04-ind-goldberg03-hard_eq_check [6] (henceforth, abbreviated to ug), sat04-ind-maris03-gripper [6] (mm), sat04-ind-velev-vliw_unsat_2.0 [9] (uv2), SAT-Race_TS_1 [10] (ms1), SAT-Race_TS_2 [10] (ms2), velev_fvp-sat.3.0 [11] (sv3), velev_fvp-unsat.3.0 [11] (uv3), velev_vliw_unsat_4.0 [9] (uv4).

For each solver, we compared the following four versions, applying: (1) no shrinking; (2) the base version of shrinking, corresponding to Eureka's version of shrinking (recall from Section 2 that Eureka's shrinking algorithm is largely similar to Chaff's: its shrinking condition is based on clause length and the sorting scheme picks variables in descending order of decision levels); (3) the base version, modified by applying activity ordering; (4) the base version, modified by using the decision-level-based shrinking condition.

Table 1 provides some statistics regarding the benchmark families as well as Eureka's results. The first column of the table contains the family name, the second column specifies whether the instances are satisfiable, unsatisfiable, or mixed, and the third column contains the number of instances in the family. Each subsequent pair of columns shows the number of instances solved by Eureka within a three hour timeout and the overall run-time for the particular version in seconds (10800 seconds, that is, three hours, is added for an unresolved benchmark). Table 2 provides Minisat's results in the same format. (A table with all the details of the experimental results appears in [12].)

Compare the empirically best shrinking algorithm versus the version without shrinking for each solver. For Eureka, shrinking (the base version) is helpful for solving seven out of eight families, and critical for solving ug, uv2, uv3 and uv4.

Table 1. Shrinking within Eureka

Family	SAT?	Inst.	No Shr. Solved	Time	Base Shr. Solved	Time	Act. Order Solved	Time	Dec. Cond. Solved	Time
ug	UNS	13	10	67005	13	12041	13	14389	12	28457
mm	MIX	10	5	66602	7	39870	7	39426	8	44404
uv2	UNS	8	1	78870	8	12129	8	10283	8	10914
ms1	MIX	50	47	51117	49	27352	48	38208	50	16279
ms2	MIX	50	42	109899	44	92813	43	96564	42	99882
sv3	SAT	20	20	767	20	1119	20	788	20	1375
uv3	UNS	6	1	62038	6	10863	6	11761	6	11251
uv4	UNS	4	0	43200	4	10874	4	9018	4	10677
Sum		161	126	479498	151	207061	149	220437	150	223239

Table 2. Shrinking within Minisat

Family	SAT?	Inst.	No Shr. Solved	Time	Base Shr. Solved	Time	Act. Order Solved	Time	Dec. Cond. Solved	Time
ug	UNS	13	7	82310	10	43007	10	43686	11	44140
mm	MIX	10	0	108000	4	71234	0	108000	4	76680
uv2	UNS	8	1	85508	8	12235	8	10817	8	11743
ms1	MIX	50	48	36771	47	37771	49	26894	49	20557
ms2	MIX	50	44	82982	41	122233	42	107147	41	107780
sv3	SAT	20	16	53968	20	9330	20	10084	20	6954
uv3	UNS	6	0	64800	3	38056	0	64800	3	39652
uv4	UNS	4	1	33370	4	15230	4	9912	4	14798
Sum		161	117	547709	137	349096	133	381340	140	322304

For Minisat, shrinking (with the decision-level-based shrinking condition) is critical for solving seven out of eight families (ms2 is an exception). Overall, shrinking enables Eureka and Minisat to solve, respectively, 25 and 23 more benchmarks within the timeout. Hence employing shrinking is highly advantageous.

Compare now our two variations of shrinking versus the base version. The effect of applying the decision-level-based shrinking condition in Minisat is clearly positive as it leads to better overall performance in terms of both the number of solved instances and the run-time. Although applying the decision-level-based ordering condition within Eureka does not lead to better results overall, the solver does perform better for four families (the gap is especially significant for ms1) than with the base version. While the impact of activity ordering is negative for Minisat overall, it performs better than best version (the version with the decision-level-based shrinking condition) for three families. Activity ordering is not helpful overall for Eureka, but is does help solve four families more quickly than the best version (the version with base shrinking). Hence it is recommended that shrinking be tuned for each specific solver and benchmark family.

An important question is whether the effect of shrinking can be achieved by applying other algorithms, proposed after shrinking. Consider the following three techniques: (1) Frequent restarts [13,2]; (2) A clause-based heuristic, such as CBH [3]; and (3) RSAT's polarity selection heuristic [14], which assigns every decision variable the last value it was assigned. Observe that the combined effect of these three techniques seems to be similar to that of shrinking. First, restarting the search when a certain condition holds corresponds to backtracking when the shrinking condition is met. Second, applying a clause-based heuristic and RSAT's polarity selection heuristic results in selecting the last conflict clause and assigning its literals the value false, similar to what happens in shrinking. It was claimed in [13] that the impact of conflict clause minimization [15,16] could be considered somewhat similar to the impact of shrinking, since minimization reduces the size of conflict clauses, as does shrinking, according to [5].

However, we have seen that shrinking is extremely useful within Eureka, which employs all the above-mentioned techniques, and Minisat with local restarts, which uses some of them. Thus empirically the effect of shrinking is not achieved by combining other techniques. Let us take a closer look at the differences between our basic version of shrinking and the combination of frequent restarts, CBH, and RSAT's polarity selection heuristic. First, the shrinking condition differs from the restart condition of any known restart strategy. Second, shrinking restarts the search only partially, in contrast to most modern restart strategies. Third, unlike clause-based heuristics, shrinking continues selecting variables from the last conflict clause, even if it is satisfied. Fourth, shrinking re-orders the variables in the last conflict clause. It is, therefore, the simultaneous effect of these features, achieved by carefully choosing the shrinking condition, the sorting scheme, and the shrinking backtrack level, that makes shrinking highly efficient.

4   Conclusion

Assignment stack shrinking is a technique that boosts the performance of modern complete SAT solvers by making them more dynamic and local, and by enhancing the interrelation of the assigned variables. We have described in detail different variations of the shrinking algorithm, including two new heuristics, one of which improves Minisat's overall performance. We have shown that shrinking is extremely efficient within Minisat and Eureka, and that its effects cannot be achieved by other modern algorithms. Shrinking is proving to be a useful concept (that is, a collective name for a family of algorithms) that can be enhanced independently of the other components of SAT solvers, such as restart strategies or decision heuristics.

Acknowledgment

The authors would like to thank Amit Palti for supporting this work and Paul Inbar for editing the paper.

References

1. Moskewicz, M.W., Madigan, C.F., Zhao, Y., Zhang, L., Malik, S.: Chaff: Engineering an efficient SAT solver. In: DAC, pp. 530–535. ACM, New York (2001)
2. Ryvchin, V., Strichman, O.: Local restarts. In: Büning, H.K., Zhao, X. (eds.) SAT 2008. LNCS, vol. 4996, pp. 271–276. Springer, Heidelberg (2008)
3. Dershowitz, N., Hanna, Z., Nadel, A.: A clause-based heuristic for SAT solvers. In: Bacchus, F., Walsh, T. (eds.) SAT 2005. LNCS, vol. 3569, pp. 46–60. Springer, Heidelberg (2005)
4. Nadel, A.: Backtrack search algorithms for propositional logic satisfiability: Review and innovations. Master's thesis, Hebrew Univeristy of Jerusalem, Jerusalem, Israel (November 2002)
5. Mahajan, Y.S., Fu, Z., Malik, S.: Zchaff2004: An efficient SAT solver. In: [17], pp. 360–375
6. Berre, D.L., Simon, L.: Fifty-five solvers in Vancouver: The SAT 2004 competition. In: [17], pp. 321–344
7. Nadel, A.: Understanding and improving a modern SAT solver. PhD thesis, Tel Aviv University, Tel Aviv, Israel (August 2009)
8. Audemard, G., Simon, L.: Predicting learnt clauses quality in modern SAT solvers. In: Boutilier, C. (ed.) IJCAI, pp. 399–404 (2009)
9. Velev, M., Bryant, R.: Effective use of Boolean satisfiability procedures in the formal verification of superscalar and VLIW microprocessors. In: Proceedings of the 38th Design Automation Conference (DAC 2001), pp. 226–231 (2001)
10. Sinz, C.: SAT-Race 2006 (2006), http://fmv.jku.at/sat-race-2006/
11. Velev, M.N.: Using rewriting rules and positive equality to formally verify wide-issue out-of-order microprocessors with a reorder buffer. In: Proc. Design, Automation and Test in Europe Conference and Exhibition, pp. 28–35 (2002)
12. Nadel, A., Ryvchin, V.: Experimental results for the SAT'10 paper Assignment stack shrinking, http://www.cs.tau.ac.il/research/alexander.nadel/sat10_ass_res.xlsx

13. Biere, A.: PicoSAT essentials. JSAT 4(2-4), 75–97 (2008)
14. Pipatsrisawat, K., Darwiche, A.: A lightweight component caching scheme for satisfiability solvers. In: Marques-Silva, J., Sakallah, K.A. (eds.) SAT 2007. LNCS, vol. 4501, pp. 294–299. Springer, Heidelberg (2007)
15. Beame, P., Kautz, H.A., Sabharwal, A.: Towards understanding and harnessing the potential of clause learning. J. Artif. Intell. Res. (JAIR) 22, 319–351 (2004)
16. Sörensson, N., Biere, A.: Minimizing learned clauses. In: Kullmann, O. (ed.) SAT 2009. LNCS, vol. 5584, pp. 237–243. Springer, Heidelberg (2009)
17. Hoos, H.H., Mitchell, D.G. (eds.): SAT 2004. LNCS, vol. 3542. Springer, Heidelberg (2005)

Simple but Hard Mixed Horn Formulas

Gayathri Namasivayam and Mirosław Truszczyński

Department of Computer Science, University of Kentucky, Lexington, KY 40506, USA
{gayathri,mirek}@cs.uky.edu

Abstract. We study simple classes of mixed Horn formulas, in which the structure of the Horn part is drastically constrained. We show that the SAT problem for formulas in these classes remains NP-complete, and demonstrate experimentally that formulas randomly generated from these classes are hard for the present SAT solvers, both complete and local-search ones.

1 Introduction

We study some simple classes of mixed Horn formulas and show that randomly generated formulas from these classes are hard for the present SAT solvers. A conjunctive normal form (CNF) formula F is a *mixed Horn formula* (an MHF, for short) if each clause in F is either a positive 2-clause (a clause of the form $a \vee b$, where a and b are propositional variables), or is a Horn clause. MHFs have received much attention recently [1,2]. Researchers proved that many NP-complete problems have simple encodings as MHFs [2], showed that the satisfiability of MHFs remains NP-compete even under additional restrictions of the structure of input MHFs [2], and developed satisfiability algorithms for MHFs with good worst-case behavior lower bounds [1,3].

Due to their simplicity on the one hand, and the expressive power on the other, MHFs are attractive as possible benchmarks for SAT solvers. Our goal in this paper is to propose some models of simple random MHFs of particularly constrained structure, and to show that these models yield instances that are hard for SAT solvers. In the process, we find interesting connections to a class of random logic programs that we recently identified as consisting of instances that are hard for answer-set solvers [4].

2 Preliminaries

Let $V = \{v_1, v_2, \ldots\}$ be a fixed set of propositional variables. We define the class $MH_n(k, m)$, where $k \geq 1$ and $m \geq 0$, to consist of MHFs F such that

1. the set of atoms occurring in F is $\{v_1, \ldots, v_n\}$
2. F contains m positive 2-clauses
3. for every $v \in V$, F contains a negative clause $C_v = \neg v \vee \neg w_1 \vee \ldots \vee \neg w_k$, where $w_1, \ldots w_k \in V$ (note: clauses C_v and C_w, $v \neq w$, need not be distinct)
4. there are no other clauses in F.

We also define $MH_n(k) = \bigcup_m MH_n(k, m)$ (here m ranges from 0 to $\binom{n}{2}$), and $MH(k) = \bigcup_{n=0}^{\infty} MH_n(k)$.

O. Strichman and S. Szeider (Eds.): SAT 2010, LNCS 6175, pp. 382–387, 2010.

Thus, formulas in $MH(k)$ contain only positive 2-clauses and negative $(k+1)$-clauses. The key aspect of the model is, though, that there is an additional constraint imposed on the set of negative $(k+1)$-clauses of a formula $F \in MH(k)$: the set of variables of F must be a system of distinct representatives for the family (with possibly repeated occurrences) of the sets of variables of $(k+1)$-clauses of F.

Other classes of MHFs we consider in the paper impose additional connections between the negative and positive parts. Namely, we consider the class $MH_n^1(k)$, which we define as follows: an MHFs $F \in MH_n(k)$ belongs to $MH_n^1(k)$ if and only if its set of positive 2-clauses is given by $\{v \vee w \mid w \in Var(C_v), \text{ where } C_v \in F\}$. In the case of MHFs in $MH_n^1(k)$, there is a strong connection between the sets of positive and negative clauses: if $F \in MH_n^1(k)$, then F is *entirely* determined by its negative part. We note that the number of 2-clauses in formulas in $MH_n^1(k)$ is not fixed and ranges between $kn/2$ and kn. We write $MH^1(k)$ for $\bigcup_{n=0}^\infty MH_n^1(k)$, and MH_n^1 for $\bigcup_{k=1}^\infty MH_n^1(k)$.

Despite constraints on MHFs that form the classes $MH(k)$ and $MH^1(k)$, for each of them deciding satisfiability remains NP-complete.

Proposition 1. *For each of the classes $MH(k)$ and $MH^1(k)$, with $k \geq 2$, the satisfiability problem restricted to that class of formulas is NP-complete.*

Proof. (Sketch) The class $MH(k)$ represents a more general class of formulas than $MH^1(k)$. Hence, it is sufficient to show the NP-completeness for the problem restricted to the class $MH^1(k)$. We do it by constructing a polynomial-time reduction from the problem to decide whether a logic program built of clauses of the form $a \leftarrow not\ b$ has a stable model — the problem which is known to be NP-complete [5].

Thus, the classes of formulas discussed above are on the one hand extremely simple, and on the other hand, as expressive as the class of (unrestricted) CNF formulas. Moreover, it is clear that in the case of each class $\mathcal{C}$ of formulas we introduced above, there are straightforward algorithms to generate formulas from $\mathcal{C}$ uniformly at random. These properties suggest that these classes be considered as possible models of random CNF formulas for use as benchmarks for SAT solvers.

3   Phase Transition for $MH_n(k, m)$

For every fixed k and sufficiently large n, the class $MH_n(k)$ shows the classical phase-transition behavior. That is, when m (we recall that m stands for the number of 2-clauses in a formula) is small, formulas in $MH_n(k)$ are almost certainly satisfiable, when m is large, they are almost certainly unsatisfiable, and the transition from satisfiability to unsatisfiability occurs rapidly (the rate of change increases with n). Figures 1(a) and (b) show the phase transition for $k = 5$ and $k = 10$, respectively, as the ratio m/n grows ($n = 225$; 200 instances). As in the case of random 3-CNF formulas, we also observe that with increase in n the phase transition gets sharper while the cross-over point remains approximately the same (we do not provide graphs due to space limits).

The approximate location of the phase-transition region expressed in terms of the ratio m/n, for which the phase transition occurs, grows with k. Our experimental results

(a)   (b)

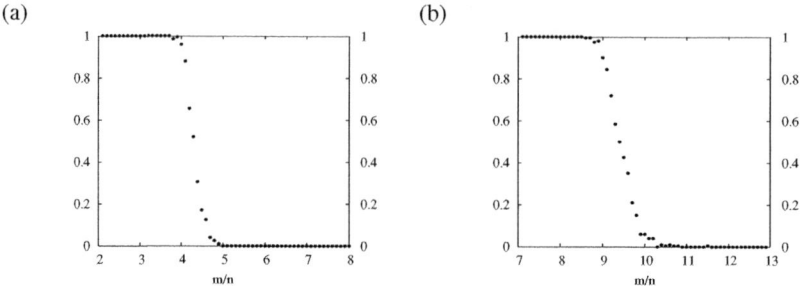

Fig. 1. The phase transition $k = 5$ and $k = 10$, $n = 225$, 200 instances

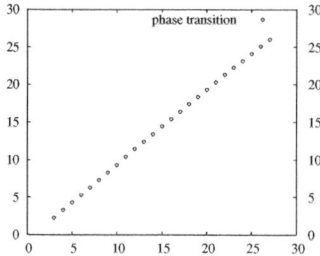

Fig. 2. The location of the phase transition in the model $MH_n(k)$ as a function of k

for $n = 200$ and $k = 3, \ldots 25$ (200 instances), show that the location of the phase transition grows slightly slower than k (Figure 2).

For each fixed k, we also observe that as the density m/n of 2-clauses grows, instances generated demonstrate the "easy-hard-easy" pattern, and those generated from the phase transition region are especially hard. The graphs given in Figure 3 for $k = 5$ are representative. They give the probability of having a model and the average running time in seconds for the SAT solver *clasp* for instances from $MH_n(k)$, for $k = 5$, $n = 200$, as the density m/n grows (for each density, 100 instances were generated).

4   Easy-Hard-Easy Behavior

We pointed out above that for a fixed k, as the number m of 2-clauses grows, instances from $MH_n(k)$ are initially getting harder and then, after passing the area of the phase transition, start getting easier again. However, the framework of the classes $MH_n(k)$ we consider reveals yet another interesting phenomenon. Being parameterized with k, it allows us to compare hardness of instances generated from the phase transition region for *different* values of k. Somewhat surprisingly, it turns out that as we increase k, the easy-hard-easy pattern emerges again. We observed an easy-hard-easy pattern using several SAT solvers that performed well in the SAT 2009 competition [6], including *precosat*, *glucose*, *clasp* and *march hi* (each was the winner in at least one of the categories of that competition). Initially, as k grows, the phase-transition instances are getting harder at an increasing rate. The hardness peaks when $k \approx 15, 16$, and from that

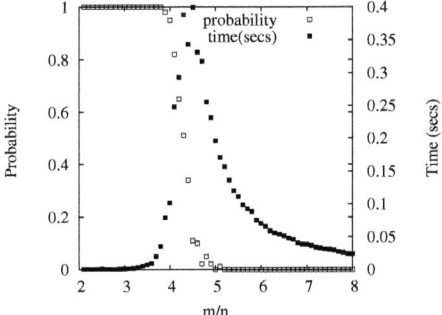

Fig. 3. The correlation of the instance hardness and the phase transition regions for the model $MH_n(k)$; $k = 5$, $n = 200$

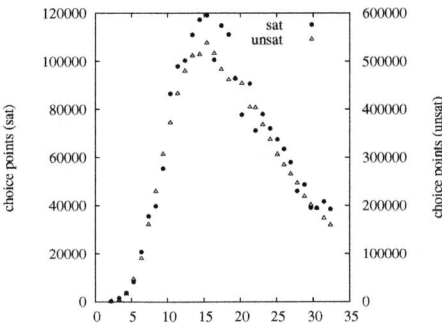

Fig. 4. The easy-hard-easy pattern of instances generated from the phase transition region of classes $MH_n(k)$ as a function of k

point on the instances are becoming increasingly easier. Figure 4 illustrates that pattern observed for *clasp* for $n = 200$, and k ranging from 3 to 34.

While it is rather natural that the hardness of instances from the phase transition in the model $MH_n(k)$ initially grows with k, it may seem surprising that at some point it peaks and then starts to decrease. Providing a formal explanation to this phenomenon is an open problem.

5 Hard Benchmarks for SAT Solvers

Our results suggest that MHFs randomly generated from the phase transition region for the class $MH_n(k)$ for $k = 15$ or 16 (located when $m \approx (k - 0.5)n$, where m stands for the number of 2-clauses) can provide challenging instances for SAT solvers. It is indeed so. We randomly generated 50 instances from $MH_n(k, m)$, with $n = 350$, $k = 15$ and $m = 14.5n$. Given the timeout limit of 1800 seconds, *clasp* and *march hi* solved fewer than 20% of the instances all of which were satisfiable and did not solve any of the unsatisfiable ones.

We stress that the instances in the set $MH_n(15, 14.5n)$ are small (350 atoms and 5425 clauses), and more importantly, that most of their clauses (5025) are 2-clauses.

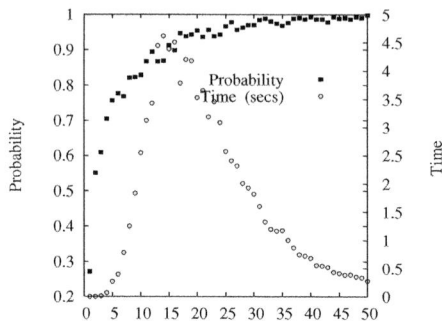

Fig. 5. The easy-hard-easy pattern for the model $MH_n^1(k)$, and the probability of satisfiability ($n = 100$)

Since they pose a challenge for the state-of-the-art complete solvers, we believe that the class $MH_n(15, 14.5n)$ is important for the design and testing solvers performance.

The classes $MH_n^1(k)$ offer even harder instances. While they can also serve as benchmarks for complete solvers, even for relatively small values of n, satisfiable instances from $MH_n^1(15)$ become very hard also for local-search solvers! The selection of $k = 15$ is not accidental. Our experiments showed that when we vary k, we observe the easy-hard-easy pattern, with the peak for $k \approx 15$. We also found (a property important below) that in the maximum hardness area, the percentage of instances that are satisfiable exceeds 90% for different N values. Figure 5 illustrates these claims.

We generated 100 random CNF formulas from each of the sets $MH_n^1(15)$, where $n = 450$ and 550. Given our experiments, the expected number of satisfiable instances in these two sets of formulas is at least 90. We ran *TNM* [6] on these formulas. *TNM* is currently one of the best local-search solvers. It won in the random category (satisfiable instances only) at the SAT 2009 competition. The solver does not require any parameters, as it adaptively selects them. We observed that for $n = 450$, *TNM* could still solve 86% of the instances in less than 1800 seconds (yet, already likely missing some satisfiable instances). The larger value of n, $n = 550$, resulted in many hard instances. Indeed, for $n = 550$, *TNM* solved only 53 of the 100 instances within 1800 seconds while we expect about 90 instances to be satisfiable in this sample.

6   Conclusions

We studied classes of simple MHF's: $MH_n(k)$ and MH_n^1. The key finding is that despite their simple form, randomly generated formulas from these classes (for the appropriate selections of parameters) are challenging benchmarks for the current generation of the state-of-the-art SAT solvers. Thus, formulas in these classes are relevant for the design of fast SAT solvers and deserve attention. We studied these classes experimentally, focusing on identifying phase transitions and hardness patterns, in order to facilitate generation of hard formulas. Interestingly, we found that the hardness of the instances from the phase transition region in the classes $MH_n(k)$ shows the easy-hard-easy pattern as a function of k, with the peak hardness for $k = 15$. We showed that instances generated from the phase transition region of $MH_n(15)$ (which occurs when the number of 2-clauses is about $14.5n$) pose a challenge instances for the current generation of SAT solvers. Similarly, the instances from MH_n^1 show the easy-hard-easy behavior (as the length k of purely negative clauses, grows), with the peak hardness when $k = 15$. The instances generated from $MH_n^1(15)$ are predominantly satisfiable (probability of a random formula generated from that class being satisfiable is at least 0.9). Those instances that are satisfiable pose a challenge for local-search solvers.

We note that the class MH_n^1 is closely related to a class R^- of logic programs studied in [4] and identified as containing programs that are especially hard for the current generation of the answer-set solvers. The class R^- consists of programs whose every clause is of the form $a \leftarrow not\ b$. The *completions* [7] of such programs (certain CNF theories whose models capture, in this particular case, the semantics of logic programs given by answer sets [8]) consist of positive 2-clauses and purely negative clauses (possibly of varying lengths). Completions of programs from R^-, in which every atom appears in the head of exactly k rules form precisely the class $MH_n^1(k)$.

This paper contains mostly experimental results. Yet it opens several interesting theoretical questions concerning tight bounds on the location of the phase transition in the model $MH_n(k)$, the properties of formulas in sets $MH_n(k)$ and $MH_n^1(k)$, when k grows with n (for instance, when $k = cn$, for some positive c), and possible reasons for the easy-hard-easy pattern demonstrated by formulas from the phase transition region of $MH_n(k)$ as k grows (or by formulas in $MH_n^1(k)$, as k grows).

Acknowledgements

The work was partially supported by the National Science Foundation under grant IIS-0913459.

References

1. Kottler, S., Kaufmann, M., Sinz, C.: A new bound for an NP hard subclass of 3-sat using backdoors. In: Kleine Büning, H., Zhao, X. (eds.) SAT 2008. LNCS, vol. 4996, pp. 161–167. Springer, Heidelberg (2008)
2. Porschen, S., Schmidt, T., Speckenmeyer, E.: On some aspects of mixed horn formulas. In: Kullmann, O. (ed.) SAT 2009. LNCS, vol. 5584, pp. 86–100. Springer, Heidelberg (2009)
3. Porschen, S., Speckenmeyer, E.: Worst case bounds for some NP-complete modified horn-sat problems. In: H. Hoos, H., Mitchell, D.G. (eds.) SAT 2004. LNCS, vol. 3542, pp. 251–262. Springer, Heidelberg (2005)
4. Namasivayam, G., Truszczyński, M.: Simple random logic programs. In: Erdem, E., Lin, F., Schaub, T. (eds.) LPNMR 2009. LNCS, vol. 5753, pp. 223–235. Springer, Heidelberg (2009)
5. Marek, W., Truszczyński, M.: Autoepistemic logic. Journal of the ACM 38, 588–619 (1991)
6. The SAT 2009 competition (2009), http://www.satcompetition.org/
7. Clark, K.: Negation as failure. In: Gallaire, H., Minker, J. (eds.) Logic and data bases, pp. 293–322. Plenum Press, New York (1978)
8. Gelfond, M., Lifschitz, V.: The stable semantics for logic programs. In: Proceedings of the 5th International Conference on Logic Programming (ICLP 1988), pp. 1070–1080. MIT Press, Cambridge (1988)

Zero-One Designs Produce
Small Hard SAT Instances

Allen Van Gelder and Ivor Spence

http://www.cse.ucsc.edu/~avg
http://www.cs.qub.ac.uk/~I.Spence

Abstract. Some basics of combinatorial block design are combined with certain constraint satisfaction problems of interest to the satisfiability community. The paper shows how such combinations lead to satisfiability problems, and shows empirically that these are some of the smallest very hard satisfiability problems ever constructed. Partially balanced $(0, 1)$ designs (PB01Ds) are introduced as an extension of balanced incomplete block designs (BIBDs) and $(0, 1)$ designs. Also, $(0, 1)$ difference sets are introduced as an extension of certain cyclical difference sets. Constructions based on $(0, 1)$ difference sets enable generation of PB01Ds over a much wider range of parameters than is possible for BIBDs. Building upon previous work of Spence, it is shown how PB01Ds lead to small, very hard, unsatisfiable formulas. A new three-dimensional form of combinatorial block design is introduced, and leads to small, very hard, satisfiable formulas. The methods are validated on solvers that performed well in the SAT 2009 and earlier competitions.

1 Introduction

Combinatorial block design addresses constraint satisfaction problems that are frequently of interest to the satisfiability community, yet many of its results are not well known in this community. This short paper shows how they combine with certain constraint satisfaction problems to create very hard satisfiability problems. The page limit forces proofs and many other details to be omitted. The Conclusion gives a URL for more information.

Our goal is to construct benchmarks with approximately 100 variables, with overall size varying linearly with the number of variables, that are very hard for all known solvers.

The main ideas are based on $(0, 1)$ designs and the newly introduced partially balanced $(0, 1)$ designs (PB01Ds), which generalize the very stringent class of balanced incomplete block designs (BIBDs). Whereas BIBDs exist for very limited sets of parameters, $(0, 1)$ designs exist and can be constructed over wider ranges of parameters, and PB01Ds extend the range even further. Techniques for constructing $(0, 1)$ designs and PB01Ds are presented, based on the newly introduced idea of $(0, 1)$ difference sets.

Building upon previous work of Spence [Spe10], Section 3 describes how to generate families of small unsatisfiable formulas with regularly varying properties. The relationship to certain restricted forms of SAT is briefly discussed.

O. Strichman and S. Szeider (Eds.): SAT 2010, LNCS 6175, pp. 388–397, 2010.

A new three-dimensional form of combinatorial block design is introduced in Section 4, which leads to a generator for families of small satisfiable formulas with regularly varying properties, again following and extending earlier work of Spence.

The generators are validated on solvers that performed well in the SAT 2009 and earlier competitions. Experiments are presented that demonstrate that these generators produce some of the smallest very hard satisfiability problems ever constructed. They are orders of magnitude smaller than random 3-CNF formulas of comparable difficulty. Whereas random formulas are often criticized for having no structure, and therefore not representative of "real world" problems, our formulas are highly structured and represent certain resource allocation problems. Therefore, we hope that they might be useful in the study of how to further improve solvers aimed at practical application problems.

2   Combinatorial Block Designs

The study of combinatorial block designs dates back to the 1840s, and has a rich literature. A book by Ian Anderson [And90] provides an accessible treatment of the field, from introductory through advanced topics, with emphasis on construction methods. Various unattributed facts about combinatorial designs in this paper may be found, together with citations, in this book. In general, we follow this book's terminology, and we thank the author for providing clarifications and further insights by email. However, we present concepts in a different order, reflecting our application of them in this paper.

Definition 1. A *balanced* $(0, 1)$*design* $((0, 1)$design for short) is an undirected bipartite graph on two collections of vertices, called *objects* and *blocks*, such that any pair of objects has edges to at most one block in common, and any pair of blocks has edges to at most one object in common. Each object is incident upon the same number of blocks, and each block is incident upon the same number of objects; however these two numbers may be different.

Blocks are often viewed as sets of the objects to which they are adjacent. (This is why we used the phrase "collection of vertices" for blocks and objects.) We shall also sometimes view an object as the set of blocks that contain it. Thus we say a pair of blocks *intersects* in certain objects, and a pair of objects *intersects* in certain blocks.

A bipartite graph is conveniently represented by its *incidence matrix* M, in which each row corresponds to an object and each column corresponds to a block. In fact, we often think of the incidence matrix as *being* the design. Entries in M are either *null* or a *flag*. When doing arithmetic on M, nulls are treated as 0 and flags as 1, but flags might contain other information in some cases. In diagrams, nulls are blank and flags are "X."

The constraint on intersections can be described using the idea of a quadrangle in a matrix M. A *quadrangle* is a set of four flags (or 1's) at locations $M(r_1, c_1)$, $M(r_1, c_2)$, $M(r_2, c_1)$, $M(r_2, c_2)$, forming a rectangle. The quadrangle is a forbidden pattern in M.

Following the conventions in the field, v denotes the number of objects, b denotes the number of blocks, k denotes the *degree* of each block (i.e., the number of objects incident upon it), and r denotes the degree of each object. Clearly, the number of non-null flags (edges) is $vr = bk$.

The design is called *symmetric* if $v = b$. Note that a symmetric design does not, in general, have a symmetric incidence matrix. In fact, the order among the rows or columns of an incidence matrix is usually immaterial, because they correspond to disjoint collections. □

Note that M^T represents a $(0,1)$ design whenever M does. Thus objects and blocks have dual roles. This duality does *not* extend to many other designs. This paper is primarily concerned with symmetric $(0,1)$ designs, and their relation to constraint satisfaction problems. When every pair of objects is incident upon *exactly* one block, the structure is called a $(v, k, 1)$ design, short for $(v, k, 1)$ *balanced incomplete block design* (BIBD). In general, a BIBD may be a (v, k, λ) design, in which every pair of objects is incident upon exactly λ blocks.

The duality enjoyed by $(0,1)$ designs is lost for BIBDs because it is not required in a BIBD that every pair of blocks is incident upon exactly the same number of objects. Duality is restored for *symmetric* $(v, k, 1)$ designs, which are also called *finite projective planes*. These designs exist for very limited combinations of v and k. However, they serve as starting points for constructing $(0,1)$ designs of many sizes. A finite projective plane of order q is a $(v, k, 1)$ BIBD, where $v = q^2 + q + 1$ and $k = q + 1$.

Definition 2. A *partially balanced* $(0,1)$ *design* (PB01D for short) is like a balanced $(0,1)$ design except that all block sizes are in a finite set K and all object sizes are in a finite set R. The *size* of a block or object is its degree in the bipartite graph for the design. In this paper, the cardinalities of K and R do not exceed two, although the definition does not require this. □

The term "partially balanced $(0,1)$ design" is apparently new in this paper. Closely related *partially balanced designs* are very important in combinatorial-design construction. However, the constraints for PB01Ds, defined here, are apparently too loose to say anything that is mathematically interesting about them.

Definition 3. A $(0,1)$ *difference set* D for Z_v (the set of integers modulo v) is a set of integers in Z_v such that each element of Z_v occurs at most once as a difference (modulo v) between an ordered pair of distinct integers in D. (Note that, for distinct $a, b \in D$, $b - a$ and $a - b$ are distinct differences.) In this context, let $k = |D|$, the cardinality of D. Since there are $k(k-1)$ ordered pairs, we have $k(k-1) \leq v - 1$. □

The term "$(0,1)$ difference set" is apparently new in this paper. As with balanced $(0,1)$ designs, the limiting case of a $(0,1)$ difference set is a more familiar concept in combinatorial design.

Definition 4. The limiting case of a $(0,1)$ difference set in which each positive integer in Z_v occurs in *exactly* one difference is called a *cyclic* $(v, k, 1)$ *difference set*. In general, a cyclic (v, k, λ) difference set requires that each positive integer in Z_v occurs as exactly λ differences. □

For cyclic $(v, k, 1)$ difference sets, we have $k(k-1) = v - 1$, not just "$\leq$". Letting $k = q + 1$, we see that $v = q^2 + q + 1$, which is the relationship present for finite projective planes. In fact, difference sets with $k = q + 1$ are very useful for constructing finite projective planes of order q.

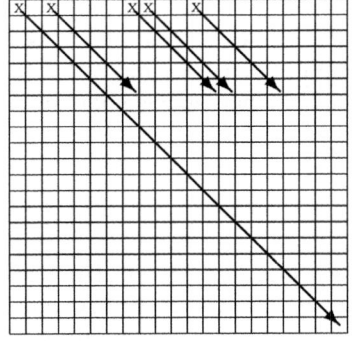
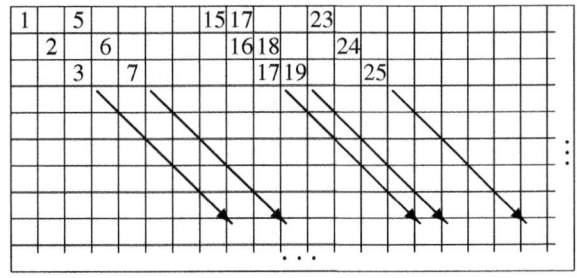

Fig. 1. (Left) Using a cyclic difference set to generate a finite projective plane. (Right) Generating a 3D $(v, 5, (0, 1))$ design, for $v = 21$ or $v \geq 23$, as described in Section 4.

Example 1. It will be shown later, in Lemma 1, that $\{0, 2, 7, 8, 11\}$ is a cyclic $(21, 5, 1)$ difference set. In a 21×21 matrix (with rows and columns indexed beginning at 0) an incidence matrix for a finite projective plane of order 4 can be generated by placing X's (for flags) in row 0, columns $\{0, 2, 7, 8, 11\}$, then continuing in a circulant fashion through succeeding rows by shifting the pattern 45° down and to the right, as suggested in Figure 1. The pattern wraps around just like integers modulo 21.

It is straightforward to verify that every row has five X's and every column has five X's also. A moment's reflection shows that, if a quadrangle occurs in this generation scheme, then the zero-th row must not be a cyclic $(21, 5, 1)$ difference set, or even a $(0, 1)$ difference set for Z_{21}.   □

The next question of interest is how to find $(0, 1)$ difference sets for various combinations of $(v, k, 1)$, for if we find such a $(0, 1)$ difference set, then generation of a *symmetric* $(0, 1)$ design for v objects and block size k becomes routine. Anderson describes an involved and expensive general procedure, but fortunately gives answers for $3 \leq k \leq 6$, attributed to Rev. T. Kirkman, 1857 [And90, Example 2.1.3, p. 50]. Values for larger k are tabulated, as well [CD96]. In each case, $k = q + 1$, $v = q^2 + q + 1$, and the difference set supports the generation of a projective plane of order q. We have found that these cyclic difference sets serve as $(0, 1)$ difference sets for many $v' > v$. The following lemma explains how.

Lemma 1. The sets D_3 through D_6 have properties as follows:

1. $D_3 = \{0, 1, 3\}$ is a $(0, 1)$ difference set for $(v, 3, 1)$, for all $v \geq 8$, and is a cyclic difference set for $(7, 3, 1)$.
2. $D_4 = \{0, 4, 5, 7\}$ is a $(0, 1)$ difference set for $(v, 4, 1)$, for all $v \geq 15$, and is a cyclic difference set for $(13, 4, 1)$.
3. $D_5 = \{0, 2, 7, 8, 11\}$ is a $(0, 1)$ difference set for $(v, 5, 1)$, for all $v \geq 23$, and is a cyclic difference set for $(21, 5, 1)$.
4. $D_6 = \{0, 4, 11, 13, 14, 19\}$ is a $(0, 1)$ difference set for $(v, 6, 1)$, for $35 \leq v \leq 37$, and for all $v \geq 39$, and is a cyclic difference set for $(31, 6, 1)$.

If D is a cyclic $(v, k, 1)$ difference set, any translation, or shift, of the elements of D, modulo v, continues to be a cyclic difference set. For example, $(0, 1, 4, 14, 16)$ is a translate of $(0, 2, 7, 8, 11)$ in Z_{21}, and it is also a cyclic difference set for $(21, 5, 1)$. Although $(0, 2, 7, 8, 11)$ is a $(0, 1)$ difference set for $(23, 5, 1)$, the same is *not* true for $(0, 1, 4, 14, 16)$ (differences of 10 occur multiply).

3 Generating Unsatisfiable Formulas

The primary goal of this research is to generate small, very hard, propositional satisfiability benchmarks. In several previous SAT competitions benchmarks have been divided into three categories, *Industrial*, *Crafted*, and *Random*. The benchmarks described in this paper fall into the crafted category.

Although crafted benchmark generators often employ some degree of randomness, their distinguishing feature is some underlying common structure, and the randomness produces variations of that structure. To have benchmarks that are solvable within our computational resources, we generated starting at 85 variables, with 40 samples at each size, and gathered statistics on several solvers. The experiments are described in a later section. Here we explain the methods of generation.

First, we need to describe how a $(0, 1)$ design or a PB01D is mapped to a constraint-satisfaction problem. Our mapping is quite different from those found in studies of *linear CNF* [Kul07, PS09]. In [PS09] a linear formula is derived from a PB01D as defined in Definition 2 by mapping objects to clauses and blocks to propositional variables. Thus there are v clauses on b variables. Our mappings all have one propositional variable per flag, giving $v r = b k$ variables for $(0, 1)$ designs. Our mappings are inspired by, and generalize, previous work of Spence [Spe10].

Recall that each $(0, 1)$ design is represented by an incidence matrix M. Let $x_{r,c}$ be a set of propositional variables corresponding to the flags of M. A resource allocation problem can be associated with an incidence matrix in a natural way. Let each row r require $need_r$ resources, and let each column c be able to supply $avail_c$ resources, both in integer units. The flag $x_{r,c}$ means that column c can supply a resource that row r can use. With slight abuse of notation, we encode clauses that are equivalent to

$$\sum_{c \in r} x_{r,c} \geq need_r, \tag{1}$$

$$\sum_{r \in c} x_{r,c} \leq avail_c. \tag{2}$$

This is the arithmetic sum, with $x_{r,c}$ interpreted as 0 or 1. Constraints of this type are in the language of integer linear programming.

For example, a company needs to manage numerous ongoing small projects with a pool of workers, and $x_{r,c}$ is defined when worker c is qualified for project r. Workers can time-share on a limited number of projects according to $avail_c$, whereas projects need staffing at levels given by $need_r$. Can all projects be staffed adequately?

For CNF, we encode the statement "each subset (r, c) of cardinality $|r| - need_r + 1$ contains a true $x_{r,c}$." Here, $|r|$ denotes the cardinality of the *set* of variables $x_{r,c}$ defined in row r. Such a set of clauses is generated for each r, $1 \leq r \leq v$. Analogously, for

Eq. 2 for each c, $1 \leq c \leq b$, we generate negative clauses to encode "each subset (r, c) of cardinality $|c| - avail_c + 1$ contains a false $x_{r,c}$."

It is evident to a human that, if $\sum_r need_r > \sum_c avail_c$, then the system is unsatisfiable. For a $(21, 5, 1)$ design it is easy to choose 2 or 3 as values for $need$, and the same for $avail$, so that $\sum_r need_r = 53$ and $\sum_c avail_c = 52$. The formula generated for the resulting constraint-satisfaction problem has 105 variables, 320 clauses, and 1060 literals. We issue a challenge for any solver to solve it in less than one day.

Although the formula just described appears to be very challenging for its size, it is one of a kind, except for permuting variable numbers and flipping some polarities. We need methods to generate formulas of varying sizes to get useful benchmarks.

The first method of generation consists of deleting flags from the 21×21 incidence matrix for the $(21, 5, 1)$ BIBD just mentioned, then encoding similar constraints that ensure that $\sum_r need_r > n/2$ and $\sum_c avail_c \leq n/2$, where n is the number of flags. Flags are deleted in a manner such that every object has 4 or 5 flags in its row, and every block has 4 or 5 flags in its column. Then $need$ and $avail$ values are assigned so that the sums come out to $\lceil (n+1)/2 \rceil$ and $\lfloor n/2 \rfloor$, respectively, ensuring unsatisfiability. Labels "sgen2del" in the experiments refer to this method.

The second method of generation is considerably simpler, but less flexible. Spence reported a procedure that uses simulated annealing to reduce a penalty function, called the *score*, computed on permutations [Spe10]. As it happens, a permutation has score 0 if and only if it corresponds to a PB01D. (If the last variable is ignored, it would be a $(v, 4, (0, 1))$ design.) For g groups it is a PB01D with $v = g$ and $R = K = \{4, 5\}$. All objects but one, and all blocks but one, have size 4. The remaining object and remaining block have size 5.

As it happens the original study began in a range where achieving score 0 would be a rare event. Also, for simplicity and speed, the program only computed *changes* in score, not actual score. Labels "orig" in the experiments refer to this method.

Using "back of the envelope" methods we calculated that in the range $v = 21$ $(v, 4, (0, 1))$ designs should be plentiful enough that random searching should have a good chance to succeed. Therefore, we modified the program to compute and check the actual score, and stop if the score reached 0. This method has so far always succeeded within about 100,000 tries, for $n \geq 85$, and group size 4 (except 5 for the last group).

Once the permutation with score 0 is found, the *need* is set to 2 for all objects of size 4 and is set to 3 for the object of size 5. The *avail* is set to 2 for all blocks of size 4 and is set to 2 for the block of size 5, as in the original study. Labels "score" in the experiments refer to this method.

4 Generating Satisfiable Formulas

Since there is no such thing as an objectively hard satisfiable formula, generating satisfiable benchmarks that are predictably difficult for a wide range of solvers is challenging, if not impossible. We investigated a 3D analogy of $(0, 1)$ designs, as well as modifying the generator reported by Spence to compute scores and stop at 0. The extension to 3D appears to be new and quite different from other combinatorial designs that involve three dimensions, such as Room squares, used to schedule leagues and tournaments.

Definition 5. We define a 3D $(0, 1)$ design directly in terms of the 3D incidence matrix, M_3, of size $v \times v \times v$. Besides objects and blocks a third collection, called *layers*, is introduced. Although there are many more matrix cells, the number of flags remains at $v\,r = b\,k$. Let ℓ denote layer, from 1 to v, so M_3 is indexed by (r, c, ℓ).

The first constraint on flags is that for any two fixed indices, at most one third index contains a flag. That is, for a fixed pair (r, c), at most one cell of the form $M_3(r, c, \ell)$ has a flag, and so forth for other index combinations. Note that a 3D $(0, 1)$ design can be represented in a 2D matrix, say M, indexed on (r, c) by letting each flag contain the value of the *layer* that it is in.

The size of an object is now found by counting the flags in that row for all columns and layers; for the size of a block, count flags in that column for all rows and layers; and for the size of a layer, count the flags in that layer, for all rows and columns. All objects, blocks, and layers are required to have size $k = r$.

The intersection constraint is most easily stated by requiring that, if any one dimension is projected out, the 2D matrix represents a $(0, 1)$ design.   □

A 3D $(v, 5, (0, 1))$ design can be generated, in some cases, by a variant of the method shown in Figure 1 (left), by having each flag contain its layer number, as shown on the right side of that figure. This method works in practice for odd $v \geq 21$.

5   Experimental Results

Benchmarks generated by the methods described were put through a computationally expensive evaluation on several multi-core 2.5 GHz Linux machines. With the goal of getting statistically meaningful results, each generation method at each size was tested on a sample of 40 benchmarks. See Figure 2 for an overview of the unsatisfiable results.

Some of the better-performing solvers in the SAT 2009 were used; we used the static binaries from the computation web page. The clause-learning solvers selected were clasp [GKNS07] and minisat2 [ES05]. We chose march_hi because it is based on

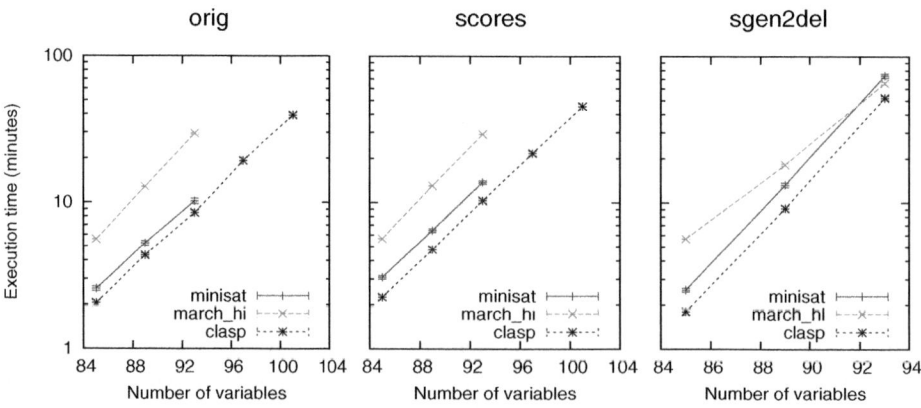

Fig. 2. Growth rates of solving times for unsat benchmarks produced by three generators

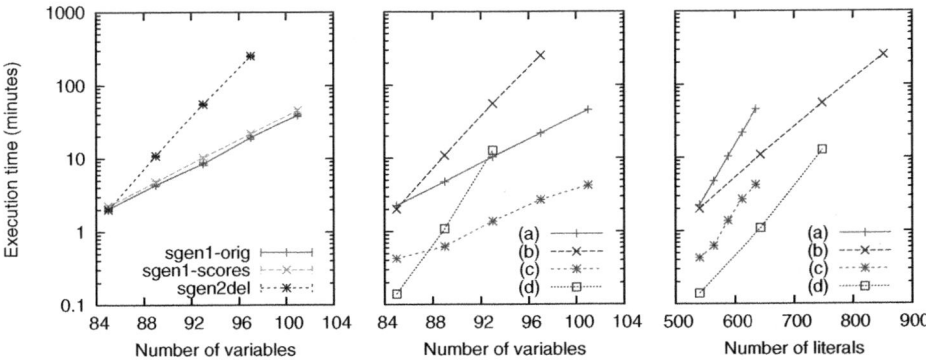

Fig. 3. (Left) Time growth for `clasp` on formulas from three generators, as a function of number of variables. (Center, Right) Time growth for two size measures, variables (center), and literals (right). `clasp` times are for "scores" (a), "sgen2del" (b). `tts` times are for "scores" (c), "sgen2del" (d).

look-ahead and is *not* in the clause-learning family. Local search was represented by `adaptg2wsat2009++` (adapt for short).

Some well-performing solvers were not used, due to limited resources. For example, `precosat` tested considerably slower than `minisat2` on our smallest benchmarks and is also in the clause-learning family, while `satzilla` is a portfolio solver from which it would be difficult to draw any conclusions.

All benchmarks were run to completion with no time-outs. Statistics computed on each sample of 40 included mean, standard error (seen as barely visible error bars in Figure 2), minimum, and maximum. The *standard error* estimates how different the calculated mean is likely to be for an independently drawn sample of the same size n.

Growth rates are estimated using two size measures: variables and literals. While variables might not be an appropriate size measure on some classes of benchmark, for ours it is reasonable because the number of literals was linear in the number of variables. Noting that a straight line on a semi-log plot represents an exponential function, we observed exponential growth rates in all solvers, for all generation methods, on the unsatisfiable benchmarks.

The original generation method of Spence [Spe10] can be seen as an approximation of our methods based on $(0, 1)$ designs. Did using $(0, 1)$ designs create harder benchmarks? The evidence is that they did, for the unsatisfiable cases. The plot on the left of Figure 3 shows that "scores" (short for "zero scores") were definitely harder than "orig" for `clasp`, with a separation of about two standard errors. Other solvers were similar.

The other type, "sgen2del," used a more general encoding scheme that allowed it to grow in difficulty the most rapidly, based on the numbers of variables, as seen in Figure 3 (left), for `clasp`. Other solvers were similar. However, Figure 3 shows that this relationship changes when growth rate is based on the number of literals. The "sgen2del" family is believed by us to contain the hardest known benchmarks with under 100 variables and under 1000 literals.

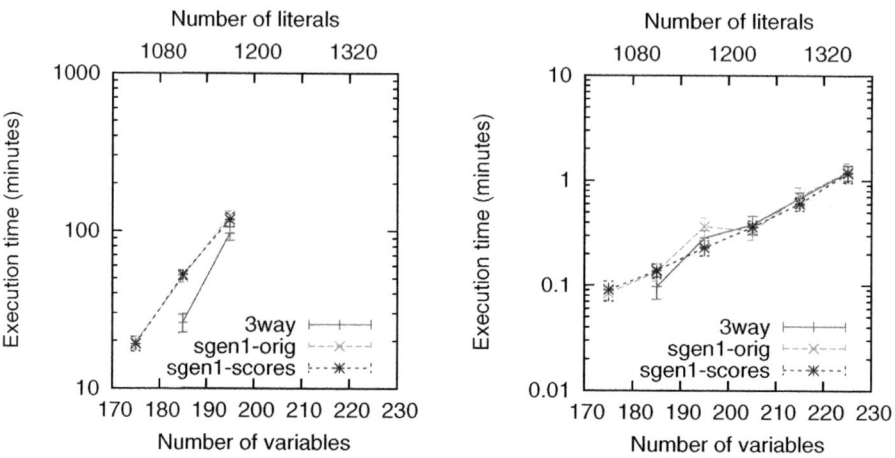

Fig. 4. Time growth for satisfiable formulas produced by three generators; (left) `clasp`, a complete solver; (right) `adaptg2wsat2009++`, an incomplete solver

A natural question, since we have a random element in our generation methods, is whether some formulas are intrinsically harder than others, across a variety of solvers. Such a correlation is difficult to see in scatter plots, but was found to be statistically significant in some cases, by analysis of variance. The greater correlation appeared between `minisat2` and `clasp` (both clause-learning solvers), while correlation between `march_hi` (a look-ahead solver) and either clause-learning solver was much lower. Apparently, our generated unsatisfiable formulas differ *randomly* in difficulty across solver styles; no "killers" stand out.

Turning to satisfiable benchmarks, results are more random, as shown by the wider error bars in Figure 4, compared to Figure 2. The incomplete `adapt` plowed through them, without much difference between generation methods, and with somewhat irregular growth in time vs. size. It was two orders of magnitude faster than the best complete solver we tested. Nevertheless the plot strongly suggests exponential growth, although at a lower rate than `clasp`. It is worth noting that in the SAT 2009 Competition [IB09], `adapt` typically solved random 3-CNF with 4000 variables and 50,000 literals in about the same same it required on our benchmarks with 225 variables and 1350 literals.

Constrained by preserving the underlying 3D $(0, 1)$ design, the `sgen2del` family has less randomness than the other families. Our tentative conclusion is that a high degree of randomness is more important than perfection of the structure (represented by score 0), for satisfiable formulas. This contrasts with our observation that generating unsatisfiable formulas with score 0 increases the difficulty by a statistically significant amount.

A specialized solver called `tts` was also tested briefly. This solver is designed for small hard formulas [Spe08]. The results appear in Figure 3. Notice that, measured by number of variables, `clasp` time is growing faster for "sgen2del," but measured by number of literals, it is growing faster for "scores." Interestingly, `tts` time growth is about the same for both generators when measured by number of literals. This shows that several measures of formula size are useful.

6   Conclusion

We introduced a less stringent class of combinatorial block designs, called PB01Ds, and a new combinatorial structure, called 3D $(0, 1)$ designs, which serve as the bases for for constraint satisfaction problems. Additional information and related files are available at http://www.cse.ucsc.edu/~avg/Sgen2/.

We showed that the CNF encodings of these unsatisfiable problems are quite small, yet extremely difficult for their sizes. The challenge formula mentioned in Section 3, and available at the above URL, has 105 variables and 1060 literals, but no solver has solved it in less than one day, so far. Smaller formulas in the same family have under 100 variables and under 1000 literals, but their difficulty (about one hour, for complete solvers) is comparable to random 3-CNF formulas with 500 variables and 6375 literals, or larger. The crafted benchmarks in the SAT 2009 competition that caused any difficulty were larger still (other than the orig family included in this paper for comparison).

Future work includes better understanding the structure of 3D $(0, 1)$ designs and investigation of whether the families presented in this paper have a provable exponential lower bound for resolution. However, it can be shown that generated instances based on the fixed-bandwidth incidence matrices of Figure 1 have polynomial-length refutations. Therefore, the random swapping used in the actual generation of instances, or a similar expansion idea, would be integral to establishment of an exponential lower bound.

Acknowledgment. We thank Ian Anderson for helpful email discussions about constructing combinatorial designs.

References

[And90]   Anderson, I.: Combinatorial Designs: Construction Methods. Ellis Horwood (1990)

[CD96]   Colbourn, C.J., Dinitz, J.H. (eds.): CRC Handbook of Combinatorial Designs. CRC Press, Boca Raton (1996)

[ES05]   Eén, N., Sörensson, N.: MiniSat – A SAT Solver with Conflict-Clause Minimization. In: Bacchus, F., Walsh, T. (eds.) SAT 2005. LNCS, vol. 3569. Springer, Heidelberg (2005)

[GKNS07] Gebser, M., Kaufmann, B., Neumann, A., Schaub, T.: Conflict-driven answer set solving. In: IJCAI, pp. 386–392. AAAI, Menlo Park (2007)

[Kul07]   Kullmann, O.: Polynomial time SAT decision for complementation-invariant clause-sets, and sign-non-singular matrices. In: Marques-Silva, J., Sakallah, K.A. (eds.) SAT 2007. LNCS, vol. 4501, pp. 314–327. Springer, Heidelberg (2007)

[lB09]   Le Berre, D.: The SAT competitions (2009), http://www.satcompetition.org/

[PS09]   Porschen, S., Schmidt, T.: On some SAT-variants over linear formulas. In: Nielsen, M., Kucera, A., Miltersen, P.B., Palamidessi, C., Tuma, P., Valencia, F.D. (eds.) SOFSEM 2009. LNCS, vol. 5404, pp. 449–460. Springer, Heidelberg (2009)

[Spe08]   Spence, I.: tts: A SAT-solver for small, difficult instances. Journal on Satisfiability, Boolean Modeling and Computation 4, 173–190 (2008)

[Spe10]   Spence, I.: sgen1: A generator of small but difficult satisfiability benchmarks. Journal of Experimental Algorithms 15, 1.1–1.15 (2010)

Author Index